Control and Automation: Devices, Systems and Engineering Applications

Control and Automation: Devices, Systems and Engineering Applications

Edited by Marques Vang

Clanrye International
www.clanryeinternational.com

Clanrye International,
750 Third Avenue, 9th Floor,
New York, NY 10017, USA

ISBN: 978-1-63240-581-4

Cataloging-in-publication Data

Control and automation : devices, systems and engineering applications / edited by Marques Vang.
p. cm.
Includes bibliographical references and index.
ISBN 978-1-63240-581-4
1. Automatic control. 2. Automation. 3. Industrial engineering. 4. Mechanization. 5. Engineering. I. Vang, Marques.
TJ213 .C66 2017
629.8--dc23

For information on all Clanrye International publications visit our website at www.clanryeinternational.com

CLANRYE INTERNATIONAL

Printed in China.

Contents

Preface

The ever-growing need of advanced technology is the reason that has fueled the research in the field of control and automation in recent times. This book is a compilation of chapters that discuss the most vital concepts and emerging trends in the field. Automation is the practice of using control systems in order to operate machinery like boilers, ovens, aircrafts, ships, etc. While reducing the cost of production, automation also helps in saving energy and time and improves quality and precision. The elements used in this area of science and engineering are of electrical, hydraulic, electronic, computer, mechanical devices, systems and engineering. Different approaches, evaluations, methodologies and advanced studies on this subject have been included in this text. It will serve as a reference to a broad spectrum of readers. This book presents the complex subjects of control and automation in the most comprehensible and easy to understand language.

This book has been the outcome of endless efforts put in by authors and researchers on various issues and topics within the field. The book is a comprehensive collection of significant researches that are addressed in a variety of chapters. It will surely enhance the knowledge of the field among readers across the globe.

It gives us an immense pleasure to thank our researchers and authors for their efforts to submit their piece of writing before the deadlines. Finally in the end, I would like to thank my family and colleagues who have been a great source of inspiration and support.

Editor

1

Process Fault Diagnosis Using Support Vector Machines with a Genetic Algorithm based Parameter Tuning

Mehdi Namdari*, Hooshang Jazayeri-Rad, Seyed-Jalaladdin Hashemi

Department of Instrumentation and Automation, Petroleum University of Technology, Ahwaz, Iran
*Corresponding author: namdari_mehdi@yahoo.com

Abstract Fault diagnosis, centered on pattern recognition techniques employing online measurements of process data, has been studied during the past decades. Amongst those techniques, artificial neural networks classifiers received an enormous attention due to some of their remarkable features. Recently, a new machine learning method based on statistical learning theory known as the Support Vector Machine (SVM) classifier is offered in the pattern recognition field. Support vector machine classifiers were originally used to solve binary classification problems. Subsequently, methods were proposed to apply support vector machine classifier to multiclass problems. Two of these mostly used methods are known as one versus one and one versus all. This paper deals with the application of the above mentioned classifiers for fault diagnosis of a chemical process containing a continuous stirred tank reactor and a heat exchanger. The results show a superior classification performance of the support vector machine versus the selected artificial neural network. In addition, the support vector machine classifier is very sensitive to the proper selection of the training parameters. It is shown that the utilization of genetic algorithm for optimal selection of these parameters is feasible and can help to improve the support vector machine classifier performance.

Keywords: *process fault diagnosis, continuous stirred tank reactor, support vector machine, artificial neural network, SVM Parameter Tuninge*

1. Introduction

Recently, Process Fault Diagnosis (PFD) systems have been developed to a high degree of quality in modern industry, because the cutting-edge faults diagnosis system has been employed to avoid severe damages. Fault diagnosis objective is to isolate and classify the detected irregularities and also to determine the origin of the observed abnormal status. Fault diagnosis is a crucial step to properly remove the fault. Besides, large volumes of data collected from chemical processes necessitate the development of the data-driven methods to convert them into proper information suitable for many fields such as PFD.

Amongst the numerous approaches that have been employed in PFD, those which utilize the pattern recognition problem are very striking. Because, they do not require any qualitative or quantitative model of the process and solely employ the historical data to design a fault diagnosis model to lay down the current process operating condition. Furthermore, these models permit dealing with the inherent nonlinearities of most chemical processes with no extra cost. In these methods, diverse operating conditions including normal and abnormal ones are considered as patterns. Then, the resulting classifier is used to examine the online measurement data and to convert them to a known class label for abnormal or normal so that the existing system condition is categorized.

Many methodologies based on the so called pattern recognition techniques have been implemented. Amongst those methods, Artificial Neural Networks (ANN) had an enormous consideration in past years because of some of their remarkable characteristics such as dealing with nonlinearity, noise tolerance and generalization capability [1,2,3,4,5]. ANNs have demonstrated to be good classifiers; however, they require a large number of samples for training, which are not every time essentially obtainable. Also, they infrequently give high generalization error because in the training process they only attempt to enhance categorization performance due to the training data. To resolve this problem, a new machine learning technique relying on statistical learning theory known as the support vector machine is suggested in pattern recognition issues to achieve high generalization capability and to deal with problems with low samples and high input features [6]. It has been shown that the SVM based classifiers are more effective than the formerly reported pattern classifiers and in recent years have been established to be extremely effective in numerous practical applications [7,8,9,10]. However, their applications to process engineering problems are still rare [11,12,13,14]. This paper makes use of SVM for fault diagnosis of a combination of the Continuous a Stirred Tank Reactor (CSTR) and a heat exchanger in a chemical process plant.

SVM was initially developed for binary classification. Several techniques have been proposed for the SVM classifier to be extended to multiclass problems. Amongst them, two methods: One Versus One (OVO) and One Versus All (OVA) are very popular. So in this work a comparison study will be performed amongst three classifiers: the SVM based on OVO (OVO-SVM); the SVM based on OVA (OVA-SVM) and the ANN classifier. A major improvement of the SVM performance can be achieved by the proper selection of the classifier parameters. Initially, a trial and error (T&E) method is used to select the parameters of both of the OVO-SVM and OVA-SVM classifiers. Then, to obtain the optimal parameter values for the SVM classifiers, the Genetic Algorithm (GA), as a well-known optimization solver is used. The result indicated improved classification performance.

2. Methods

2.1. ANN

ANNs were inspired from the study of the human brain which consists of millions of organized neurons. ANNs were developed to imitate the computational structures of these neurons. They contain numerous linked artificial processing neurons known as nodes which are joined together in layers making a network. A node obtains weighted inputs from other nodes or sources, and directs the summation of them to a transfer function known as the activation function. Activation function is a kind of mathematical function. It is typically selected to be a constrained differentiable function such as the sigmoid.

Of all the arrangements of ANNs, the Multi-Layer Perceptron (MLP) neural network is the most prevalent one. As an example, a MLP network with three layers is presented in Figure 1. The circles signify neurons organized in three layers as input, hidden and output. Each hidden neuron is linked to each input and output units. The number of nodes in the input and output layers are determined by the nature of the problem to be unraveled and the number of input and output variables required to define the problem. The number of hidden layers and the nodes within each hidden layer is typically found by the T&E method.

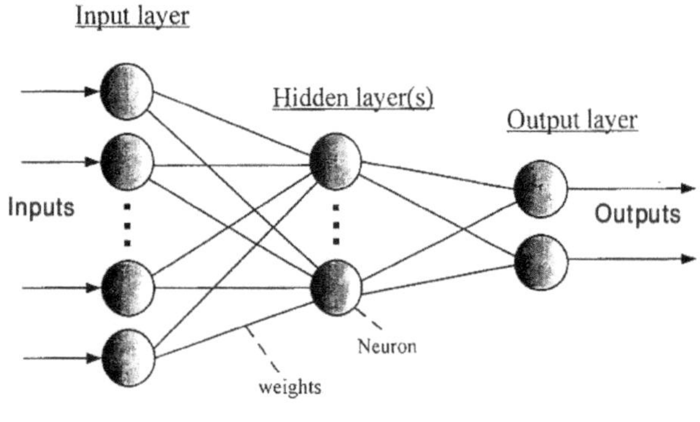

Figure 1. MLP neural network with three layers

Training is used to obtain the values of the signal weights. The network is trained on a collection of data known as the training set, acquired from the process to be modeled. This method is known as the "learning" procedure. The learning procedure is basically an optimization algorithm. As soon as a network is trained, it can be suitably employed as a model to characterize the system for many different purposes.

2.2. SVM

SVM was initially developed for pattern recognition tasks. Pattern recognition or classification is to label some object into one of the specified categories called classes. Conventional classifiers such as neural networks minimize the training data set error called Empirical Risk Minimization (ERM), while SVM relies on the Structural Risk Minimization (SRM) principle founded on the statistical learning theory, which enhances generalization capabilities.

2.2.1. Two-Class SVM

SVM, in its simple form, is a binary classifier which gives rise to a linear hyper-plane which distinguishes a set of positive examples from a set of negative examples with extreme margin (the margin is outlined by the distance of the hyper-plane to the bordering positive and negative examples as shown in Figure 2). This maximum margin hyper-plane is known as the optimal separating hyper-plane. Because the theories behind the statistical learning endorse that considering maximum margin in training procedure ends up to acquiring better generalization capability [15]. The nearest data points which are employed to define the margin are known as support vectors. The number of support vectors rises with the enormity of the problem.

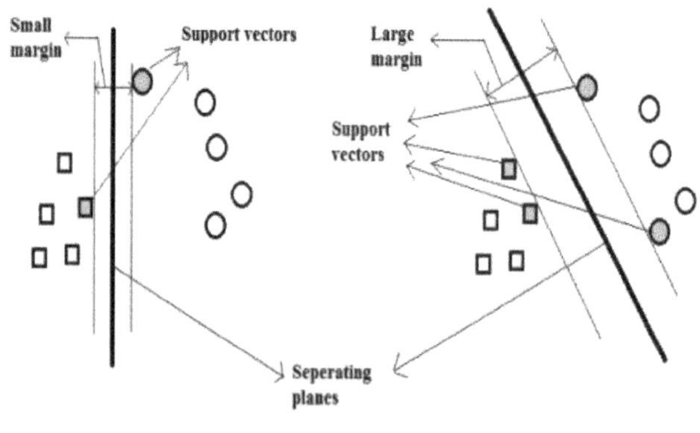

Figure 2. Separating hyper-planes with small and large margins in binary classification

Assume there is a known training sample set $G = \{(x_i, y_i), i=1...M\}$, where M is the number of samples and each sample $x_i \in R^d$ fits into a class by $y_i \in \{+1, -1\}$. In the case of a linearly separable data set, it is possible to se parate the given data into the two classes using the hyper-plane:

$$\mathbf{W}^T.\mathbf{x} + b = 0 \quad (1)$$

where **W** is an M-dimensional vector and b is a scalar bias term. The vector **W** and scalar b are employed to describe the position of the separating hyper-plane. The following decision function, $D(x_i)$, can be employed to categorize input data into either positive class or negative class. So for a known input data x_i:

$$D(x_i) = \mathbf{W}^T.\mathbf{x}_i + b \begin{cases} > 0 for y_i = +1 \\ < 0 for y_i = -1 \end{cases} \quad (2)$$

It is possible to demonstrate that maximization of the margin can be achieved through minimization of **W**. In order to obtain a hyper-plane with larger margin which results in better generalization ability, SVM defines a slack variable ζ_i for each training sample and allows some of them to be misclassified. So the optimal hyper-plane separating the data can be determined as a solution to the following constrained quadratic optimization problem:

$$Minimise \frac{1}{2}\left\|\mathbf{W}^2\right\| + c\sum_{i=1}^{M} \zeta i \qquad (3)$$

$$Subject\ to y_i\left[\mathbf{W}.\mathbf{x}_i + b\right] \geq 1 - \zeta i, i = 1, \ldots M \qquad (4)$$

where c is the regularization parameter defined as the trade-off between the margin maximization and classification error minimization. Hence, it can be set to prevent the over-fitting phenomenon.

If the input data are not linearly separable, the original input space is mapped into a high-dimensional space called the feature space. It is then possible to determine a hyper-plane that allows linear separation in the feature space. SVM employs kernels to transform the data into higher dimensions. The most common kernel function is Radial Basis Functions (RBF) which is defined as:

$$K(x, y) = \exp(-\gamma\|x - y\|^2) \qquad (5)$$

2.2.2. Multi-Class SVM

The discussion above handles binary classification where the class labels can take only two values: 1 and -1. In practice, however, we find more than two classes for examples in fault diagnosis. Unfortunately there is no unique way to employ SVM for multiclass problems. At this time, there are two general methodologies for multiclass strategies. One is creating and combining several binary classifiers while the other is by directly employing all data in an optimization problem. Two techniques which are used in this paper employ the first approach to tackle the multiclass problem. These are the OVA and OVO methods. The OVA method constructs k binary classifiers where k is the number of classes. Each binary classifier splits the training samples of one class from all other classes. That is why it is known as the one versus all. So in OVA for the k binary SVMs, all training samples of all classes are used. However, the second strategy, OVO, creates $k*(k-1)/2$ binary classifiers where each binary classifier separates the training samples of one class from another and so in training of each binary SVM, training samples of only two classes are employed. Further details of these two strategies and other existing methods for the multiclass SVM can be found in [16].

2.3. GA

The genetic algorithm is a technique for resolving optimization problems that relies on natural selection which is a process motivating biological evolution. The genetic algorithm continually adjusts the population of individual solutions. At each step, the genetic algorithm picks individuals at random from the current population to be parents and uses them to generate the children for the next generation. Over consecutive generations, the population evolves toward an optimal solution. GA can handle large search spaces effectively, and therefore has less chance to get into local optimal solutions than other algorithms. The genetic algorithm generates three types of children for the next generation:

- Elite children are the individuals in the existing generation with the optimal fitness values. These individuals are routinely carried on to the next generation.
- Crossover children are generated by joining the vectors of a pair of parents.
- Mutation children are generated by producing random variations, or mutations, into a single parent.

Figure 3 illustrates the genetic operations of crossover and mutation. Detailed description of GA can be found in [17].

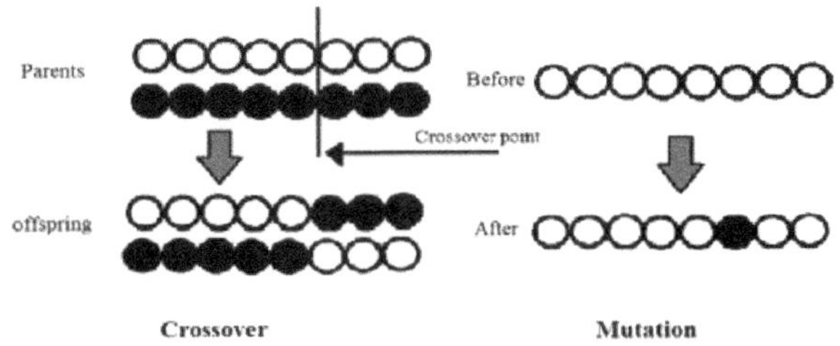

Figure 3. Genetic crossover and mutation operations

3. Case Study

The process used in this paper is shown in Figure 2. This process is made up of a heat exchanger and a CSTR where an irreversible first-order reaction $A \rightarrow B$ occurs. The reaction is catalytic and exothermic. The temperature of the reactor (T_m) is regulated by pumping a part of the reactor outlet stream back to the reactor through the heat exchanger where the temperature (T_R) of the recycle flow (F_R) is cooled by an external flow (F_w). F_R is maintained using a flow controller. The process employs two other feedback controllers (CL and CT) that keep the level of the reactor (L_m) and the temperature of the reactor at the set point temperatures of L_s and T_s, respectively. This process is selected because it reveals the most common features appearing in industrial processes. The fundamental model of the process can be found in [18].

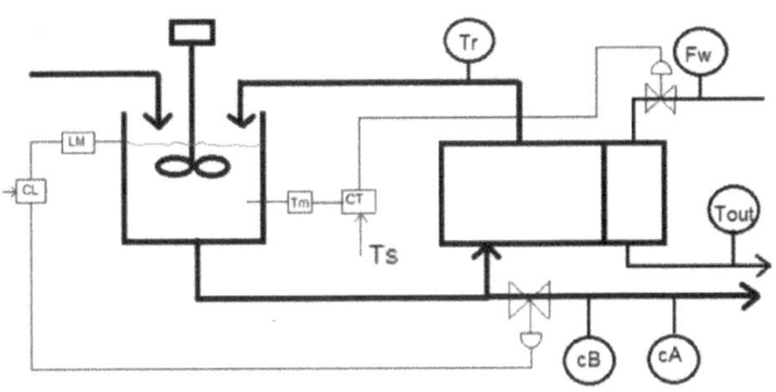

Figure 4. The simulated process

Altogether, five variables (F_w, T_r, T_{out}, c_A, *and* c_B) are measured from the process (the encircled variables in Figure 4). T_{out} is the output temperature of the coolant fluid; c_A and c_B are the concentrations of the components A and B, respectively. Eight typical fault situations have been chosen for this study. These faults are listed in Table 1. Each fault in the heat exchanger-CSTR system corresponds to fluctuations in one related faulty variable. For each fault situation, different step changes within the range of 0-10% in the corresponding faulty variable generated different faulty samples. For each faulty case, 100 data samples were generated for the network training and 100 samples were generated for the network testing. Gaussian noise has been added to both of the training and testing data. Also, some outlier samples are added to the generated data.

Table 1. Fault Situations in the Example Process

Faults	Description
Fault 1	Low feed concentration
Fault 2	Low recycle flow
Fault 3	Bias error in the T_2 sensor
Fault 4	Fouled heat exchanger
Fault 5	Deactivated catalyst
Fault 6	Low inlet flow
Fault 7	Low feed temperature
Fault 8	Low cooling water temperature

It is very difficult to comprehend the geometrical properties of different patterns in a 5-dimensional measurement space. Therefore, we need to use a dimensionality reduction technique to visualize the samples in a lower dimensional space. The Principal Component Analysis (PCA) technique is used for this purpose. A detailed description of PCA is beyond the scope of this work. Briefly, PCA reduces the dimension of the data to a number of principal components which explain different proportion of the variance within the original data. Typically, the first three components account for the majority of the variations, and may be used to represent the data in a lower dimension. Figure 5 presents the generated samples for eight fault situations with the first and second principal components. It should be noticed that PCA here is used only for the visualization purpose and geometrical comprehension about different generated faulty samples. The original 5-dimensional patterns will be used to train the classifiers.

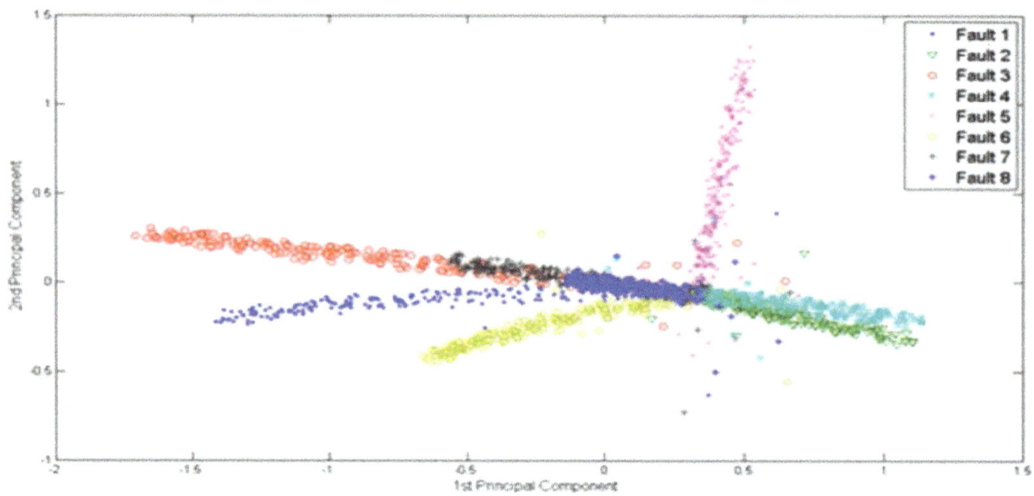

Figure 5. The fault situations of example process presented with the first and second principal components

4. Results and Discussion

4.1. Classification Performance Comparison

For training of both of the ANN and SVM classifiers, training and testing data were normalized within the range [0, 1]. For the ANN a two layers feed-forward MLP network with sigmoid activation function for both hidden and output layers has been chosen. The number of neurons in the output layer is fixed and is equal to the defined number of fault situations which is equal to eight. The number of neurons in the hidden layer should be determined. This highly affects the classification performance of ANNs. So the ANN is trained for different numbers of hidden neurons and then the network with the least average misclassification rate is selected. In the training cycle for each fault situation, its output value is set to 1 and the output values corresponding to other fault situations are set to 0. In testing the network, for an arbitrary input, a particular fault will be detected if its corresponding output value is greater than the values at other outputs.

For training the ANN the Scaled Conjugate Gradient (SCG) algorithm has been used. The MATLAB 7.12 neural network toolbox has been used for this purpose. The network is trained several times with different random initial weight values. The number of training iteration is set to 200. Then, the network with the best classification performance is chosen. The classification performance of the trained ANN is reported in Table 2 and shown in Figure 6.

For training the SVM classifiers first a suitable kernel function should be chosen. Usually the RBF kernel is a reasonable first choice because it only needs two parameters (c, γ) which need to be determined and has fewer numerical difficulties than the polynomial and sigmoid kernels [19]. So, RBF kernel has been selected for this work. The two parameters (c, γ) are determined using the T&E method. For training the SVM classifiers, the Libsvm toolbox developed by Chang and Lin has been used [20]. Libsvm considers a Sequential Minimal Optimization (SMO) type decomposition method for training the SVM classifiers. The classification performances of both of the OVA-SVM and OVO-SVM classifiers are reported in Table 2 and shown in Figure 6.

As seen in Table 2, the magnitudes of average misclassification rates indicate that the classification performances of the OVO-SVM and OVA-SVM classifiers outperform the ANN classifier. Furthermore, although a fast training technique such as the SCG algorithm has been used to train the ANN classifier, the training rate of each of the OVA-SVM and OVO-SVM classifiers was much faster than the ANN classifier (about three times faster). A major difficulty in using neural network is that it may entrap in local minima during the training cycle. To resolve this problem, the network should be trained with different initial weights. However, the SVM classifier always converges to a unique solution in the training phase. These major advantages of the SVM classifier over the ANN classifier make it a better tool for pattern recognition and process fault diagnosis tasks. Also, the performance of the OVO-SVM is better than the OVA-SVM with respect to the classification rate. Therefore, the OVO technique is better than the OVA scheme for the multiclass SVM as was reported in [16].

Table 2. Average Misclassification Rates of the Trained Classifiers

Classifier	Parameter	Misclassification Rate (test data) (%)	Misclassification Rate (training data) (%)
ANN	hidden neurons= 6	14	12.625
OVA-SVM	(c, γ)= (2^3, 1)	12.75	10.875
OVO-SVM	(c, γ)= (2^3, 1)	11.875	9.875

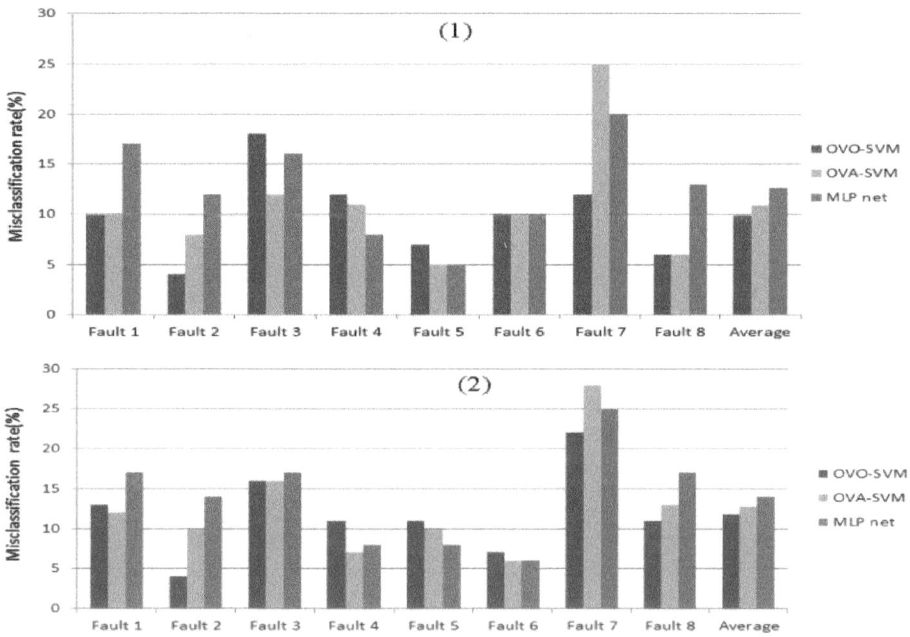

Figure 6. Misclassification rates of different fault situations forthe training (1) and test (2) data

4.2. GA-based Parameter Tuning

Classification performance of the SVM classifier to a large extent depends on its regularization parameter *c* and the kernel function parameters. The common approach to find the optimal values for these hyper-parameters is to use a grid search with cross-validation [19]. In the grid search, various pairs of (c, γ) values are tried and the one with the best cross-validation accuracy is selected. In the v-fold cross-validation, the training set is divided into v subsets of equal sizes. Sequentially, one subset is tested using the classifier trained on the remaining v-1 subsets. Thus, each instance of the whole training set is predicted once and the cross-validation accuracy is the percentage of data which are *incorrectly* classified.

The application of grid search combined with the cross-validation for a wide range of the parameters (c, γ) can be very time consuming and may not result in the optimal hyper-parameters. So in this work, a GA-based parameter tuning method is used and proposed to automatically obtain the optimal hyper-parameters of the SVM classifier. The classifier is trained and validated using the v-fold cross-validation based on the training parameters proposed by the GA. The classification rate of the training data is the fitness function of the GA. Therefore, the optimum tuning is achieved when a minimum misclassification is detected by the classifier. This procedure is shown in Figure 7.

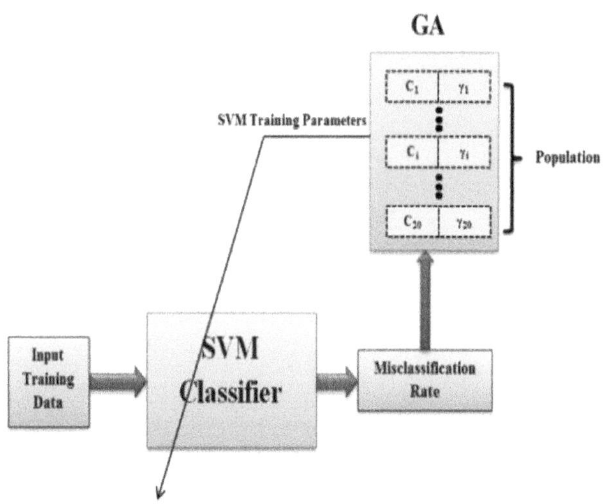

Figure 7. GA-based parameter tuning procedure

For both the OVO and OVA methods, the fitness function is the classification rate of a 10-fold cross-validation procedure and the population size is considered to be equal to 20. The initial range is taken to be within [0,

2] for all individuals. Figure 8 (1) and Figure 8 (2) present the optimization procedures and as seen in each of these figures the fitness function has been minimized under 30 iterations. Table 3 and Figure 9 show the classification performance of both of the OVA and OVO methods using the T&E method of parameter selection and the GA-based parameter tuning. As seen in Table 3, the classification performances of both of the OVA and OVO methods, when compared against the T&E scheme, are improved using the GA-parameter tuning. These results show the advantage of using the GA for the parameter tuning of the SVM classifier.

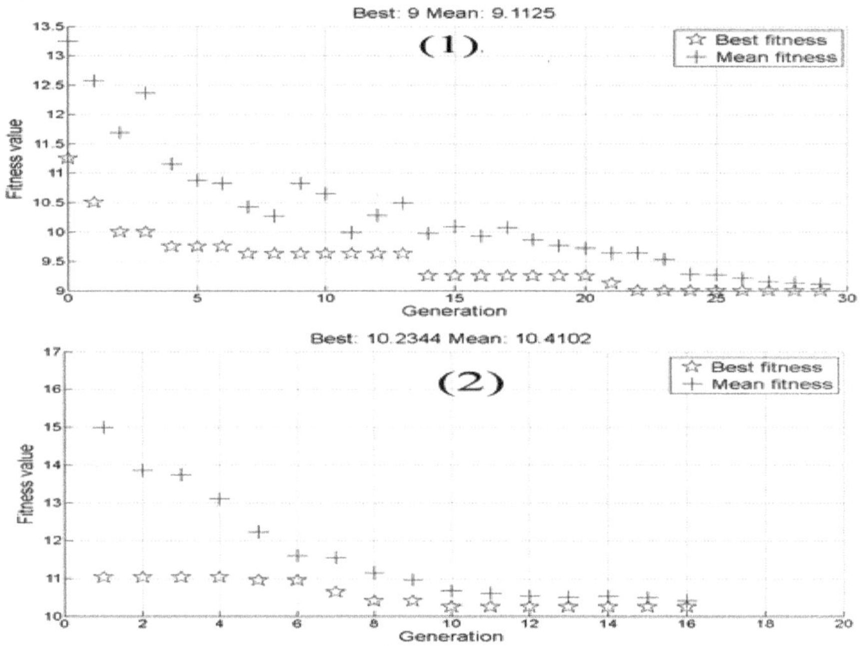

Figure 8. GA-based optimization procedures for the OVO-SVM (1) and OVA-SVM (2)

Table 3. Average Misclassification Rate with the Trial and Error (T&E)- and GA-based Parameter Selections

Classifier	Parameter (c, γ)	Misclassification Rate (testing data) (%)	Misclassification Rate (training data) (%)
T&E-OVA	(2^3, 1)	12.75	10.875
GA-OVA	(43.9,18.3)	12	3.625
T&E-OVO	(2^3, 1)	11.875	9.875
GA-OVO	(1.5,103.9)	10.875	4.25

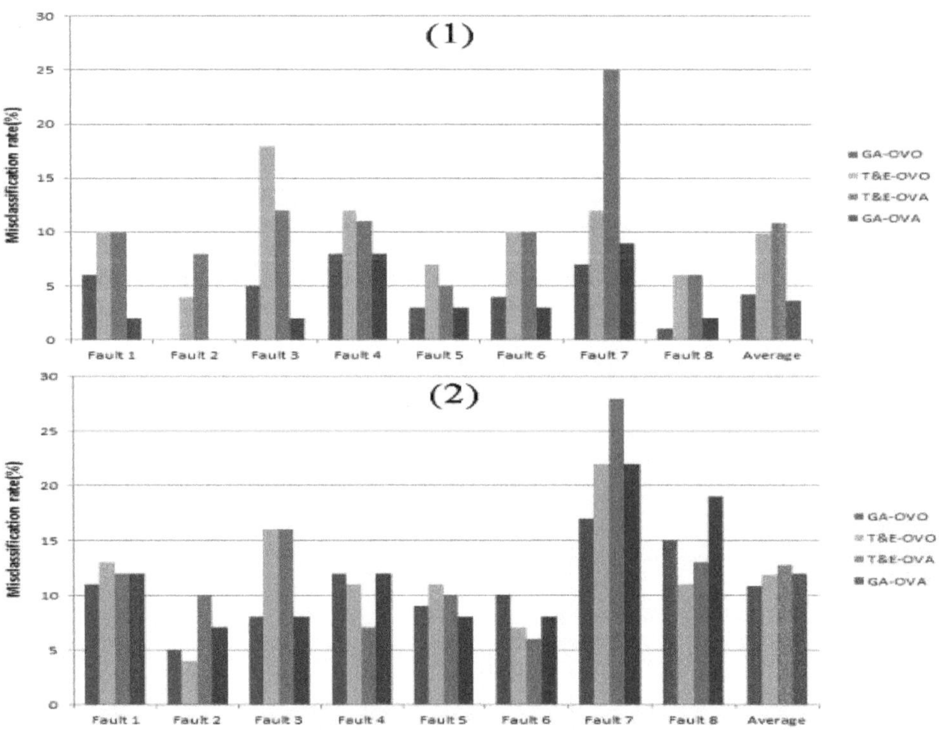

Figure 9. Misclassification rates of different fault situations with trial and error (T&E)- and GA-based parameter selections for the training (1) and test(2) data

5. Conclusions

In this work the SVM and ANN classifiers are applied to the fault diagnosis of a chemical process containing a CSTR and a heat exchanger. As expected, the results show a superior classification performance for the SVM classifier against the ANN classifier. The performance measures were based on the misclassification rates. Also the entrapment in local minima problem for ANNs, sometimes necessitate the training procedure to be repeated several times with different initial weights to achieve acceptable results. In contrast, the objective function formulation in SVM gives a quadratic problem with linear constraint that has a distinctive solution, therefore, training of the SVM classifier always leads to a unique result. These major advantages of the SVM classifier over the ANN classifier make it a superior tool for pattern recognition and process fault diagnosis tasks. In addition, the performances of the OVO-SVM and OVA-SVM approaches in multiclass problems have been compared and the results show that the OVO scheme outperformed the OVA method for the multiclass SVM. Finally, it has been shown that the use of the T&E method for the selection of the training parameters provides an improved classification performance for the SVM classifier against the ANN classifier. In addition, the results show that a GA-based parameter tuning for the SVM classifier gives rise to further enhancement of the classification performance of the SVM classifier.

References

[1] Watanabe, K., Matsuura, I., Abe, M., Kubota, M. and Himmelblau, D. M., "Incipient fault diagnosis of chemical processes via artificial neural networks," *AIChE Journal*, 35 (11), 1803-1812, 1989.

[2] Koivo, H. N., "Artificial neural networks in fault diagnosis and control," *Control Engineering Practice*, 2 (1), 89-101, 1994.

[3] Sharma, R., Singh, K., Singhal, D. and Ghosh, R., "Neural network applications for detecting process faults in packed towers," *Chemical Engineering and Processing: Process Intensification*, 43 (7), 841-847, 2004.

[4] Behbahani, R. M., Jazayeri-Rad, H. and Hajmirzaee, S, "Fault detection and diagnosis in a sour gas absorption column using neural networks," *Chemical engineering & technology*, 32 (5), 840-845, 2009.

[5] Eslamloueyan, R., "Designing a hierarchical neural network based on fuzzy clustering for fault diagnosis of the Tennessee–Eastman process," *Applied Soft Computing*, 11 (1), 1407-1415, 2011.

[6] Vapnik, V., *The nature of statistical learning theory*. New York, Springer, 1999.

[7] Widodo, A. and Yang, B.S. "Support vector machine in machine condition monitoring and fault diagnosis," *Mechanical Systems and Signal Processing*, 21 (6), 2560-2574, 2007.

[8] Tyagi, C. S., "A comparative study of SVM classifiers and artificial neural networks application for rolling element bearing fault diagnosis using wavelet transform preprocessing," *Neuron*, 1, 309-317, 2008.

[9] Lee, M. C. and To, C., "Comparison of support vector machine and back propagation neural network in evaluating the enterprise financial distress," *International Journal of Artificial Intelligence & Applications*, 1(3), 31-43, 2010.

[10] Azar, A. and El-Said, S., "Performance analysis of support vector machines classifiers in breast cancer mammography recognition," *Neural Computing and Applications*, 1-15, 2013.

[11] Chiang, L. H., Kotanchek, M.E. and Kordon, A. K., "Fault diagnosis based on Fisher discriminant analysis and support vector machines," *Computers & chemical engineering*, 28 (8), 1389-1401, 2004.

[12] Yélamos, I., Graells, M., Puigjaner, L. and Escudero, G., "Simultaneous fault diagnosis in chemical plants using a multilabel approach," *AIChE Journal*, 53 (11), 2871-2884, 2007.

[13] Mao, Y., Xia, Z., Yin, Z., Sun, Y. and Wan, Z., "Fault diagnosis based on fuzzy support vector machine with parameter tuning and feature selection," *Chinese Journal of Chemical Engineering*, 15(2), 233-239, 2007.

[14] Yélamos, I., Escudero, G., Graells, M. and Puigjaner, L., "Performance assessment of a novel fault diagnosis system based on support vector machines," *Computers & Chemical Engineering*, 33 (1), 244-255, 2009.

[15] Abe, S., *Support vector machines for pattern classification*. New York, Springer, 2005.

[16] Hsu, C. W. and Lin, C. J., "A comparison of methods for multiclass support vector machines," *Neural Networks, IEEE Transactions on*, 13(2), 415-425, 2002.

[17] Whitley, D., "A genetic algorithm tutorial," *Statistics and computing*, 4(2), 65-85, 1994.

[18] Sorsa, T., Koivo, H. N. and Koivisto, H., "Neural networks in process fault diagnosis," *Systems, Man and Cybernetics, IEEE Transactions on*, 21 (4), 815-825, 1991.

[19] Hsu, C. W., Chang, C.C. and Lin, C. J., "A practical guide to support vector classification," 2010.

[20] Chang, C.C. and Lin, C. J., "LIBSVM: a library for support vector machines," *ACM Transactions on Intelligent Systems and Technology (TIST)*, 2(3), 2011.

Iris Verification and ANOVA for Iris Image Quality

Lidong Wang*

Department of Engineering Technology, Mississippi Valley State University, Itta Bena, USA
*Corresponding author: lwang22@students.tntech.edu

Abstract Iris recognition has been acknowledged as the most accurate method in biometrics that is one of main automated identification technologies. The iris image quality and iris verification of eyes in dark brown, regular brown, hazel, green, and blue was tested. The effects of eyeglasses and contact lenses on the iris image quality score and iris verification were investigated, respectively. The investigated results indicate that the iris verifications with eyeglasses or contact lenses can still be successful although both eyeglasses and contact lenses decrease iris image quality. Analysis of variance (ANOVA) for the iris image quality of three eye colors (dark brown, hazel, and blue) was conducted to study the difference in the image quality due to the eye colors. The ANOVA results show there is no significant difference in the iris image quality of eyes in dark brown, hazel, and blue at the 0.05 level of significance.

Keywords: *iris image quality, iris verification, biometrics, automated identification, ANOVA, level of significance*

1. Introduction

The human iris is rich with features and can be used to quantitatively distinguish one eye from another. The iris contains many collagen fibers, contraction furrows, coronas, crypts, color, serpentine vasculature, striations, freckles, rifts, and pits. Measuring the patterns of these features and their spatial relationships to each other provides other quantifiable parameters for identification processes [1]. To capture a rich iris texture, the iris recognition system should have high resolution, good sharpness, and good lighting conditions [2]. Iris recognition technology has been reported to be more stable and reliable than other biometric technologies; however, not all of the iris images acquired from the device are in-focus and sharp enough for recognition [3].

Poor quality images increase false rejection rate (FRR) and false accept rate (FAR), significantly degrade the performance of iris recognition systems [4]. It has been shown that the matching performance of iris recognition systems can be improved when the iris image quality is improved [3]. Best image quality optimizes segmentation and recognition performance, especially the false non-match rate (FNMR). Quality predicts match performance [5]. Therefore, iris image quality assessment is very important for an iris recognition system.

Several quality factors have an effect on the iris image quality. They are: defocus blur, motion blur, occlusion, off-angle, specular reflection, lighting, and pixel-counts [6,7,8,9]. It is a complex problem to assess the overall quality of an iris image. Defocus blur, motion blur, occlusion, and off-angle are the main factors among these factors [3,6].

Defocus blur images can result from many sources, but in general, defocus occurs when the focal point is outside the 'depth of field' of the object to be captured. The further an object is from this depth of field, the higher the degree of defocus. Depth of field is affected by the aperture size (i.e. the smaller the aperture size, the greater the depth of field) [3]. Motion blur can result from either the relative motion of an object or relative motion of the camera during exposure time [6]. In general, there are two types of motion blur: linear and non-linear. Linear motion blur can be considered as smearing in only one direction, while non-linear can be considered smearing in multiple directions at different strengths. An occlusion image is the case that a part of iris area is covered by eyelid and eyelashes [3]. Off-angle can result from non-cooperative users or when capturing iris's at a distance [6].

Specular reflections come in two forms: diffuse-source reflections, caused by scattered ambient light, and point-source reflections (PSRs), usually caused by highly localized sources such as desk lamps or overhead lighting. With diffuse-source reflections the underlying image remains partially visible and the affected area can be quite large. PSRs are usually smaller high-intensity regions with semi-transparent borders [10]. The position of specular reflection relative to pupil boundary provides indication of gaze angle; the size of specular reflection indicates focus quality and motion blur; pupil and iris edge contrast/sharpness indicate focus quality; the distance between upper and lower lid can be compared to iris diameter to estimate iris exposure; and the presence of specular reflections outside pupil may indicate obscuration of the iris area [5].

Several kinds of iris image quality metrics were presented. Cambier (2007) [5] presented the quality metrics including: precise segmentation and determination of iris area based on eyelids, eyelashes, specular reflections, etc.; the focus assessment based on spatial frequency content – may be limited to iris area; and the measurement of pupil/iris diameter ratio. McConnon et al. (2010) [10] presented their quality metrics to assess the iris image characteristics with regard to focus, entropy, reflections, pupil constriction and pupillary boundary contrast. Li et al. (2011) [9] proposed three approaches to estimate the quality metrics of defocus, motion blur, and off-angle in an iris image, respectively. A fusion method based on likelihood ratio was proposed to combine the six quality factors (defocus, motion blur, off-angle, occlusion, deformation, and light variation) into a unified quality score; and a statistical quantization method based on *t-test* was proposed to adaptively classify the iris images in a database into a number of quality levels. The relationship between iris recognition results and the quality level of iris images could be explicitly formulated.

Chaskar et al. (2012) [11] though that iris images usually got affected from wide range of qualities like dilation, iris resolution, specular reflection, motion blur, camera diffusion, the presence of eyelids and eyelashes, head rotation, camera angle, contrast, luminosity etc. Nine quality factors were assessed. They are: dilation measure (DM), ideal iris resolution(IIR), actual iris resolution (AIR), processable iris resolution (PIR), signal to noise ratio (SNR), occlusion measure (OM), specular reflection (SR), eccentric distance measure (EDM), and angular assessment (θ). The dilation of a pupil can affect the recognition accuracy. If the iris is too dilated, there is a possibility of losing the information; this may not serve the adequate necessary information for recognition. IIR is the resolution of the iris obtained ideally in the absence of noise; AIR is the resolution of the iris obtained in the presence of noise like an eyelid or an eyelash; PIR is the available part of iris from which features can be extracted for further processing and the ratio of AIR to IIR gives the PIR; SNR should be as high as possible and it is high if the noise is low; OM measures how much percentage of the iris area is invalid due to eyelids, eyelashes, and other noise; Once eyelid occlusions are estimated, occlusions resulting from specular reflection (SR) are evaluated on the remaining iris portion unaffected by the eyelids; EDM measures the position of the iris and the pupil with respect to each other using the coordinates of their centers; and θ is the evaluation of the iris center relative to the pupil center.

Research on iris recognition was conducted based on iris images acquired under varying illumination, at the visible wavelength, and using less constrained imaging protocols. The term "less constrained iris imaging conditions" refers to the acquisition of iris images at larger distances, under dynamic lighting conditions, and without requiring any active participation of the subjects. This extends the use of iris recognition technology to domains where the subjects' cooperation cannot be expected. However, it is highly probable that the captured images are off-angle or contain partial or blurred irises (due to bad focus or motion) [12].

As wireless technology has advanced, many applications have been developed on mobile devices. Therefore, many mobile iris recognition devices, such as the Pier™ 2.3 system and the HIIDE ™ Series 4, were commercialized recently. A motion-and-optical blurred image can be sometimes captured because a user captures the iris images of a subject by holding the recognition device and the user's hand-shaking happens. Motion-and-optical blurred images reduce iris recognition accuracy. To overcome this problem, a new method of restoring motion-and-optical blurred iris images at the same time was proposed [13].

In general, greater capture distance has convenience and hygiene value as the device need not be taken very close to eyes during the capture. Sources of motion can be: body motion in three dimensions, the head's independent motion, and the eye's independent motion. Many professionals typically thought: 1) both the eyes and the head's position must be controlled and eye motion during exposure causes image blurring; 2) the iris capture process was sensitive to lighting conditions present in the testing room and that no direct or artificial light should directly reflect off the enrollee's eyes [6]; 3) image quality can be affected under changes in the stand-off distance (capture distance). The author in this paper studied the effects of the three factors on iris image quality using the IrisAccess™ 4000 system [14,15] and the following conclusions were reached [16]: 1) there were almost no motion-induced effects on iris image quality scores and the iris verification was successful when a person shook his/her head, nodded, or had eye motion during the enrollment and the verification; 2) while the lights were on and off in the lab where iris images were captured, there was almost no difference in iris image scores; and 3) distance changes within the capture range (18-25cm) of the IrisAccess™ 4000 system did not affect the iris image quality score.

Dark brown irises often recode lighter with infrared photography than blue ones. For blue irises, infrared illumination may produce less iris detail than visible light. Under visible illumination, blues irises are lighter than dark brown irises [17]. The iris has a complex texture. As all enrollment forms suggest, the iris color (usually called eye color) is a recognized discriminant among people. Yet iris recognition methods do not use color. They use structure (shape) [18]. For Asian people, brown irises are dominant. Their irises have a lower contrast texture [19]. The utility of texture and color for iris recognition systems was proposed. It contributes for improvement of system accuracy with a reduction of false acceptance rate (FAR) and false rejection rate (FRR). Experimental results indicated that the proposed method using only color achieves 99.9993 accuracy, 0.0160 FAR, and 0.0813 FRR [20]. The effect of eye color on texture-based iris recognition in both the near -IR and visible bands was examined. The findings of this work showed that the matching performance associated with texture-based encoding varies with eye color and the specific band of the iris template [21]. Iris recognition performance as a function of eye color was studied. The performance was described in the form of relative operating characteristic (ROC) curve. It was shown that there was statistically significant difference in performance between blue eyes and brown eyes [22,25].

In this paper, the following study has been conducted: 1) iris images and their quality scores of people with different eye colors (dark brown, regular brown, hazel, and blue); 2) the effect of eyeglasses on the iris image quality score and iris verification; 3) the effect of contact lenses on the iris image quality score and iris verification; and 4) statistical analysis for the image quality of three eye colors (dark brown, hazel, and blue) based on the parametric method ANOVA to study the difference in the iris image quality of eyes due to the eye colors.

2. The Iris Recognition System and the Experimental Method

An IrisAccess™ recognition system developed by the LG Electronics – Iris Technology Division, was used in this study. The EAC Software v 3.00.14 was installed in the iris recognition system. The EAC Software was designed to operate using both IrisAccess™ 3000 and IrisAccess™ 4000 series hardware. An iris camera iCAM4000 in the IrisAccess™ 4000 series was used to capture the subjects' iris images. The iCAM4000 is a two-eye iris camera which includes an alignment indicator behind the mirror. The iris system can be used for enrollment and verification [14,15]. The system displays the iris image and the iris image quality score as soon as an iris scanning is completed. The iris image quality score called "IrisCode™ quality" is from 0 to 100. The lowest value is fixed at 0 and the highest value is fixed at 100. To obtain iris images with a higher quality, a user should keep both eyes wide open and look into the rectangular mirror aligning the colored dot between the eyes until the audio message of "we finish taking pictures of your eyes" plays. The user should not rotate, pan, or tilt his/her face. To test the effect of eyeglasses or contact lenses on the iris image quality score and iris verification, eyeglasses or contact lenses are worn during both the iris enrollment and the iris verification.

The user can choose to scan the right eye, the left eye, or both though a setting in the system. When both the right eye and the left eye are chosen, the iris image quality score is a combined score for both.

3. Experimental Results and Analysis

3.1. Image Quality Scores and Iris Verification for Different Eye Colors and Races

Four people's iris images were captured using the IrisAccess™ 4000 system in the Automated Identification Technology Lab at Mississippi Valley State University, USA, in April, 2013. The four people had hazel, dark brown, regular brown, and blue eyes, respectively. None of them wore eyeglasses. The stand-off distance from the camera to the individual was 21 cm. Figure 1 shows the four individuals' right iris images and the image quality scores after the enrollment process. The iris image acquisition yielded images of the irises and the surrounding eye regions. All of the four images were used for the each individual verification and the verifications were successful, although Figure 1 (d) is an image with a relatively low quality score due to the lower eyelid occlusion.

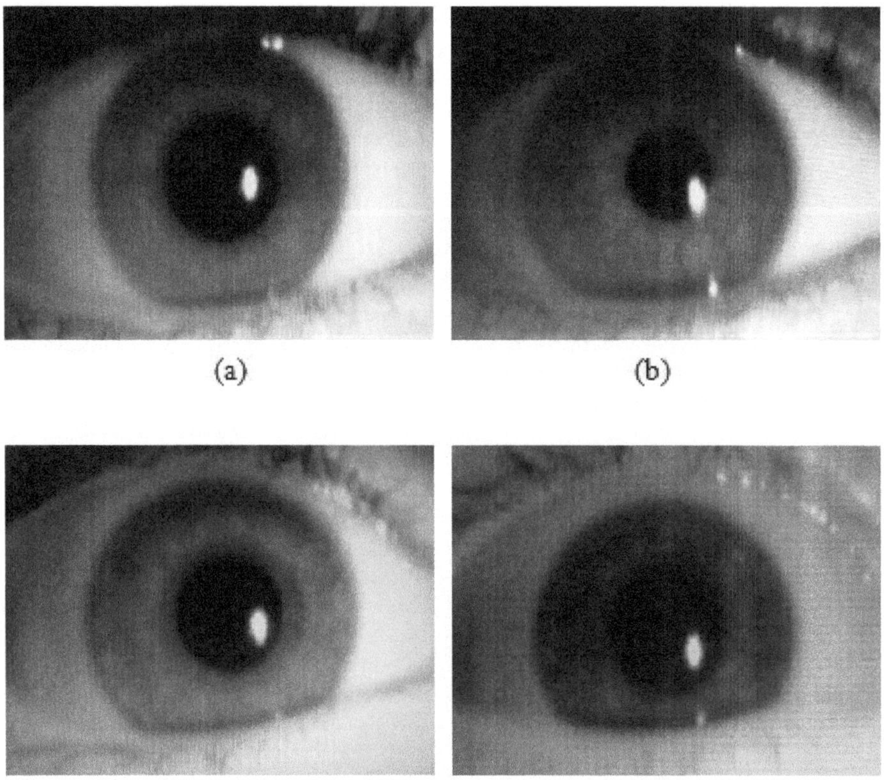

(a) (b) (c) (d)

Figure 1. Iris images and image quality scores for four subjects: (a). African American; Male; Eye color: hazel; Quality score—99.3; (b). African American; Male; Eye color: dark brown; Quality score—99.2; (c). Russian; Caucasian; Male; Eye color: regular brown; Image with light occlusion; Quality score—98.7; (d). American; Caucasian; Female; Eye color: blue; Image with occlusion; Quality score— 89.6

3.2. Iris Images, Image Quality Scores and Iris Verification with Eyeglasses/Contact Lenses

Figure 2 (a) shows a Chinese male's brown iris image and the quality score during the iris enrollment with eyeglasses. It indicates that there were reflections due to his eyeglasses; therefore, the iris quality score decreased. Figure 2 (b) shows an African American male's dark brown iris image and the quality score during the iris enrollment with contact lenses. It shows a contact lens shifted over the lower right part of the iris region and slightly off-center of the pupil.

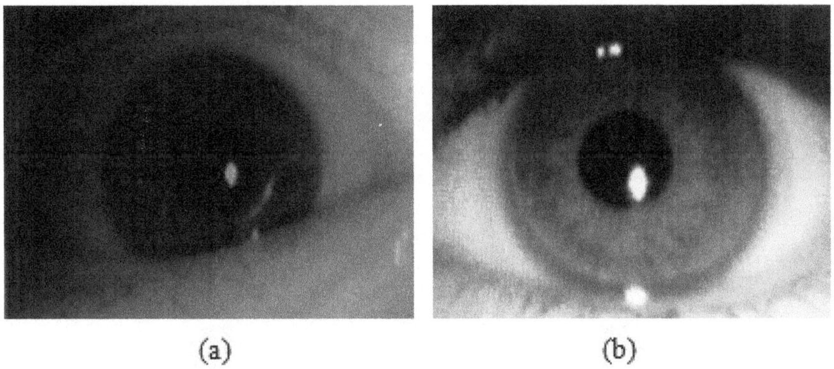

Figure 2. Iris images and image quality scores with eyeglasses/contact lenses: (a). Chinese; Male; Eye color: brown; With eyeglasses; Right eye iris image with occlusion; Quality score—82.2; (b). African American; Male; Eye color: dark brown, With contact lens; Left eye iris image; Quality score—98.9

Six people were tested when they wore eyeglasses and when they did not wear eyeglasses using the IrisAccess^{TM} 4000 system. Ten people were tested when they wore contact lenses and when they did not wear contact lenses using the system. The stand-off distance was still 21 cm. Table 1 lists the iris testing results, including the iris image quality scores of the enrollment and verification outcomes under the conditions with/without eyeglasses. Table 2 lists the iris testing results, including the iris image quality scores of the enrollment and verification outcomes under the conditions with/without contact lenses. Because the verification was conducted right after the enrollment process (almost at the same time), the iris image score during the verification was regarded as almost the same as the score during the enrollment. Table 1 indicates that eyeglasses did not affect the success in iris verification although they can decrease the iris image quality scores. Table 2 indicates that contact lenses can decrease the iris image quality scores; however, they did not affect the success in iris verification.

Table 1. The effects of eyeglasses on iris image scores and verification outcomes (without occlusion; the right eye tested)

No	Eye colors	Enrollment *With glasses?*	Verification *With glasses?*	Image scores	Verification outcomes
1	Dark Brown	No	No	99.4	Success
		Yes	Yes	89.6	Success
2	Dark Brown	No	No	98.3	Success
		Yes	Yes	87.4	Success
3	Regular Brown	No	No	98.6	Success
		Yes	Yes	89.5	Success
4	Blue	No	No	98.1	Success
		Yes	Yes	87.9	Success
5	Green	No	No	97.8	Success
		Yes	Yes	87.1	Success
6	Hazel	No	No	97.7	Success
		Yes	Yes	86.8	Success

Table 2. The effects of contact lenses on iris image scores and verification outcomes (without occlusion; the right eye tested)

No	Gender and Eye colors	Enrollment *With contact lenses?*	Verification *With contact lenses?*	Iris image scores	Verification outcomes
1	Male, Dark Brown	No	No	98.5	Success
		Yes	Yes	97.9	Success
2	Male, Dark Brown	No	No	99.2	Success
		Yes	Yes	98.9	Success
3	Male, Dark Brown	No	No	99.2	Success
		Yes	Yes	99.0	Success
4	Female, Dark Brown	No	No	99.3	Success
		Yes	Yes	98.9	Success
5	Female, Dark Brown	No	No	98.9	Success
		Yes	Yes	98.7	Success
6	Male, Regular Brown	No	No	98.6	Success
		Yes	Yes	98.0	Success
7	Female, Blue	No	No	98.1	Success
		Yes	Yes	97.3	Success
8	Female, Green	No	No	97.8	Success
		Yes	Yes	97.1	Success
9	Male, Hazel	No	No	97.7	Success
		Yes	Yes	96.6	Success
10	Female, Hazel	No	No	98.5	Success
		Yes	Yes	97.1	Success

3.3. Parametric Analysis for the Image Quality of Three Eye Colors Based on ANOVA

In addition to the above subjects tested, 98 additional people were invited to participate in iris enrollment and verification tests in April, 2013 to study the difference in iris image quality among three types of eyes in dark brown, hazel, and blue. The individuals were 17 to 28 years old.

Thirty four of them are African Americans with dark brown eyes; and 33 are African Americans with hazel eyes; and 31 are Caucasians with blue eyes. The mean of the 98 people's iris image quality scores is $\overline{x} = 98.79$; the median m= 98.8; and the standard deviation s= 0.393. Figure 3 shows the frequency distribution of these people's iris image quality scores.

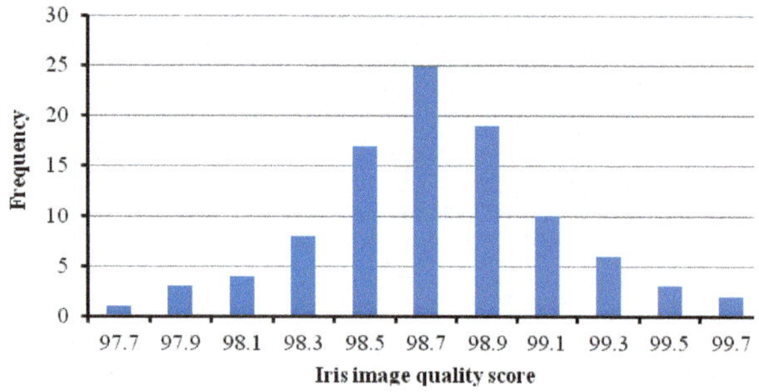

Figure 3. Frequency distribution for 98 subjects' iris image quality scores

The Pearsonian coefficient of skewness [23] is given by

$$SK = \frac{3*(\overline{x}-m)}{s} \tag{1}$$

Substituting into the formula the values of the values of the mean, the median, and standard deviation, hence, *SK* = - 0.0856.

This *SK* value reflects that the distribution is of an approximately symmetrical distribution with a very slight negative skewness. Therefore, the distribution can be assumed to be approximated closely with a normal distribution.

Table 3 shows the means and standard deviations of the iris image quality scores for people in three eye colors. The means and the standard deviations for the people in dark brown eyes, hazel eyes, and blue eyes are very close, respectively.

Table 3. The means and the standard deviations of the iris image quality scores in three eye color groups

Eye Group No	Eye Color	Number of Subjects	Mean $\overline{x}$	Standard Deviation s
1	Dark Brown	34	98.800	0.355
2	Hazel	33	98.824	0.392
3	Blue	31	98.739	0.439

Analysis of variance (ANOVA) for the means the iris image quality scores for the people in the three eye colors was conducted. During the ANOVA procedures, the total sum of squares (*SST*), the treatment of squares (*SS(Tr)*), the error sum of squares (*SSE*), the treatment mean square (*MS(Tr)*), the error mean square (*MSE*), the F statistic were calculated using the following formula [23]:

$$SST = \sum_{i=1}^{k}\sum_{j=1}^{n_j} x_{ij}^2 - \frac{1}{N}T_{..}^2 \tag{2}$$

$$SS(Tr) = \sum_{i=1}^{k}\frac{T_i^2}{n_i} - \frac{1}{N}T_{..}^2 \tag{3}$$

$$SSE = SST - SS(Tr) \tag{4}$$

$$MS(Tr) = \frac{SS(Tr)}{k-1} \tag{5}$$

$$MSE = \frac{SSE}{k(n-1)} \tag{6}$$

$$F = \frac{MS(Tr)}{MSE} \tag{7}$$

Where x_{ij} is the *j*th observation of the *i*th sample; $N = n_1 + n_2 + \cdots + n_k$; T_i denotes the sum of the values in the *i*th sample; and $T_{..}$ denotes the grand total of all the data in the *k* samples. In this study, $k = 3$; $n_1 = 34$; $n_2 = 33$; $n_3 = 31$; $N = 98$; $T_{1.} = 3359.2$; $T_{2.} = 3261.2$; $T_{2.} = 3060.9$; and $T_{..} = 9681.3$.

The author formulated the following null hypothesis:

There is no statistically significant difference in means of the iris image quality scores for the people in the three eye colors (dark brown, hazel, and blue). The outcome is: the hypothesis is accepted or rejected at $\alpha = 0.05$. α is the level of significance. It can be other values such as 0.01. The ANOVA test follows the following criterion: If F exceeds the value $F_{0.05}$, the null hypothesis must be rejected.

All the results for the people in the three eye colors are shown in the following analysis-of-variance table:

Table 4. ANOVA table

Source of Variation	Degrees of Freedom	Sum of Squares	Mean Square	F
Treatments	2	0.13	0.0650	0.4153
Error	95	14.87	0.1565	
Total	97	15.00		

Since $F = 0.4153$; the value is less than $F_{0.05} = 3.09$ obtained from the literature [23,24], the null hypothesis must be accepted. In other words, there is no significant difference in the iris image quality scores of eyes in dark brown, hazel, and blue when the level of significance is 0.05.

4. Conclusions and Future Work

Iris image quality scores for five kinds of eyes (dark brown, regular brown, hazel, green, and blue) without eyeglasses were obtained through the IrisAccess™ 4000 system. The iris verifications for the five kinds of eyes were successful although occlusion reduced the image quality score of the blue eye.

The iris verifications with eyeglasses can be still successful although eyeglasses decrease iris image quality scores. Contact lenses decrease the iris image quality scores; however, they do not affect the success in iris verification.

The results obtained from the parametric method based on ANOVA indicate that there is no significant difference in the iris image quality of eyes in dark brown, hazel, and blue when the level of significance α is 0.05.

For the author's future work, the experimental datasets should include more iris image quality data, in particular the subjects from different races. The effect of choosing the left eye, the right eye, or both on the performance (such as the number of trials or attempts to capture images or complete verification, the successful rate) during iris enrollment and iris verification will be evaluated. This needs a lot of subjects to participate in testing. More research will be conducted on iris recognition in less-controlled conditions, and improving recognition or identification performance for people wearing eyeglasses or contact lenses.

References

[1] Devireddy, Srinivasa Kumar. "An Accurate Human Identification Through Iris Recognition." *Computer Science and Telecommunications*, vol. 23, no. 6, 2009, pp. 22-29.

[2] Ezhilarasan, M., Jacthish, R., Subramanian, Ganabathy K.S. and Umapathy, R. "Iris Recognition Based on Its Texture Patterns." *International Journal on Computer Science and Engineering*, vol. 2, no. 9, 2010, pp. 3071-3074.

[3] Lee, J-C, Su, Y., Tu, T-M. and Chang C-P. "A novel approach to image quality assessment in iris recognition systems." *The Imaging Science Journal*, vol. 58, 2010, pp. 136-145.

[4] Wei, Zhuoshi, Tan, Tieniu, Sun, Zhenan and Cui, Jiali. "Robust and Fast Assessment of Iris Image Quality." *Proceedings of the 2006 International Conference on Advances in Biometrics*, Springer-Verlag Berlin, Heidelberg, 2006, pp. 464-471.

[5] Cambier, Jim. "Iris Image Quality Metrics", Company Confidential and Proprietary, Technical Report, November, 2007.

[6] Kalka, Nathan D., Zuo, Jinyu, Schmid, Natalia A. Cukic, Bojan. "Image quality assessment for iris biometric." *Proc. of 2006 SPIE Conf. on Biometric Technology for Human Identification III*, Orlando, FL, USA, April 17-18, 2006, vol. 6202, pp. pp. 445-452.

[7] Bowyer, Kevin W., Hollingsworth, Karen and Flynn, Patrick J. "Image Understanding for Iris Biometrics: A Survey." *Computer Vision and Image Understanding*, vol. 110, no. 2, May 2008, pp. 281-307.

[8] Kang, Byung Jun, Park, Kang Ryoung. "A new multi-unit iris authentication based on quality assessment and score level fusion for mobile phones." *Machine Vision and Applications*, no. 21, 2010, pp. 541-553.

[9] Li, Xingguang, Sun, Zhenan, Tan, Tieniu. "Comprehensive assessment of iris image quality." 18th IEEE International Conference on Image Processing, Brussels, Belgium, September 11-14, 2011, pp. 3117-3120.

[10] McConnon, G. et al. "A Survey of Point-Source Specular Reflections in Noisy Iris Images." International Conference on Emerging Security Technologies, Canterbury, United Kingdom, September 6-7, 2010, pp. 13-17.

[11] Chaskar, U. M., Sutaone, M. S., Shah, N. S., Jaison. T. "Iris Image Quality Assessment for Biometric Application." *International Journal of Computer Science Issues*, vol. 9, no. 3, May 2012, pp. 474-478.

[12] Proenca, Hugo. "An iris recognition approach through structural pattern analysis methods." *Expert Systems*, February 2010, vol. 27, no. 1, pp. 6-16.

[13] Kang, Byung Jun, Park, Kang Ryoung. "A Study on Restoration of Iris Images with Motion-and-Optical Blur on Mobile Iris Recognition Devices." *International Journal of Imaging Systems & Technology*, vol. 19, 2009, pp. 323-331.

[14] LG Electronics - Iris Technology Division. IrisAccess™ Software User Manual, Version 3.00, New Jersey, USA, December13, 2007.

[15] LG Electronics - Iris Technology Division. IrisAccess™ 4000 Hardware Manual, New Jersey, USA, 2008.

[16] Wang, Lidong. "Iris Image Quality Testing and Iris Verification." *International Journal of Electrical and Computer Engineering*, vol. 3, no. 4, 2013, pp. 1-7.

[17] Wasserman, Philip D. "Digital Image Quality for Iris Recognition." Biometric Image Quality Workshop, National Institute of Standards and Technology, USA, March 8-9, 2006.

[18] Fu, Jian, Caulfield, H. John, Yoo, Seong-Moo, Atluri, Venkata, "Use of Artificial Color filtering to improve iris recognition and searching." *Pattern Recognition Letters*, vol. 26, 2005, pp. 2244-2251.

[19] Wang, Changyu, Song, Shangling. "An iris recognition algorithm based on fractal dimension." *Acta Automatica Sinica*, vol. 33, no. 7, 2007, pp.608-702.

[20] Birgale, L., Kokare, M. "Comparison of Color and Texture for Iris Recognition." *International Journal of Pattern Recognition and Artificial Intelligence*, vol. 26, no. 3, 1256007, 2012.

[21] Monaco, Matthew K. "Effect of Eye Color on Iris Recognition." Thesis for the degree of Master of Science, Virginia University, 2007.

[22] Smith, Kelly N. "Analysis of Pigmentation and Wavefront Coding(TM) Acquisition in Iris Recognition." January 1, 2007, ProQuest Publisher.

[23] Freund, J. E. and Perles, B. M. *Statistics: A First Course.* (8th Ed.), Pearson Prentice Hall, New Jersey, 2004.

[24] Merrington M. and Thompson, C. M. "Tables of percentage points of the inverted beta (F) distribution." *Biometrika*, vol. 33, 1943.

[25] Sulem, P. et al. "Genetic determinants of hair, eye and skin pigmentation in Europeans." *Nature Genetics*, vol. 39, Dec. 2007, pp. 1443-1452.

3

Robust Identification of Hydrocarbon Debutanizer Unit using Radial Basis Function Neural Networks (RBFNNs)

Masih Vafaee Ayouri[1,*], Mehdi Shahbazian[1], Bahman Moslemi[2], Mahboobeh Taheri[3]

[1]Department of Instrumentation and Automation Engineering, Petroleum University of Technology, Ahwaz, Iran
[2]Department of Basic Science, Petroleum University of Technology, Ahwaz, Iran
[3]Senior expert in R&D, Sarkhoon & Qeshm Gas treating Company, Bandar Abbas, Iran
*Corresponding author: masih.vafaee@gmail.com

Abstract Radial Basis Function Neural Network (RBFNN) is considered as a good applicant for the prediction problems due to it's fast convergence speed and rapid capacity of learning, therefore, has been applied successfully to nonlinear system identification. The traditional RBF networks have two primary problems. The first one is that the network performance is very likely to be affected by noise and outliers. The second problem is about the determination of the parameters of hidden nodes. In this paper, a novel method for robust nonlinear system identification is constructed to overcome the problems of traditional RBFNNs. This method based on using Support Vector Regression (SVR) approach as a robust procedure for determining the initial structure of RBF Neural Network. Using Genetic Algorithm (GA) for training SVR and select the best parameters as an initialization of RBFNNs. In the training stage an Annealing Robust Learning Algorithm (ARLA) has been used for make the networks robust against noise and outliers. The next step is the implementation of the proposed method on the Hydrocarbon Debutanizer unit for prediction of n-butane (C4) content. The performance of the proposed method (ARLA-RBFNNs) has been compared with the conventional RBF Neural Network approach. The simulation results show the superiority of ARLA-RBFNNs for process identification with uncertainty.

Keywords: *robust system identification, RBF Neural Networks, hydrocarbon debutanizer unit, support vector regression, genetic algorithm, annealing robust learning algorithm*

1. Introduction

Artificial neural networks are recognized as major tools for optimization, pattern recognition and nonlinear system identification because of their learning and modeling abilities [1]. System identification is the science of building mathematical models of dynamic systems from Input–Output(I/O) pairs. It can be realized as the interface between the real world of applications and the mathematical world of model abstractions [2]. Robust identification is a method to determine the parameters of neural network when training data contaminated with noise and outliers. The intuitive definition of an outlier (Hawkins, 1980) is "*an observation which deviates so much from other observations as to arouse suspicions that it is generated by a different mechanism*" [3,4]. However, outliers may occur due to erroneous measurements in measurement device (sensors). When outlier exist in training data set, approximation of predicted models with neural networks is largely deteriorated [5,6].

Between the existing neural networks architectures, the Radial Basis Function Neural Networks (RBFNNs) is considered as a good candidate for approximation and prediction due to its rapid learning capacity and simpler network structure [7]. Accordingly, the RBFNNs is a popular alternative to the Multilayer Perceptron (MLP). In contrast to MLPs, RBF networks use a localized representation of information. RBFNNs were introduced into the neural network literature by Broomhead and Lowe (1988). Radial basis function (RBF) networks typically have three layers which include an input layer, a hidden layer with a non-linear activation function and a linear output layer [3,4,7,8]. The training procedure of RBF networks is accomplished through the estimation of three kind of parameters, namely the centers and the width of basis function and finally, the synaptic weights [8]. Center selection is an important role in performance of RBFNNs. There are two kinds of methods exist in this scope. One of them is fixed selection of hidden node center (such as random selection from training data and fuzzy c-means clustering) and the other chooses this parameter systematically. SVR approach is a systematic way to define the initial structure of RBFNNs. supervised and unsupervised learning are common learning methods in RBF networks. Unsupervised learning is used to initialize the network parameters, and supervised learning is usually used for the fine tuning of the network parameters [9].

Many learning algorithms in literature have been proposed for training RBFNNs such as hybrid fuzzy clustering approach [8], orthogonal least squares (OLS) [10], gradient descent [11] and extended kalman filter (EKF) [12]. When training data contaminated with noise and outliers, traditional learning methods (least squares based) show poor performance therefore we must approach robust learning algorithm. The following robust training methods are proposed in literature for training RBFNNs: In 1995, sanchez presented Scaled Robust Loss Function (SRLF) and conjugate gradient method for Robust learning RBFNNs [5]. Chien-Cheng Lee et al. derived new robust objective function from robust statistics (type of Hample's M-estimator) and back propagation as learning algorithm in order to reduce the influence of outliers in training patterns [13]. Chen-Chia Chuang et al. proposed Annealing Robust Backpropagation Learning Algorithm (ARBP) that adopts the annealing concept into the robust learning algorithms to overcome the problem of training network with noise and outliers [14]. Mei-juan Su et al. (2006) illustrate that by using new SRLF and gradient descent as training method and shows strength robustness against outliers.

In this paper we use ε-SVR with Gaussian kernel function for systematic determination of an initial structure of RBFNNs. Another advantage of this method is robust initialization of RBFNNs against data uncertainties. When the initial structure of RBFNNs determined with ε-SVR, the next step is training the SVR parameter using Genetic Algorithm to find optimal parameters from SVR. Constructing the network using this parameter and using Annealing Robust Learning Algorithm (ARLA) in training phase. In this stage we used as M-estimator [15] as a popular method to fix the problem of parameter estimation when data contain uncertainty. The least squares based estimators residuals deviated from zero when we faced outliers in the data set therefore M-estimators are used as a robust estimation method to decrease the effect of outliers in the data set. At the point When the residuals goes outside a threshold, the M-estimator suppresses the response instead. Therefore, the M-estimator based error function is more robust to noise and outliers than the Least Mean Square (LMS) based error function. This robustness can be achieved by replacing the MSE criterion with M-estimator [6]. In this study Logistic loss function was adopted to develop a robust RBF network. Thus, the cost function is a special case of the logistic function with properly setting constants in the theory of the M-estimator. We have adopted the concepts of M-estimator and annealing to develop a feedforward network with a robust backpropagation learning algorithm. At the same time, the proposed approach has fast convergence speed and robust against outliers on the identification of the nonlinear dynamic systems contain noise and outliers.

The proposed method will be implemented on the Hydrocarbon Debutanizer unit and the capability of this method for prediction of n-butane (C4) content will be investigated. This paper is organized as follows: In Section 2 full description of the method is proposed. The implementing and its results of the proposed method on a Hydrocarbon Debutanizer unit together with a short description of the process are presented in section 3 and section 4 presents conclusions.

2. Methodology Description

2.1. Architecture of RBFNNs for Identification of Nonlinear Systems

Assume that the unknown nonlinear dynamical system is expressed by:

$$y(t+1)=f\begin{pmatrix}u(t),u(t-1),\dots,u(t-m),y(t),\\ y(t-1),\dots,y(t-n)\end{pmatrix} \quad (1)$$

Where m is input lag, n is output lag and f(.) is an unknown nonlinear function needed to be identified. The goal of the identification problems is to find a suitable prediction model $\hat{y}(t+1)$.

$$\hat{y}(t+1)=\hat{f}\begin{pmatrix}u(t),u(t-1),\dots,u(t-m),y(t),\\ y(t-1),\dots,y(t-n)\end{pmatrix} \quad (2)$$

Such that $\hat{f}$ predicts f(.).

A Radial Basis Function Neural Networks (RBFNNs) have three layers, the input layer, the hidden layer with nonlinear activation function, and the linear output layer. The architecture of RBFNNs with Gaussian basis function for identification problem is expressed in Figure 1.

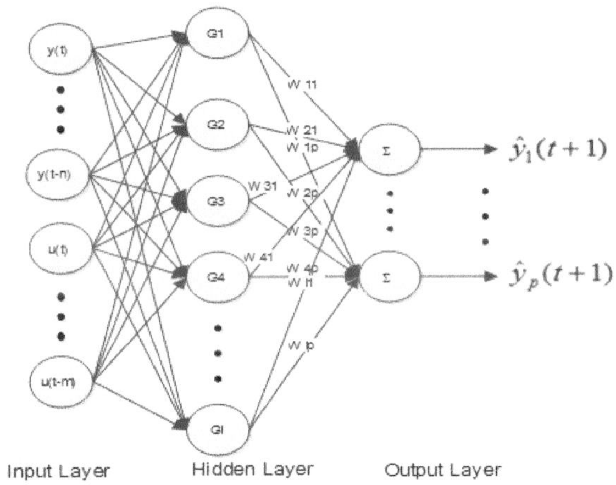

Figure 1. Architecture of Radial Basis Function Neural Networks for Identification Problem

A RBFNNs would be shown in the form:

$$\hat{y}_o(t+1)=\sum_{j=0}^{L}W_{jo}G_j=\sum_{j=0}^{L}W_{jo}\exp(-\frac{X-m_j^{\ 2}}{2\sigma_j^2}) \quad (3)$$

For $0\le j\le L$ and $1\le o\le p$.

Where $X=\begin{pmatrix}u(t),u(t-1),\dots,u(t-m),\\ y(t),y(t-1),\dots,y(t-n)\end{pmatrix}$ is the input vector, $\hat{y}_o(t)$ is the oth network predicted output, W_{jo} is the synaptic weight, G_j is the Gaussian function at the jth hidden layer neuron, m_j and σ_j are the center and width of Gaussian function, respectively, and L is equal to the number of hidden layer neurons [16].

When utilizing an RBFNNs for the identification of nonlinear dynamical system, our objective is to determine the values of RBFNNs parameters (L, W_{jo} , m_j) to minimize the following objective function:

$$\sum_{i=1}^{N}\left\| y_i - \hat{y}_i \right\|^2 \quad (4)$$

Where N is the number of training data pairs (I/O pairs), y_i is the output of unknown nonlinear dynamical systems and $\hat{y}_i$ is the prediction value from identified model. For determining better combination of optimal parameters, iterative methods can be applied. When constructing a RBFNNs, it is significant to define the net structure and initializes the network parameters. Most of the conventional RBFNNs approaches (Least squares based) are easily influenced by long tail noise and outliers , therefore, robust approaches are given to overcome the problem of traditional RBFNNs. These robust radial basis functions approaches mainly focus on robust learning algorithms. These algorithms assume the concept of robust estimators in the training phase. For construction the network structure we use $\varepsilon-SVR$ approach for determining initial network structure and parameter of $\varepsilon-SVR$ method tuned with Genetic Algorithm (GA) and using iterative methods for training RBFNNs parameters. How to use the $\varepsilon-SVR-GA$ method to get the optimal structure of RBFNNs will be exemplified in the succeeding segment.

2.2. ε–*SVR*–*GA* Approach for Structure Selection of RBFNNs

Suppose we are given training I/O pair data which includes $\{(x_1,y_1),(x_2,y_2),\ldots,(x_N,y_N)\}$ where X denotes the input space pattern and Y means output data. That is, regression function in the SVR approach is approximated by the following function as:

$$f(x,\theta)=\sum_{i=1}^{N}\theta_k\Phi(x_k)+b \quad (5)$$

In 1995 Vapnik [17] proposed $\varepsilon-SVR$ as a solution for the problem is to find $f(x,\theta)$ that minimize the subsequent risk function

$$R(\theta)=\frac{1}{N}\sum_{i=1}^{N}L_\varepsilon(y_i-f(x,\theta)) \quad (6)$$

Subject to the constraint

$$\theta^2 < C, \quad (7)$$

Where C is a constant and $L_\varepsilon(.)$ is the ε -insensitive loss function defined as:

$$L_\varepsilon(e)=\begin{cases} o & for\ |e|<\varepsilon \\ |e|-\varepsilon & otherwise \end{cases} \quad (8)$$

Where $e=y(i)-\hat{y}(i)$ is the error between ith desired output and ith output of RBFNNs and ε is a nonnegative number.

Since an SVR approach with the insensitive loss function provides an estimated function within zone, the initial construction of the RBFNNs can be obtained by the SVR approach that evenhandedly supplies better initialization to learning algorithm. Three kinds of parameter which can be chosen appropriately to determine the best initialization parameters of RBFNNs consist of penalty factor (C), epsilon (coefficient of the SVR loss function) and sigma (width of kernel function). Moreover, an SVR approach with the ε -insensitive loss function can make use of a small subset of the training data, called the support vectors (SVs), to approximate the unknown functions within a tolerance band ε The numbers of SVs are controlled by the values of tolerance band ε [17,18,19].

Shows the situation graphically. Only the points outside the shaded region contribute to the cost insofar, as the deviations are penalized in a linear fashion. It turns out that the optimization problem can be solved more easily in its dual formulation and solving this problem [19] with the Lagrange multipliers method and minimized above loss function leads to following dual optimization problem [4,16,17,19].

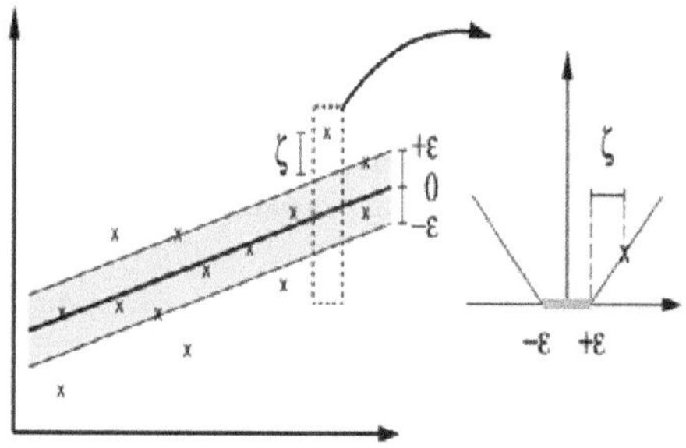

Figure 2. The soft margin loss setting corresponds for a linear SVR [19]

Minimize

$$Q(\alpha,\alpha^*)=\varepsilon\sum_{r=1}^{N}(\alpha_r+\alpha_r^*)-\sum_{r=1}^{N}y_r(\alpha_r-\alpha_r^*) + \frac{1}{2}\sum_{r,s=1}^{N}(\alpha_r^*-\alpha_r)(\alpha_s^*-\alpha_s)\ll\Phi(x_r).\Phi(x_s)\gg \quad (9)$$

Subject to the constraint

$$\sum_{r=1}^{N}\alpha_r^*=\sum_{r=1}^{N}\alpha_r, 0<\alpha_r^*,\alpha_r<C\ for\ r=1,2,\ldots,N \quad (10)$$

Where Q is a cost function in the $\varepsilon-SVR$, $\alpha,\alpha^*,\alpha_r,\alpha_r^*,\alpha_s,\alpha_s^*$ are all the nonnegative Lagrange multipliers, r and s are all indexes, X is input and Y is output. The inner product of basis function is replaced via kernel function

$$K(x_r,x_s)=\ll\Phi(x_r).\Phi(x_s)\gg \quad (11)$$

The kernel function determines the smoothness properties of solutions and should reflect a previous knowledge of the data. In the literature Gaussian kernel function often used. Hence equation (9) rewritten as

$$Q(\alpha,\alpha^*)=\varepsilon\sum_{r=1}^{N}(\alpha_r+\alpha_r^*)-\sum_{r=1}^{N}y_r(\alpha_r-\alpha_r^*) + \frac{1}{2}\sum_{r,s=1}^{N}(\alpha_r^*-\alpha_r)(\alpha_s^*-\alpha_s)K(x_r,x_s) \quad (12)$$

Therefore, the solution of the SVR method [15] is in the form of the following linear expansion of kernel function:

$$f\left(x,\alpha,\alpha^*\right)=\sum_{k=1}^{N}(\alpha_k^*-\alpha_k)K\left(x,x_k\right)+b \qquad (13)$$

Note that only some of $\left(\alpha_k^*-\alpha_k\right)$s are not zeros and the corresponding vectors x_k are termed support vectors (SVs). That is, $\#SVs \le N$, SVs is number of SVs. In this paper, we use the Gaussian kernel function is used in kernel of $\varepsilon-SVR$ and then relation (13) can be rewritten as

$$f\left(x,\lambda\right)=\sum_{k=1}^{\#SVs}\lambda_k \exp\left(-\frac{\left\|x-x_k\right\|^2}{2\sigma_k^2}\right)+b \qquad (14)$$

Where $\lambda_k=(\alpha_k^*-\alpha_k)\neq 0$ and x_k are SVs and b is bias of the network.

To design an effective SVR model, the values of SVR parameters have to be chosen carefully. For this purpose, we use Genetic Algorithm as an advanced tool to find the best solution. The concept of GA was developed by Holland and his coworkers in the 1960s and 1970s [20]. Genetic algorithms (GAs) are well appropriate for searching global optimal values in complex search space (multi-modal, multi-objective, non-linear, discontinuous, and highly constrained space), combined with the fact that they work with raw objectives only when compared with conventional techniques [21,22].

The procedure of GA for finding the best solution can be summarized below:

1. Choose a randomly generated population.
2. Calculate the fitness of each chromosome in the population.
3. Create the offspring by the genetic operators: selection, crossover and mutation.
4. Check the termination condition. If the new population does not satisfy the termination condition, repeat steps 2 up to 4 for the generated offspring as a new starting population

The SVR parameters are as follows:

- Regularization parameter C, which determines the tradeoff cost between minimizing the training error and minimizing the complexity of the model.
- Spread parameter (σ^2) of the kernel function which defines the width of Gaussian kernel function.
- Epsilon parameter (ε) of the loss function which determine the number of SVs. A small ε value allows more points to be outside the ε-tube and results in more SVs and a large ε value results in less SVs and probably in a smoother regression function.

2.3. Annealing Robust Learning Algorithm (ARLA) for Training RBFNNs

An ARLA is proposed as a learning algorithm for training RBFNNs. An important feature of ARLA that adopts the annealing concept in the cost function of the robust backpropagation learning algorithm is suggested in [4]. Therefore, the ARLA can overcome the existing problems in the traditional backpropagation learning algorithm when data contaminated with outliers. The intuitive definition of an outlier (Hawkins, 1980) is "*an observation which deviates so much from other observations as to arouse suspicions that it is generated by a different mechanism*" [3,4]. However, outliers may occur due to erroneous measurements in measurement device(sensors). Outliers in training data set can cause substantial deterioration of the approximation realized by a neural network architecture [5,6]. Statistical techniques are often used which are sensitive to such outliers, and negative results may have been affected by them, and the most robust and resistant methods have been developed since 1960 and less sensitive to outliers. Robustness is the key issue for system identification.

A cost function for ARLA is defined as

$$E_{ARLA}\left(t\right)=\frac{1}{N}\sum_{i=1}^{N}\rho[e_i\left(t\right);\beta(t)] \qquad (15)$$

Where

$$e_i\left(t\right)=y_i-f\left(x_i\right)=y_i-\sum_{j=1}^{L}W_j\exp\{-\frac{\left\|x-m_j\right\|^2}{2\sigma_j^2}\} \qquad (16)$$

t is number of epochs, $e_i\left(t\right)$ is the error between the ith desired output and the ith output of the trained network at epoch t, $\beta\left(t\right)$ is a deterministic annealing schedule acting like the cut-off points (scale estimator) and $\rho(.)$ is Logistic loss function and defined as

$$\rho\left(e_i,\beta\right)=\frac{\beta}{2}\ln(1+\frac{e_i^2}{\beta}) \qquad (17)$$

Our objective function defined as

$$J\left(h\right)=\frac{1}{N}\sum_{i=1}^{N}\rho[e_i\left(t\right);\beta\left(t\right)] \qquad (18)$$

In a Logistic loss function using annealing schedule (Scale Estimator) as a threshold for the rejection of outliers. Usually, the scale estimator can be chosen in two ways. One is to obtain the scale estimator based on some robust statistic theories such as the median errors and the median of absolute deviation (MAD) and the other way of defining the scale estimator is to count out the pre-specified percentage of points [14]. This loss function with scale estimator degrade the effect of noise and outliers in dataset.

Based on Gradient-Descent as a type of learning algorithm the parameters of RBFNNs (W_{jo},m_j,σ_j) updated as

$$\Delta W_{jo}=-\eta\frac{\partial J}{\partial W_{jo}}=-\eta\sum_{i=1}^{N}\varphi(e_i,\beta)\frac{\partial e_i}{\partial W_{jo}} \qquad (19)$$

$$\Delta m_j=-\eta\frac{\partial J}{\partial m_j}=-\eta\sum_{i=1}^{N}\varphi(e_i,\beta)\frac{\partial e_i}{\partial m_j} \qquad (20)$$

$$\Delta \sigma_j=-\eta\frac{\partial J}{\partial \sigma_j}=-\eta\sum_{i=1}^{N}\varphi(e_i,\beta)\frac{\partial e_i}{\partial \sigma_j} \qquad (21)$$

$$\varphi\left(e_i;\beta\right)=\frac{\partial\rho(e_i,\beta)}{\partial e_i}=\frac{e_i}{1+\left(e_i\right)^2/\beta(h)} \qquad (22)$$

Where η is a learning rate, $\varphi(.)$ is usually called the influence function. When outliers exist in dataset, they have a major impact on the predicted results. In the ARLA, the properties of the annealing schedule $\beta(h)$ have [12]:

- $\beta_{initial}, \beta(h)$ has large values for the first epochs;
- $\beta(h) \to 0^{+}$ for $h \to \infty$;
- $\beta(h) = \dfrac{k}{h}$ for any h epoch, where k is a constant.

The procedure of robust learning algorithm for RBFNNs described as follows:

2.3.1. Robust learning Procedure

Step 1: Initializing the structure of RBFNNs using $\varepsilon - SVR$ -GA as shown in section 2.2.

Step 2: Compute the predicted output of network and its error for training data.

Step 3: Find the values of annealing schedule $\beta(h) = \dfrac{k}{h}$ for each epoch, where $k = 2{*}\max\left\{|e_i|_{initial}\right\}$.

Step 4: Update the network parameters such as (synaptic weights (W_{jo}), width of Gaussian function (σ_j) and the centers (m_j)).

Step 5: Compute the robust cost function explain in (18).

Step 6: If the stopping criteria are not satisfied, then go to Step 2; otherwise terminate the training stage.

3. Case Study, Implementation and Results

3.1. Hydrocarbon Debutanizer Process

Debutanizer column is an essential component of the gas recovery unit in the petroleum refineries. It is used to recover butane from the light end product containing C2–C8 hydrocarbons [23]. Nevertheless, the debutanizer column demonstrations high dimensional coupling with severe nonlinearity and cooperative set of operational constraints. A debutanizer is a type of fractional distillation column used to recover light gases (C1-C4) and Liquefied Petroleum Gas (LPG) from the overhead distillate before producing light naphtha during the refining process. Distillation is the process of heating a liquid to vapor and condensing the vapors back to liquid in order to separate or purify the liquid. The main equipment of this process consist of a distillation column, reboiler and condenser. The debutanizer condenser condenses the overhead vapor and the debutanizer overhead pressure control valves controls the overhead system. The reflux from the top of the debutanizer consists of the collected condensed hydrocarbon (light hydrocarbon). There are three manipulated variables for the distillation column which are the feed flow rate, reflux flow rate and reboiler duty. The feed flow rate controls the feed to the column, the debutanizer reboiler control valve controls the reboiler temperature while the debutanizer bottom level controller controls the bottom product (Heavier Hydrocarbon) level. The debutanizer reflux control valve controls the ratio of the liquid and distillate flow rate at the top of the column. This column is a challenging process because a highly nonlinear multivariable process, involves a great deal of interactions between the variables, has lag in many variable of the control system (Dynamic System), all of which makes it difficult system to be modeled by linear techniques [24]. Therefore we use nonlinear models for the prediction of properties and identification of this process. At this time, the composition of the debutanizer products is measured through tedious and time consuming laboratory measurements. Consequently, prediction of product quality is an important issue for solving these problems. To maintain the product quality at a desired level, it is necessary to predict the top and bottom compositions of the debutanizer column quickly with a high degree of precision[25].Therefore robustness issues are important for contaminated data. The debutanizer column considered in this study is a fifteen-stage multi-component distillation column fed by two input streams consisting of a mixture of light hydrocarbons [26]. The two input feed streams to debutanizer are compositions of the light hydrocarbons containing i-butane, n-butane, i-butene, i-pentane, n-pentane, n-hexane, n-heptane and n-octane. The Process Flow Diagram (PFD) of this process presented in the Figure 3.

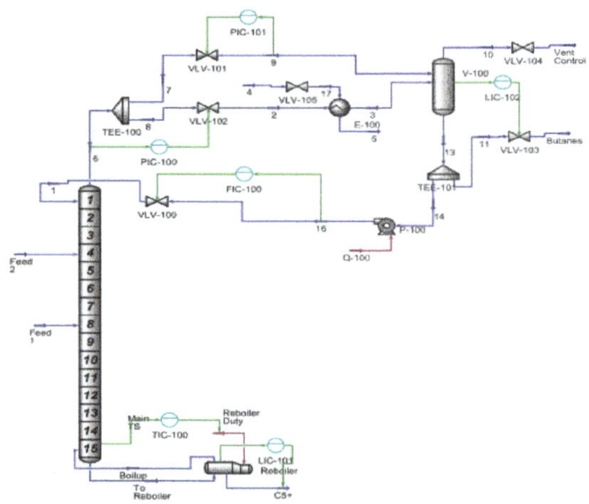

Figure 3. PFD of Debutanizer column

3.2. Results and Discussion

To demonstrate the validity of the proposed ARLA-RBFNNs identification method in practical modeling applications, the proposed method has been used for identification of Debutanizer unit.

In the identification problems, the training input-output data are obtained by feeding a signal $x(t)$ to the MISO system and measuring corresponding outputs $y(t+1)$. An objective of the debutanizer is minimizing the n-butane (C4) content in the debutanizer bottom flow, this concentration is chosen as the output variable which must be estimated by the proposed algorithm.

After gathering 1000 data samples from the debutanizer, 60 percent of given data selected randomly picked up as the training data (600 sample) for modeling and the remaining samples were used for testing purpose. some of the original data set are replaced by artificially generated Gaussian noise, Cauchy noise and outliers for evaluating the robustness of the proposed algorithm against uncertainty. The following Figures Demonstrates the data set contaminated with artificially Gaussian noise and Cauchy noise with outliers.

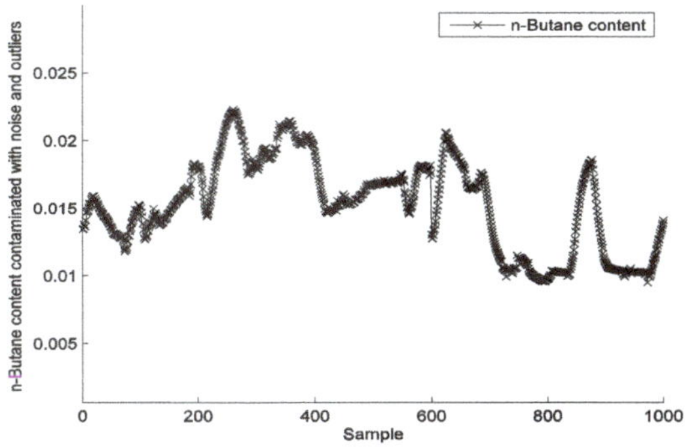

Figure 4. Contaminated data with Gaussian noise

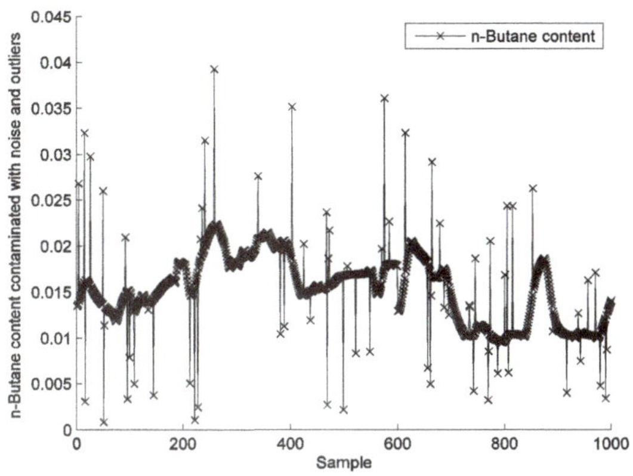

Figure 5. Contaminated data with Gaussian noise and 7% outlier (STD=5)

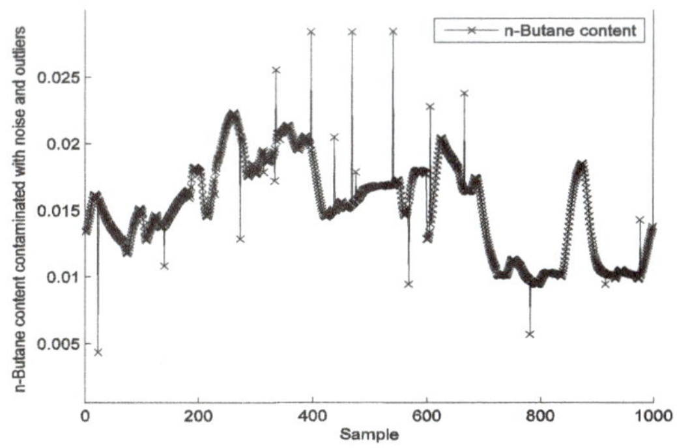

Figure 6. Contaminated data with Cauchy noise

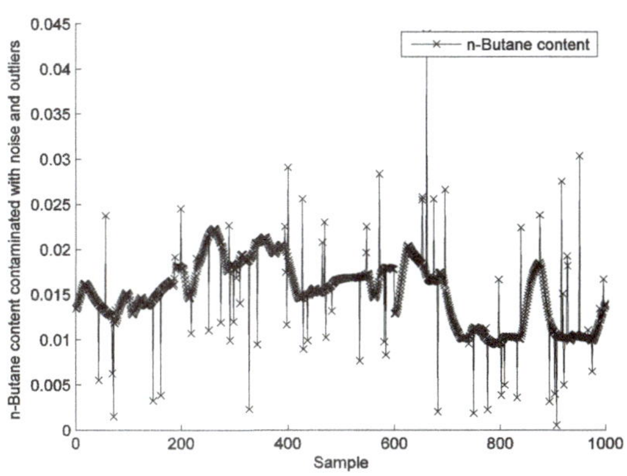

Figure 7. Contaminated data with Cauchy noise and 7% outlier (STD=5)

Figure 4, shows contaminated data with Gaussian noise, Figure 5, illustrates contaminated data with Gaussian noise and 7% outlier when standard deviation (STD) is equal to 5, Figure 6, represents Contaminated data with Cauchy and Figure 7, shows the Contaminated data with Cauchy noise and 7% outlier when STD is equal to 5.

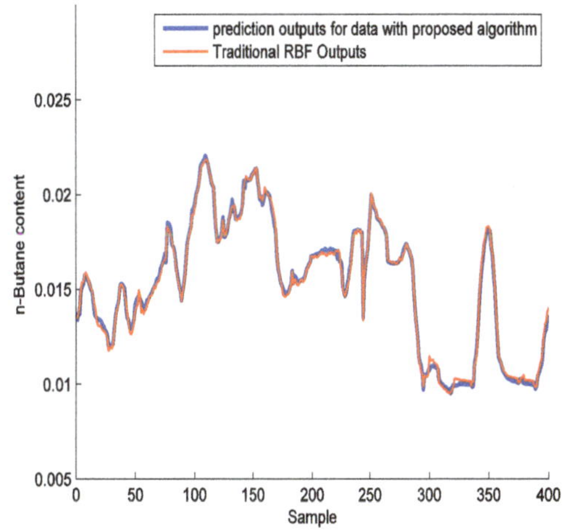

Figure 8. Output prediction with the proposed algorithm versus traditional RBF for Gaussian noise

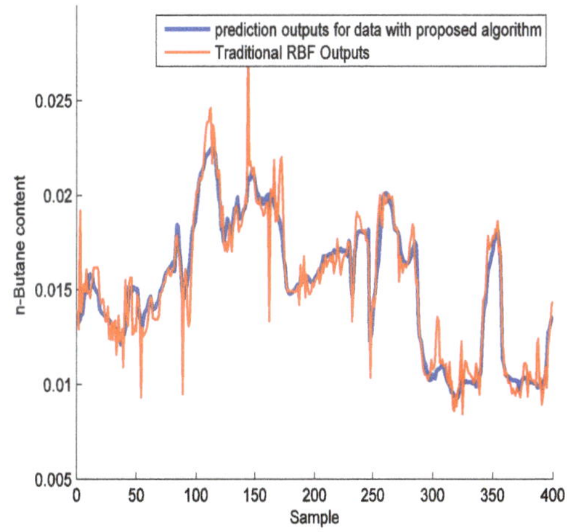

Figure 9. Output prediction with the proposed algorithm versus traditional RBF for Gaussian noise and 7% outliers (STD=5)

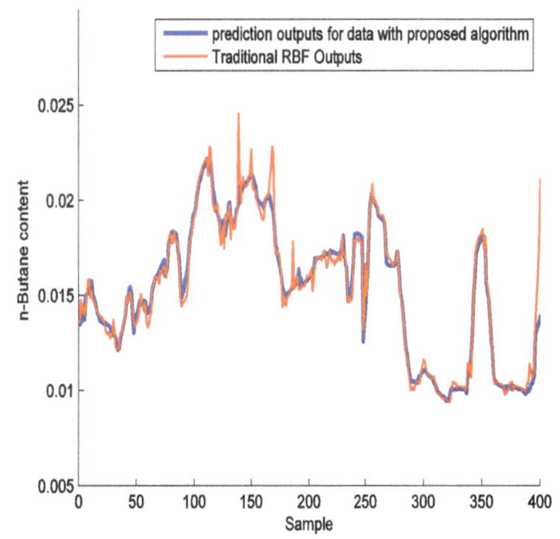

Figure 10. Output prediction with the proposed algorithm versus traditional RBF for Cauchy noise

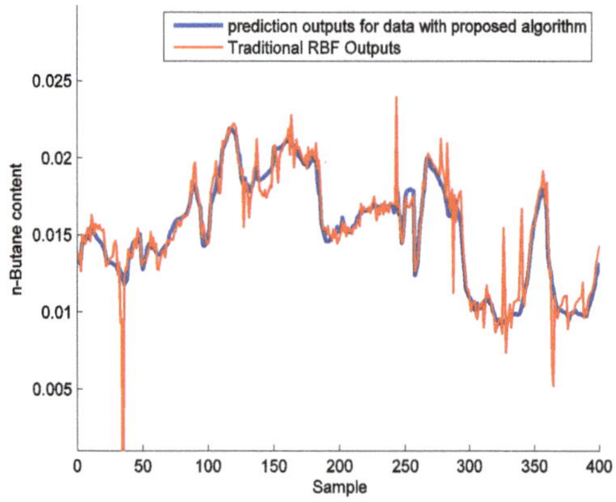

Figure 11. Output prediction with the proposed algorithm versus traditional RBF for Cauchy noise and 7% outliers (STD=5)

Inputs applied to $\varepsilon - SVR$ which parameters optimized with GA and initial parameters for RBF structure selection are determined. After this step, using robust learning algorithm to train the proposed network and then determined a prediction outputs. Simulation results show the effectiveness of the proposed algorithm versus traditional RBF. Traditional RBF consists of **MATLAB** Toolbox for creating RBFNNs with **newrb** function. The width of basis functions and number of neurons in this function replace with calculated value from GA-SVR (initialization method).

The Figure 8 - Figure 11 depicts the comparison results between proposed algorithm and traditional RBF.

The simulation results show that the proposed approach significantly improves the robustness against outliers versus traditional methods and follows real value very smoother than conventional approach.

The Root Mean Square Error (RMSE) and Correlation Coefficient (CC) of test data are used to measure the performance of the learned networks. The RMSE is defined as:

$$RMSE = \sqrt{\frac{1}{N-1}\sum_{i=1}^{N}(y_i - \hat{y}_i)^2} \quad (23)$$

Where y_i is the desired output and $\hat{y}_i$ is the output of the proposed method. And the value of the Correlation Coefficient(CC) is given by:

$$CC = \frac{\sum_{i=1}^{N}(\hat{y}_i - \hat{Y})(y_i - \overline{y})}{\sqrt{\sum_{i=1}^{N}(\hat{y}_i - \hat{Y})^2(y_i - \overline{y})^2}} \quad (24)$$

Where $\overline{y}$ is the mean desired output and $\hat{Y}$ is the mean predicted output of the proposed method. Table 1 shows the comparison result of the prediction performance for proposed network and traditional RBFNNs.

Table 1. Comparison results of the prediction error for the proposed method and traditional RBF in various case studies

	Proposed method		Traditional RBFNNs	
	RMSE	CC	RMSE	CC
Output data contaminated with Gaussian noise	0.00022869	0.9977	**0.00012979**	**0.9992**
Output data contaminated with Gaussian noise and 7% outliers with STD=5	**0.00025760**	**0.9972**	0.0011	0.9530
Output data contaminated with Cauchy noise	**0.00025001**	**0.9977**	0.00068912	0.9757
Output data contaminated with Cauchy noise and 7% outliers with STD=5	**0.00028294**	**0.9978**	0.0014	0.9191

The results from the demonstrate that when the data set is contaminated with Gaussian noise the proposed method reveals a good performance. Although the conventional RBF approach (least square estimators) has slightly better performance; this is because the conventional approach is an optimal estimator (Maximum Likelihood Estimator) for Gaussian noise. Since, the data set contain non-Gaussian (Cauchy) noise and outliers, performance of the conventional least squares estimator is getting worse and robust training methods have been used for achieving better performance.

Evidently, the proposed method (ARLA-RBFNNs) has superior performance and is more robust against outliers than traditional RBF.

4. Conclusion

In this paper, the radial basis function networks with the support vector regression and the robust learning algorithm is developed for the system identification of nonlinear plant with outliers. To design an effective SVR model, suitable values of parameters in SVR have been chosen using GA. The SVR approach has been used to determine the number of hidden nodes, the initial parameters of the kernel, and the initial weights of the proposed neural networks. Using the annealing robust learning algorithm to adjust the parameters of the model, the successful results indicated that the proposed ARLA-RBFNNs method can be used as a reliable technique for the system identification from data contamination with outliers. This algorithm has been implemented on Hydrocarbon Debutanizer unit for prediction of normal butane (C4) concentration. The simulation results show that the proposed approach overcome the problems of identification with outliers and have more accurate and smoother results than the conventional approaches.

Acknowledgement

The authors would like to thank the technical and financial supports provided by the National Iranian Gas Company–Sarkhoon & Qeshm Gas treating Company (NIGC-SQGC).

References

[1] G. Apostolikas and S. Tzafestas, "On-line RBFNN based identification of rapidly time-varying nonlinear systems with optimal structure-adaptation," *Mathematics and Computers in Simulation*, vol. 63, pp. 1-13, 2003.

[2] L. Ljung, "Perspectives on system identification," *Annual Reviews in Control*, vol. 34, pp. 1-12, 2010.

[3] C.-N. Ko, "Identification of nonlinear systems with outliers using wavelet neural networks based on annealing dynamical learning

algorithm," *Engineering Applications of Artificial Intelligence,* vol. 25, pp. 533-543, 2012.

[4] C.-C. Chuang, J.-T. Jeng, and P.-T. Lin, "Annealing robust radial basis function networks for function approximation with outliers," *Neurocomputing,* vol. 56, pp. 123-139, 2004.

[5] A. Sánchez and V. David, "Robustization of a learning method for RBF networks," *Neurocomputing,* vol. 9, pp. 85-94, 1995.

[6] C.-C. Lee, Y.-C. Chiang, C.-Y. Shih, and C.-L. Tsai, "Noisy time series prediction using M-estimator based robust radial basis function neural networks with growing and pruning techniques," *Expert Systems with Applications,* vol. 36, pp. 4717-4724, 2009.

[7] D. S. Broomhead and D. Lowe, "Radial basis functions, multi-variable functional interpolation and adaptive networks," DTIC Document 1988.

[8] A. D. Niros and G. E. Tsekouras, "A novel training algorithm for RBF neural network using a hybrid fuzzy clustering approach," *Fuzzy Sets and Systems,* vol. 193, pp. 62-84, 2012.

[9] K.-L. Du and M. N. Swamy, *Neural networks in a softcomputing framework*: Springer, 2006.

[10] S. Chen, C. F. Cowan, and P. M. Grant, "Orthogonal least squares learning algorithm for radial basis function networks," *Neural Networks, IEEE Transactions on,* vol. 2, pp. 302-309, 1991.

[11] N. B. Karayiannis, "Gradient descent learning of radial basis neural networks," in *Neural Networks, 1997., International Conference on*, 1997, pp. 1815-1820.

[12] D. Simon, "Training radial basis neural networks with the extended Kalman filter," *Neurocomputing,* vol. 48, pp. 455-475, 2002.

[13] C.-C. Lee, P.-C. Chung, J.-R. Tsai, and C.-I. Chang, "Robust radial basis function neural networks," *Systems, Man, and Cybernetics, Part B: Cybernetics, IEEE Transactions on,* vol. 29, pp. 674-685, 1999.

[14] C.-C. Chuang, S.-F. Su, and C.-C. Hsiao, "The annealing robust backpropagation (ARBP) learning algorithm," *Neural Networks, IEEE Transactions on,* vol. 11, pp. 1067-1077, 2000.

[15] P. J. Huber, *Robust statistics*: Springer, 2011.

[16] Y.-Y. Fu, C.-J. Wu, C.-N. Ko, and J.-T. Jeng, "Radial basis function networks with hybrid learning for system identification with outliers," *Applied Soft Computing,* vol. 11, pp. 3083-3092, 2011.

[17] V. N. Vapnik and V. Vapnik, *Statistical learning theory* vol. 2: Wiley New York, 1998.

[18] A. J. Smola and B. Schölkopf, "A tutorial on support vector regression," *Statistics and computing,* vol. 14, pp. 199-222, 2004.

[19] J. H. Holland, *Adaptation in natural and artificial systems: An introductory analysis with applications to biology, control, and artificial intelligence*: U Michigan Press, 1975.

[20] A. Konak, D. W. Coit, and A. E. Smith, "Multi-objective optimization using genetic algorithms: A tutorial," *Reliability Engineering & System Safety,* vol. 91, pp. 992-1007, 2006.

[21] B. Üstün, W. Melssen, M. Oudenhuijzen, and L. Buydens, "Determination of optimal support vector regression parameters by genetic algorithms and simplex optimization," *Analytica Chimica Acta,* vol. 544, pp. 292-305, 2005.

[22] A. K. Jana, A. N. Samanta, and S. Ganguly, "Nonlinear state estimation and control of a refinery debutanizer column," *Computers & Chemical Engineering,* vol. 33, pp. 1484-1490, 2009.

[23] N. Mohamed Ramli, M. Hussain, B. Mohamed Jan, and B. Abdullah, "Composition Prediction of a Debutanizer Column using Equation Based Artificial Neural Network Model," *Neurocomputing,* vol. 131, pp. 59-76, 2014.

[24] M. Behnasr and H. Jazayeri-Rad, "Robust data-driven soft sensor based on iteratively weighted least squares support vector regression optimized by the cuckoo optimization algorithm," *Journal of Natural Gas Science and Engineering,* vol. 22, pp. 35-41, 2015.

[25] S. Ferrer-Nadal, I. Yélamos-Ruiz, M. Graells, and L. Puigjaner, "On-line fault diagnosis support for real time evolution applied to multi-component distillation," *Computer Aided Chemical Engineering,* vol. 20, pp. 961-966, 2005.

4

Comparing the Fault Diagnosis Performances of Single Neural Networks and Two Ensemble Neural Networks Based on the Boosting Methods

Pooria Karimi*, Hooshang Jazayeri-Rad

Department of Instrumentation and Automation, Petroleum University of Technology, Ahwaz, Iran
*Corresponding author: pooria.k@gmail.com

Abstract This work employs potential and distinct features of artificial neural networks, in particular the feed-forward multilayer structure, to achieve fault diagnosis in chemical plants. Artificial neural networks can automatically store information by learning from historical fault data without using any qualitative or quantitative model of the system. However, different limitations are encountered in utilizing a single neural network classifier in fault diagnosis of complex large-scale chemical processes. These limitations are mostly originated from the low reliability and also high generalization error of a single neural network classifier which may be faced in real applications. An attractive remedy for these issues is to use an ensemble of different neural network models instead of relying on a single one. In this work, boosting as a general well-known approach to improve the learning performance of learning algorithms by building ensemble models is investigated. In particular, an efficient boosting method called the Stage-wise Additive Modeling using a Multi-class Exponential Loss Function (SAMMELF) is utilized. This technique can effectively overcome the main limitations of the early boosting methods for fault diagnosis applications. The potential of the SAMMELF algorithm is proved through its application to the challenging diagnosis problem of the Tennessee-Eastman Process (TEP) where the whole sets of the predefined TEP faults are considered in the classification problem.

Keywords: *artificial neural network, process fault diagnosis, adaptive boosting, Tennessee-Eastman Process*

1. Introduction

Changes in the physical conditions of process units, control systems or external conditions may lead to what are generally referred to as faults. Faults in the broadest sense include symptoms resulting from physical changes, such as deviations of temperature or pressure from their normal operating ranges, as well as physical changes themselves such as scaling, foaming, leaks and wear. Even changes in unmeasured process parameters such as heat or mass transfer coefficients can be considered to be faults. Generally speaking, fault is defined as any departure from an acceptable range of an observed variable or a calculated parameter of the process [1]. Faults in process equipment can give rise to off-specification production, increased operating expenses, the likelihood of line shutdown and an increasing risk of damaging environments. Hence, the growing demand for performance, efficiency, reliability and safety of industrial systems, is encouraging a growing interest both from industry and academic circles in fault diagnosis.

Process monitoring associated to fault management mission is essentially a four-stage process: fault detection, fault identification, fault diagnosis, and process recovery. (i) The task of fault detection is to determine when abnormal process conditions occur on-line. (ii) Fault identification is identifying the observation variables most relevant to diagnosing the fault. The purpose of this procedure is to focus the plant operator's and engineer's attention on the subsystems most pertinent to the diagnosis of the fault, so that the effect of the fault can be eliminated in a more efficient manner. (iii) Fault diagnosis which is the purpose of this article, is determining which fault occurred, in other words, determining the root cause of the observed out-of-control status. The fault diagnosis procedure is essential to the counteraction or elimination of the fault. (iv) Process recovery, also called intervention, is removing the effect of the fault [2].

Numerous techniques have been proposed for process fault detection and diagnosis in the past few decades. According to [3] these techniques can be broadly classified as model based and process history-based approaches. Model based approaches generally utilize results from the field of control theory and are based on parameter estimation or state estimation [4]. This approach is based on the fact that a fault will cause changes in certain physical parameters which in turn will lead to changes in some model parameters or states. It is then possible to detect and diagnose faults by monitoring the estimated model parameters or states. When using this approach, it is essential to have the knowledge about the relationships between faults and model parameters or states. Furthermore, quite accurate models are required. The main problem with these techniques is that obtaining an exact process model is not always possible or economical in industrial systems. In addition, errors in the model can be interpreted as faults thus yielding false alarms, or can prevent faults from being detected when they occur, especially in nonlinear or uncertain systems. Hence, many process model based quantitative techniques are only applicable to linear systems with limited availability of methods for nonlinear chemical processes [5].

The process history-based methods make use of the large amount of process data obtained from recorded measured variables of the process during abnormal and normal operations. Due to the recent advances in sensing and data collection technology which provides large amounts of data collected in many processes at chemical industries, these methods are more popular than model-based approaches in real applications. Many multivariate statistical techniques for analyzing these massive datasets have been developed including Principal Component Analysis (PCA), Partial Least Squares (PLS), and Fisher Discriminant Analysis (FDA) [6]. Although most of these techniques are well designed for the fault detection, one of the most relevant techniques for the diagnosis is the supervised classification. In this technique, different operating conditions including normal and abnormal ones are treated as patterns or classes. Given that there are many datasets in the historical database, each related to a different abnormal condition (root cause); the aim is then to allocate the on-line out-of-control measurements to the most narrowly associated fault class. Classification tools are one of the most extended fault diagnosis systems in literature. They have gotten ahead because of their ease of implementation, so they do not necessitate process operators' experience or process first principles information. In addition, they can handle the inherent nonlinearity of variables existed in most chemical processes at no extra cost.

Numerous methods have been developed for supervised classification in machine learning area. These techniques which were successfully applied to fault diagnosis applications include the *k*-Nearest Neighborhood (kNN) [7], FDA [8], the Bayesian Networks (BNs) [9,10] and the Support Vector Machines (SVMs) in more recent works [11,12,13,14]. In particular, amongst the classification techniques, Artificial Neural Networks (ANNs) had an enormous attention in past years. ANNs have many useful properties concerning process fault diagnosis. They can handle nonlinear and undetermined processes where no process model is available and learn the diagnosis by means of the information of the learning data. ANNs are very noise tolerant and work well with noisy measurements. The ability to generalize the knowledge as well as the ability to adapt during their use is one of their very interesting properties. The use of ANNs in fault diagnosis is very straightforward [15]. Multi-Layer Perceptron (MLP) networks are the most popular ANNs which have been used extensively for fault diagnosis [16-21].

Despite simplicity and versatility of MLP networks, their application to process fault diagnosis in plant-wide scale is not without difficulties. One of the most important issues of employing MLP networks in process fault diagnosis is related to their over-fitting, meaning that the network shows very good performance on the training data, however behaves poorly with unseen data. Moreover, The MLP neural networks do not necessarily converge to a unique solution during the training phase and they suffer from low robustness especially when data available for training the network are not abundant. Besides, often the input-output relationship represented by a large set of fault data dispersed on a wide input feature space might not be captured adequately by a single network and so the obtained generalization performance may not be satisfactory. Hence, it is often difficult, if not impossible, to build a perfect single MLP network model for fault classification tasks in real industrial problems involving many types of faults. To encounter these issues, use of an ensemble of neural networks is proposed in literature. Ensemble or committee of neural networks is achieved through information fusion by combining different single neural network outputs by some kind of voting scheme to get a single output [22]. Using an ensemble based method may result in improving the performances of any learning system [23]. For neural networks mainly, if one takes up that the output of an individual neural network of the ensemble comprises of a true output combined with a random error constituent with zero average, then the grouping of the outputs from the individual networks lead to averaging of the random error constituents. Hence, it reduces the estimation error. Also, an ensemble of networks might reduce the over-fitting, by combining different networks with different architectures. Moreover, the risk of an unfortunate selection of a poorly performing classifier is reduced. Use of ensemble of neural networks as a general way to improve the generalization performance is quite common in neural network literature [24,25,26]. Although the advantages of ensemble network were recognized by several researchers, it is believed that no attempt except by [27] was made to apply an ensemble network for process fault diagnosis. In that work, to develop a diverse range of individual networks, each individual network is trained on a replication of the original training data generated through bootstrap resampling with replacement. The power of fusion approach in increasing neural network robustness and reliability and leading to an improved on-line diagnosis system was also demonstrated [27]. However, a quite simple chemical process comprising a Continuous Stirred Tank Reactor (CSTR) and a heat exchanger is employed to investigate the performances of different diagnosis systems.

In this work, the possibility of improvement in performance of the MLP neural network classifiers based on ensemble architecture achieved by boosting techniques is investigated through the application to the TEP fault diagnosis problem. Boosting in machine learning area refers to the general techniques attempt to improve the performance of learning methods. In particular, two powerful techniques for ensemble classification, called adaptive boosting, or AdaBoost for short, and the SAMMELF algorithm are examined in this work. Whole sets of the TEP faults are considered in this work to provide a challenging diagnosis problem. The remainder of this paper is organized as follows. Section 2 deliberates about the ensemble based learning principles and provides a brief explanation for AdaBoost, SAMMELF and MLP neural network classification methods. The TEP fault diagnosis problem is detailed in section 3. Fault diagnosis results obtained by application to the TEP problem are presented and discussed in section 4. Finally, section 5 concludes this paper.

2. Methods

2.1. MLP Neural Network Classifier

ANNs were motivated from the examination of the human brain and consist of many basic computational elements (neurons) locally affecting other neurons through connections. Neurons or nodes in ANNs are very basic computational elements originated from their biological equivalents. A neuron can be defined by a basic function providing a conversion from n-dimensional space (inputs acquired from other neurons) to 1-dimensional space (an output value that it directs to other neurons). There are weights on each of the interconnections and it is these weights that are updated during the training process to ensure that the inputs produce an output which is close to its actual value, using an appropriate training rule being applied to adjust the weights. Through the training process, the network learns from examples and in doing so acquires some capabilities for generalization beyond the training data. The architecture of these models is specified by the node characteristics, network topology and learning algorithm.

The classification of process data can be carried out with the information about different classes. Then we know that certain measurement pattern correspond to normal operation and each other measurement pattern correspond to each faulty operation. The training of the neural network using this kind of information is called supervised. The MLP network is the most prevalent architecture of ANNs which can be used for supervised classification. The network contains one input layer, one or more hidden layers and one output layer. Figure 1 depicts a simple structure of the MLP network with one hidden layer where the circles represent neurons arranged in the hidden and output layers. Each connection has a weight associated with it. For input layer, the input value is forwarded straight to the hidden layer. The hidden and output layers carry out two calculations: first a weighted sum of the inputs and then the output is calculated using a non-decreasing and differentiable transfer function. Usually a sigmoidal function is used. The information propagates in one direction only through the network (feed-forward structure).

For fault diagnosis, the input pattern represents the important variables that are affected by the existing faults and the output pattern represents the fault to be identified. The input pattern is directly fed to the network input layer and the output pattern corresponds to each node in the network output layer. Then, an output node target would be either 0 (no fault) or 1 (a particular fault). Once the weights on the connections are finally adjusted in the training phase, a new vector of sensor measurements can be sent to the input nodes of the network to be classified. Very little computation time is needed for this step. The degree of misclassification that occurs is a function of how well the knowledge stored in the connections and weights.

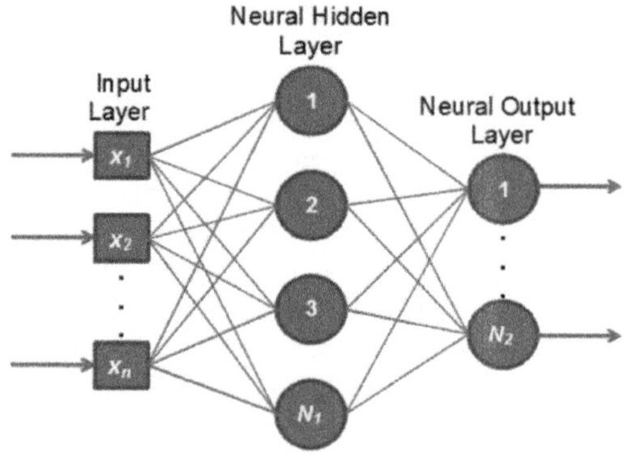

Figure 1. MLP neural network with one hidden layer

2.2. Ensemble Models for Classification

The idea of building ensembles of classifiers has gained interest in the last decade. Instead of building a single complex classifier, one aims at designing a combination of several simple (weak) classifiers (see Figure 2). For instance, instead of training a large neural network, we train several simpler neural networks and combine their individual outputs in order to produce the final output. The basic idea is to generate several classifiers and pool them in such a way as to improve on the performance of any single one. Even in some cases, even though the ensemble may not result in better than the optimum single classifier, it could reduce or remove the risk of choosing an inappropriate single classifier.

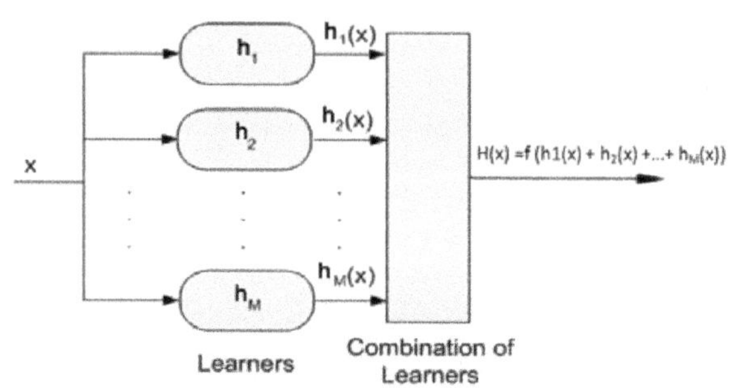

Figure 2. The concept of an ensemble of classifiers

All ensemble models are built on two key components: first, a strategy is needed to build diverse classifiers; second, a strategy, is needed to combine the outputs of the individual classifiers that make up the ensemble in such a way that the correct decisions are amplified, and the incorrect ones are cancelled out. Diversity means the base classifiers of an ensemble disagree with each other as much as possible particularly with respect to misclassified instances. Specifically, we need classifiers whose decision boundaries are sufficiently different from those of others. Various techniques have been suggested for obtaining diversity in the base models of an ensemble, e.g. using different training parameters [22], different training patterns [28], different feature subsets [29] and different learning methods for each classifier of the ensemble [30]. For combining the individual decisions, the most popular (and simplest) method is the majority voting rule, although there exist other more complex schemes (e.g., average, minority, medium and product of votes) [31]. Ensemble based classification methods have been widely studied in such diverse fields as image segmentation, data mining from noisy data streams, credit card fraud detection, sensor networks, image/speech/handwritten recognition, fault diagnosis, etc [32].

Given: $(x_1, y_1), \ldots, (x_m, y_m)$, *Where* $x_i \in X$, $y_i \in Y = \{1, 2, \ldots, K\}$

1. *Initialize the weights* $w_i^1 = 1/m$, $i = 1, 2, \ldots, m$
2. *For t=1 to T*
   - *(a)* *Fit a classifier* $h^t(x)$ *to the training data using weights* w_i^t
   - *(b)* *Compute*

$$err^t = Pr_{i \sim w_i^t}[h^t(x_i) \neq y_i] = \sum_{i=1,\, h^t(x_i) \neq y_i}^{m} w_i^t \Big/ \sum_{i=1}^{m} w_i^t$$

If $err^t > 1/2$, *then t=T-1 and abort loop.*

   - *(c)* *Compute*

$$\alpha^t = \log \frac{1 - err^t}{err^t}$$

   - *(d)* *Set*

$$w_i^t \leftarrow \begin{cases} w_i^t . \exp(\alpha^t) & if \quad h^t(x_i) \neq y_i \\ w_i^t & otherwise \end{cases} \quad i = 1, 2, \ldots, m$$

   - *(e)* *Renormalize* w_i^t
3. *Output*

$$H(x) = \arg\max_{y \in Y} \sum_{t=1,\, h^t(x) = y}^{T} \alpha^t$$

Figure 3. The boosting algorithm AdaBoost

2.2.1. Multi-Class Boosting

Among the different methods for constructing diverse classifiers in an ensemble classification system, most techniques are based on using different training patterns for each classifier. One of the most popular approaches for this purposes are based on boosting techniques which have considerable success in various applications [33]. Boosting encompasses a family of methods which attempts to "boost" the accuracy of any given learning algorithm. The well-known AdaBoost algorithm which is the focus of this paper is introduced in 1995 by Freund and Schapire [34] has become widespread due to its capability of solving many of the practical difficulties of the earlier boosting algorithms.

A Pseudocode for AdaBoost is given in Figure 3. The algorithm takes as input a training set $(x_1, y_1), \ldots, (x_m, y_m)$ where each x_i belongs to some domain or instance space X, and each label y_i is in some label set $Y=\{1, 2, \ldots, K\}$ where K is the number of classes. The goal is to find a classification rule $H(\boldsymbol{x})$ from the training data, so that when given a new input $\boldsymbol{x}$, we can assign it a class label y from $\{1, \ldots, K\}$. AdaBoost constructs a composite strong classifier by sequentially training weak or base classifier classifiers repeatedly in a series of rounds $t = 1, \ldots, T$. One of the main ideas of the algorithm is to maintain a distribution or set of weights over the training set. Starting with the un-weighted training sample, AdaBoost builds a classifier that produces class labels. If a training data point is misclassified, the weight of that training data point is increased (boosted) so that the weak learner is forced to focus on the hard examples in the training set. This is achieved by the rule represented in Figure 3. This rule is based on the weak classifier error rate. Notice that the error is measured with respect to the weights of the training data on which the weak learner was trained (step 2b in Figure 3). A second classifier is built using the new weights which are no longer equal. Again, misclassified training data have their weights boosted and the procedure is repeated. A score relating to the goodness of a classifier measured by its error is assigned to each classifier (α^t). The final classifier is defined as a weighted majority vote of the T weak classifiers. It should be noted that that this boosting algorithm, is sometimes denoted as adaboost.M1 in literature to emphasize on the capability of handling multi-class classification problems [33]. However, we simply refer to this algorithm as AdaBoost through this paper.

In practice, the weak learner may be an algorithm that can be trained directly based on the weights on the training examples. This method of boosting is called reweighting. However, some learning algorithms don't permit training with respect to a weighted cost function and require an un-weighted set of examples. In this case, resampling with replacement (using the probability distribution w) can be used to approximate a weighted cost function. In this case, examples with high probability occur more often than those with low probability, while some examples may not occur in the sample at all, although their probability is not zero. Neural networks can be trained directly with respect to a distribution over the learning data by weighting the cost function. However, using weighted cost function may cause numerical problems [35]. Hence, in this work, using neural network as the weak learner, the resampling boosting is utilized.

Essentially, AdaBoost has many benefits. It is fast, straightforward and easy to be implemented. It has no parameters to be tuned (except for the number of iterations). It needs no previous information about the weak learner and therefore can be easily joined to any method for finding weak hypotheses. AdaBoost has very interesting theoretical properties; in particular it can be shown that the error of the composite classifier on the training data decreases exponentially to zero as the number of combined classifier is increased provided that each classifier yields a weighted error that is less than 50% [34]. More importantly, however, bounds on the generalization error of such a system have been formulated in [36]. These are based on a notion of margin of classification, defined as the difference between the score of the correct class and the best score of a wrong class. It has been shown empirically in various applications that AdaBoost tends to be robust to over-fitting [37,38,39].

Given: $(x_1, y_1), \ldots, (x_m, y_m)$, *Where* $x_i \in X$, $y_i \in Y = \{1, 2, \ldots, K\}$

4. *Initialize the weights* $w_i^1 = 1/m$, $i = 1, 2, \ldots, m$
5. *For t=1 to T*
   - *(f)* *Fit a classifier* $h^t(x)$ *to the training data using weights* w_i^t
   - *(g)* *Compute*

$$err^t = Pr_{i \sim w_i^t}[h^t(x_i) \neq y_i] = \sum_{i=1,\, h^t(x_i) \neq y_i}^{m} w_i^t \Big/ \sum_{i=1}^{m} w_i^t$$

If $err^t > 1 - 1/K$, *then t=T-1 and abort loop.*

   - *(h)* *Compute*

$$\alpha^t = \log \frac{1 - err^t}{err^t} + \mathbf{log}(\boldsymbol{K} - \mathbf{1})$$

   - *(i)* *Set*

$$w_i^t \leftarrow \begin{cases} w_i^t . \exp(\alpha^t) & if \quad h^t(x_i) \neq y_i \\ w_i^t & otherwise \end{cases} \quad i = 1, 2, \ldots, m$$

   - *(j)* *Renormalize* w_i^t
6. *Output*

$$H(x) = \arg\max_{y \in Y} \sum_{t=1,\, h^t(x) = y}^{T} \alpha^t$$

Figure 4. The boosting algorithm SAMMELF

The main disadvantage of the AdaBoost is that it requires the weak classifier have prediction error less than 1/2 (with respect to the distribution on which it was trained). The expected error of a hypothesis which randomly guesses the label is 1-1/K, where K is the number of possible labels. Thus the AdaBoost requirement for $K = 2$ is that the prediction is just slightly better than random guessing. However, when $K>2$, the requirement of AdaBoost is much stronger than that, and might be hard to meet. To solve this issue, a new modification of AdaBoost is proposed by [40] in which the new algorithm only requires the performance of each weak classifier just to be better than random guessing similar to AdaBoost in the two class case. The new algorithm is called SAMMELF. They provided a statistical justification for the algorithm by using a novel multi-class exponential loss function and the forward stage-wise additive modeling technique. The SAMMELF algorithm is presented in Figure 4. Notice that SAMMELF is very similar to AdaBoost with the only difference in weight updating rule as presented in the step 2c of the algorithm shown in Figure 4. Now in order for α^t to be positive, we only need $(1-err^t) > 1/K$, or the accuracy of each weak classifier to be better than random guessing rather than 1/2. As a consequence, the new algorithm puts more weight on the misclassified data points when compared to AdaBoost. It is worth noting that when $K = 2$, SAMMELF reduces to AdaBoost.

An experimental comparison between SAMMELF and AdaBoost methods is reported in [40] based on a simple three-class simulation example. The results are provided in Figure 5. This figure (upper row) shows how AdaBoost breaks using ten-terminal node trees as weak classifiers. As we can see (upper right panel), the test error of AdaBoost decreases for a few iterations, then levels of around 0.53. What has happened can be understood from the upper left and upper middle panels: the err^t starts below 0.5; after a few iterations, it overshoots 0.5 (α^t below 0), then quickly hinges onto 0.5. Once err^t is equal to 0.5, the weights of the training samples do not get updated ($\alpha^t = 0$), hence the same weak classifier is fitted over and over again but is not added to the existing fit, and the test error rate stays the same.

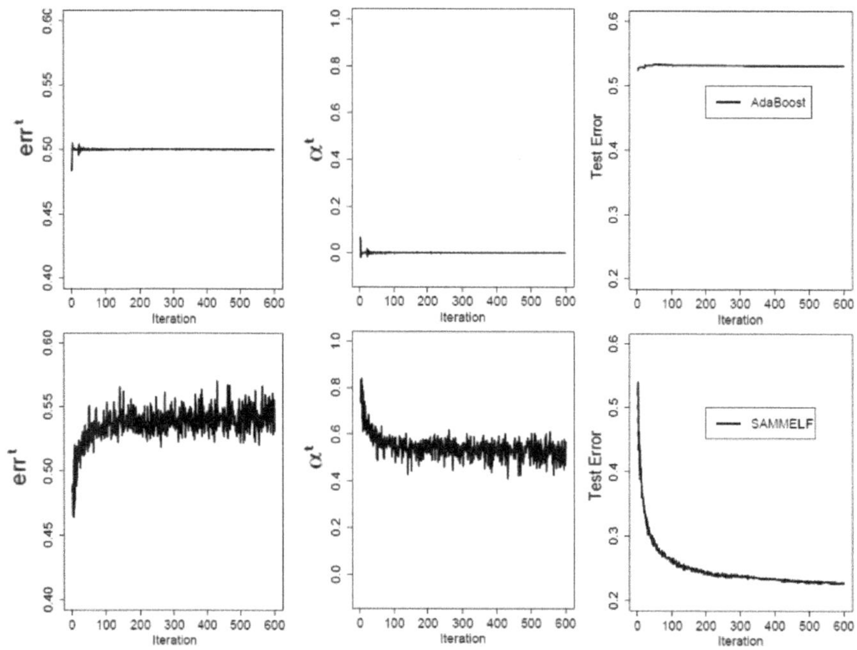

Figure 5. Comparison of AdaBoost and the new algorithm SAMMELF on a simple three-class simulation example as reported in [40]

3. Application: Tennessee-Eastman Process

The TEP problem was created by the Eastman chemical company to provide a realistic industrial process in order to evaluate process control and monitoring methods [41]. A simplified process flow sheet of TEP is plotted in Figure 6. TEP is a chemical process composed of five major operation units: a reactor, a condenser, a compressor, a stripper and a separator. Four gaseous reactants A, C, D, E and an inert B are fed to the reactor where the liquid products F, G and H are formed. The TEP represents a challenging real diagnosis problem that has been widely used for many research experiments in fault detection and control. Main diagnosis difficulties in this problem come from the number of different faults to be managed and the wide faults diversity. Random and step biases as well as soft drifts on the process variables must be managed in this complex diagnosis case study. Besides, some of the considered disturbances cause insignificant biases in the variables because of the adjusted control strategy, which makes very difficult to find out the biases source causes [12].

The TEP is unstable in open-loop. Chiang et al. [8] simulated the plant by using a closed-loop control structure adopted in [42]. Their generated simulation data which can be accessed in *http:rrbrahms.scs.uiuc.edu* were employed in this work. The process contains 41 measured variables (*XMEAS*) and 12 manipulated variables (*XMV*). However, only 52 variables are employed in this problem because an input variable (the reactor agitator speed) is constant. All the process measurements include Gaussian noise. An observation vector at a particular time instant is given by:

$$x = \left[XMEAS_1 \ldots XMEAS_{41} XMV_1 \ldots XMV_{11}\right] \qquad (1)$$

The TEP simulation contains 21 preprogrammed faults. Sixteen of these faults are known and five are unknown. These faults are listed in Table 1. Two sets of training data and test data for 22 kinds of operation conditions, i.e. 21 kinds of faulty and the other one of normal status were generated. Each faulty data set started with no faults, and the faults were then introduced for the duration of 1 and 8 simulation hours into the run, respectively, for the training and testing data sets. The data were sampled every 3 min. The total number of observations generated for each run was 500 and 960 for the training and testing data sets, respectively. However, only 480 and 800 observations were correspondingly collected after the introduction of the fault.

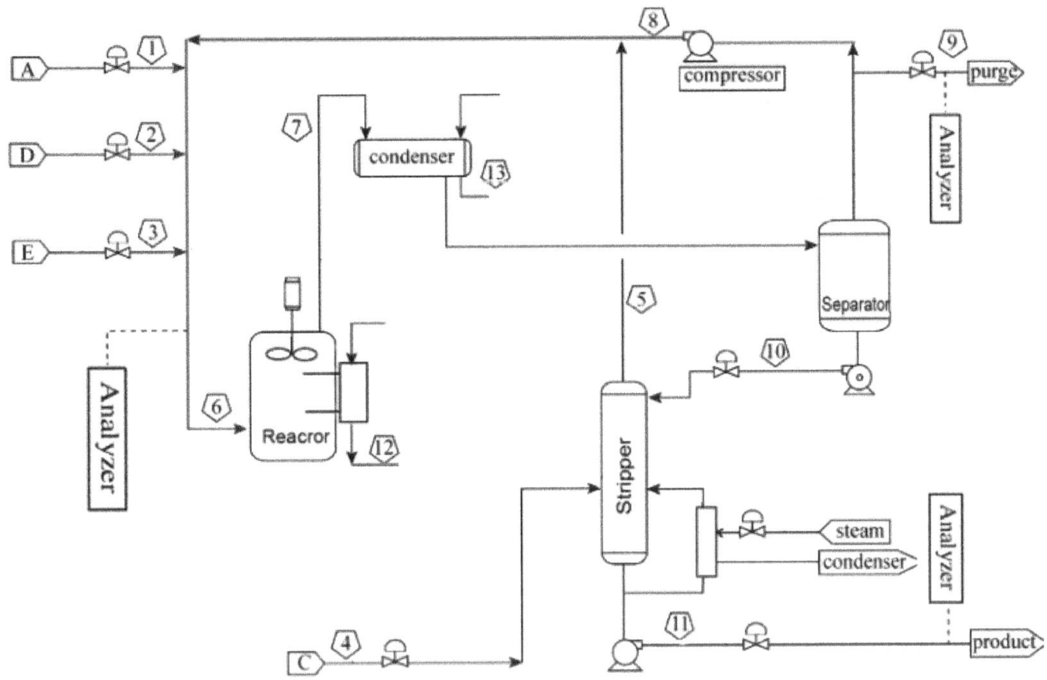

Figure 6. The process flow sheet of TEP

Table 1. Types of faults in the TE process

Fault	Description	Type
1	A/C feed ratio, B composition constant (stream 4)	Step
2	B composition, A/C ratio constant (stream 4)	Step
3	D feed temperature (stream 2)	Step
4	Reactor cooling water inlet temperature	Step
5	Condenser cooling water inlet temperature	Step
6	A feed loss (stream 1)	Step
7	C header pressure loss-reduced availability (stream 4)	Step
8	A, B, C feed composition (stream 4)	Random
9	D feed temperature (stream 2)	Random
10	C feed temperature (stream 4)	Random
11	Reactor cooling water inlet temperature	Random
12	Condenser cooling water inlet temperature	Random
13	Reaction kinetics	Slow drift
14	Reactor cooling water valve	Sticking
15	Condenser cooling water valve	Sticking
16	Unknown	Unknown
17	Unknown	Unknown
18	Unknown	Unknown
19	Unknown	Unknown
20	Unknown	Unknown
21	The valve for Stream 4 was fixed at the steady state position	Constant position

4. Results and Discussion

4.1. Single Neural Network Classifier

The raw data obtained from the TEP simulation was used to construct the different classifies discussed in this section. First, in order to get a deeper knowledge on the ANN performance, different single neural network classifiers are evaluated under different conditions. Different neural network classifiers are trained when the number of fault classes is progressively increased as shown by the fault numbering listed in Table 1. The networks used here are single hidden layer feed-forward neural networks. For each case, the number of output neurons is fixed. Each output neuron corresponds to one normal state and the matching considered fault situation. The number of hidden neurons however needs be determined. We always utilized 20 hidden neurons which experimentally found to be an optimal choice between the network training time and the classification performances in different cases. The results obtained are depicted in Figure 7 for the training and also test data. The diagnosis performance is indicated as the error rate which is simply defined as the rate of misclassified points in the whole data set. Here, misclassification indicates a point whose class is determined incorrectly. As expected, Figure 7 shows how the error rate globally increases when more classes are considered and the size of the network increases.

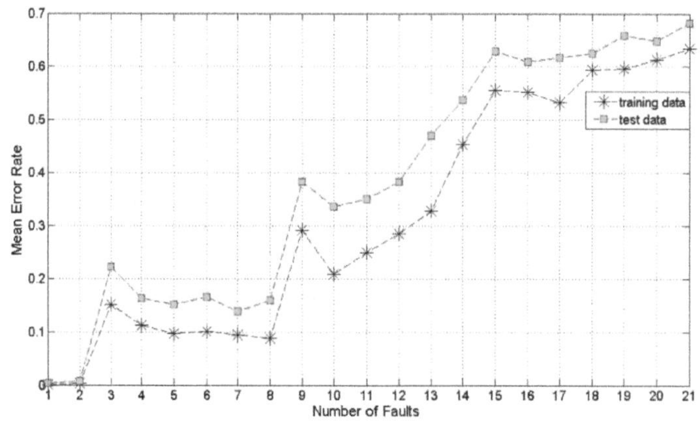

Figure 7. Error rate versus the number of considered faults. Number of faults indicate that faults 0+1+...+n are considered

In the employed MLP neural networks, both of the hidden and output layer neurons utilize the sigmoid activation functions. The sigmoid activation function naturally bounds the neural network outputs within the range [0,1]. In the training phase, a target value of 1 is considered for the output node corresponding to the label of the current training dataset, while the target values for all other output nodes are set to 0. For a new testing example, the labeled prediction corresponds to the node with the highest confidence. It should be noted that the Scaled Conjugate Gradient (SCG) algorithm is used in this paper for the training purpose of the MLP network [43]. In order to deal with the over-fitting issue in the training step, the early stopping technique is employed. For this purpose, the training data set is randomly divided into two subsets: the learning and the validation sets with proportion of 7 to 3, respectively. The network is trained using the learning data-set; however, its performance is monitored on the validation data set. The validation error normally decreases during the initial phase of training, as does the learning set error. However, when the network begins to over-fit the data, the error on the validation set typically begins to rise. Hence, the network training is stopped when the minimum error is reached on the validation data, i.e. when the validation error begins to rise. Otherwise, a maximum of 200 epochs of training must be utilized. Also, the entrapment in local minima problem for ANNs, sometimes necessitates the training procedure to be repeated several times each time employing different initial weights to increase the chance of selecting the best model. This procedure is considered for each case and consequently, the network with the best classification result has been chosen.

4.2. Boosting Based on AdaBoost

As mentioned before in section 2, the basic AdaBoost algorithm converges (leads to reduced training error) if the basic classifier yields a weighted error that is less than 0.5. Hence, the network trained based on the whole fault sets of the TEP problem (21 fault cases in Figure 7) cannot be directly boosted by the basic AdaBoost algorithm because it always produces error more than 0.5. Hence, we apply and evaluate the AdaBoost method to the largest network having error less than 0.5, i.e. the network which diagnoses 14 fault classes. For this purpose, first a uniform weighting distribution is assigned to the original training samples. We consider the trained network as the first round classifier of the boosting ensemble and based on its observed errors for the original training samples, the weight distribution is updated (step 2d in Figure 3). The new weighting distribution is used to generate a new training set by randomly picking samples with replacement from the current training set. The new dataset size obtained by resampling is equal to the size of the current training data set. Because the new distribution uses more weight on the training samples which were wrongly classified before, the new training set obtained by resampling on the current training set contains more difficult samples in classification view. The second network is then trained using the new training set which is more focused on the classification of the hard examples. Then all the training samples are passed through this network and the weighted error is computed and used to obtain new modification for the weighting distribution of the third round. This procedure is repeated in sequential rounds. Meanwhile, because the final networks are trained using more difficult data values than the previous ones, the weighted error rate on the training dataset normally increases. Finally the boosting procedure is stopped when the obtained weighted error exceeds the value of 0.5. In addition, in order to take into account the fact that a network may become trapped into a local minimum of the cost function in each round, the training will be restarted with new random initial weights if the current weighted training error exceeds 0.5. The maximum number of restarts for a given network is taken to be five, after which the boosting algorithm terminates. Note that, each training task is performed based on the early stopping method; i.e. the training is stopped when the network either reaches the minimum mean squared error in the validation data set or passes 200 epochs as described before.

The result obtained in this part is demonstrated in Figure 8. In this figure, the error rates on the training and test samples of the boosted classifiers, as the number of networks is increased, are presented. In each round of boosting, the outputs of current trained classifier and all other previous classifiers are combined through the majority voting method described in the step 3 of the algorithm shown in Figure 3. As can be observed in Figure 8, the training error and the test error in sequential rounds are normally decreased. This implies the general success of the boosting mission. However, after the fifth round the weighted error goes beyond 0.5, hence the boosting procedure is aborted. The final error rate of 0.28 and 0.397 are respectively obtained for the training and testing data by the ensemble classifiers compared to the corresponding error rates of 0.453 and 0.538 generated by the single un-boosted network.

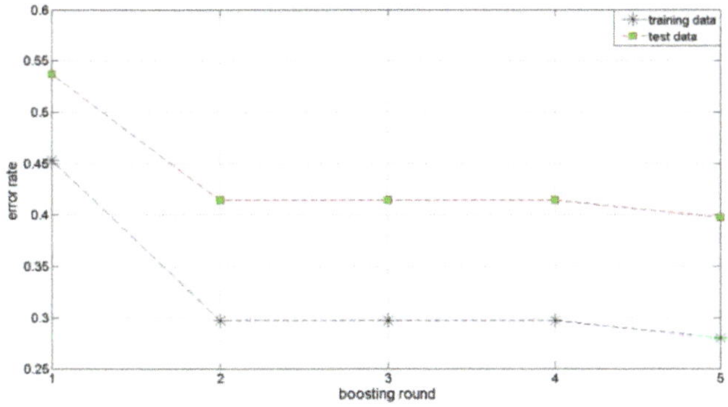

Figure 8. Boosting procedure for the network diagnoses of faults 1 to 14 of TEP obtained by AdaBoost

A remarkable feature of the ensemble networks reported in literature is their resistance to over-fitting. In other words, while the error on the training set is decreased in sequential rounds the generalization error is always bounded in the ensemble classifiers without concerning whether the single classifiers are over-fitted or not. In many experiments, it has been observed that the generalization error continues to decrease toward an apparent asymptote after the training error has reached zero. Schapire et al. [36] suggested a possible explanation for these behaviors based on the definition of margin of classification. After that, many empirical evaluations of AdaBoost also provided an analysis of the so-called margin distribution. Consider an ensemble classification scheme obtained by a majority voting combination of different single classifiers. Suppose that the weights assigned to the different base classifiers are normalized. The classification margin for each example is defined as the difference between the weight assigned to the correct label and the maximal weight assigned to any single incorrect label. It is easy to see that the margin is a number within the range [-1,1] and that an example is classified correctly if and only if its margin is positive. Moreover, the magnitude of the margin can be interpreted as a measure of confidence in the prediction. Schapire et al. [36] proved that larger margins on the training set translate into a superior upper bound on the generalization error. They also showed that that boosting is particularly aggressive at reducing the margin (in a quantifiable sense) since it concentrates on the examples with the smallest margins (either positive or negative).

Figure 9 shows the margin cumulative distributions (MCD) obtained in our experiment. MCD represents the fraction of examples in the training data whose margin is at most x as a function of $x \in [-1,1]$. The margin distribution is plotted after two rounds of boosting and also for the final obtained ensemble which is obtained after the 5th round. It is clear that the number of examples with high margin increases when more classifiers are combined by boosting. This result also confirms the success of boosting in our application.

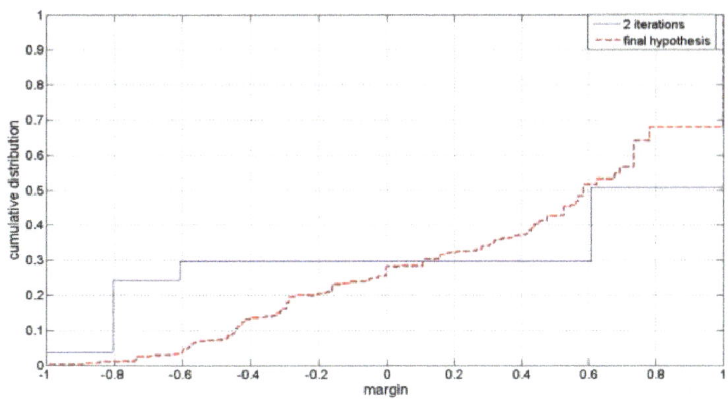

Figure 9. The margin distribution of the training examples obtained in the boosting of the network diagnoses of faults 1 to 14 of TEP by AdaBoost

4.3. Boosting Based on the SAMMELF Method

The AdaBoost algorithm cannot be employed for the boosting of the networks which classifies large sets of TEP faults. Also, for the largest possible classification problem, the AdaBoost based boosting procedure could not be executed for more than 5 rounds. In this section, we first apply the SAMMELF boosting method on the same network boosted by AdaBoost to obtain a comparative view. The SAMMELF boosting algorithm is described in Figure 4. As stated before, the SAMMELF algorithm is similar to AdaBoost. The only difference originates from the weight updating function. The one which is suggested by the SAMMELF method requires the weak learner to perform just slightly better than random guessing.

Whereas, for the AdaBoost algorithm a weighted error less than 0.5 is always required. In implementation of the SAMMELF algorithm, similar to AdaBoost, the trained network with 20 hidden layers is used as the first round classifier and based on its observed training error, the weighting distribution is updated for the next round. Also, the same procedure of network learning in each round, based on early stopping technique, is carried out again. The final classifier is obtained by the majority voting technique expressed in the step 3 of the algorithm shown in Figure 4. The consequent result is shown in Figure 10. Despite the fact that the boosting algorithm basically reduces the errors obtained on the training data, however the errors of the test data are also improved in the same manner. The resistance of the SAMMELF method to over-fitting is also exhibited in this figure. The boosting procedure can be proceeded to higher rounds, compared to AdaBoost, without being aborted due to large weighted errors. However, after 20 rounds, no improvement in training error is seen by this method. The final results of the training and generalization errors are presented in Table 2 which indicates a superior performance for the SAMMELF boosting method. The MCD curve obtained by the SAMMELF algorithm is also depicted in Figure 11 for three different rounds (iterations). Similar to the results shown in Figure 9 for the AdaBoost method, Figure 11 shows the achieved performance of the SAMMELF method in boosting the margin of the training data.

Table 2. Classification performances of different methods in diagnosis cases of faults 1 to 14 of TEP

	Single un-boosted network	Boosting by AdaBoost	Boosting by SAMMELF
Training error	0.453	0.280	0.182
Test error	0.538	0.398	0.382

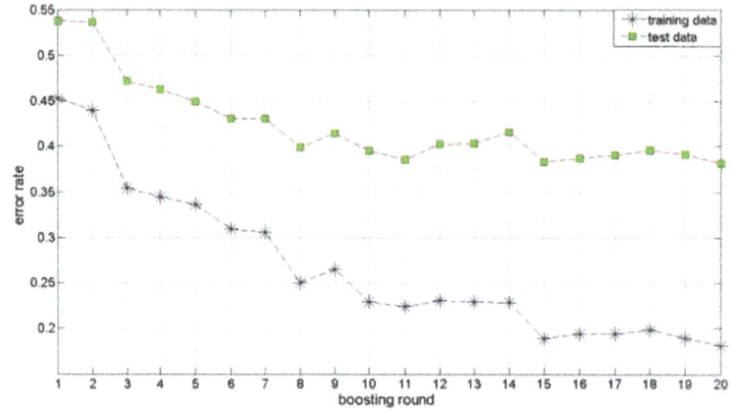

Figure 10. Boosting procedure for the network diagnoses of faults 1 to 14 of TEP obtained by SAMMELF

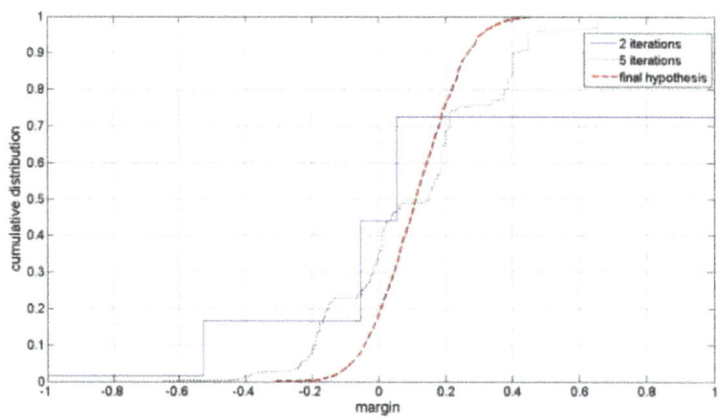

Figure 11. The margin distribution of the training examples obtained in the boosting of the network diagnoses of faults 1 to 14 of TEP by SAMMELF

We applied the SAMMELF boosting method to the largest neural networks diagnosing all the TEP faults because the generated training errors by this networks are less than the random guessing (less than 1-1/K where K is equal to 22 classes of data). The corresponding result obtained is presented in Figure 12. As seen in this figure, both the training and the generalization errors continue to decrease to their associated apparent asymptotic levels. The training and test errors are 0.349 and 0.536, respectively. The interrelated values achieved for the single network are 0.634 and 0.682. The errors for the single network were obtained after 20 iterations (no improvement was observed after the round 20).

The MCD obtained by the SAMMELF algorithm applied to the whole TEP faults is also depicted in Figure 13 for three different rounds. As a general conclusion, due to the fact that an error less than the random guessing can always be obtainable in any large classification problem, the SAMMELF method demonstrates a superior substitute for the conventional AdaBoost algorithm in real diagnosis problems.

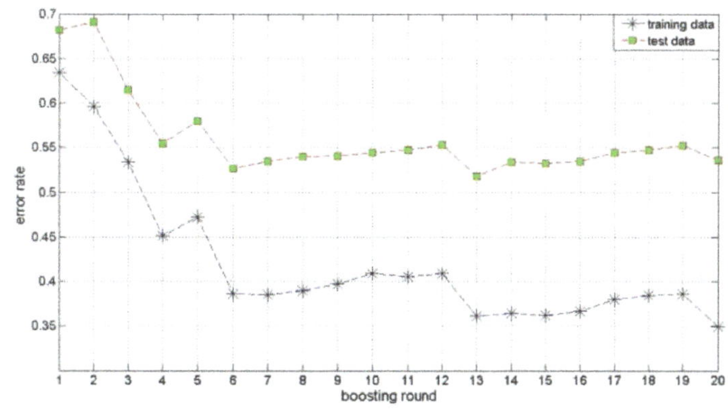

Figure 12. Boosting procedure for the network diagnoses of the whole TEP faults obtained by SAMMELF

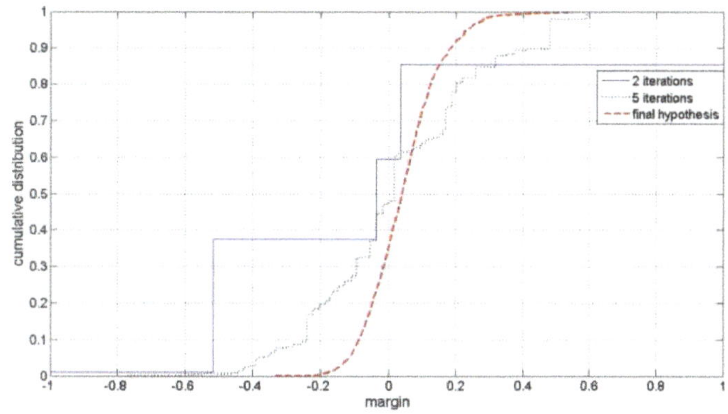

Figure 13. The margin distribution of the training examples obtained in the boosting of the network diagnoses of the whole TEP faults by SAMMELF

Using one fault pattern at a time, the single network and the boosted network obtained by the SAMMELF method are tested to determine a general view of the capabilities of these two diagnosis systems under different fault conditions. Here, the whole TEP diagnosis problem is considered. Each test pattern is applied to both of the diagnosis methods and consequently the corresponding misclassification errors are computed. The results are presented in Figure 14. As seen in this figure, some faults cannot be diagnosed by the single neural network classifiers at all. These include the challenging faults of 3, 9 and 15 which were also referred to as unobserved faults in some previous works [44]. On the other hand, the boosting improvement are more noticeable for the poorly identified faults. This reveals the fact the boosting algorithm heavily concentrates on the hard-classified examples. However, in cases such as for the faults 1, 6, 17, 18, and 20 the diagnosis performance is decreased in the boosted network. This indicates that the application of the boosting method, sacrifices the diagnosis strength of some of the well-diagnosed faults to attain a minimum level of diagnosis ability for each fault. However, the overall average error is reduced as shown in Figure 14.

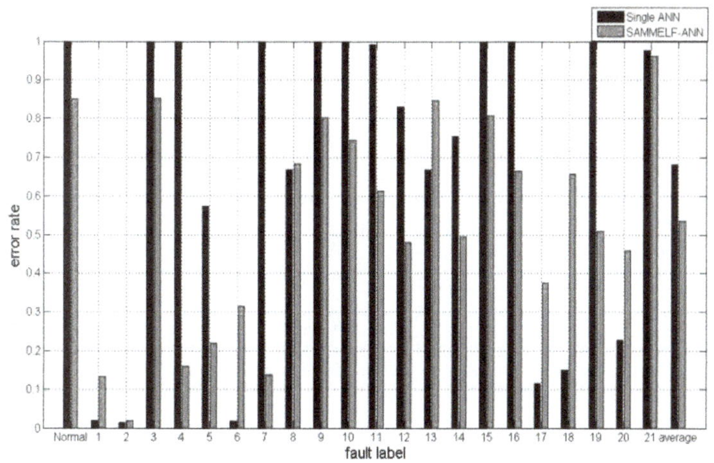

Figure 14. Classification performances of the single and boosted networks obtained by the SAMMELF algorithm on the test data under different fault conditions

5. Conclusion

In this work, three networks used for fault diagnosis of a chemical plant are compared to each other. Boosting as a general method employed to improve the classification performance of learning algorithms is utilized. The main idea in the boosting method is to enforce the basic classifier to work on the hard examples in a sequential rounds and then to combine the obtained single classifiers in an ensemble system. AdaBoost algorithm as the most popular boosting method is employed and its advantages and limitations in practical applications of process fault diagnosis are discussed. AdaBoost guarantees that the error of the composite classifier on the training data decreases exponentially to a very small value. However, the convergence is achieved if the weak learner (MLP neural networks in our application) has always prediction error less than 0.5 in each boosting stage. Whereas, this condition may not always be obtainable especially in applications where a large number of faults is encountered, a substitute boosting method called SAMMELF is also utilized. The SAMMELF algorithm follows closely the philosophy of boosting and shares the same simple modular structure of AdaBoost. However, its convergence is achieved by imposing a relatively relaxed condition on each weak learner prediction error which normally requires error to be less than a random guessing.

The TEP simulator as a well-known challenging diagnosis benchmark is utilized in this work to assess three different fault classification methodologies. The performances of the single neural network and the ensemble model obtained by the AdaBoost and SAMMELF algorithms are compared against each other under different conditions. The results showed that using a boosting based ensemble method always led to an improved diagnosis performance. The employed boosting methods were able to exponentially reduce the training error without generating over-fitted classification models. In addition, the generalization error is improved in all cases. However, the AdaBoost algorithm failed when boosting the network dealing with the whole TEP diagnosis problem because this network always produced errors more than 0.5 on the training data set. Consequently, the conventional AdaBoost algorithm may not always be applied to complex diagnosis problems where numerous process faults are presented. On the other hand, the SAMMELF algorithm performed well in these difficult diagnosis situations and always improved the fault diagnosis performance in the sequential rounds of boosting. Hence, it presents a proficient boosting scheme which is more suitable for industrial process fault diagnosis.

References

[1] Himmelblau, D.M., *Fault detection and diagnosis in chemical and petrochemical processes*, Elsevier Scientific Publishing Company, Amsterdam, 1978.

[2] Chiang, L.H., Braatz, R.D. and Russell, E.L., *Fault detection and diagnosis in industrial systems*, Springer, New York, 2001.

[3] Venkatasubramanian, V., Rengaswamy, Yin, R.K. and Kavuri, S.N., "A review of process fault detection and diagnosis: Part I: Quantitative model-based methods," *Computers & chemical engineering*, 27 (3). 293-311. 2003.

[4] Isermann, R., "Process fault detection based on modeling and estimation methods-a survey," *Automatica*, 20 (4). 387-404. 1984.

[5] Dash, S. and Venkatasubramanian, V., "Challenges in the industrial applications of fault diagnostic systems," *Computers & chemical engineering*, 24 (2). 785-791. 2000.

[6] Kresta, J.V., Macgregor, J.F. and Marlin, T.E., "Multivariate statistical monitoring of process operating performance," *The Canadian journal of chemical engineering*, 69 (1). 35-47. 1991.

[7] He, Q.P. and Wang, J., "Fault detection using the k-nearest neighbor rule for semiconductor manufacturing processes," *Semiconductor manufacturing, IEEE transactions on*, 20 (4). 345-354. 2007.

[8] Chiang, L.H., Kotanchek, M.E. and Kordon, A.K., "Fault diagnosis based on Fisher discriminant analysis and support vector machines," *Computers & chemical engineering*, 28 (8). 1389-1401. 2004.

[9] Verron, S., Tiplica, T. and Kobi, A., "Bayesian networks and mutual information for fault diagnosis of industrial systems," in *Workshop on Advanced Control and Diagnosis (ACD'06)*, 2006.

[10] Verron, S., Tiplica, T. and Kobi, A., "Fault diagnosis of industrial systems by conditional Gaussian network including a distance rejection criterion," *Engineering applications of artificial intelligence*, 23 (7). 1229-1235. 2010.

[11] Chen, K.Y., Chen, L.S., Chen, M.C. and Lee, C.L., "Using SVM based method for equipment fault detection in a thermal power plant," *Computers in industry*, 62 (1). 42-50. 2011.

[12] Yélamos, I., Escudero, G., Graells, M. and Puigjaner, L., "Performance assessment of a novel fault diagnosis system based on support vector machines," *Computers & chemical engineering*, 33 (1). 244-255. 2009.

[13] Kulkarni, A., Jayaraman, V.K. and Kulkarni, B.D., "Knowledge incorporated support vector machines to detect faults in Tennessee Eastman process," *Computers & chemical engineering*, 29 (10). 2128-2133. 2005.

[14] Xu, J., Zhao, J., Ma, B. and Hu, S., "Fault diagnosis of complex industrial process using KICA and sparse SVM," *Mathematical problems in engineering*, 2013. 1-6. 2013.

[15] Hussain, M. A., Hassan, C.R.C., Loh, K.S. and Mah, K.W., "Application of artificial intelligence technique in process fault diagnosis," *Journal of engineering science and technology*, 2 (3). 260-270. 2007.

[16] Watanabe, K., Matsuura, I., Abe, M., Kubota, M. and Himmelblau, D.M., "Incipient fault diagnosis of chemical processes via artificial neural networks," *AIChE journal*, 35 (11). 1803-1812. 1989.

[17] Koivo, H.N., "Artificial neural networks in fault diagnosis and control," *Control engineering practice*, 2 (1). 89-101. 1994.

[18] Himmelblau, D.M., "Applications of artificial neural networks in chemical engineering," *Korean journal of chemical engineering*, 17 (4). 373-392. 2000.

[19] Sharma, R., Singh, K., Singhal, D. and Ghosh, R., "Neural network applications for detecting process faults in packed towers," *Chemical engineering and processing: process intensification*, 43 (7). 841-847. 2004.

[20] Eslamloueyan, R., "Designing a hierarchical neural network based on fuzzy clustering for fault diagnosis of the Tennessee–Eastman process," *Applied soft computing*, 11 (1). 1407-1415. 2011.

[21] Behbahani, M.R., Jazayeri-Rad, H. and Hajmirzaee, S., "Fault detection and diagnosis in a sour gas absorption column using neural networks," *Chemical engineering & technology*, 32 (5). 840-845. 2009.

[22] Hansen, L.K. and Salamon, P., "Neural network ensembles," *Pattern analysis and machine intelligence, IEEE transactions on*, 12 (10). 993-1001. 1990.

[23] Polikar, R., "Ensemble based systems in decision making," *Circuits and systems magazine, IEEE*, 6 (3). 21-45. 2006.

[24] Baxt, W.G., "Improving the accuracy of an artificial neural network using multiple differently trained networks," *Neural computation*, 4 (5). 772-780. 1992.

[25] Ali, K.M. and Pazzani, M.J., "Error reduction through learning multiple descriptions," *Machine learning*, 24 (3). 173-202. 1996.

[26] Valdovinos, R.M. and Sanchez, J. S., "Ensembles of multilayer perceptron and modular neural networks for fast and accurate learning," in *Artificial Intelligence, MICAI'06. Fifth Mexican International Conference on*, 2006, 229-236.

[27] Zhang, J., "Improved on-line process fault diagnosis through information fusion in multiple neural networks," *Computers & chemical engineering*, 30 (3). 558-571. 2006.

[28] Bauer, E. and Kohavi, R., "An empirical comparison of voting classification algorithms: bagging, boosting, and variants," *Machine learning*, 36 (1). 105-139. 1999.

[29] Zio, E., Baraldi, P. and Gola, G., "Feature-based classifier ensembles for diagnosing multiple faults in rotating machinery," *Applied soft computing*, 8 (4). 1365-1380. 2008.

[30] Xu, L., Krzyzak, A. and Suen, C.Y., "Methods of combining multiple classifiers and their applications to handwriting recognition," *Systems, man and cybernetics, IEEE transactions on*, 22 (3). 418-435. 1992.

[31] Kuncheva, L.I., Bezdek, J.C. and Duin, R.P.W., "Decision templates for multiple classifier fusion: an experimental comparison," *Pattern recognition*, 34 (2), 299-314, 2001.

[32] Ruta, D. and Gabrys, B., "An overview of classifier fusion methods," *Computing and Information systems*, 7 (1). 1-10. 2000.

[33] Freund, Y. and Schapire, R.E., "Experiments with a new boosting algorithm," in *ICML*, 1996, 148-156.

[34] Freund, Y. and Schapire, R.E., "A desicion-theoretic generalization of on-line learning and an application to boosting," in *Computational learning theory*, 1995, 23-37.

[35] Schwenk, H., and Bengio, Y., "Boosting neural networks," *Neural Computation*, 12 (8). 1869-1887. 2000.

[36] Schapire, R.E., Freund, Y., Bartlett, P. and Lee, W.S., "Boosting the margin: A new explanation for the effectiveness of voting methods," *The annals of statistics*, 26 (5). 1651-1686. 1998.

[37] Samanta, B., Bandopadhyay, S., Ganguli, R. and Dutta, S., "A comparative study of the performance of single neural network vs. adaboost algorithm based combination of multiple neural networks for mineral resource estimation," *Journal of South African institute of mining and metallurgy*, 105 (4). 237-246. 2005.

[38] Schwenk, H. and Bengio, Y., "Adaboosting neural networks: Application to on-line character recognition," in *Artificial Neural Networks-ICANN'97*, Springer, 1997, 967-972.

[39] Canty, M.J., "Boosting a fast neural network for supervised land cover classification," *Computers & geosciences*, 35 (6). 1280-1295. 2009.

[40] Zhu, J., Zou, H., Rosset, S. and Hastie, T., "Multi-class adaboost," *Statistics and Its Interface*, 2. 349-360. 2009.

[41] Downs, J.J. and Vogel, E.F., "A plant-wide industrial process control problem," *Computers & chemical engineering*, 17 (3). 245-255. 1993.

[42] Lyman, P.R. and Georgakis, C., "Plant-wide control of the Tennessee Eastman problem," *Computers & chemical engineering*, 19 (3). 321-331. 1995.

[43] Møller, M.F., "A scaled conjugate gradient algorithm for fast supervised learning," *Neural networks*, 6 (4). 525-533. 1993.

[44] Mahadevan, S. and Shah, S.L., "Fault detection and diagnosis in process data using one-class support vector machines," *Journal of process control*, 19 (10). 1627-1639. 2009.

5

Distance Measurement via Using of Ultrasonic Sensor

Michal Kelemen[1,*], Ivan Virgala[1], Tatiana Kelemenová[2], Ľubica Miková[1], Peter Frankovský[1], Tomáš Lipták[1], Milan Lörinc[1]

[1]Department of Mechatronics, Technical University of Kosice, Faculty of Mechanical Engineering, Kosice, Slovak Republic
[2]Department of Biomedical Engineering and Measurement, Technical University of Kosice, Faculty of Mechanical Engineering, Kosice, Slovak Republic
*Corresponding author: michal.kelemen@tuke.sk

Abstract Paper deals with ultrasonic sensors used for distance measurement. Selected ultrasonic sensor has been tested and results of experiments are shown in the paper. Uncertainty analysis also has been realized.

Keywords: *ultrasonic, distance measurement, sensor, uncertainty*

1. Introduction

Ultrasonic distance sensors are designed for non-contact distance measurement and these types consist of transmitter and receiver or transceiver which is able to transmit and to receive ultrasonic sound (Figure 1). Main idea is to measure time to fly of ultrasonic sound wave from sensor to detected object. An ultrasonic transmitter sends a sound frequency of above 18 kHz in the air at the speed of 344 meter per second (at 20°C) and the receiver receives the reflected sound from the object. Distance between the transmitter and the object can be calculated by simple calculation by considering the time taken by the ultrasonic wave to travel from transmitter and received back (reflected) by the receiver. Measurement range is up to several meters.

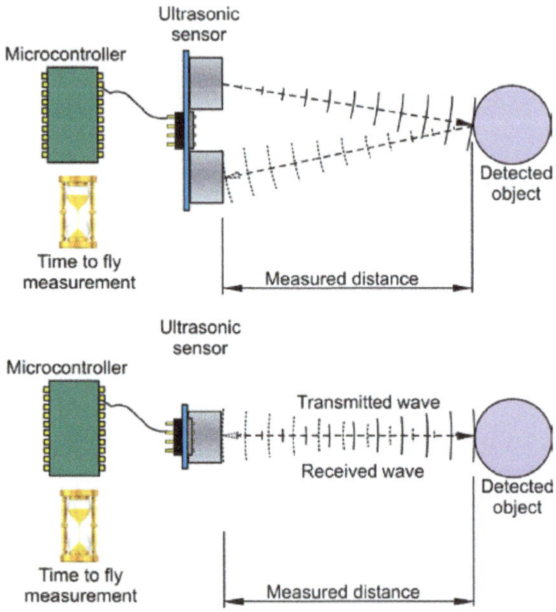

Figure 1. Ultrasonic sensor working principle

Almost all materials reflect sound waves, so ultrasonic sensors are a fine choice for many tasks. Excellence in the detection and measurement of films, transparent objects, and liquids separate these sensors from their photoelectric counterparts. Target color or frequent color changes also have no effect on ultrasonic sensors.

Due to their use of sound waves, ultrasonic sensors also perform well in dusty, dirty environments. However, they do not operate well with small targets against large backgrounds or targets such as foam batting that are excellent for absorbing sound waves [1,2,3,4].

A typical ultrasonic sensor (Figure 2) comprises a clock (signal) generator and a controller to excite the transducer, then a processor and output amplifier to handle the return signal [4].

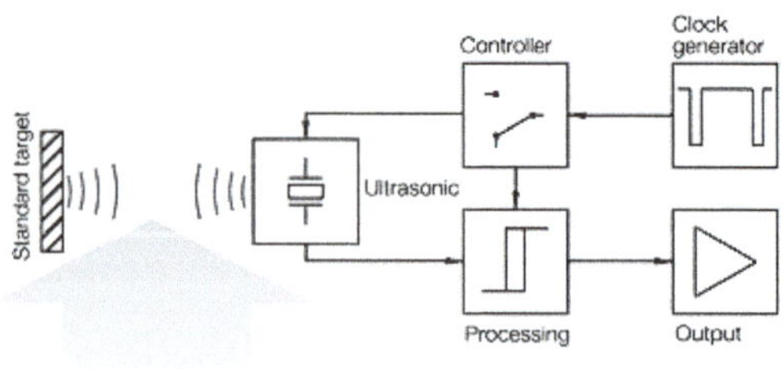

Figure 2. Ultrasonic sensor block scheme [4]

Besides the time of flight principle also they are used physical principles based on Doppler effect and the attenuation of sound waves.

Frequently application is as navigation sensor for mobile robots for obstacle avoiding [5].

Figure 3. Ultrasonic sensor on for mobile robot navigation

Ultrasonic sensors have been used in Tukebot robot (Figure 3, Figure 4) built for "puck collecting competition" at RobotChallenge. RobotChallenge is one of the biggest competitions for self-made, autonomous and mobile robots worldwide. Competitive robots have to collect small discs ("pucks") on the field according to color. To robots compete against each other on a 250 x 250 cm field. The aim is to collect all pucks of the assigned color and carry them to the own home base. The first robot which collects all the assigned pucks wins [5,6].

Locomotion microcontroller obtains signals from infrared distance sensor, from collision bumper touch sensors and from ultrasonic distance sensors (Figure 4). On the base of these sensors, locomotion microcontroller plans next locomotion and control it through drives of both wheels. These sensors enable to recognize, where the other rival robot is and where these assigned pucks are. It means that robot is still looking for assigned pucks and it avoids the rival robot [5,6].

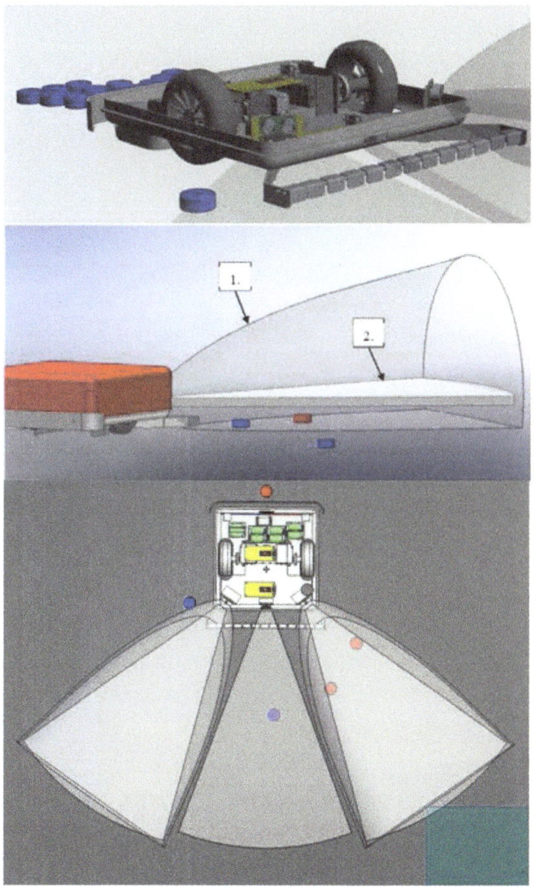

Figure 4. Sensor for puck collecting robot (1 – ultrasonic sensor active field), (2 – infrared sensor active field)

2. Experimental testing of ultrasonic distance sensor

Ultrasonic distance sensor with analogue output 0 – 10V has been selected for testing. The sensor uses 300 kHz sound frequency. Measurement range is from 120 mm up to 1000 mm and it has linear characteristic. Repeat accuracy is ±0.15% and resolution is 0.037 mm.

Next step is to examine properties of the sensor. Experimental stand (Figure 5) has been designed for testing and set of length gauges (Figure 6) have been used for sensor testing.

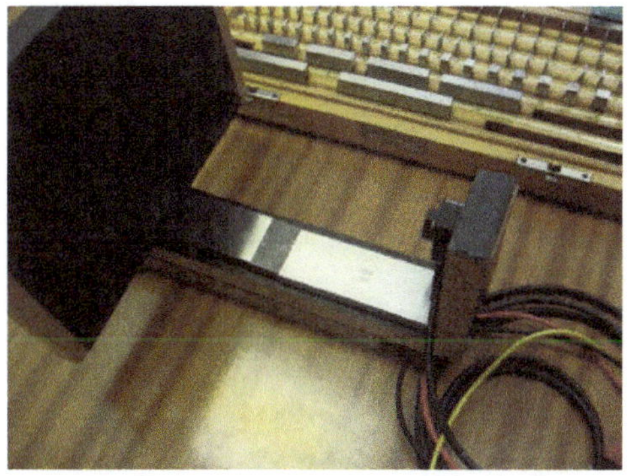

Figure 5. Testing of ultrasonic distance sensor

Figure 6. Gauge length for testing of sensor

Gauge lengths - blocks have been used as etalon of length and desired value of length or distance has been composed from these blocks. Two sets of gauges have been used for testing and it is possible to compose every possible etalon of length with resolution 0.001 mm in range 0 to 1000 mm.

Some materials are not suitable for distance measurement. Vibration absorbents may cause problems with measurement. For this reason testing of distance with this material has been executed.

Figure 7. Testing of ultrasonic sensor with compliant obstacle

Experimental testing has been executed with solid and compliant (vibration absorbent) material. Results from these testing are visible on Figure 8.

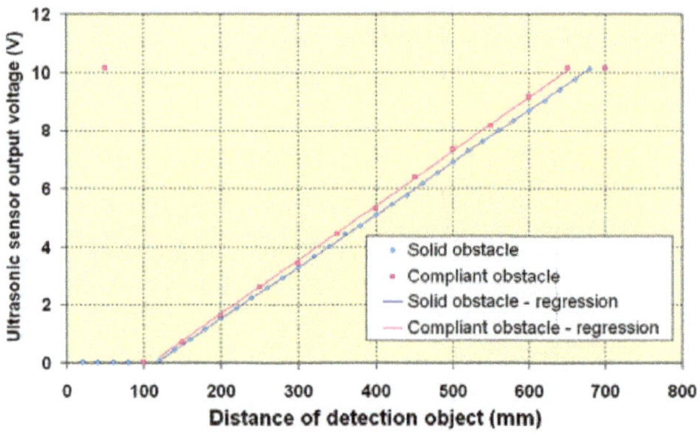

Figure 8. Measurement result of ultrasonic sensor – transformation characteristic

As it is visible (Figure 8), the sensor has linear characteristic with dead zone, it means that it is not able to measure distances less then 120 mm. Total measurement range is up to 700 mm and after this value sensor has constant value of output voltage. There is also visible difference (change of slope of the characteristic) between the measurement realized with solid material and compliant material.

Calibration characteristic from the measurement to solid obstacle is shown on Figure 9.

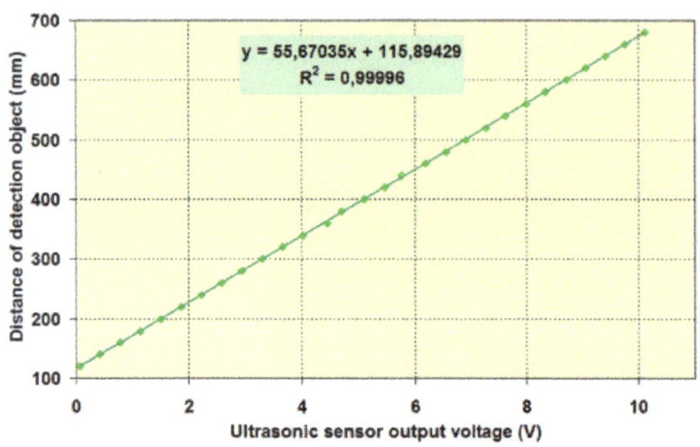

Figure 9. Measurement result of ultrasonic sensor – calibration characteristic

Math model obtained from regression of calibration characteristic define the equation (Figure 9), which can be used for recalculation of measured output sensor voltage to searched information about position measurement. For practical using of the sensor, it is necessary to analyse the uncertainty of measurement.

3. Uncertainty of Measurement

Calibration characteristic (Figure 9) is made for set of values (x_i - sensor output voltage, y_i – distance of detected object). The calibration curve is approximated with linear model $y = b_0 + b_1 \cdot x$, where regression coefficients have also uncertainty of determination expressed with equations:

$$u^2_{(b1)} = \frac{n}{n\sum_{i=1}^{n} x_i^2 - \left(\sum_{i=1}^{n} x_i\right)^2} \cdot \sigma^2 \qquad (1)$$

$$u^2_{(b0)} = \frac{\sum_{i=1}^{n} x_i^2}{n\sum_{i=1}^{n} x_i^2 - \left(\sum_{i=1}^{n} x_i\right)^2} \cdot \sigma^2 \qquad (2)$$

Covariance between these regression coefficients estimation is defined with equation:

$$u_{b0,b1} = \mathrm{cov}(b0, b1) = \frac{-\sum_{i=1}^{n} x_i}{n\sum_{i=1}^{n} x_i^2 - \left(\sum_{i=1}^{n} x_i\right)^2} \cdot \sigma^2 \qquad (3)$$

Where σ is standard deviation of distance (y_i) is possible to estimate with residual variance:

$$\sigma^2_{MSE} = \frac{1}{n-2}\sum_{i=1}^{n}\left[w_i - \left(b_1 \cdot x_i + b_0\right)\right] \qquad (4)$$

For general math model described with polynom of p degree $y = b_0 + b_1 \cdot x + b_2 \cdot x^2 + ... + b_p \cdot x^p$ overall standard uncertainty is defined as:

$$u_y = \sum_{j=0}^{p} x^{2 \cdot j} \cdot u_{bj} + \left(\sum_{j=1}^{p} j \cdot x^{j-1} \cdot b_j\right)^2 \cdot u_x^2 + \\ +2 \cdot \sum_{j=0}^{p-1} \sum_{k=j+1}^{p} x^j x^k u_{bj,bk} \qquad (5)$$

For our linear model equation (5) can be simplified:

$$u_y = \left(u_{b0}^2 + x^2 \cdot u_{b1}^2\right) + b_1 \cdot u_x^2 + 2\left(x \cdot u_{b0,b1}\right) \qquad (6)$$

Standard uncertainty (Figure 10) is obtained applying the equation (4) for measurement chain with ultrasonic distance sensor. These values represent together uncertainty for all parts of measuring chain (ultrasonic sensor, multimeter, length gauges, positioning table etc.)

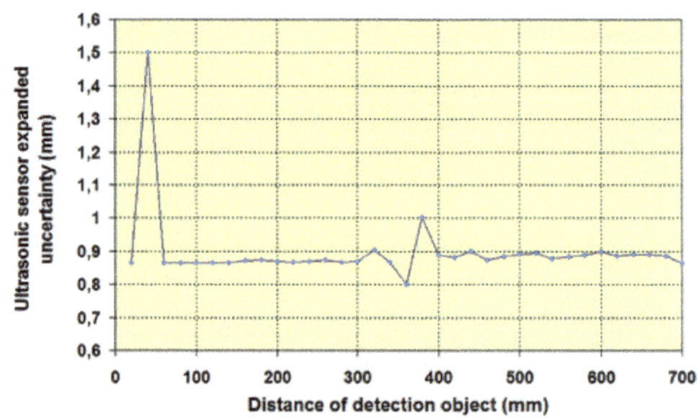

Figure 10. Standard uncertainty of measurement for measurement chain with ultrasonic distance sensor

Larger value of uncertainty occurs because it is out of the useful range of the sensor. All others uncertainties are less than 1 mm. Consequently standard uncertainty depends also on more factors (not only on sensor).

4. Conclusion

Ultrasonic sensors have variety application as distance measurement, obstacle avoiding and anti-collision detection, robot navigation, measurement in automotive parking assistance systems, measurement of air flow velocity - anemometer, medical ultrasonography, non-destructive testing, piezoelectric transducers, level measurement, pallet detection on forklifts, vehicle detection in barrier systems etc.

Ultrasonic sensors are non-intrusive in that they do not require physical contact with their target, and can detect certain clear or shiny targets otherwise obscured to some vision-based sensors. On the other hand, their measurements are very sensitive to temperature and to the angle of the target. Temperature and humidity affect the speed of sound in air. Therefore, range finders may need to be recalibrated to make accurate measurements in a new environment. Temperature variations and air currents can create invisible boundaries that will reflect ultrasonic waves, so care must be taken to avoid these. For the transmitted wave to echo back to the receiver, the target surface must be perpendicular to the transmitter. Round objects are therefore most easily sensed since they always show some perpendicular face. When targeting a flat object, care must be taken to ensure that its angle with respect to the sensor does not exceed a particular range.

Ultrasonic sensors typically have a "dead zone" immediately in front of them in which objects cannot be detected because they deflect the wave back before the receiver is operational. (This is because reverberations from the transmitter force the receiver to pause a moment before beginning to listen for the echo). Some materials are more absorbent than others, and these will reflect less ultrasound. This complicates using the attenuation method to measure the distance of arbitrary objects [7-29].

Acknowledgement

The work has been accomplished under the research projects No. KEGA 048TUKE-4/2014, VEGA 1/0182/15 and KEGA No. 014STU-4/2015 financed by the Slovak Ministry of Education.

References

[1] *Smart Sensors and Applications Student Guide.* VERSION 1.0. Parallax Inc., https://www.parallax.com/sites/default/files/downloads/28029-Smart-Sensors-Text-v1.0.pdf

[2] *PING))) Ultrasonic Distance Sensor.* Parallax Inc., https://www.parallax.com/product/28015

[3] *PING))) Ultrasonic Distance Sensor* (#28015). https://www.parallax.com/sites/default/files/downloads/28015-PING-Sensor-Product-Guide-v2.0.pdf

[4] *The Search for a Better Proximity Sensor Starts Here* Published in Design World, February 2010. Balluff Inc. http://www.balluff.com/balluff/MUS/en/news/search-for-better-proximity-sensor.jsp

[5] M. Kelemen et all, Puck collecting robot, 2014. In: *Applied Mechanics and Materials.* Vol. 611 (2014), p. 256-264.

[6] *RobotChallenge. International championship for self-made, autonomous, and mobile robots.* Available online. cited 06-112013. http://www.robotchallenge.org/

[7] jwilans on SensorWiki.org, *Ultrasound.* Last modified: 2011/04/11, cited 2015/25/11. Available online: http://www.sensorwiki.org/doku.php/sensors/ultrasound

[8] J. Wilson, 2005. *Sensor Technology Handbook.* Oxford, UK.: Elsevier.

[9] R. Pallàs-Areny, , and J. Webster. 2001. *Sensors and Signal Conditioning*, 2nd ed. Toronto, Ontario: John Wiley & Sons.

[10] J. J. Carr. 1993. *Sensors and Circuits.* Upper Saddle River, NJ: PTR Prentice-Hall, Inc.

[11] Li-Hong Juang, Ming-Ni Wu, Zhi-Zhong Weng, Object identification using mobile devices, *Measurement*, Volume 51, May 2014, Pages 100-111, (2014).

[12] M. Dekan, F. Duchoň, L. Jurišica, A. Vitko, A. Babinec, iRobot Create Used in Education, *Journal of Mechanics Engineering and Automation.* Vol. 3, Iss. 4, 2013, pages 197-202, (2013).

[13] Z. Liu, F. Li, G. Zhang, An external parameter calibration method for multiple cameras based on laser rangefinder, *Measurement*, Volume 47, January 2014, Pages 954-962 (2014).

[14] Y. Xue et al., A New Approach for Autonomous Robot Obstacle Avoidance Using PSD Infrared Sensor Combined with Digital Compass, *Applied Mechanics and Materials*, Volume 511-512, February, 2014, pages 101-104, (2014).

[15] N. Kelly, *A Guide to Ultrasonic Sensor Set Up and Testing Instructions, Limitations, and Sample Applications.* Application Note. Michigan State University. available online: http://www.egr.msu.edu/classes/ece480/capstone/fall09/group05/docs/ece480_dt5_application_note_nkelly.pdf

[16] A. Carullo; M. Parvis (2001). *An Ultrasonic Sensor for Distance Measurement in Automotive Applications.* In: IEEE SENSORS JOURNAL, vol. 1 n. 2, pp. 143-147.

[17] Z. Yi, H. Y. Khing ; Ch. Ch. Seng ; Z. X. Wei, Multi-ultrasonic sensor fusion for mobile robots, *Intelligent Vehicles Symposium*, 2000. IV 2000. Proceedings of the IEEE, pages 387-391.

[18] D. Han, S. Kim, S. Park, Two-dimensional ultrasonic anemometer using the directivity angle of an ultrasonic sensor, *Microelectronics Journal*, Volume 39, Issue 10, October 2008, Pages 1195-1199.

[19] M.Y.I. Idris et all, Parking Guidance System Utilizing Wireless Sensor Network and Ultrasonic Sensor, *Information Technology Journal* 8 (2), 2009, pages 138-146.

[20] B. Henning , J. Rautenberg, Process monitoring using ultrasonic sensor systems, *Ultrasonics*, Volume 44, Supplement, 22 December 2006, Pages e1395-e1399.

[21] D. Koniar, L. Hargaš, M. Hrianka, P. Špánik, Frequency and Trajectory Objects Evaluation in Frames Sequence, *Advances in Electrical and Electronic Engineering*, 2011.

[22] G. V. Blessing ; D. G. Eitzen, Ultrasonic Sensor For Measuring Surface Roughness, *Proc. SPIE 1009, Surface Measurement and Characterization*, 281 (March 21, 1989).

[23] P. G. Kim et all, Obstacle Avoidance of a Mobile Robot Using Vision System and Ultrasonic Sensor, *Advanced Intelligent Computing Theories and Applications.* With Aspects of Theoretical and Methodological Issues, Volume 4681 of the series Lecture Notes in Computer Science pp 545-553.

[24] A. Babinec, M. Dekan, F. Duchoň, A. Vitko, Modifications of VFH Navigation Methods for Mobile Robots, *Procedia Engineering*, Volume 48, 2012, Pages 10-14.

[25] F. Fang, X. Ma, X. Dai, A Multi-sensor Fusion SLAM Approach for Mobile Robots. In: *Proceedings of the IEEE International Conference on Mechatronics and Automation*, Niagara Falls, Canada (2005).

[26] J. Yun-ho, *A Real Time Obstacle Avoidance for Mobile Robot Using Nearness Diagram*, Limit-cycle and Vector Field Method. Korea Aerospace University (2006).

[27] F. Duchoň, P. Hubinský, J. Hanzel, A. Babinec, M. Tölgyessy, Intelligent Vehicles as the Robotic Applications, *Procedia Engineering*, Volume 48, 2012, Pages 105-114.

[28] D. Koniar, , L. Hargaš, S. Štofan, Segmentation of Motion Regions for Biomechanical Systems, *Procedia Engineering*, Volume 48, 2012, Pages 304-311.

[29] Sukop M., Hajduk M., Varga J., Proposal Hardware and Software Modules for Multi-agent System of Robosoccer, Acta Mechanica Slovaca. Volume 18, Issue 2, Pages 20-27.

6

The Block Diagram and Equations of State of the Bond Graph Example

Darina Hroncová*, Alexander Gmiterko, Tomáš Lipták

Department of Mechatronics, Faculty of Mechanical Engineering, Technical University of Košice, Letná 9, 042 00 Košice, Slovak republic

*Corresponding author: darina.hroncova@tuke.sk

Abstract The work shows the use the methodology of Bond Graph for modeling electric system of simple RLC circuit. Electrical model is solved by this approach at the level of its physical behavior. In this paper the method of generation of state equations system is discussed. Model of a simple electrical RLC circuit consisting of a resistor, an inductor, and a capacitor is taken. The differential equations describing the dynamics of the system are obtained in terms of the states of the system.

Keywords: *Modeling of dynamic systems, Bond Graphs, energy modeling, state equation*

1. Introduction

The concept of bond graphs was originated by Paynter. The idea was further developed by Karnopp and Rosenberg in their textbooks, such that it could be used in practice by Thoma, Van Dixhoorn, Breedveld and by others [1,2,3,4].

The language of bond graphs aspires to express general class physical systems through power interactions. The factors of power i.e., effort and flow, have different interpretations in different physical domains. Yet, power can always be used as a generalized coordinate to model coupled systems residing in several energy domains.

A causal bond graph contains all information to derive the set of state equations. The procedure to derive these equations is covered by bond graph software like Enport (Rosenberg, 1974), MS1 (Lorenz, 1997), CAMP (Granda, 1985), and 20-SIM (Broenink, 1990, 1995, 1997, 1999; Broenink and Kleijn, 1999). Therefore, in practice, generation of equation need not be done by hand. However, we discuss the generation of equations to indicate what exactly has to be done.

2. Bond Graph of the Electrical System - Description of the Model

The language of bond graphs aspires to express general class physical systems through power interactions. The factors of power i.e., effort and flow, have different interpretations in different physical domains. Yet, power can always be used as a generalized coordinate to model coupled systems residing in several energy domains. To demonstrate the bond graph methodology as an example an electrical model of RLC system is analyzed Figure 1a. An RLC circuit is an electrical circuit consisting of a resistor, an inductor, and a capacitor. The RLC part of the name is due to those letters being the usual electrical symbols for resistance, inductance and capacitance respectively [5].

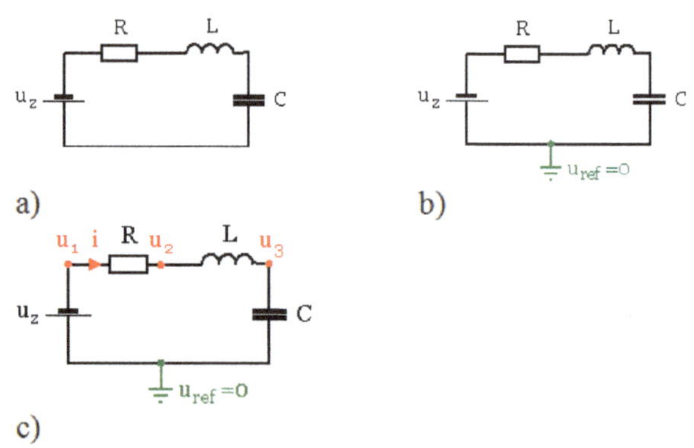

Figure 1. Electrical system: a) electrical model of RLC system; b) reference voltage u_{ref} with positive direction; c) efforts (voltages) with unique names u_1, u_2, u_3

3. Systematic Procedure to Derive a Bond Graph of the Electrical System

We have discussed the basic bond-graph elements and the bond, so we can transform a domain-dependent ideal-physical model, written in domain-dependent symbols, into a bond graph. For this transformation, there is a systematic procedure, which is presented here. This electrical system contains a voltage source effort SE (SE:uz), a resistor R (R:R), an inductor I (I:L) and a capacitor C (C:1/C). In the **step 1** we determine which physical domains exist in the system and identify all basic elements like C (capacitor), I (inductor), R (rezistor), SE (source of the effort), SF (source of the flow), TF (transformer) and GY (gyrator) [6,7,8].

This system contains a electrical domain part with the inductance L of the inductor (I:L), the resistance R of the resistor (R:R) and a capacity of the condensator C (C:C). Voltage u_z is considered as an effort source (SE: u_z).

In the **step 2** we indicate in the ideal-physical model per domain a reference source – effort voltage u_{ref} (reference voltage with positive direction), Figure 1b. The references are indicated in the ideal physical model: the ideal velocity $u_{ref} = 0$. In the next **step 3** we identify all other efforts (voltages) and give them unique names u_1, u_2, u_3, Figure 1c). In **step 4** we draw these efforts (electrical: voltages), graphically by 0-junctions in Figure 2a).

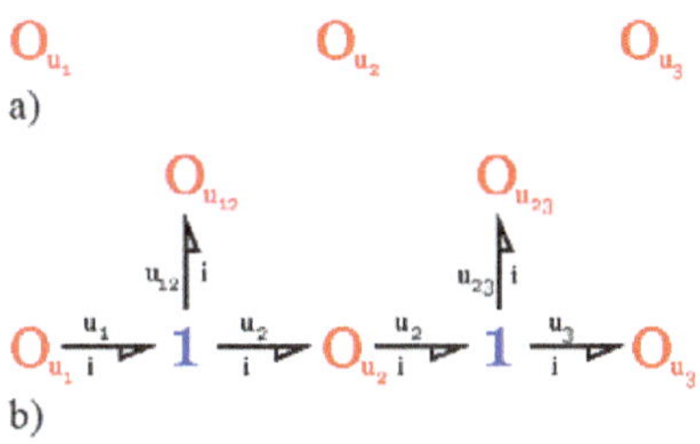

Figure 2. Electrical system (a) voltages are shown as 0-junctions (b) connecting 0-junctions with 1-junctions

We identify in **step 5** all effort differences (electrical: voltage = effort) needed to connect the ports of all elements enumerated in step 1 to the junction structure. When checking all ports of the elements found in step 1 linear voltage differences, u_{12} and u_{23} are identified:

$$u_1 - u_2 = u_{12},\, u_2 - u_3 = u_{23}. \tag{1}$$

In the **step 6** we construct the voltages differences using a 1-junction and draw them as such in the graph Figure 2b). The junction structure is now ready and the elements can be connected.

In the **step 7** we connect the port of all elements found at step 1 with the 1-junction of the corresponding efforts or efforts differences in Figure 3a).

We simplify in the **step 8** the resulting graph by applying the simplification rules in Figure 3b) and in Figure 4a).

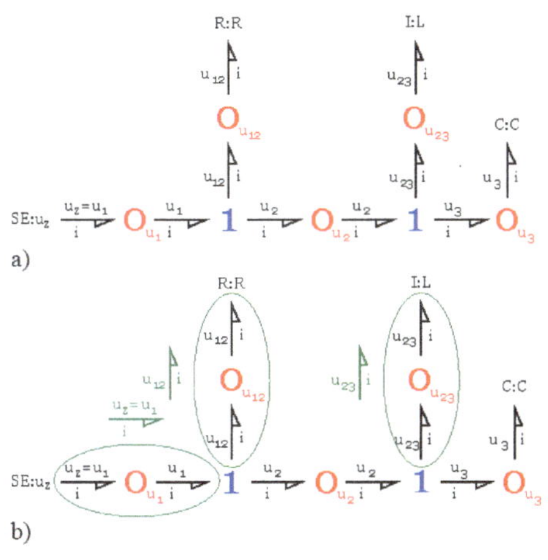

Figure 3. Electrical system (a) connecting elements R, I, C, SE with 0-junctions (b) simplify the resulting graph by applying the simplification rules

A junction between two bonds can be left out, if the bonds have a through power direction (one bond incoming, the other outgoing). A bond between two the same junctions can be left out, and the junctions can join into one junction. Two separately constructed identical effort or flow differences can join into one effort or flow difference.

In the **step 9** we determine the signal direction and causality in Figure 4 b). Causality establishes the cause and effect relationships between the factors of power.

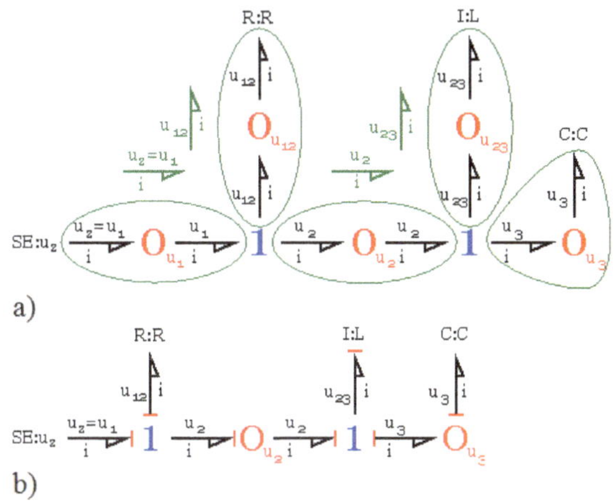

Figure 4. Simplify the resulting graph by applying the simplification rules (a) the 0-junctions with voltages u_2 and u_3 (b) determine the causality

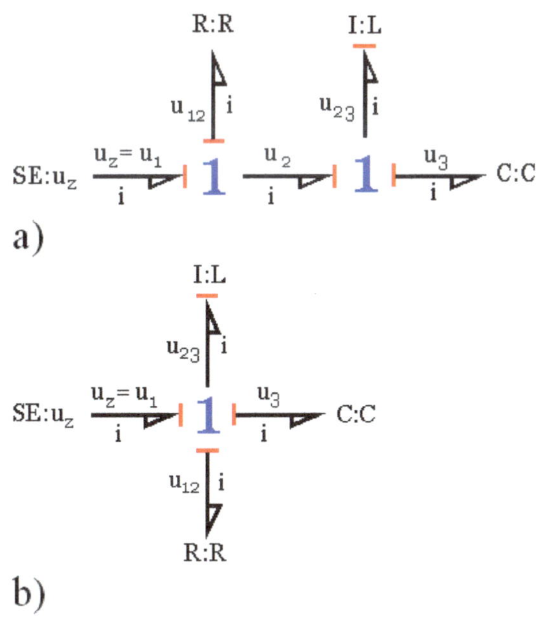

Figure 5. Simplify the resulting graph by applying the simplification rules (a) the 0-junctions reduction and causality and (b) the final bond graph with signal direction and causality

In bond graphs, the inputs and the outputs are characterized by the causal stroke. The causal stroke indicates the direction in which the effort signal is directed (by implication, the end of the bond that does not have a causal stroke is the end towards which the flow signal is directed).

There are two ways of describing an element's behavior (e.g. effort in, flow out vs. flow in, effort out) as different causal forms. Note that the two alternative causal forms may, in general, require quite different mathematical operations. The causal form we use, i.e. which variable we select as input and which we select as output, can make a lot of difference. For example, the required mathematical operations may be well defined in one causal form, but not defined at all in the other.

The causal bond graph of this system can be derived, in which the inputs and the outputs are characterized by the causal stroke. This is the starting point, from which we continue toward the differential equations describing the dynamics of the system.

4. Expansion to Block Diagrams

The equations are derived in the steps described below.

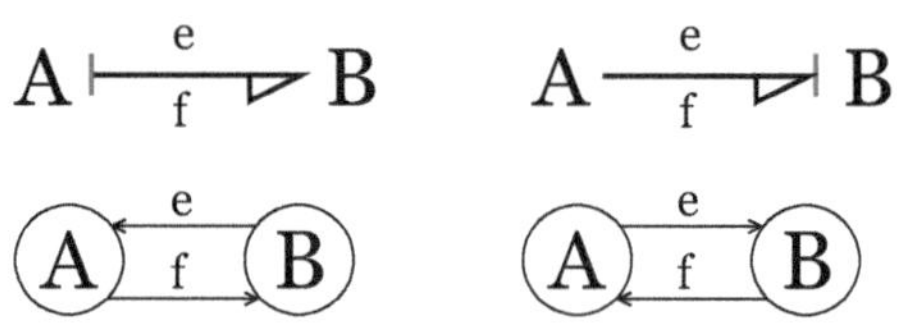

Figure 6. The bond graph example – bilateral signal flows between ports and definition of the causal stroke

A causal bond graph in Figure 5b) can be expanded into a block diagram in three steps.

All node symbols in the **step 1** are encircled in Figure 7.

Step 2 – All bonds are expanded into bilateral signal flows according to the assigned causality in Figure 7.

In **step 3** all constitutive relations of each node are written into block diagram form, according to the assigned causality of each port; one-junction is represented by a signal-node for the flows and a summation for the efforts (Figure 8).

All signals entering a summation resulting from a junction are given a sign corresponding to the half-arrow direction: if, while traveling from causal input to causal output, the bond orientation does not change (this does not exclude an orientation opposite to the signal direction), then a plus sign is added representing a positive contribution to the summation; by contrast if the bond orientation does change, then a minus sign is added representing a negative contribution to the summation. In principle, a complete block diagram is obtained at this point.

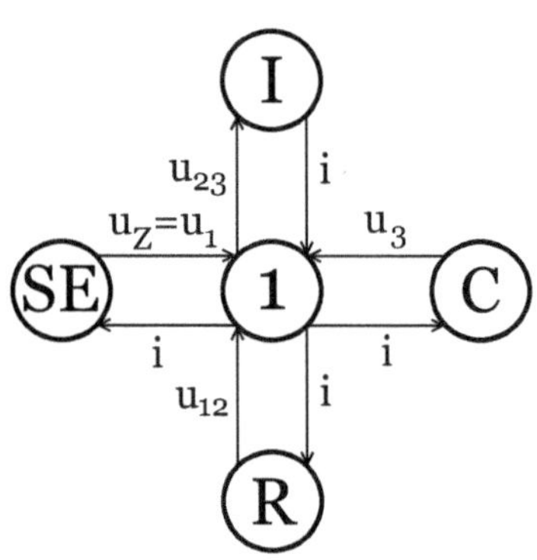

Figure 7. The bond graphs element layout and bond graph with bilateral signals flow

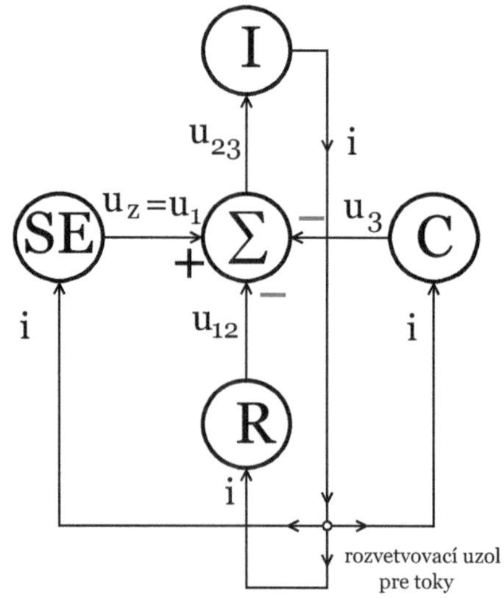

Figure 8. The block diagram with the line signal direction

The procedure to obtain a signal flow graph is completely analogous to the above procedure as all operations represented by blocks, including the signs of the summations, are combined as much as possible and then written next to an edge, while all summations become nodes, as signal nodes can be distinguished from signal summation points by observing the signal directions (signal node has only one input, summation has only one output).

Table 1. Element in block diagram

Bond graph	Type of element	Element in block diagram	Equations
u23 i I:L	u23 i I	u23 ∫ Φ(t) i 1/L	$i(t)=\frac{1}{L}.\Phi(t)$ $\Phi(t)=\int_{t_0}^{t}u(\tau)d\tau$
u12 i R:R	u12 i R	i R i u12	$u(t)=R.i(t)$
u12 i R:R	u12 i R	i R i u12	$u(t)=R.i(t)$
u3 i C:C	u3 i C	u3 1/C i ∫ Q(t	$u(t)=\frac{1}{C}.Q(t)$ $Q(t)=\int_{t_0}^{t}i(\tau)d\tau$

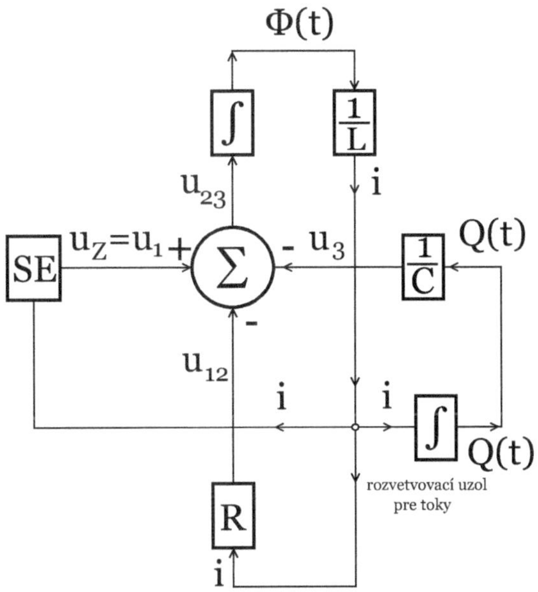

Figure 9. Bond graph expanded to a block diagram (a) the block diagram with source effort voltage u_z

5. Generation of the Equations

The two state equations derived from a bond graph diagram (Figure 9) are:

$$\frac{d\Phi(t)}{dt}=u_{23}, \tag{2}$$

$$\frac{dQ(t)}{dt}=i, \tag{3}$$

where state variables Φ (t) and Q (t) are magnetic flux and electric charge respectively, , and voltage u_{23} is:

$$\begin{aligned} u_{23} &= -u_3-u_{12}+u_Z = \\ &= -\frac{1}{C}.Q(t)-R.\frac{1}{L}.\Phi(t)+u_Z \end{aligned} \tag{4}$$

Next equations are:

$$u_3=\frac{1}{C}.Q(t), \tag{5}$$

$$\frac{d\Phi(t)}{dt}=u_{23}, \tag{6}$$

$$\frac{dQ(t)}{dt}=i, \tag{7}$$

$$\frac{1}{L}.\Phi(t)=i, \tag{8}$$

$$R.i=u_{12}, \tag{9}$$

After substitution we get:

$$\frac{d\Phi(t)}{dt}=-\frac{1}{C}.Q(t)-R.\frac{1}{L}.\Phi(t)+u_Z \tag{10}$$

$$\frac{dQ(t)}{dt}=\frac{1}{L}.\Phi(t) \tag{11}$$

or in matrix form:

$$\begin{bmatrix} \frac{d\Phi(t)}{dt} \\ \frac{dQ(t)}{dt} \end{bmatrix} = \begin{bmatrix} -R.\frac{1}{L} & -\frac{1}{C} \\ \frac{1}{L} & 0 \end{bmatrix} . \begin{bmatrix} \Phi(t) \\ Q(t) \end{bmatrix} + \begin{bmatrix} 1 \\ 0 \end{bmatrix} . u_Z \tag{12}$$

By choosing different state variables i electric current and u_3 electric charge, after substitution the state equations are in following form:

$$\frac{d(L.i)}{dt}=-\frac{1}{C}.(C.u_3)-R.\frac{1}{L}.(L.i)+u_Z, \tag{13}$$

$$\frac{d(C.u_3)}{dt}=\frac{1}{L}.(L.i), \tag{14}$$

and finally:

$$\frac{di}{dt}=-\frac{1}{L}.u_3-R.\frac{1}{L}.i+\frac{1}{L}u_Z, \tag{15}$$

$$\frac{du_3}{dt}=\frac{1}{C}.i, \tag{16}$$

or in matrix form:

$$\begin{bmatrix} \frac{di}{dt} \\ \frac{du_3}{dt} \end{bmatrix} = \begin{bmatrix} -\frac{R}{L} & -\frac{1}{L} \\ \frac{1}{C} & 0 \end{bmatrix} . \begin{bmatrix} i \\ u_3 \end{bmatrix} + \begin{bmatrix} \frac{1}{L} \\ 0 \end{bmatrix} . u_Z \tag{17}$$

The equation can be solved in Matlab/Simulink [9].

6. State Equation in Matlab/Simulink

Solution of the non-homogenous system of differential equations of a mechanical system with two degrees of freedom is done in Matlab/Simulink using State-Space. The most general state-space representation of a linear system with *u(t)* inputs, *y(t)* outputs and *n* state variables is written in the following form:

$$\dot{\mathbf{x}}(t)=\mathbf{A}.\mathbf{x}(t)+\mathbf{B}.\mathbf{u}(t), \tag{18}$$

$$\mathbf{y}(t)=\mathbf{C}.\mathbf{x}(t)+\mathbf{D}.\mathbf{u}(t). \tag{19}$$

We obtain:

$$\begin{bmatrix} \dot{i} \\ \dot{u}_3 \end{bmatrix} = \mathbf{A}.\begin{bmatrix} i \\ u_3 \end{bmatrix} + \mathbf{B}.u_Z \tag{20}$$

$$y=\mathbf{C}.\begin{bmatrix} i \\ u_3 \end{bmatrix} + \mathbf{D}.u_Z \tag{21}$$

where **A**, **B**, **C**, **D** are the respective matrices of the mechanical system defined as follows:

$$\mathbf{A}=\begin{bmatrix} -\frac{R}{L} & -\frac{1}{L} \\ \frac{1}{C} & 0 \end{bmatrix}, \mathbf{B}=\begin{bmatrix} \frac{1}{L} \\ 0 \end{bmatrix}, \mathbf{C}=\begin{bmatrix} 1 & 0 \end{bmatrix}, \mathbf{D}=[0] \tag{22}$$

The input **u**(t) is represented by the source effort voltage u_Z :

$$\begin{bmatrix} \dot{i} \\ \dot{u}_3 \end{bmatrix} = \begin{bmatrix} -\dfrac{R}{L} & -\dfrac{1}{L} \\ \dfrac{1}{C} & 0 \end{bmatrix} \cdot \begin{bmatrix} i \\ u_3 \end{bmatrix} + \begin{bmatrix} \dfrac{1}{L} \\ 0 \end{bmatrix} . u_Z \quad (23)$$

And the output **y**(t) is represented by the electric current *i*:

$$y = \begin{bmatrix} 1 & 0 \end{bmatrix} . \begin{bmatrix} i \\ u_3 \end{bmatrix} + [0] . u_Z \quad (24)$$

We determine the matrix and block diagram in Matlab/Simulink shown in Figure 10:

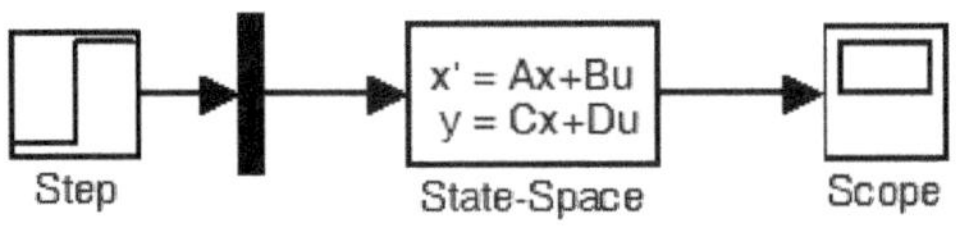

Figure 10. The block diagram in Simulink

Parameters used in the process of solution of the RLC circuit electrical system (Figure 1a) in Matlab/Simulink [10] are: R = 5 (Ω), L =0.1 (H), C = 100 (μF) and input voltage u_Z = 10 (V).

Output parameters of the electrical current i and voltages u_C ($u_C = u_3$), u_L ($u_L = u_{23}$), u_R ($u_R = u_{12}$) of the RLC circuit electrical system in Matlab [11] are shown in Figure 11.

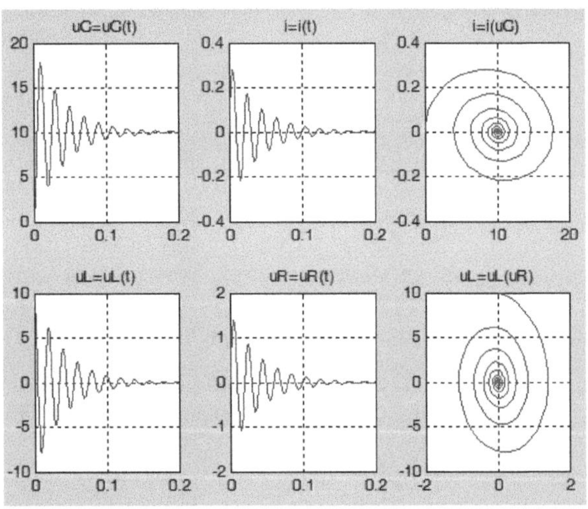

Figure 11. The results in Matlab $u_C=u_C(t)$, $i=i(t)$, $i=i(u_C)$, $u_L=u_L(t)$, $u_R=u_R(t)$, $u_L=u_L(u_R)$

7. Conclusion

Bond graphs represent a convenient tool for physical system analysis. We presented a method to systematically build a bond graph starting from an ideal physical model. Causal analysis gives, besides the computational direction of the signals at the bonds, also information about the correctness of the model. A practical example of an electrical model is presented as the application of this methodology.

The differential equations describing the dynamics of the system in terms of the states of the system were derived from a bond graph diagram of a simple electrical system. The results correspond with equations obtained using traditional method, where the equations for individual components are created first and then the simulation scheme is derived on their basis, although the described method uses the reverse procedure.

Acknowledgement

This work was supported by grant projects VEGA No. 1/1205/12 and grant projects VEGA No. 1/0937/12.

References

[1] Breedveld, P., C., Rosenberg, R., C., Zhou, T., 1991. *Bibliography of Bond Graph Theory and Application*, Journal of the Franklin Institute, 328(5/6), p.1067-1109.

[2] Breedveld, P., C. *Modeling and Simulation of Dynamic Systems Using Bond Graphs*, http://eolss.net.

[3] Broenink, J., F. *Introduction to Physical Systems Modelling with Bond Graphs*, http://www.ce.utwente.nl.

[4] Karnopp, D., C., Margolis, D., L., Rosenberg, R., C., 2005. *Modeling and simulation of mechatronic systems*. John Willey & Sons, Inc., Hoboken, New Jersey.

[5] Horáček, P., 1999. *Systems and models*, Nakladatelství ČVUT, Praha, Česko.

[6] Gmiterko, A., 2004. *Mechatronics*, Emilenia, Košice, Slovakia.

[7] Gmiterko, A., Šarga, P., Hroncová, D., 2012. *Mechatronika I*, Technická univerzita Košice, Košice, Slovakia.

[8] Hroncová, D., 2012. *Studies of compatible software tools for abstract and concrete design of mechatronic systems and linking them for the purpose of logical and physical modeling*, Dissertation thesis, Košice, Slovakia.

[9] Karban, P., 2006. *Výpočty a simulace v programech Matlab a Simulink*. Computer Press, Brno, Česko.

[10] Virgala, I., Frankovský, P., Kenderová, M.: *Friction Effect Analysis of a DC Motor*. In: American Journal of Mechanical Engineering. 2013. Vol. 1, no. 1 (2013), p. 1-5

[11] Vittek, J., 1997. *Matlab pre elektrické pohony*, Nakladateľstvo Žilinskej univerzity, Žilina, Slovensko, p. 35-36.

Maximum Power Point Tracking for Solar Photovoltaic System Using Genetic Programming Toolbox for Identification of Physical System

Sajjad Aliesfahani*, Mehdi Shahbazian

Department of Instrumentation and Automation Engineering, Petroleum University of Technology, Ahwaz, Iran
*Corresponding author: sajjadaliesfahani@gmail.com

Abstract Due to the increasing demands for energy in recent years and declining fossil fuel resources and the pollution that these energy sources are created, the solar energy produced by sun, has attracted much attention. Solar cells are elements that convert the solar energy into electricity. One of the challenges that we encounter is to obtain the most of the solar resource potential in various conditions of radiation, temperature and wind speed. A major task is to find an efficient and fast algorithm for tracking the maximum power point. In this paper, a new method for maximum power point tracking of photovoltaic (PV) cell that based on genetic programming toolbox of identification of physical system (GPTIPS), has been proposed. The maximum values of the voltage and the current of the solar cell is predicted by the GPTIPS algorithm and then the optimum duty cycle is produced for the chopper. Which cause maximum power deliver to the load. It was observed that the mean square error is reached to order of 10^{-5} approximately, that in similar circumstances, the mean square error and offline training time is less than a multi-layer perceptron of neural network. Simulation results show that the proposed GPTIPS approach is superior to the neural network method.

Keywords: *solar cell, maximum power point tracking, GPTIPS, optimal control*

1. Introduction

Due to depleting fossil fuel resources and the pollution that these are transfer into the environment, tend to produce clean and renewable energy is much higher. one of these energy sources is solar energy that is always available in everywhere. [1,2] Solar cells are components that produce electricity by solar energy. Performance of this elements is affected by environmental factors such as radiation, air temperature and wind speed. Voltage, current and power of solar cells varies nonlinearity with changing environmental conditions. [3,4] It is desirable that at any moment and in any circumstances, the maximum power generated by the solar cell and transfer to load. several of algorithms available for the MPPT including perturb and observe, incremental conductance, parasitic capacitance, constant voltage and fuzzy logic algorithms [5,6,7,8]. These methods have disadvantages like costly, difficult to implement and non-stable. For this purpose genetic programming toolbox for identification of physical system (GPTIPS) come with a solution. GPTIPS sometimes (usually when only a few input variables are involved) lags behind a neural network model in terms of raw predictive performance but the equivalent GP models are often simpler, shorter and may be open to physical interpretation. It's not always an easy question to answer.

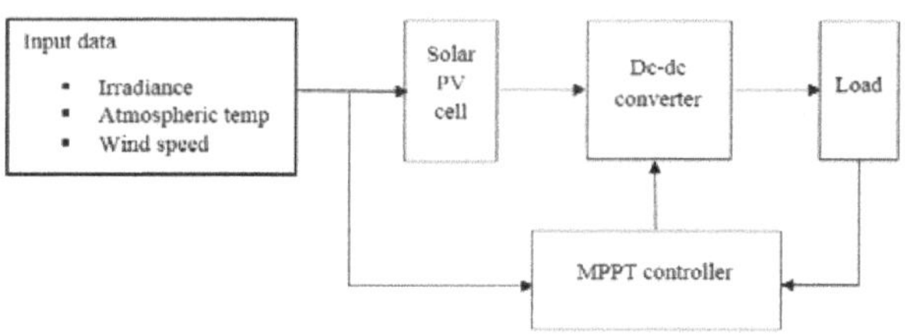

Figure 1. Photovoltaic system

At any rate, on practical applications GPTIPS normally significantly outperforms neural network models in terms of model performance, interpretability, deployability and robustness.also GPTIPS is better than the PLS. In my opinion, PLS regression is an overused and overvalued analytical tool. PLS regression is also hard to understand (well at least I find it hard to understand) and the models it produces are complex, fragile, hard to deploy and difficult to interpret.For impedance matching and obtain maximum power, a dc-dc converter located between the solar cell and load. A maximum power point tracker required for optimal control of duty cycle of dc-dc converter. [9,10] There are many techniques for maximum power point tracking. In this paper for generation and control the duty cycle used GPTIPS. It has several advantages such as high accuracy and faster performance compared to neural network and fuzzy logic control. [11,12] View block diagram of the solar photovoltaic system, we can see the Figure 1.

2. Model of Solar PV Cell

A simple model of a PV cell is shown as an equivalent circuit in Figure 2 that consists of an ideal current source in parallel with an ideal diode, with a parallel shunt resistance R_{sh} and series resistance R_s [13,14].

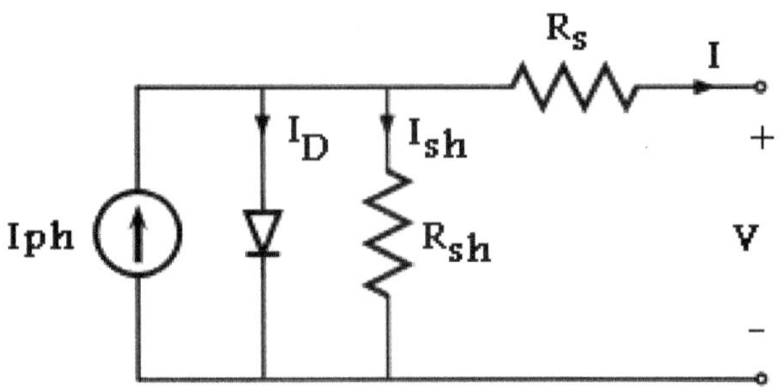

Figure 2. PV cell equivalent circuit

Based on the equivalent circuit, the current to the load can be given as [15,16,17]:

$$I = I_{ph} - I_s\left(\exp\frac{q(V+R_sI)}{NKT} - 1\right) - \frac{(V+R_sI)}{R_{sh}} \quad (1)$$

In this equation, I_{ph}, is the photocurrent, I_s, is the reverse saturation current of the diode, q is the electron charge, V is the voltage across the diode, K is the Boltzmann's constant, T is the junction temperature, N is the ideality factor of the diode, and R_s and R_{sh} are the series and shunt resistors of the cell, respectively. As a result, the complete physical behavior of the PV cell depends on I_{ph}, I_s, R_s and R_{sh} from one hand and with three environmental parameters as the temperature, the solar radiation and wind speed on the other hand. The above equation is a function of two equations:

One that calculates the PV cell photocurrent which depends on the radiation and the temperature according to equation [18].

$$I_{ph} = \left[I_{sc} + K_i(T-298)\right]\frac{\beta}{1000} \quad (2)$$

Where $K_i = 0.0017A/^\circ C$ is the cell's short circuit current temperature coefficient, I_{sc} short circuit current of solar cell (A) and β is the solar radiation (W/m). The actual cell temperature (T) has been found to be dependent on ambient air temperature, radiation and wind speed, which has been modeled by the following relationship [19]:

$$T(^\circ C) = 0.943T_A + 0.028(irradiance) \\ -1.528(wing\ speed) + 4.3 \quad (3)$$

Where T_A is the ambient air temperature given in °C, irradiance in W/m^2 and wind speed in m/s. The above solar PV cell model has been implemented in MATLAB and the model's accuracy is also validated by a comparison between it and datasheet of Q6LMXP3-G3.

Simulated P-V and I-V characteristics of solar cell are shown in Figure 3.

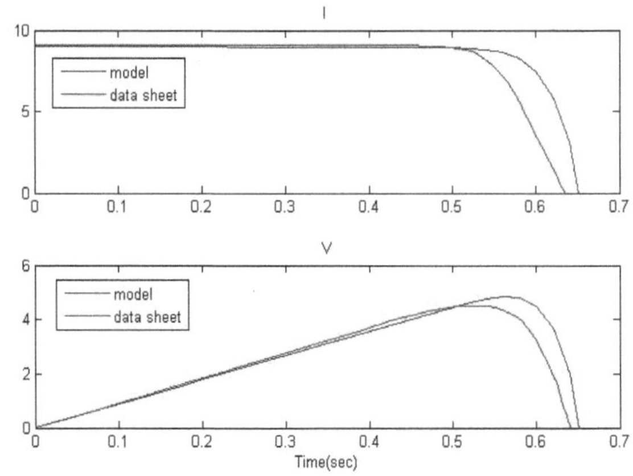

Figure 3. Comparison of the simulated P-V and I-V characteristics of solar cell at irradiance level 1000 W/m^2 and ambient air temperature 25°C and the Data sheet of Q6LMXP3-G3

3. Methodology and MPPT Controller

Genetic programming is an approach to identification and estimation the output of a system with respect to itself inputs. In this method, set the number of generations and functions that are used to identify the system will be done. Then having the inputs and output data, mathematical model of the system is constructed by this method. In fact, a function or a equation is produced respect to inputs and output of the system. According to this equation, the output of the system with respect to new inputs can be estimated. It means after training the system, the output can be achieved in a short time. [20].

Table 1. comparison between GPTIPS & NN

Methods	Number of train data	Number of test data	Offline training time	Mean square error for Voltage(test data)	Mean square error for Current(test data)
GPTIPS	870	372	0.214s	0.000035	0.0000346
MLP of NN	870	372	3s	0.0000728	0.0000758

The model uses the cell operating temperature and radiation as input values and estimates the corresponding maximum power voltage and current values. In Figure the actual data values and the GPTIPS-estimated values of maximum power current I_{max} and voltage V_{max} for training data and testing data have been compared. The values I_{max} and V_{max} correspond to solar intensity variation between 100 and 1000 W/m^2 and the cell temperatures of 10 to 50°C, whose ranges are relevant to practical situation. One can notice the good agreement between all sets of results. It was observed that the mean square error is reached to order of 10^{-5} approximately, that in similar circumstances, the mean square error and offline training time is less than a multi-layer perceptron of neural network. See Table 1 and Figure 4 and Figure 5.

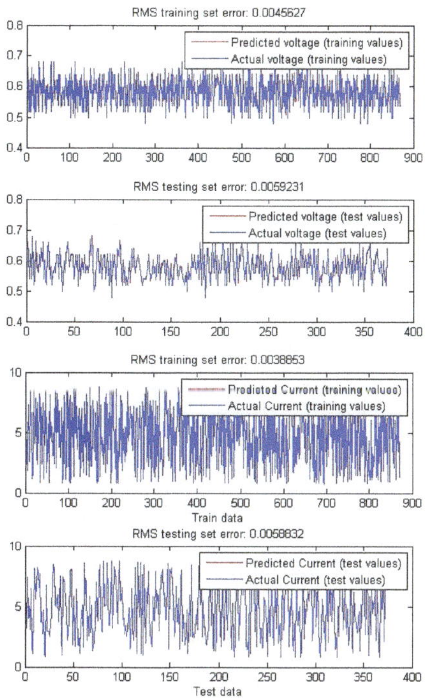

Figure 4. Comparison of actual max power current and voltage values with the maximum power current and voltage values obtained by GPTIPS predictor.

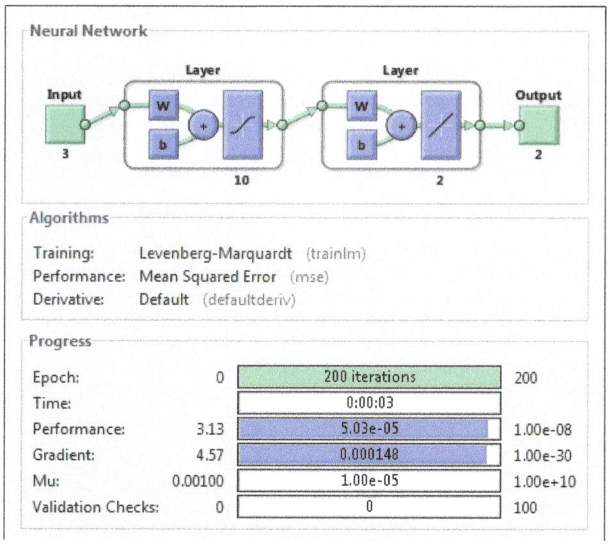

Figure 5. neural network

4. Extract of Maximum Power

A buck & boost chopper is used for impedance matching between the solar cell and load. The GPTIPS predictor estimate maximum power current and maximum power voltage and then the optimum duty cycle generates by using equation (1) for feeding the chopper[21].

$$D_{opt} = \frac{1}{1+\sqrt{V_P / R_L I_P}} \quad (4)$$

5. Results and Discussion

Output voltage, current and power at constant atmospheric temperature and wind speed with different irradiation display of the SIMULINK model of the complete SPV system is illustrated in Figure 6.

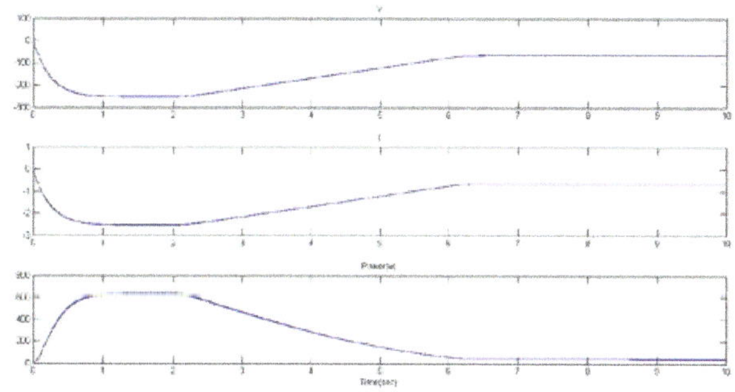

Figure 6. Output voltage, current and power display of SIMULINK model of complete SPV system

6. Simulation of Maximum Power Point Tracker Using SIMULINK

A SIMULINK model of a complete SPV system is shown in Figure 7 realizing the LSSVM based maximum power point controller is. This model consists of six subsystems, namely Inputs block, SPV cell Model, prediction system, Dopt block, PWM block and Chopper. The Inputs subsystem contains the input parameters for the SPV cell subsystem are incident solar radiation; atmospheric temperature and wind speed. The SPV cell Model subsystem is characterized by Eqs. (1)-(3); The predictor system subsystem gives the maximum voltage and current for all practical values of radiation and temperature. The Dopt block calculates the optimum duty cycle using Eq. (4) and it produces into the PWM subsystem. The variation of maximum power output P_{max} corresponding atmospheric ambient conditions is shown in Figure 8.

A ramp load has been considered to show that the optimum power delivered is constant at particular radiation and temperature at variable conditions of load.

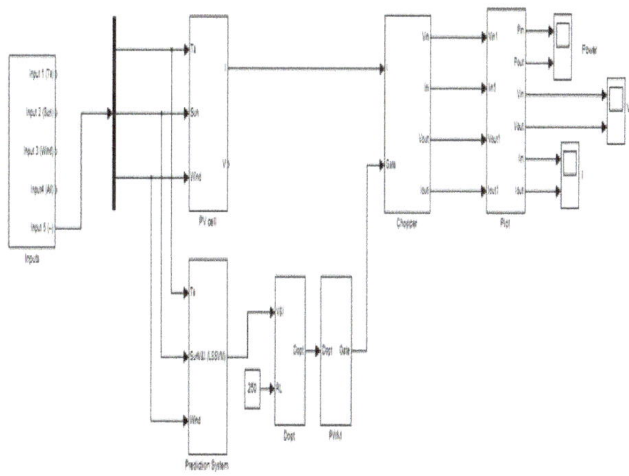

Figure 7. SIMULINK model of the complete SPV system

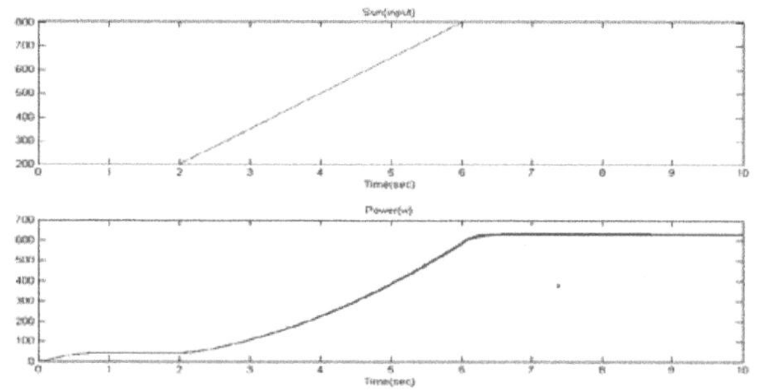

Figure 8. The variation of maximum power output P_{max} corresponding the variation of irradiation and constant temperature and wind speed

7. Conclusion

A new method based on the genetic programming for MPPT of PV energy harvesting system has been developed. A complete PV system with GPTIPS MPPT controller was modeled in MATLAB/Simulink Simulation It was observed that the mean square error is reached to order of 10^{-5} approximately, that in similar circumstances, the mean square error and offline training time is less than a multi-layer perceptron of neural network. results show that the MPPT controller using the GPTIPS method is robust and accurate compared to the other MPPT methods and that it successfully tracks the global maximum power point of a PV module. This system can be applied in the building integrated PV or grid connected PV systems.

Nomenclature

A: Ampere
D: duty cycle of the converter
D_{opt}: optimum duty cycle
E: electron charge $1.6X10^{-19}C$
E_g: energy band gap (eV)
I_D: diode current (A)
I_{in}: converter input current (A)
I_{max}: maximum power current (A)
I_P: current estimated by LSSVM predictor corresponding to peak power point (A)
$I_s(t)$: reverse saturation current (A)
I_s: reverse saturation current reference (A)
I_{out}: converter output current (A)
I_{ph}: photo-generated current (A)
I_{sh}: shunt current (A)
K: Boltzmann constant, $1.38X10^{-23}J/K$
R_L: load resistance (Ω)
R_{Lin}: input resistance of the converter (Ω)
R_s: series resistance (Ω)
R_{sh}: shunt resistance (Ω)
T_A: ambient air temperature (1C)
V: voltage
V_{max}: maximum power voltage (volt)
V_{in}: converter input voltage (volt)
V_P: voltage estimated by LSSVM predictor corresponding to peak power point (volt)
V_{out}: converter output voltage (volt)
N: diode ideality factor

References

[1] F. Blaabjerg, Z. Chen, and R. Teodorescu, "Renewable energy systems in the power electronics curriculum," in *Power Electronics Education, 2005. IEEE Workshop*, 2005, pp. 58-68.

[2] T. Khatib, A. Mohamed, and K. Sopian, "Optimization of a PV/wind micro-grid for rural housing electrification using a hybrid iterative/genetic algorithm: case study of Kuala Terengganu, Malaysia," *Energy and Buildings*, vol. 47, pp. 321-331, 2012.

[3] D. Furkan and M. Mehmet Emin, "Critical factors that affecting efficiency of solar cells," *Smart Grid and Renewable Energy*, vol. 2010, 2010.

[4] T. Esram and P. L. Chapman, "Comparison of photovoltaic array maximum power point tracking techniques," *IEEE TRANSACTIONS ON ENERGY CONVERSION EC*, vol. 22, p. 439, 2007.

[5] C. Hua and C. Shen, "Comparative study of peak power tracking techniques for solar storage system," in *Applied Power Electronics Conference and Exposition, 1998. APEC'98. Conference Proceedings 1998., Thirteenth Annual*, 1998, pp. 679-685.

[6] K. Hussein, I. Muta, T. Hoshino, and M. Osakada, "Maximum photovoltaic power tracking: an algorithm for rapidly changing atmospheric conditions," *IEE Proceedings-Generation, Transmission and Distribution*, vol. 142, pp. 59-64, 1995.

[7] R. Ramaprabha and B. Mathur, "Intelligent controller based maximum power point tracking for solar PV system," *International Journal of Computer Applications*, vol. 12, pp. 37-42, 2011.

[8] M. M. Algazar, H. Al-Monier, H. A. EL-halim, and M. E. E. K. Salem, "Maximum power point tracking using fuzzy logic control," *International Journal of Electrical Power & Energy Systems*, vol. 39, pp. 21-28, 2012.

[9] C.-F. Lu, C.-C. Liu, and C.-J. Wu, "Dynamic modelling of battery energy storage system and application to power system stability," *IEE Proceedings-Generation, Transmission and Distribution*, vol. 142, pp. 429-435, 1995.

[10] A. F. Boehringer, "Self-Adapting dc Converter for Solar Spacecraft Power Supply Selbstanpassender Gleichstromwandler fÃ¼r die Energieversorgung eines Sonnensatelliten," *Aerospace and Electronic Systems, IEEE Transactions on*, pp. 102-111, 1968.

[11] H. J. Möller, *Semiconductors for solar cells*: Artech House Boston^ eMA MA, 1993.

[12] R. Gottschalg, M. Rommel, D. Infield, and H. Ryssel, "Comparison of different methods for the parameter determination of the solar cell's double exponential equation," in *Proceedings of the 14th European Photovoltaic Solar Energy Conference (PVSEC), Barcelona, Spain*, 1997, pp. 321-324.

[13] F. M. González-Longatt, "Model of photovoltaic module in Matlab," *II CIBELEC*, vol. 2005, pp. 1-5, 2005.

[14] H.-L. Tsai, C.-S. Tu, and Y.-J. Su, "Development of generalized photovoltaic model using MATLAB/SIMULINK," in *Proceedings of the world congress on Engineering and computer science*, 2008, pp. 1-6.

[15] I. Altas and A. Sharaf, "A photovoltaic array simulation model for Matlab-Simulink GUI environment," in *Clean Electrical Power, 2007. ICCEP'07. International Conference on*, 2007, pp. 341-345.

[16] K. Kachhiya, M. Lokhande, and M. Patel, "MATLAB/Simulink model of solar PV module and MPPT algorithm," in *Proceedings of the National Conference on Recent Trends in Engineering and Technology*, 2011.

[17] T. Salmi, M. Bouzguenda, A. Gastli, and A. Masmoudi, "Matlab/simulink based modeling of photovoltaic cell," *International Journal of Renewable Energy Research (IJRER)*, vol. 2, pp. 213-218, 2012.

[18] S. Nema, R. Nema, and G. Agnihotri, "MATLAB/Simulink based study of photovoltaic cells/modules/array and their experimental verification," *International journal of Energy and Environment*, vol. 1, pp. 487-500, 2010.

[19] R. Chenni, M. Makhlouf, T. Kerbache, and A. Bouzid, "A detailed modeling method for photovoltaic cells," *Energy*, vol. 32, pp. 1724-1730, 2007.

[20] D. P. Searson, D. E. Leahy, and M. J. Willis, "GPTIPS: an open source genetic programming toolbox for multigene symbolic regression," in *Proceedings of the International multiconference of engineers and computer scientists*, 2010.

[21] N. Mohan and T. M. Undeland, *Power electronics: converters, applications, and design*: John Wiley & Sons, 2007.

8

Decoupled PWM Based Direct Torque Control of Four Level Inverter Fed Open End Winding Induction Motor Drive

Satheesh G[1,*], T Bramhananda Reddy[2], CH Sai Babu[2]

[1]Electrical and Electronics Engineering Department, G.Pulla Reddy Engineering College, Kurnool, India
[2]Electrical and Electronics Engineering Department, University college of Engineering Kakinada, JNTUK, Kakinada, India
*Corresponding author: gsatish.eee@gmail.com

Abstract A direct torque control (DTC) of open end winding induction motor (OEWIM) with decoupled PWM is proposed in this present paper. In this approach the four level voltages are generated by two conventional two level inverters which are fed with the unequal dc link voltages. These two inverters are feeding the induction motor from either ends. However, the proposed DTC scheme does not require the sector information of the estimated fundamental stator voltage vector and its relative position with respect to the stator flux vector. Simulation results clearly demonstrate a better dynamic and simplicity in numerical calculations of the proposed method.

Keywords: *decoupled SVPWM Algorithm, DTC, four level SVPWM, Multilevel Inverters, OEWIM, zero sequence voltages*

1. Introduction

High power capability and multi level inverters controlling and simplicity in the controlling of the drive operation of the open end winding induction motor is attracting many researchers now a days. A high power configuration for the induction motor drive is proposed in [3] and later many researchers proposed many control configurations for the induction motor drives. Generation of three level voltages using the two conventional two-level inverters for open end winding induction motor is proposed in [2]. After that many multi level configurations for open end winding induction motor are proposed in [4,5,6]. Two new approaches with decoupled strategy and Nearest Sub Hexagonal Centre strategies with the multi level output are proposed in [9]. This configuration employed same amount of DC link voltage to feed the two two-level inverters. Another new Space Vector Pulse Width Modulation (SVPWM) is presented in [6] feeding the two two-level inverters with asymmetric DC link voltages. This method is the simple extension to the work proposed in [9].

A numerous direct torque controlling techniques are presented for the induction motor drives and a new approach with SVPWM based multi level inverter fed open end winding induction motor DTC control is presented in [13,14,15]. In [14] employed a decoupled strategy for the controlling of the open end winding induction motor.

In this paper, new DTC approach for the OEWIM with four level configuration is presented; this approach employs strategy specified in [7]. Transient and steady state and step change in load simulation results are discussed.

2. OEWIM with Four Level Configuration

OEWIM drive with four level configuration is shown in Figure 1, where in each inverter is fed with an individual DC supply. If the DC-link voltages are equal, then three-level inversion is obtained [9]. But in the proposed configuration, in order to generate four-level inversion Inverter-1 is having a DC voltage of $(1/3)*V_{dc}$, while inverter-2 is having DC voltage of $(2/3)*V_{dc}$.

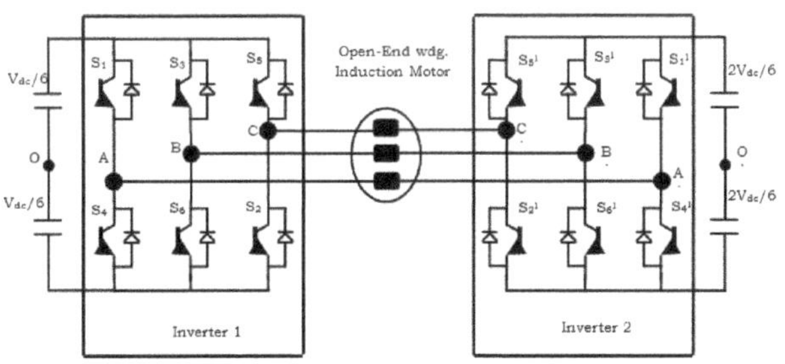

Figure 1. The OEWIM configuration with 2 isolated unequal power supplies for four level configurations

The SV locations of the inverters 1 and 2 are depicted in Figure 2. In Table 1, it is shown that, how the four level are observed in line to line voltages with the specified configuration along with the pole voltages of individual inverters. The SV locations of the inverters 1 and 2 are depicted in Figure 2. In Table 1, it is shown that, how the four level are observed in line to line voltages with the specified configuration along with the pole voltages of individual inverters. As the inverters are two level inverters, then each inverter is having two values in the pole voltages. The inverter-1 outputs are either $V_{dc}/3$ or 0.

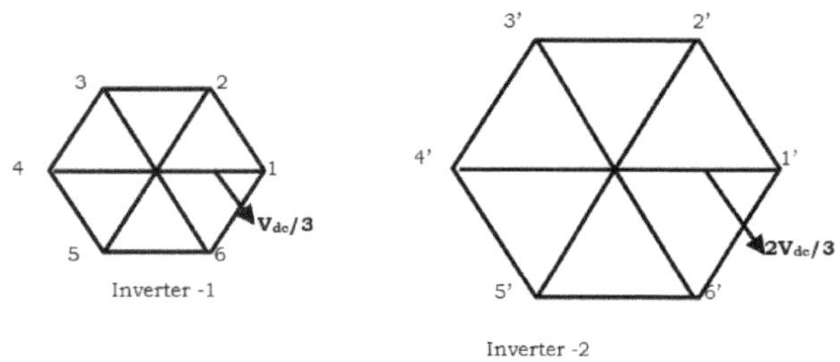

Figure 2. SV positions of inverter-1(left) and inverter-2 (Right)

Table 1. Pole and Line voltages of the 4-level inverter

Pole-voltage of inverter-1 (V_{ao})	Pole-voltage of inverter-2 ($V_{a'o}$)	Line to Line Voltage of the Motor $V_{aa'}= V_{a'o}-V_{ao}$
0	0	0
0	$2*V_{dc}/3$	$2*V_{dc}/3$
$V_{dc}/3$	$2*V_{dc}/3$	$V_{dc}/3$
$V_{dc}/3$	0	$-V_{dc}/3$

Similarly, inverter 2 outputs are either $2*V_{dc}/3$ or 0. The difference of pole voltages of inverter 2 and 1 are shown in Table 1. The same can be visualized in top trace of the voltages in Figure 7. As there is no change in the configuration, except the magnitudes of the DC voltages, there is no path for the zero sequence currents.

3. Proposed DCPWM Algorithm for Four – Level OEWIM Configuration

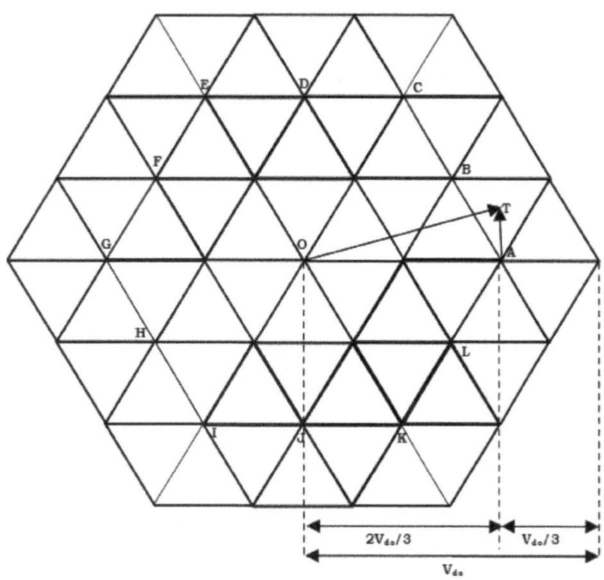

Figure 3. The DCPWM algorithm for a four-level OEWIM configuration

A DCPWM algorithm, which is proposed in [7] for OEWIM configuration with three level output, is now used for the four-level output. In this PWM algorithm the reference vector V_{ref} is constructed by subtracting the second voltage vector ($-V_{ref}/3$) from the first voltage vector ($2V_{ref}/3$). This method of PWM algorithm is shown in Figure 5.

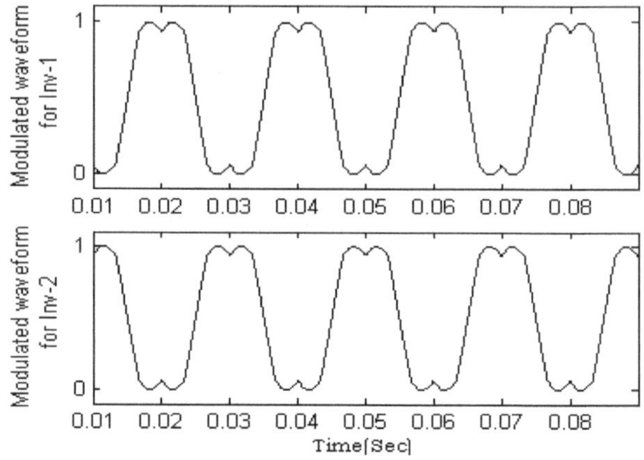

Figure 4. Modulated waveforms for the individual inverters. Top trace modulated waveform for the first inverter, bottom trace modulated waveform for the second inverter with 180 degrees phase shift

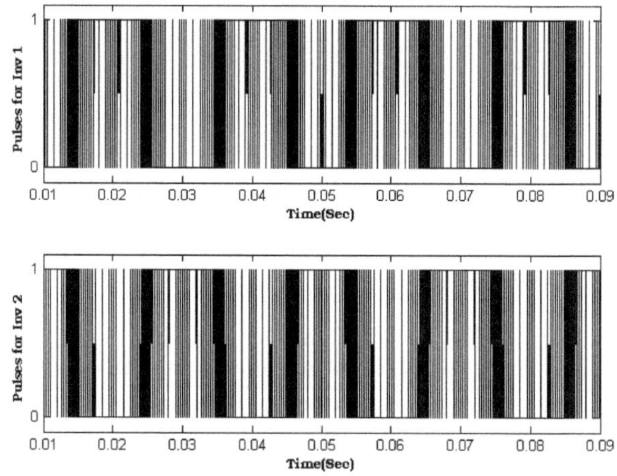

Figure 5. Individual inverter pulses for four-level OEWIM configuration

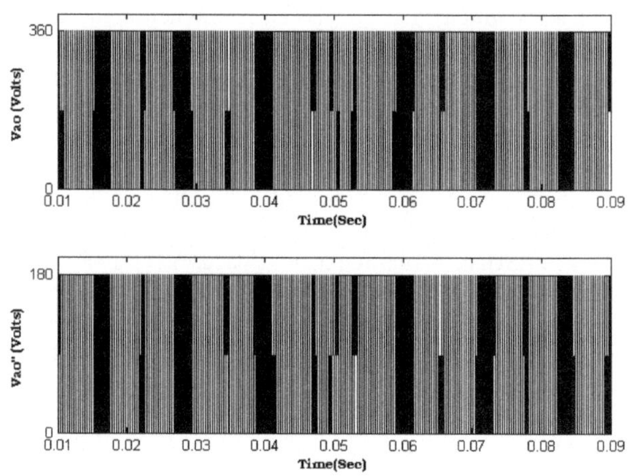

Figure 6. Pole voltages of inverter-2 (top) and inverter-1 (bottom)

With four-level OEWIM configuration the reference vector OT with a magnitude of $|V_{dc}|\angle\alpha$ is generated which is shown in Figure 3. Now this reference vector is resolved into two vectors OA (|$2V_{dc}/3$ |∠α) and AT with (|$V_{dc}/3$ | ∠180° + α). However, in this approach, both inverters are to be switched.

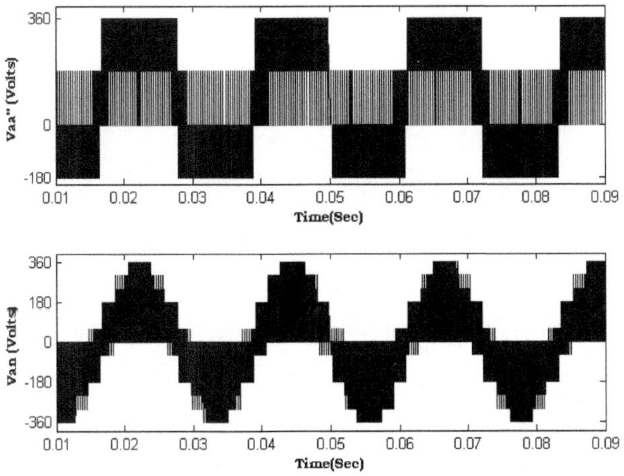

Figure 7. Difference in pole voltages (top) and effective phase voltage (bottom) of the OEWIM drive

The proposed algorithm for the four-level OEWIM drive is validated by using MATLAB simulation. The modulating signals for two inverters are depicted in Figure 4. These modulation signals are operated with 180 degrees phase shift. The generated pulses for two inverters are depicted in Figure 5.

4. DTC of Four-level OEWIM

The electromagnet torque of the open-end winding configuration of induction motor is calculated in a same manner as for normal induction motor using the standard torque equation as:

$$T_e = \frac{3}{2} P\left(\psi_{sd} I_{sq} - I_{sd}\psi_{sq}\right)$$

Where P is the number of poles, ψ_{sd} & ψ_{sq} are the stator d-q axis flux and I_{sd} & I_{sq} are the d-q axis stator currents in the stationary reference frame. The schematic of the proposed method is as shown in Figure 8. In this method, speed of the reference stator flux vector $|\psi^*{}_s|$ is derived by the addition of slip speed and actual rotor speed. The actual synchronous speed of the stator flux vector $|\psi_s|$ is calculated from the adaptive motor model. After each sampling interval, actual stator flux vector $|\psi_s|$ is corrected by the error and it tries to attain the reference flux space vector $|\psi^*{}_s|$. Thus the flux error is minimized in each sampling interval. In this paper, the direct axis and qudrature axis components of the reference voltage vector are created by corresponding direct axis and quadrature axes stator flux error components respectively. They are derived as follows Reference value of the direct axis and quadrature axis stator fluxes and actual value of the direct axis and quadrature axis stator fluxes are compared in the reference voltage vector calculator block and hence the error in the direct and quadrature axes stator flux vectors is obtained as

$$\Delta\psi_{ds} = \psi_{ds}^{\ *} - \psi_{ds}$$

$$\Delta\psi_{qs} = \psi_{qs}^{\ *} - \psi_{qs}$$

The knowledge of flux error and stator resistive drop allows the determination of appropriate reference voltage space vectors along direct axis and quadrature axis, which is given as

$$V_{ds}^{\ *} = R_s I_{ds} + \frac{\Delta\psi_{ds}}{T_s},\ V_{qs}^{\ *} = R_s I_{qs} + \frac{\Delta\psi_{qs}}{T_s}$$

The above derived direct & quadrature components of the reference voltage vector are then fed to the SVPWM block, from where the gating pulses for two inverters are generated.

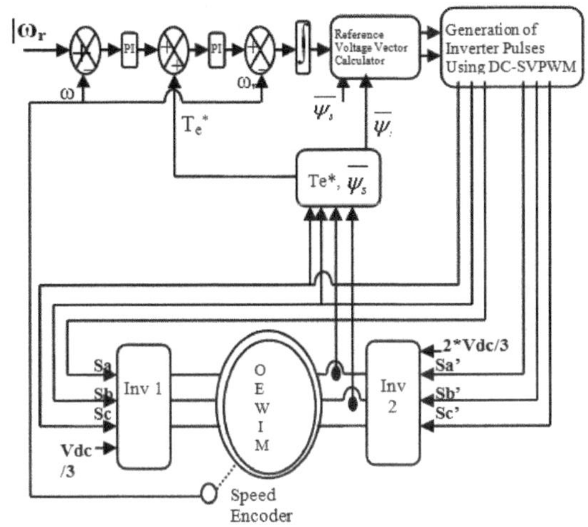

Figure 8. block diagram of the proposed 4-level OEW Induction Motor drive

5. Results and Discussion

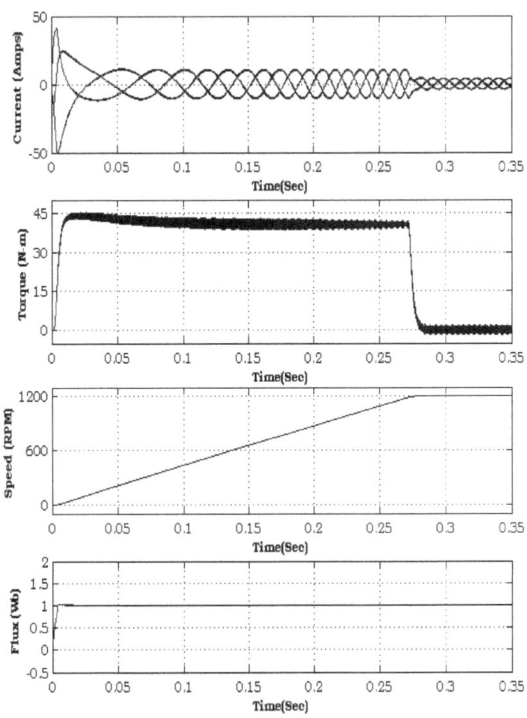

Figure 9. Starting response of DTC of DCPWM based four-level OEWIM drive

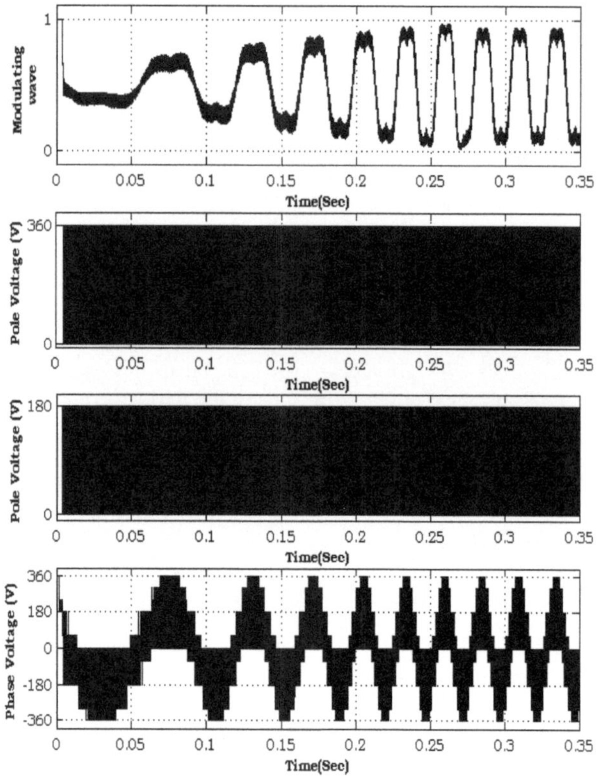

Figure 10. Starting response in modulating waveform, phase and line voltages of DTC of DCPWM based four-level OEWIM drive

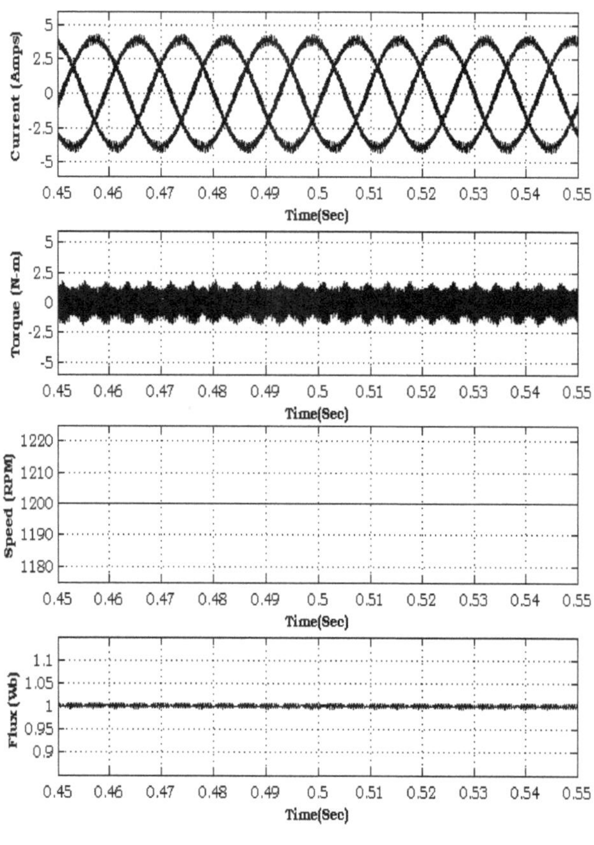

Figure 11. Steady state response of DTC of DCPWM based four-level OEWIM drive

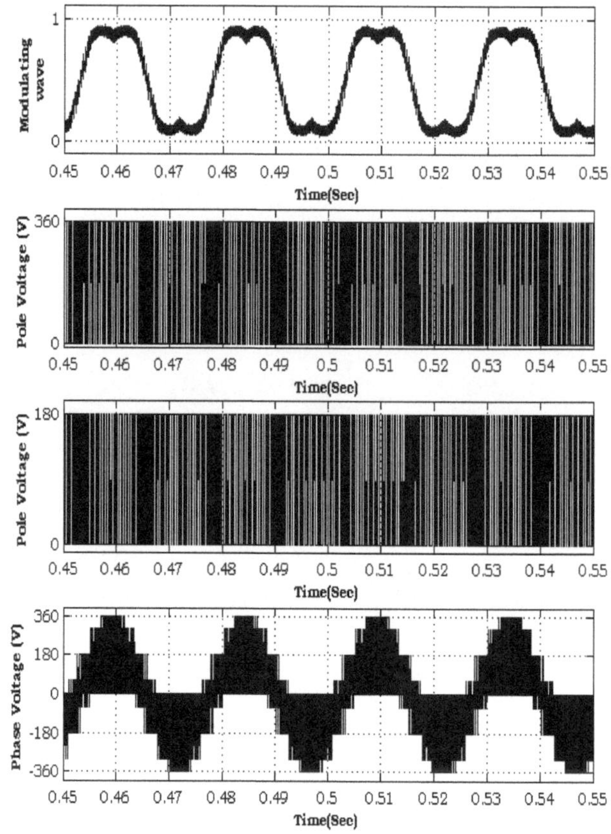

Figure 12. Steady state responses in modulating waveform and voltage waveforms of DTC of DCPWM based four-level OEWIM drive

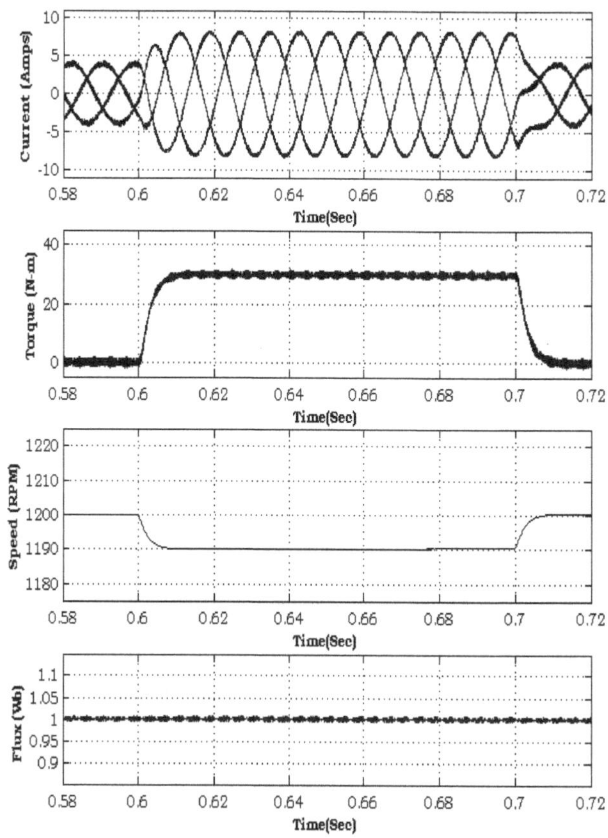

Figure 13. Response during load change between 0.6 - 0.7 sec of DTC of DCPWM based four-level OEWIM drive

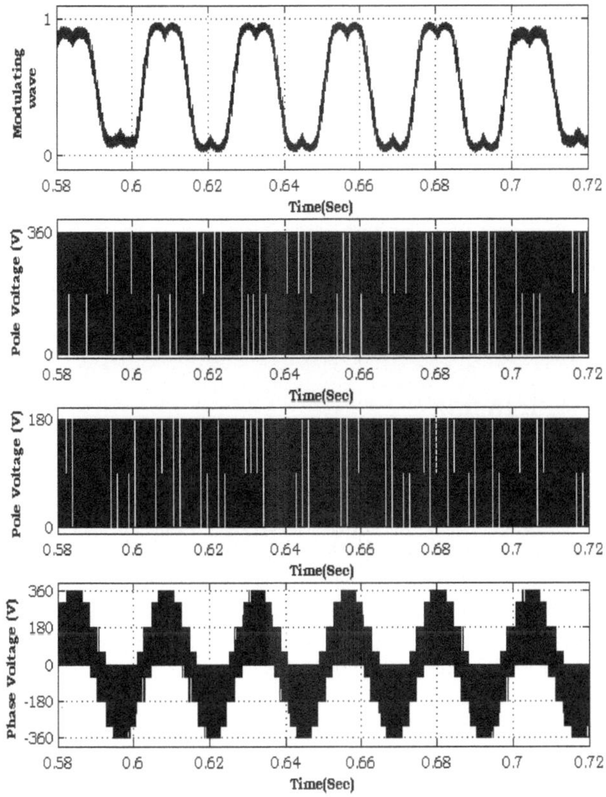

Figure 14. Responses in modulating waveform, voltage waveforms of DTC of DCPWM based four-level OEWIM drive during load change

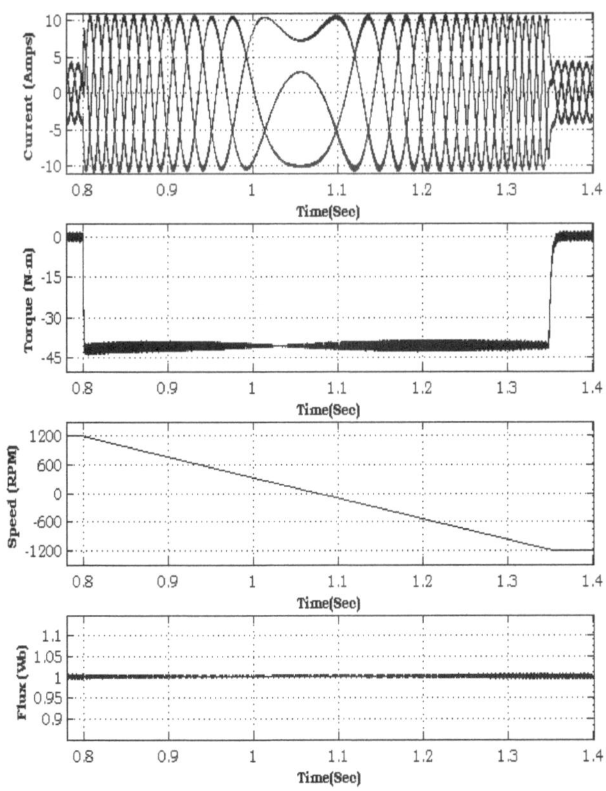

Figure 15. Responses during speed reversal of DTC of DCPWM based four-level OEWIM drive

Several simulation studies are performed to validate the proposed DTC of four-level OEWIM drive. Different operative conditions like starting, steady state, change in load, reversal of speed etc. are observed. Figure 9 to Figure 16 shows the simulation results for DTC of DCPWM based four-level OEWIM drive.

To validate the proposed method, simulation studies have been carried out by using MATLAB /SIMULINK. The motor parameters are as follows 4 KW, 400V, 30 N-m, 1470 rpm, 4-pole, 50 Hz, 3-phase, stator resistance Rs = 1.57 ohm, rotor resistance Rr = 1.21 ohm, stator inductance Ls=0.17 H, rotor inductance Lr=0.17 H, mutual inductance Lm=0.165H, moment of inertia J=0.089 Kg-m^2.

The starting response of the motor phase currents, speed, torque and flux are shown in Figure 9 in Figure 10 the modulating waveform, pole voltages of individual inverters and effective phase voltage is shown. The steady state operating period conditions and the responses of currents, torque, speed and voltage are depicted in Figure 11 and Figure 12 respectively.

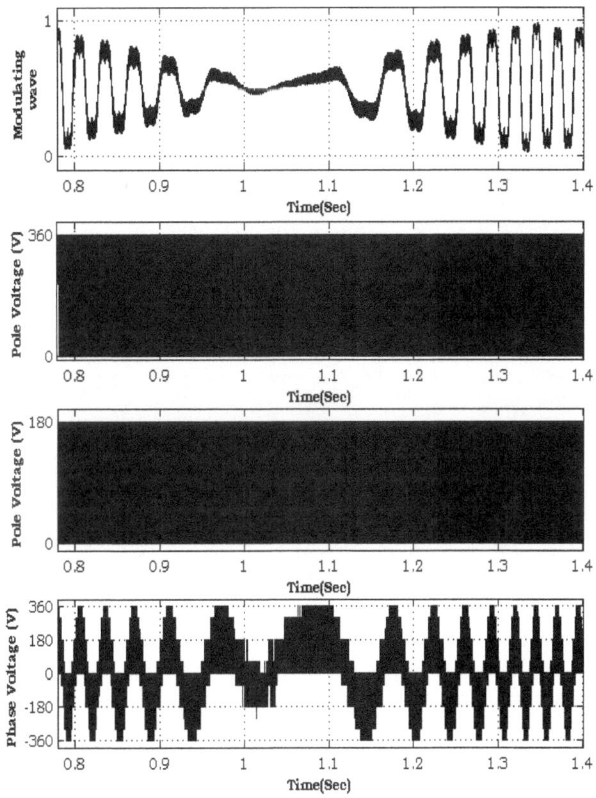

Figure 16. Responses in modulating waveform, voltage waveforms of DTC of DCPWM based four-level OEWIM drive during speed reversal

Response of motor currents, torque, speed and stator flux during change in load which is instigated at 0.6 sec and released at 0.7 sec of DTC of DCPWM based four-level OEWIM drive is observed in Figure 13. With the application of load, the current increases and the torque value also increases with a small drop in the motor speed. It can be observed that, with the increment of both current and torque values the flux value is retained at 1 Wb. The pole voltages of the two inverters and effective motor phase voltage is shown in Figure 14.

The variations in the motor phase currents and voltages during the speed reversal from 1200 rpm to -1200 are shown in Figure 15 and Figure 16 respectively. Here a speed change is instigated at 0.8 sec.

5.1. Analysis of the Current and Torque

The THD is calculated for one cycle of the motor phase current. From the Figure 17, The fundamental harmonic magnitude is 100% and the harmonics at the half of the switching frequency, at the switching frequency and twice the switching frequency with sidebands can be viewed. It is observed that high peaks of the harmonics with sidebands can be viewed exactly at the switching frequency of the inverters.

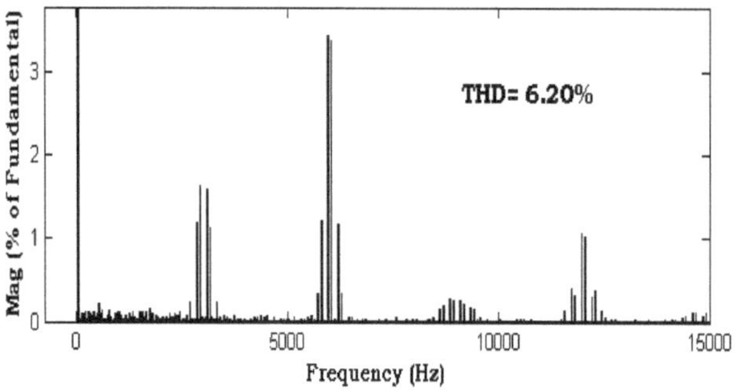

Figure 17. %THD of motor current of DTC of DCPWM based four-level OEWIM

Figure 18 and Figure 19 shows the electromagnetic torque responses of the OEWIM drive during the start-up and steady state respectively for the DTC of OEWIM drive. From the statistical analysis of the torque errors the distribution of the torque errors of start-up and steady state are listed in Table 2 and Table 3 respectively. The time range for statistics is from 0 to 0.1 sec and 0.45 to 0.55 sec for start-up and steady state respectively. There are totally 10,000 points.

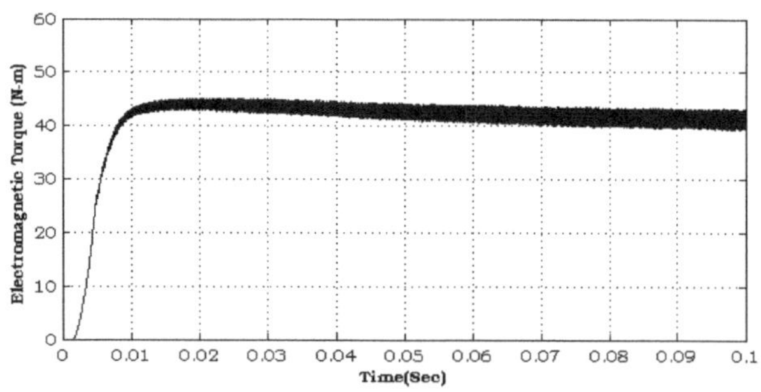

Figure 18. Torque responses during the start-up of DTC of four level OEWIM drive based on DC-PWM algorithm

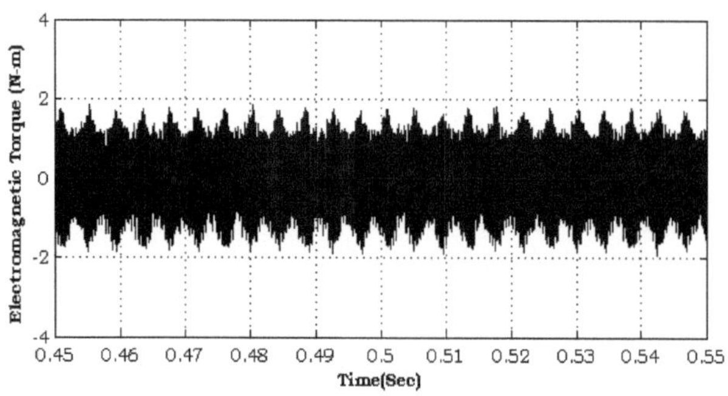

Figure 19. Steady state torque responses of DTC of four level OEWIM drive based on DC-PWM algorithm

During the start-up of the induction motor, the proposed DC-PWM produced 43.76% torque error for greater than two. During steady state running period of the induction motor, the torque errors are distributed with magnitude of less than two only.

Table 2. Comparison of torque errors during 0 to 0.1 sec

	$\lvert\Delta T_e\rvert$ (N-m)			
	0 to 0.5	0.5 to 1	1 to 2	>2
4-level DC-PWM	13.81%	11.46%	30.95%	43.76%

Table 3. Comparison of torque errors during 0.45 to 0.55 sec

	$\lvert\Delta T_e\rvert$ (N-m)			
	0 to 0.5	0.5 to 1	1 to 2	>2
4-Level DC-PWM	32.85%	36.30%	30.83%	-

6. Conclusion

A simple, effective multi level configuration for the direct torque control of the open end winding induction motor drive with asymmetrical dc link voltages is presented in this paper. This proposed method is very simple and is extension of the DC-SVPWM three level OEWIM configurations. The described PWM scheme uses only the three instantaneous phase reference voltages for the implementation and do not require either sector identification or lookup tables. This strategy will greatly reduce the maintenance cost due to the advantages associated with the induction motor. Direct torque control strategy will improve the speed control capability and dynamic response. There are two inverters connected across both the ends of the stator winding. This has another advantage of controlling the induction motor by using only one inverter in the light load condition or in emergency.

References

[1] Joohn-Sheok Kim and Seung-Ki Sul, "A novel voltage modulation technique of the space vector PWM", in Conf. Rec. IPEC'95, Yokohama, Japan, 1995, pp. 742-747.

[2] E. G. Shivakumar, K.Gopakumar, S.K. Sinha, Andre Pittet, V.T. Ranganathan, "Space Vector Control of Dual Inverter Fed Open-end Winding Induction Motor Drive", *EPE Journal*, Vol. 12, No. 1, pp. 9-18 (2002).

[3] H.Stemmler, P. Guggenbach," Configurations of High-Power Voltage Source Inverter Drives", *Proc. EPE Conf.*, pp. 7-14, (1993).

[4] M. R. Baiju, K. Gopakumar, K. K. Mohapatra, V. T.Somasekhar, L. Umanand, "Five-level inverter voltage space phasor generation for an open-end winding induction motor drive", *IEE Proc. on Electric Power Applications*, Vol. 150, No. 5, pp. 531-538, (2003).

[5] M. R. Baiju, K. Gopakumar, K.K. Mohapatra, V. T.Somasekhar L. Umanand, "A high resolution multilevel voltage space phasor generation for an open-end winding induction motor drive", *EPE Journal*, Vol. 13, No. 4, pp. 29-37, (2003).

[6] B. Venugopal Reddy, V.T.Somasekhar, Y. Kalyan, "Decoupled Space-Vector PWM strategies for a Four-Level Asymmetrical Open-End Winding Induction Motor Drive With waveform Symmetries" IEEE TRANS on INDUSTRIAL ELECTRONICS, VOL. 58, NO. 11, NOVEMBER 2011 pp. 5130-5141.

[7] G.Satheesh, T. Bramhananda Reddy and Ch. SaiBabu, "Novel SVPWM Algorithm for Open end Winding Induction Motor Drive Using the Concept of Imaginary switching Times" IJAST, Vol. 2, No. 4, 2011, pp 44-92.

[8] Arbind Kumar, B.G. Fernandes and K. Chatterjee, "Direct Torque Control of Open-end Winding Induction Motor Drive Using the Concept of Imaginary Switching Times for Marine Propulsion Systems", IEEE, pp. 1504-1509.

[9] S. Srinivas and V.T. Somasekhar, "Space-vector-based PWM switching strategies for a three-level dual-inverter-fed open-end

winding induction motor drive and their comparative evaluation" IEEE Trans, IET Electr. Power Appl., 2008, 2, (1), pp. 19-31.

[10] G. Buja, D. Casadei, and G. Serra, "Direct stator flux and torque control of an induction motor: theoretical analysis and experimental results," in *Proc. 24th Annu. Conf. IEEE Ind. Electron. Soc.*, Aachen, Germany, Sep. 31, 1998, vol. 1, pp. T50-T64.

[11] J. Quan and J.Holtz, "Sensorless vector control of induction motors at very low speed using a nonlinear inverter model and parameter identification," *IEEE Trans. Ind. Appl.*, vol. 38, no. 4, pp. 1087-1095, Aug. 2002.

[12] Arhind Kurnar, BG Fernandes, K Chatterjee. "Simplified Hybrid SVM Based Direct Torque Control of Thee Phase Induction Motor," National conference on CCIS 2W. *Goa* (India) Vol. l, pp 137-142. 2004.

[13] Arbind Kumar, BG Fernandes, K Chatterjee, "Direct Torque Control of Three Phase Induction Motor Using SVPWM With-out Sector and Angle Determination", EPE-PEMC 2004, Paper No. A-71121.

[14] Arbind Kumar, BG Fernandes, K Chatterjee, "DTC of Open-End Winding Induction Motor Drive Using Space Vector Modulation With Reduced Switching Frequency," IEEE-PESC, 2004, pp 1214-1219.

[15] G.Satheesh, T. Bramhananda Reddy and Ch. Sai Babu, "DTC of Open End Winding Induction Motor fed by Two Space-Vector-Modulated Inverters", IEEE-INDICON, 2011.

[16] G.satheesh, T. Bramhananda reddy and CH. Sai babu." Space-vector based pwm switching strategy for a four-level dual inverter fed open-end winding induction motor drive", ICAESM-2012, India, pp: 111-115.

Experimental Verification of Hall Effect Sensor Properties

Tatiana Kelemenová[1], Michal Kelemen[2,*], Ivan Virgala[2], Ľubica Miková[2], Peter Frankovský[2], Milan Lörinc[2], Tomáš Lipták[2], Peter Sedlačko[1]

[1]Department of Mechatronics, Technical University of Kosice, Faculty of Mechanical Engineering, Kosice, Slovak Republic
[2]Department of Biomedical Engineering and Measurement, Technical University of Kosice, Faculty of Mechanical Engineering, Kosice, Slovak Republic
*Corresponding author: michal.kelemen@tuke.sk

Abstract The Hall element is constructed from a thin sheet of conductive material with output connections perpendicular to the direction of current flow. Hall Effect devices are included in many products, ranging from computers to sewing machines, automobiles to aircraft, and machine tools to medical equipment. The paper deals with identification of the static characteristic of this sensor for i.e. dependence of the output voltage on permanent magnet distance from sensor. The result of measurement is application recommendation.

Keywords: *calibration, hall effect, sensor*

1. Introduction

The Hall effect has been known for over one hundred years, but has only been put to noticeable use in the last three decades. The first practical application (outside of laboratory experiments) was in the 1950s as a microwave power sensor. With the mass production of semiconductors, it became feasible to use the Hall effect in high volume products. For the first time, a Hall effect sensing element and its associated electronics were combined in a single integrated circuit. Today, Hall effect devices are included in many products, ranging from computers to sewing machines, automobiles to aircraft, and machine tools to medical equipment [1,2,3].

2. Basic Principle of the Hall Sensor

The Hall Effect is named after Edwin Hall, who in 1879 discovered that a voltage potential develops across a current-carrying conductive plate when a magnetic field passes through the plate in a direction perpendicular to the plane of the plate. Hall-effect (magnetic field) sensing applications have become practical recently through advancements in supporting technologies. Hall Effect sensors may require analog circuitry to be interfaced to microprocessors. These interfaces may include input diagnostics, fault protection for transient conditions, and short/open circuit detection. It may also provide and monitor the current to the Hall Effect sensor itself. There are precision IC products available to handle these features.

The Hall effect is an ideal sensing technology. The Hall element is constructed from a thin sheet of conductive material with output connections perpendicular to the direction of current flow. When subjected to a magnetic field, it responds with an output voltage proportional to the magnetic field strength. The voltage output is very small (μV) and requires additional electronics to achieve useful voltage levels. When the Hall element is combined with the associated electronics, it forms a Hall effect sensor.

Although the Hall effect sensor is a magnetic field sensor, it can be used as the principle component in many other types of sensing devices (current, temperature, pressure, position, etc.). Hall effect sensors can be applied in many types of sensing devices. If the quantity (parameter) to be sensed incorporates or can incorporate a magnetic field, a Hall sensor will perform the task [1-7].

In this generalized sensing device, the Hall sensor senses the field produced by the magnetic system. The magnetic system responds to the physical quantity to be sensed (temperature, pressure, position, etc.) through the input interface. The output interface converts the electrical signal from the Hall sensor to a signal that meets the requirements of the application.

General features of Hall effect based sensing devices are: True solid state, long life (30 billion operations in a continuing keyboard module test program), high speed operation - over 100 kHz possible, operates with stationary input (zero speed), no moving parts, logic compatible input and output, broad temperature range (-40 to +150°C), highly repeatable operation [1-7].

Hall Effect Sensors are available with either

- linear output or
- digital output.

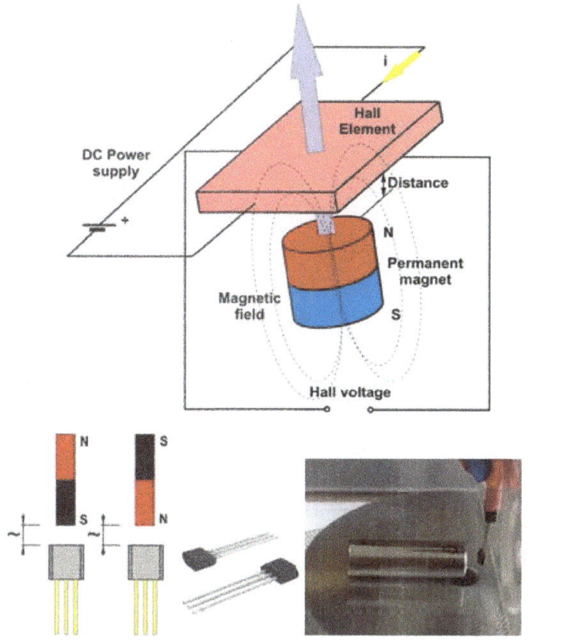

Figure 1. Principle of Hall sensor operation for distance measurement

The output signal for linear (analogue) sensors is taken directly from the output of the operational amplifier with the output voltage being directly proportional to the magnetic field passing through the Hall sensor. This output Hall voltage is given as:

$$V_H = \frac{k_h}{d} \cdot I \cdot B \qquad (1)$$

Where:

V_H is the Hall Voltage in volts,

k_h is the Hall Effect co-efficient,

I is the current flow through the sensor in amps,

d is the thickness of the sensor in mm,

B is the Magnetic Flux density in Teslas.

Linear Hall sensor (Analog ratiometric linear sensors) returns analogue Hall voltage proportional to the applied magnetic flux density. Change of the magnetic density flux can be caused via change of distance between the Hall sensor and permanent magnet connected with measured object. This type is used mainly in application, where linear distance is necessary to be measured.

Digital Hall sensor works as magnetic sensitive switch with output in form of logical levels zero "0" or logical level one "1". They just know only two states ON/OFF. These types is frequently used in application as limit switch, keyboard switch, velocity rotation sensor, brushless DC motor sensor etc.

Output current of the sensor is small only several miliampers and cannot be used direct to control of any plants. The transistor switch is necessary to use for connecting to other devices [1-7].

3. Linear Hall Sensor

When interfacing with analog output sensors, it is important to consider the effect of the load. The load must

- provide a path to ground
- limit the current through the output transistor to the rated output current for all operating conditions.

This sensor has been selected for experimental testing for position measurement with NdFeB permanent disc magnet with anisotropic magnetization (diameter 4 mm and length 10 mm). Experimental apparatus have been designed and realised for this reason (Figure 2). The position is measured with resistive sensor and parallel length gauges have been used for calibration of this sensor.

Calibration process was executed for every 10th millimetres of overall range of resistive sensor. Result is static transformation characteristic, which can be described with math model obtained from regression (Figure 4).

Figure 2. Principle of Hall sensor operation for distance measurement

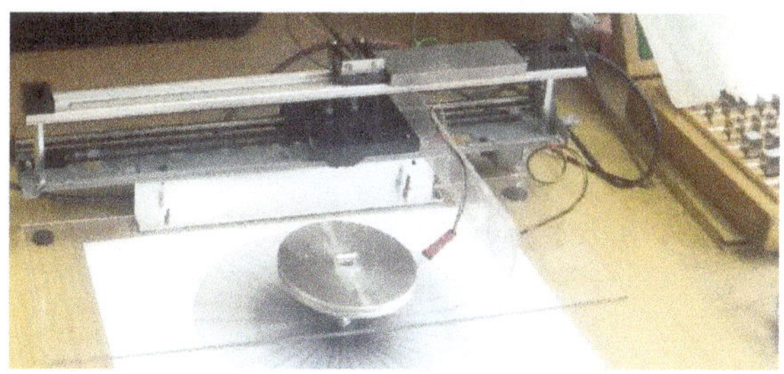

Figure 3. Calibration of resistive position sensor with length gauges

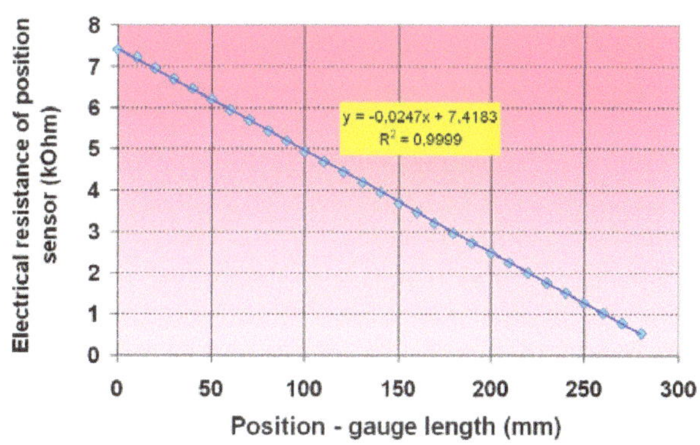

Figure 4. Static transformation characteristic of resistive position sensor

Results of calibration can be processed into calibration characteristic, which can be used for recalculation of position sensor resistance to information about actual position in millimetres (Figure 5).

Calibration model enables to measure position of Hall sensor from permanent magnet.

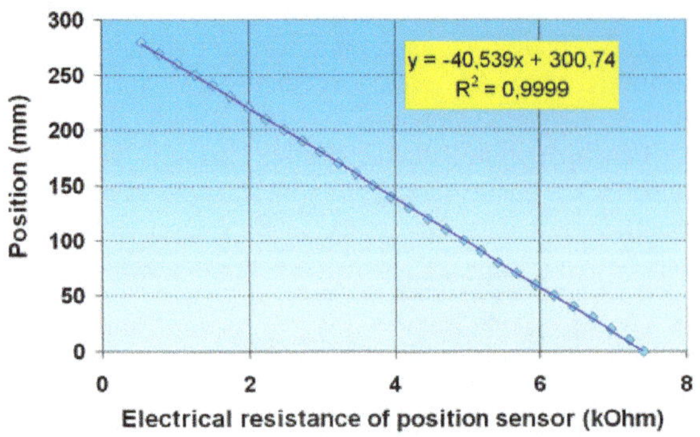

Figure 5. Calibration characteristic of resistive position sensor

4. Experimental Test of Hall Sensor

Aim of testing is to identify suitable orientation of permanent magnet and sensor. Very important is to recognize of working interval, where sensor has good sensitivity to position sensing. Magnetic sensor could be a sensitive also to direction of moving to or from permanent magnet which can occurs as hysteresis on static characteristic.

It is possible to arrange four states for position measurement:

- South pole of permanent magnet is toward to smaller surface of sensor (Figure 6).
- South pole of permanent magnet is toward to larger surface of sensor (Figure 7).
- North pole of permanent magnet is toward to smaller surface of sensor (Figure 8).
- North pole of permanent magnet is toward to larger surface of sensor (Figure 9).

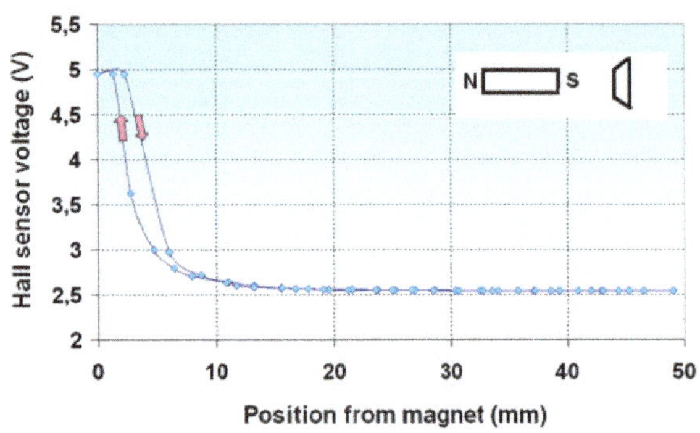

Figure 6. Calibration characteristic of resistive position sensor

Situation on Figure 6 shows the strong hysteresis and there is only small working area for position measurement 8 mm – 15 mm. There is a no change of voltage after this range.

Figure 7 also shows hysteresis, but there is also useful interval for position measurement in range 4 mm to 16 mm.

Figure 8 also has hysteresis and useful working area is too small 17mm – 22mm.

Best situation is on Figure 9, where hysteresis is negligible. Useful working interval is 3mm – 20 mm. Working range of measurement is 17 mm, which can be used for position measurement.

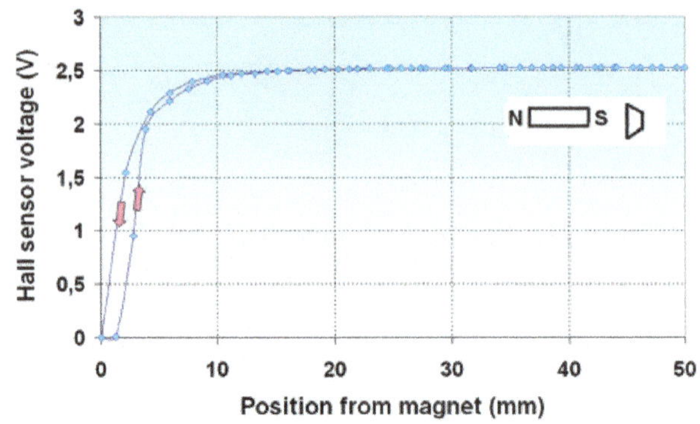

Figure 7. Calibration characteristic of resistive position sensor

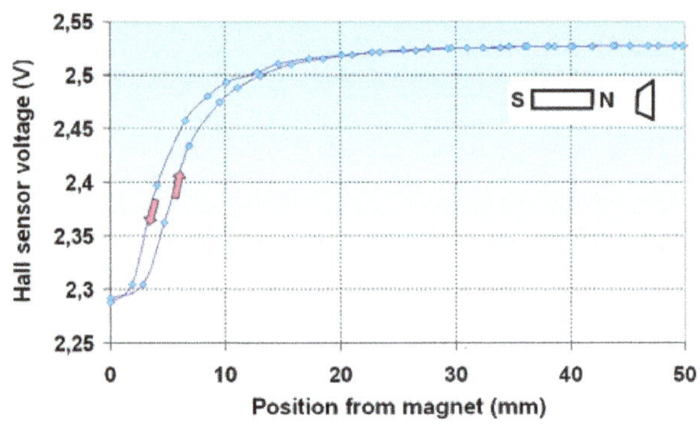

Figure 8. Calibration characteristic of resistive position sensor

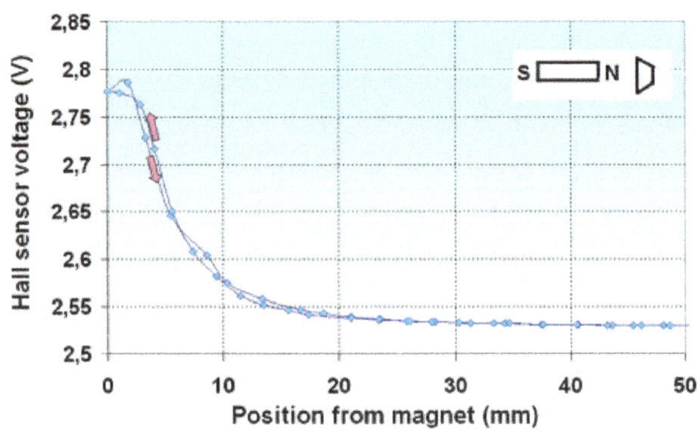

Figure 9. Calibration characteristic of resistive position sensor

This type of Hall sensor has optimal configuration when North pole of permanent magnet is toward to larger surface of sensor. This is recommendation for using of this sensor in practical application.

Our designed actuator tester includes also Hall sensor for position measurement (Figure 10).

Results from Hall sensor testing have been used for design of sensor configuration with aim to obtain maximum working range without or with minimum hysteresis effect.

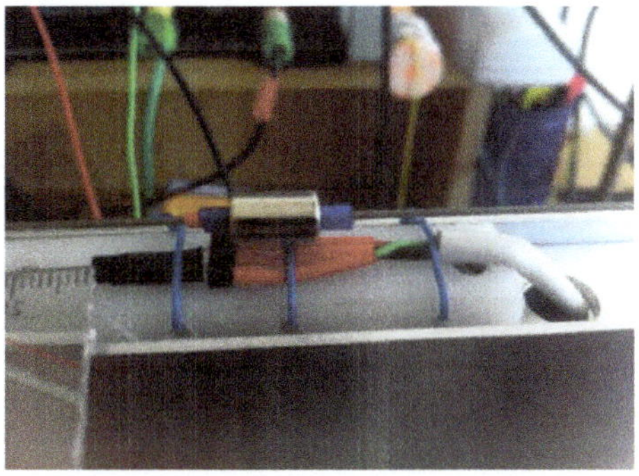

Figure 10. Hall sensor for position measurement

5. Conclusion

Currently, the Hall effect sensor application includes current sensor, gear tooth sensor, Hall effect based temperature sensor, Hall effect pressure sensor, speed sensor (tachometers and RPM sensor), position sensor (automotive sensor - valve, piston, shaft, crankshaft, ignition sensor, throttle angle sensor, anti-skid sensor, transmission sensor), door sensor (security sensor), brushless DC motor sensor, flow meter, relays, proximity sensor, encoder sensor, metal detector, vibration sensor, level and tilt sensor, etc. [8-24]

Acknowledgement

The work has been accomplished under the research projects No. KEGA 048TUKE-4/2014, VEGA 1/0182/15 and KEGA č. 014STU-4/2015 financed by the Slovak Ministry of Education.

References

[1] W. Storr, "Electronics Tutorial about Hall Effect Magnetic Sensor", *Basic Electronics Tutorials by Wayne Storr*. Last updated: November 2013, available online: http://www.electronics-tutorials.ws/electromagnetism/hall-effect.html

[2] J. Jezný, & M. Čurilla, (2013). Position Measurement with Hall Effect Sensors. American Journal of Mechanical Engineering, 1(7), 231-235.

[3] E. Ramsden, "*Hall-effect sensors. Theory and Applications*", Second edition, Newness, Elsevier. 272 pages. 2006.

[4] Interactive Catalog Replaces Catalog Pages. *Solid State Sensors SS490 Series Miniature Ratiometric Linear*. Sensing and Control. Honeywell Inc. 11 West Spring Street, Freeport, Illinois 61032. available online <http://sensing.honeywell.com/index.cfm/ci_id/16046/la_id/1.htm>. Cited july 11th 2008.

[5] SS490 Series Standard. *Miniature Ratiometric Linear Hall-Effect Sensor; radial lead IC package*. Sensing and Control. Honeywell Inc. 11 West Spring Street, Freeport, Illinois 61032. available online: http://sensing.honeywell.com/product%20page?pr_id=36221.

[6] F. Duchon, "*Sensors in mobile robotics*", 1st ed. Bratislava, Publishing STU, 2012. 97 pages.

[7] D. Koniar, L. Hargaš and M. Hrianka, "Application of standard DICOM in LabVIEW", *Proc. of 7th conf. Trends in Biomedical Engineering*, Kladno 11-13. 9. 2007.

[8] S. Milano "*Allegro Hall-Effect Sensor ICs*", Allegro MicroSystems, LLC, available online: http://www.allegromicro.com/en/Design-Center/Technical-Documents/Hall-Effect-Sensor-IC-Publications/Allegro-Hall-Effect-Sensor-ICs.aspx, 2013.

[9] G. Pepka, "*Position and Level Sensing Using Hall-Effect Sensing Technology*", Allegro MicroSystems, LLC, available online: http://www.allegromicro.com/en/Design-Center/Technical-Documents/Hall-Effect-Sensor-IC-Publications/Position-and-Level-Sensing-Using-Hall-Effect-Sensing-Technology.aspx, 2013.

[10] Swagatam, Lamar Stonecypher, "*Types, Working Principle of Hall Effect Ics*", Bright Hub Inc., available online: http://www.brighthubengineering.com/diy-electronics-devices/107450-types-working-principle-of-hall-effect-ics/.

[11] Claude Gudel, LEM SA, "Taking Transducer Technology to the Limits of Current Measurement Accuracy", *PowerGuru – Power Electronics Information Portal*. Available online: http://www.powerguru.org/taking-transducer-technology-to-the-limits-of-current-measurement-accuracy/.

[12] M. Dovica, T. Kelemenová, M. Kelemen, "Experimental apparatus for SMA actuator testing", *Journal of Automation, Mobile Robotics & Intelligent Systems*. Vol. 6, no. 3 (2012), p. 37-39. 2012

[13] R. S. Popovic, Z. Randjelovic, D. Manic, "Integrated Hall-effect magnetic sensors", *Sensors and Actuators A*: Physical, Volume 91, Issues 1-2, 5 June 2001, Pages 46-50, 2001.

[14] J. Edberg, "Selecting Hall-effect sensors for brushless DC motors", Honeywell Sensing and Control, *ECN Magazine Advantage Business Media*, Available online: http://www.ecnmag.com/articles/2012/10/selecting-hall-effect-sensors-brushless-dc-motors

[15] J. Kammerer, L. Hebrard, V. Frick, P. Poure, F. Braun, "Horizontal Hall effect sensor with high maximum absolute sensitivity", *Sensors Journal, IEEE*, Volume 3 , Issue 6, p. 700-707. 2003.

[16] F. Montaigne, A. Schuhl, F. N. Van Dau, A. Encinas "Development of magnetoresistive sensors based on planar Hall effect for applications to microcompass", *Sensors and Actuators A*: Physical, Volume 81, Issues 1-3, 1 April 2000, Pages 324-327. 2000

[17] F. Burger, P. A. Besse, R. S. Popovic, "New fully integrated 3-D silicon Hall sensor for precise angular-position measurements", *Sensors and Actuators A*: Physical, Volume 67, Issues 1-3, 15 May 1998, Pages 72-76. 1998.

[18] C. Leube, ; O. Paul, "Novel compact two-dimensional CMOS vertical Hall sensor", *Solid-State Sensors*, Actuators and Microsystems (TRANSDUCERS), 2015 Transducers - 2015 18th International Conference on, 21-25 June 2015, Pages 1164-1167.

[19] X. Wang, D. Meng, Y. Xu, "Efficiency Optimization Control for Three - Phase Induction Motors with Hall Sensor", *International Journal of Control and Automation*, Vol. 8, No. 1 (2015), pp. 1-8.

[20] E. Cardelli, A. Faba, "Vector hysteresis measurements of not oriented grain SiFe steels by a biaxial hall sensors array", *Physica B: Condensed Matter*, Volume 435, 15 February 2014, Pages 34-39.

[21] O. Petruk , R. Szewczyk, T. Ciuk, W. Strupiński, J. Salach, M. Nowicki, I. Pasternak, W. Winiarski, K. Trzcinka, "Sensitivity and Offset Voltage Testing in the Hall-Effect Sensors Made of Graphene", *Recent Advances in Automation, Robotics and Measuring Techniques*, Volume 267 of the series Advances in Intelligent Systems and Computing pp 631-640.

[22] S. Tang; H. Liu; X. Wei; Y. Lin; Y. Cao, "Analysis and design of Hall speed sensor detection system", *Journal of Science of Teachers' College and University*, 2014, Vol. 02.

[23] Y. Li, "Research about Miniature Vibration Sensor based on Hall Effect", *International Power, Electronics and Materials Engineering Conference (IPEMEC 2015)*, Published by Atlantis Press, pp. 969-973.

[24] Maciej Bodnicki, Janusz Grzybowski, "Encoder with Miniature Hall Effect Sensors – Increasing of Measurement Resolution", *Machine Dynamics Research*, Vol. 38, No. 1, 2014, pp. 13-20.

10

Proposal of Mobile Application for Rehabilitation

Rudolf Jánoš*

Department of robotics, Technical university of Kosice, Kosice, Slovakia
*Corresponding author: rudolf.janos@tuke.sk

Abstract The article describes the first version of application designed for virtual rehabilitation patients. The aim of this application is to facilitate, access and helps patients get regular exercise. The basic requirement is to perform exercises without special equipment. Therefore, our idea emerged to specialize in smartphone and use sensors incorporated therein. The article describes two kinds of sensors normally based in a smartphone with examples how it could be used. The application was developed for the Android operating system because of the broader representation in our region.

Keywords: *rehabilitation, upper limb, android, virtual rehab*

1. Introduction

Virtual rehabilitation brings several advantages over conventional rehabilitation practice. Level can be dynamically changed according to specific needs coming from previous exercise. Interactive games and activities bring greater motivation than conventional exercises hospital. Perhaps the most important is the recording of the results in real time, resulting in a better understanding of the progress of the treatment.

2. Virtual Reality for Rehabilitation

The main reasons for the introduction of virtual reality in rehabilitation practice is more attractive rehabilitation process as well as increase capacities and capabilities rehabilitation workers.

Popularizing the rehabilitation process is psychologically significant impact on speeding up the process and improve the overall results of healing. Degradation, or minimize the impact of the hospital environment has a positive effect on the psyche of patients. In a domestic environment in a circle of family members treated defect loses its priority status within the current priorities of the patient. It also deals with other activities that enable it to "forget" their health problems. [6]

This aspect is particularly significant in pediatric patients for whom the traditional rehabilitation exercises like something important and necessary. Transformation of classical rehabilitation exercises in various degrees of control computer applications and games can be suppressed perception of rehabilitation as medical acts. A similar effect has appropriately chosen a rehabilitation program for the elderly in which the patient suppresses the feeling of dependence on another person. Home rehabilitation allows for flexibility and easy access. Patients are able to carry out their rehabilitation programs in order to follow their individual plans. Depending on the severity of the defect, patients may participate in the intensive level of therapy several hours per week or follow a less demanding regime. Possibility home rehabilitation is usually used by individuals who are not able or willing to travel frequently, or require treatment of only one type of rehabilitation therapist. The main drawback is the lack of a home rehabilitation specialized equipment. However, the treatment received at home gives people the advantage to practice their skills and create your own strategy to reduce their disability in the space of their own environment. [2,11,12,13]

The second aspect is to increase the capacity possibilities of rehabilitation workers. As mentioned in Europe there is an aging population. This increases the number of residents who to speed up the recovery process as well as the most successful cure urgently need rehabilitation support. Increasingly, as rising demand for rehabilitation capacities. These may be increased movement of individual rehabilitation exercises in the home. On the physiotherapist would thus remain burden settings appropriate exercise and control. This could be realized by remote access, data sharing and automatic transmission of results exercise. The patient is not excluded from the review process, although the quality and clarity of transferred exercises at home is not completely controllable. For this reason, should be given the choice of appropriate exercises and different categories of eligible patients increased attention. [6,9,10]

In the implementation of concrete solutions rehabilitation exercises so far carried out two principles of control of interactive virtual environment:

- Position control and measurement components of the body / in hand users - joystick, interactive controls (Wii, PlayStation), virtual helmet, interactive jackets and gloves,
- Extracorporeal placement of components - cameras sensing movement microphones. [6]

Outside, it was not only in the virtual rehabilitation using other sensors to collect information on the current status and condition of the patient - infrared temperature sensors, Heart Rate, Blood Pressure and under. Said sensors allow application deployment adjust the load and speed exercises adequately state the user. This feedback minimizes the risk of overloading, but adapts the load sufficiently, which helps the patient to quicker healing process. [6]

3. Design and Implementation of Application Virtual Rehab

In order to minimize the complexity of the structure and the resulting application was at the beginning of each exercise should be classified based on how they are measured by internal sensor mobile device. This classification at a later stage helped unify the various sub-programs of exercise and thus greatly help to reduce the final size of the application. [1]

The distribution of individual exercises is based on the choice of measuring parameters of each sensor as well as the method for measuring the specific exercises. On balance, I suggest six basic categories which, in the current version of App Inventor 2 environment and the possibilities of mobile devices designed cover my rehabilitation exercises, suitable for application solutions in program Virtual Rehab and on exercises using:

- Touch screen
- Sensor lateral inclination (roll)
- Sensor lateral tilting of the retaining phase (Roll)
- Sensor vertical inclination (Pitch)
- compass sensor (Azimuth)
- A combination of lateral and vertical sensor.

When designing the basic structure of Virtual Rehab I came from the design flow chart that describes the structure of the program, each stage of the selection rehabilitated area and the final rehabilitation practice. Finally, the structure reflects the proposed category 6 exercises which are schematically noted of such subprograms of the application Figure 1. [6]

The actual exercise starts calibration phase. This phase is used to define the initial position of the mobile device before exercise and at the same time is used to determine the average dispersion of the current measured values precisely measured parameter sensor. The need for calibration is based on the uncertainty of measurements of individual sensors. Elaboration of such variance values should absolutely avoid relevant measurement and evaluation of results. Therefore, I have the judgment and decided to test several test sample as relevant for the determination of the average take into account the 100 measurements. At the same time expressing the determination of the average current location of the mobile device it is in the calibration of determining the average dispersion values. This variance in further measurements determines the range of values. It expresses the interval in which it is possible to expect change the value of the measured parameter without changing the position of a mobile device. After this phase comes the actual calibration exercises that course and structure for each program is different, so the following subprograms are described in separate sections. [3]

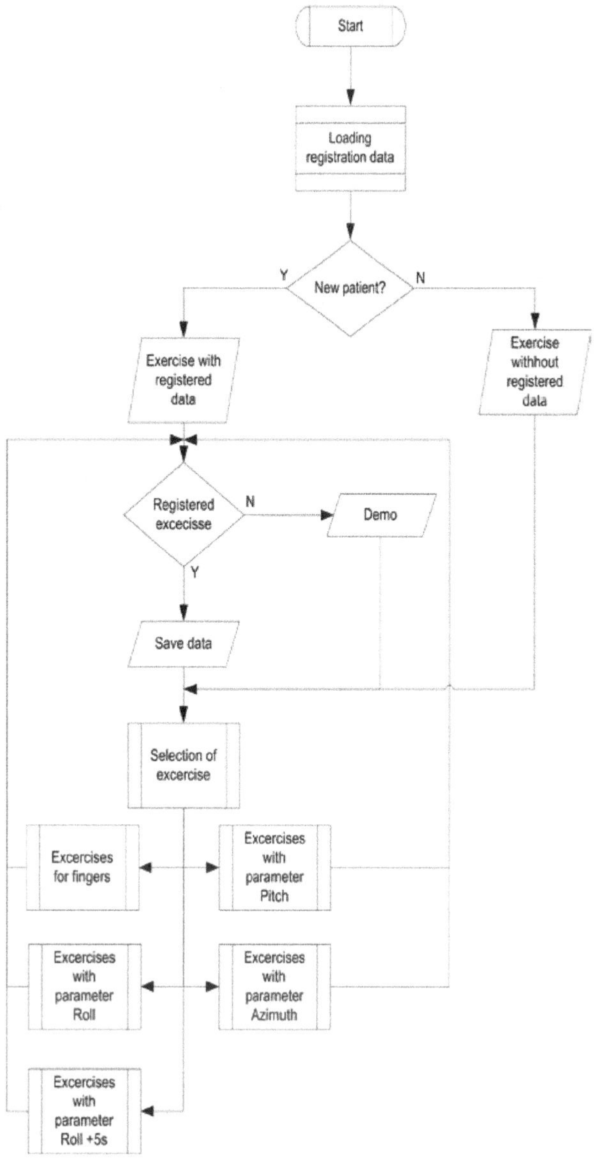

Figure 1. Basic structures of application

Exercises using azimuth sensor

To verify the functionality of the program I chose to exercise the extension of the wrist mobility - simple backward movement from the open position to the maximum flexion and then to the maximum extension. As previously mentioned, measurement principle is based on measuring of position smartphone to the magnetic field of the Earth. The purpose of this type of exercise is practicing mobility particular rehabilitated areas, while improving range of motion stretching and lengthening tendons withdrawn after long-term fixation in the plate, respectively, after surgery. The aim is to obtain a measurement value of angle, which reflect the scope of the free movement of limbs rehabilitated. At the same time during exercise extreme values they recorded maximum range of motion. This ensures that the recording not only the range of motion, but in the case of reduced mobility and the area in which this reduction occurs, Figure 2. [7,8,9]

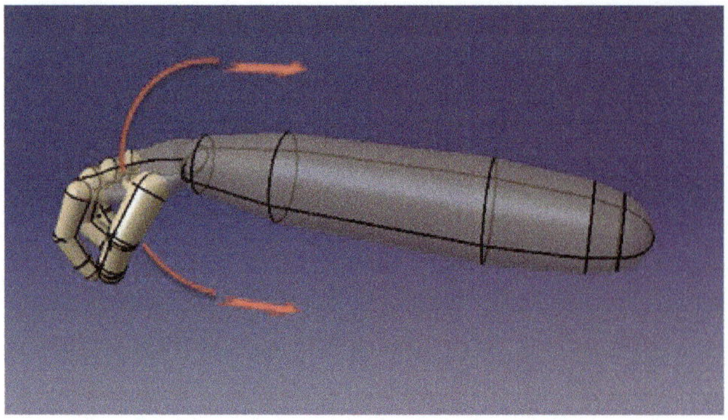

Figure 2. Wrist movement in the median plane

Then passed the initial phase of the program to measurements on which the user is alerted short vibrating. Initial phase measurement I have proposed in order to prevent distortion of the results. I suppose that, despite the clear and understandable instructions to instructional videos, are found several users who start during exercise in the opposite direction. In this part of the program therefore will wait to move in the right direction and to subsequently undergo a phase of exercise itself, which records the number of exercises as well as range of motion applied by the user.

Detecting the direction of movement is carried out by comparing the difference in the measured values of the last and the current angle. During the motion, when the differential value changes from positive to negative (or vice versa), the limit values of the last measured angle recorded as the maximum angle in a given direction. To simplify the conditions for recognizing the value of difference, I suggested a coefficient k = -1 (initial value). The coefficient at each change of direction multiplied by the number (-1). This ensures that the differential value is multiplied by the always / m at most equal to the number 0. At the same time we coefficient indicates the direction of movement (k = -1 - direction; k = 1 - direction). On this basis, it is possible to determine the end of the cycle, which phase of the exercise, the user is recorded and possibly complete a repetition of exercise (worth repeating count assign a value to count = count + 1).

Exercise cycle is repeated as far as count variable takes the value of 10, which passes through the score subroutine phase. In this phase, the display shows the measurement results achieved with color-coded values. I suggested three categories of colors.

• Red - Mobility of up to 60% compared to a standard mobility

• Orange - mobility in the range of 61-85%

• Green - mobility better than 86%

If the user does not share their results with their exercise rehabilitation specialists, application subsequently transmit measurement data to a Web database accessible only to the appropriate specialist. The current version of the transmitted data:

• date of the exercise

• particular kind of practice

• The average amount of exercise

• The maximum and minimum values of angles in both directions of movement. [7]

Exercises utilizing sensor lateral tilting with the retaining phase

In the first step was also mentioned calibration of the smartphone and the calculation of confidence.

The aim of the exercise is to improve the mobility impaired and weakened tendons, this category is a priority objective of strengthening not only the tendon apparatus, but also to improve muscle tone. To this end, it is necessary to measure the time priority of the patient and the angle in the retention phase and measurement of the range of movement of secondary importance. Limb movement after the retention phase is for relaxation, not on trying to achieve the maximum deflection, Figure 3.

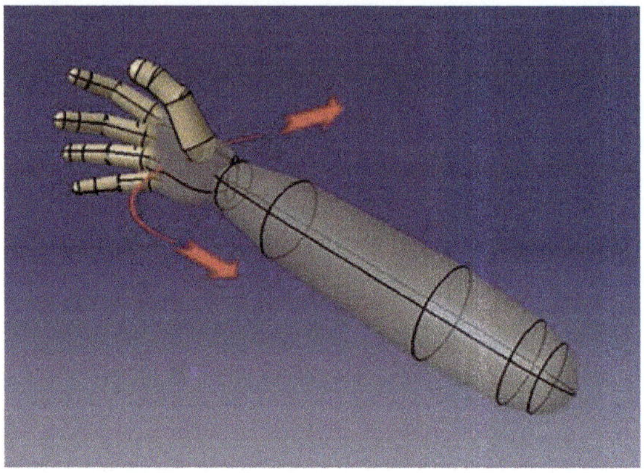

Figure 3. The basal plane and the wrist plane are perpendicular

This category has the most numerous exercises exercises (up to six). For a preview of the functionality I because of the integrity of the application chosen as in the previous subprogram area with two wrist exercises on extension and flexion. I tried to point out the possibility of using a category of sub-program for the use of multiple exercises, no writing exercises and results are presented separately.

The structure of a subroutine is nevertheless significantly different from previous practice. The main difference is the initial position of the limbs and the initial direction of motion. The previous practice, the initial position in the open position (balanced wrist). When exercise the retention phase it is based on the relaxed limit position to the maximum deflection, which leads to the arrest of movement, which aims to strengthen the ligaments and muscles in a stretched mode. [4]

The first phase is carried out initial calibration and measurement identical to the former category. From the concept and structure of the program, this step is identical for all the suggested categories. The difference will consist only in the measured parameters.

Direction of motion detection is performed similarly by comparing the difference in the measured values of the last and the current angle. When changing direction, however, records the maximum range of motion, that is the angle only in extreme retention phase. Similarly, and as with other proposed exercise is the basic number of repetitions of 10 exercises in the series. If necessary, increase rehabilitation doctor advises you to exercise several times a day. Rarely a larger series. This strategy aims to minimize the instantaneous overload. [13] The end of this phase of the exercise is similar to display the results, while the results of the color scale is divided into categories:

• Red - achieving maximum angle of 60% compared with a standard value or time in the retention phase is less than 2.5 seconds

• Orange - get the maximum angle at the level of 61-85% while retaining the time phase at least 2.5 seconds

• Green - get the maximum angle at a level higher than 85%, while the maximum time position of at least 4.5 seconds

Writing the results of measurements it is required for all measurements other exercises for a better comparison of each exercise. Shared data are entered into the website database.

Acknowledgement

Testing of proposed applications are confirmed despite slight lack of opportunity to deploy Virtual Rehab in rehabilitation practice. A prerequisite for the realization is to implement the remaining 4 exercise categories to final solution applications. A secondary objective is to solve the question of writing the results of the exercise. As I mentioned, the sharing of data on a database project, App Inventor is easy and fast, also a way of creating new databases for rehabilitation workers and their patients is relatively simple and easily manageable even for less computer-savvy people. Processing and presentation of results is stored in this way but rather limited. The possibility of sharing data by using the tool "Fusiontables", also in the web interface of Google. Shared data are clearly listed in the tables, but the setting is a bit more challenging. There would be in place to verify the possibility of automated creation of spreadsheets, or creating multi-purpose table that could be used as a template for all future users. [10]

The continuation of the project will be devoted to verify operation on the test specimen. The data obtained from the measurements and the observations incorporated into application solutions that will output of the project.

This contribution is the result of the project implementation: 059TUKE-4-2014 Rozvoj kvality života, tvorivosti a motoriky hendikepovaných a starších osôb s podporou robotických zariadení.

References

[1] CAMEIRAO, Mónica S. et al.: The Rehabilitation Gaming System: a Virtual Reality Based System for the Evaluation and Rehabilitation of Motor Deficits. In: Virtual Rehabilitation 2007: Zborník z medzinárodnej konferencie, Benátky: IEEE, 2007. s. 29-33.

[2] CHARLTON, Paula C. et al.: Reliability and concurrent validity of a Smartphone, bubble inclinometer and motion analysis system for measurement of hip joint range of motion. In: Journal of Science and Medicine in Sport. Roč. 18, č. 3 (2015), s. 262-267.

[3] MILANI, Patrizia et al.: Mobile Smartphone Applications for Body Position Measurement in Rehabilitation: A Review of Goniometric Tools. In: PM&R Journal. Roč. 6, č. 11 (2014), s. 1038-1043.

[4] HAJDECKER Alexander.: Návrh virtuálnej robotickej rehabilitácie. Diplomová práca. 2015.

[5] KUTTUVA, Manjuladevi et al.: The Rutgers Arm: An Upper-Extremity Rehabilitation System in Virtual Reality. In: Fourth Int. Workshop on Virtual Rehabilitation 2005. [online]. Catalina Island: Rutgers, 2005. [cit. 2015-04-08]. Dostupné na internete: <http://www.ti.rutgers.edu/publications/papers/2005_iwvr_kuttuva.pdf>.

[6] JÁNOŠ Rudolf, SUKOP Marek, HAJDECKER Alexander: Transfer inovácií. Č. 31 (2015), s. 85-87.

[7] Yoon, Jungwon et al.: Control Optimization of the Rutgers Ankle Rehabilitation Interface. In: Proceedings of Advances in Robot Dynamics and Control Symposium 2002, ASME International Mechanical Engineering Congress and Exposition, Louisiana: 2002. IMECE2002-32655.

[8] MILANI, Patrizia et al.: Mobile Smartphone Applications for ;Body Position Measurement in Rehabilitation: A Review of Goniometric Tools. In: PM&R Journal. Roč. 6, č. 11 (2014), s. 1038-1043.

[9] FERRIERO, Giorgio et al.: Reliability of a New Application for Smartphones (DrGoniometer) for Elbow Angle Measurement. In: PM&R Journal. Roč. 3, č. 12 (2011), s. 1153-1154.

[10] MILANESE, Steven et al.: Reliability and concurrent validity of knee angle measurement: Smartphone app versus universal goniometer used by experienced and novice clinicians. In: Manual Therapy. Roč. 19, č.6 (2014), s. 569-574.

[11] KRANZ, Matthias et al.: The mobile fitness coach: Towards individualized skill assessment using personalized mobile devices. In: Pervasive and Mobile Computing. Roč. 9, č. 2 (2013), s. 203-215.

[12] JESENSKÝ, Ján: Uvedení do rehabilitace zdravotně postižených. Praha: Karolinum, 1995. 159 s.

[13] SMRČKA, Václav – DYLEVSKÝ, Ivan: Flexory ruky. Brno: IDV PZ, 1999. 162 s.

11

A Novel Approach in Design of Model-free Fuzzy Sliding Mode Controller to SISO Chemical Processes

M. Khodadadi[1,*], M. Shahbazian[1], M. Aghajani[2]

[1]Department of Instrumentation and Automation Engineering, Ahwaz Faculty of Petroleum, PUT, Ahwaz, Iran
[2]Department of Chemical Engineering, Ahwaz Faculty of Petroleum, PUT, Ahwaz, Iran
*Corresponding author: kh.automation@yahoo.com

Abstract Sliding Mode Control (SMC) is a powerful methodology in the field of nonlinear feedback control with feature of proven robustness, performance and stability in the face of system nonlinearity and modeling uncertainties. Unfortunately, pure SMC controller suffers from two important problems; chattering phenomenon and formulation of equivalent dynamics. The aim of this paper is to design a novel model-free Fuzzy Sliding Mode Control (FSMC) to solve these drawbacks. In this method the equivalent component of the SMC is estimated using an auto-adjustable fuzzy inference system in order to remedy the model dependency problem and robustness improvement. Also, in this method the conventional switching component of the SMC is developed using additional saturation function to eliminate the chattering completely. Eventually, for verification, the proposed controller is implemented on an isothermal van de vusse reactor as a high nonlinear chemical process, subjected by unknown disturbance. It is revealed that the purposed method can improve system's robustness and transient performance effectively in terms of complete suppression of the chattering and disturbance, specially, in the face of extra disturbances and reduction in settling time and percentage of overshoot in comparison with the two other model-free FSMC and Classical PID control.

Keywords: *model-free Fuzzy Sliding Mode Control, chattering elimination, model dependency, unknown disturbance, robustness, transient performance*

1. Introduction

Sliding Mode Control (SMC) is known because of its salient feature in robustness in the face of high nonlinearity, modeling uncertainties and disturbances with quick response [1,2,3]. Hence, this methodology has been developed during last three decades in the nonlinear control area. The SMC control law consists of two component; continuous and switching (discontinuous) controllers [2]. The discontinuity of the switching component causes the structure of the SMC control action to be consisted of a set of smooth functions. The controller's structure is allowed to change with infinite frequency at any instant between these smooth members in accordance with the situation of state vector with respect to the predefined state dependent sliding surface [3,4]. The purpose of this switching is to enforce and derive the system's state trajectory to reach the sliding surface in finite time (reaching mode or approaching regime) [5]; and then, enforce the trajectory to remain and slide on this surface as a manifold toward set-point (sliding mode) [6]. During the sliding mode, the motion of the system's state trajectory is sustained under modeling errors and disturbances [2]. However, *chattering* and *requirement of SMC to model of the plant* are known as main problems which can restrict applications of the original SMC [7]. "*Chattering phenomenon*" is the high frequency oscillation of the state trajectory near the sliding surface which is created in the presence constrained frequency switching in original SMC [2]. This harmful phenomenon produces high heat in electronic circuits and extreme erosion in moving parts of mechanical devices [8]. To overcome the chattering, commonly, the boundary layer (BL) approach is used as a signal filter for alleviation of the control signal in the thin layer around the sliding surface. However, it results in the quasi sliding mode [5] and loss of the asymptotic tracking. Specially, by consideration of the large enough boundary layer thickness and high switching gain in the presence of extra bound of modeling uncertainties and disturbances, the B.L. may cause a considerable tracking error [9,10] even, the chattering on the boundary [11]. To eliminate the drawbacks of the B.L., several methods has been proposed such as varying boundary layer and switching gain for instance [9-13]. The continuous part of the SMC is *model dependent,* thus in the presence of the complex or ill-defined process or large range of uncertainties, this dependency can degrade the system's performance [7], [14]. Fuzzy Logic Control methodology (FLC) based on fuzzy set theory [15] is designed in a form of inferential rules based on some qualitative expert knowledge about plant under control. Because of its model free feature, it is implemented generally on ill-modeled or parameter-fluctuated plants to improve control system's performance [16]. Integrating of the FLC with the SMC which collects the excellent features of both original SMC and FLC [17] capable to solve the problems of the original SMC (chattering, model dependency). This integrating is done commonly in two ways [18]: Sliding Mode Fuzzy Control ***(SMFC)*** and Fuzzy Sliding Mode Control ***(FSMC)*** which have been widely used in recent years such as [7,19]. In this paper, a novel model free-FSMC is proposed which can be specially implemented for process control to improve the robustness. In this new strategy, the continuous control is estimated and updated online by fuzzy methodology to remedy the model dependency problem and removing of steady state error. Also, the traditional switching control is developed using additional filter-like saturation function to avoid the chattering. This paper is organized as follows: In section 2 the SMC is presented; the new model-free FSMC methodology is illustrated in section 3; in section 4, the proposed FSMC is implemented and its capability is verified; eventually, in the section 5 the paper is concluded.

2. Pure Sliding Mode Control

Let an nth order single input nonlinear system defined as [2]

$$x^{(n)} = f(X,t) + b(X,t)u(t) + d(t) \tag{1}$$

Where, $X(t) = \left[x, \dot{x}, \ddot{x}, \dddot{x} \ldots, x^{n-1}\right]^T$ is the state vector and scalar x is the control variable. Scalar u(t) is the control action. $f(X,t), b(X,t)$ are generally nonlinear functions which are uncertain and bounded by continuous functions $F(X,t)$, $b_{\min}(X,t)$, $b_{\max}(X,t)$ respectively and are estimated as $\hat{f}, \hat{b}$ as follows:

$$\left|\hat{f}(X,t) - f(X,t)\right| \le F(X,t) \tag{2}$$

$$0 \le b_{\min}(X,t) \le b(X,t) \le b_{\max}(X,t) \tag{3}$$

$$\hat{b}(X,t) = \sqrt{b_{\min}(X,t)\, b_{\max}(X,t)} \tag{4}$$

$$\beta(X,t) = \frac{b_{\max}(X,t)}{b_{\min}(X,t)} \tag{5}$$

Also, the disturbance d(t) is upper bounded by D.

$$0 \le |d(t)| \le D \tag{6}$$

Switching (sliding) function or sliding variable [5,11] is defined in the state space $R^{(n)}$ as:

$$s(X,t) = \left(\frac{d}{dt} + \lambda\right)^{(n-1)} e(t) \tag{7}$$

$$E(t) = X(t) - X_d(t) = \left[e, \dot{e}, \ddot{e} \ldots, e^{n-1}\right]^T \tag{8}$$

Here, $E(t)$ is the tracking error vector in the output state where $X_d(t)$ is the reference state vector and tuning factor λ is strictly positive. For second order systems (n=2), Eq. (7) gives the switching function as:

$$s = \dot{e} + \lambda e \tag{9}$$

By choosing the Lyapunov function as:

$$V(s) = \frac{1}{2}s^2 \tag{10}$$

For stability and to achieve invariant and null value sliding variable, it gives the switching (reaching) condition as:

$$\dot{V}(s) = s.\dot{s} \le -\eta|s| \tag{11}$$

Here, η is a positive fixed parameter.

2.1. SMC's Control Law

In the sliding mode, the controller satisfies the confinement of scalar sliding surface at zero:

$$\dot{s} = \ddot{e} + \lambda\dot{e} = 0 \tag{12}$$

Substituting Eq. (7) in Eq. (1), gives the approximation of continuous control law as [20]:

$$u_{eq} = \frac{-\hat{f} + \ddot{x}_d - \lambda\dot{e}}{\hat{b}} \tag{13}$$

Here, u_{eq} is the continuous (equivalent) control (element) [2] which is the system's dynamic in the sliding mode [21]. By substituting Eq. (9) into the reaching condition Eq. (11) and use of Eq. (1) the control law is obtained as [20]:

$$u = \frac{-\hat{f} + \ddot{x}_d - \lambda\dot{e}}{\hat{b}} - k_s \operatorname{sgn}(s) \tag{14}$$

$$u = u_{eq} + u_s \tag{15}$$

Thereby, the reaching mode will be assured. Here, u_s is the discontinuous (switching, relay or robustness) control (term) [11] and k_s is the positive switching gain. The severity of the reaching mode depended on this gain [22] which is bounded as follows:

$$k_s(X,t) \ge \beta(X,t)(F(X,t) + \eta) + (\beta(X,t) - 1)|u_{eq}| \tag{16}$$

2.2. Boundary Layer (BL) Technique and its Problems

In pure SMC, there is restricted frequency and non-instantaneous control switching in the sliding regime thus instead of ideal sliding mode, the state trajectory oscillates and chatters around the sliding surface with high frequency as depicted in Figure 1. The main causes of this non ideal switching are discretized (digital) implementation of the SMC's control law or neglected high frequency un-modeled dynamics (such as actuator's dynamics and time delays) [1,2,5,8,21]. The Boundary Layer (B.L) devise is the common filtering techniques to produce the smoothed control signal within a thin layer about the sliding surface [2] to suppress the chattering.

This is done using a continuous approximated saturation function such as linear type sat (instead of the discontinuous signum function sgn (as:

$$B(t)=\{X,||s(X,t)|\le \varnothing\} \tag{17}$$

$$sat\left(\frac{s}{\varnothing}\right)=\begin{cases}\operatorname{sgn}(s) & if\,|s|\ge\varnothing\\ \frac{s}{\varnothing} & if\,|s|<\varnothing\end{cases} \tag{18}$$

Where $\varnothing$ is thickness of the boundary layer $B(t)$. However, this technique causes the tracking error which is limited to the precision of ϵ [2,23].

$$\epsilon=\frac{\varnothing}{\lambda^{n-1}} \tag{19}$$

Here, ϵ is named the width of the boundary layer

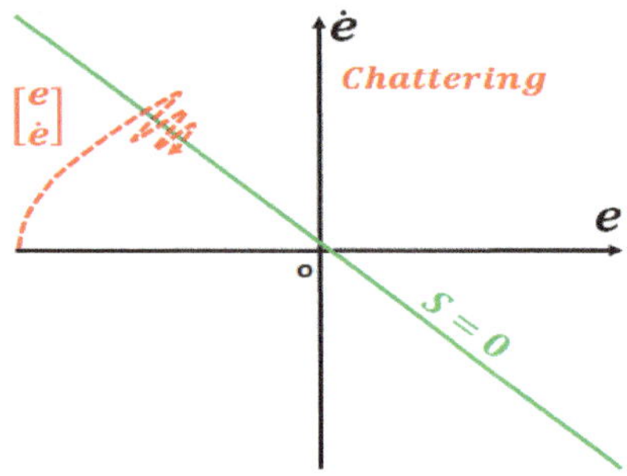

Figure 1. Chattering phenomenon as a result of non-ideal switching [23]

It is notable that, if the modeling uncertainties or disturbances vary in large range in the system, it should be assigned the *high enough switching gain* to improve tracking error and afterward, in order to prevent the chattering due to high gain, the *thickness of the boundary layer* should be assigned *large enough*, but, this devise may result in off-set considerably [9]. Furthermore, it may replace the chattering on the boundary [11]. To solve these drawbacks, time varying thickness has been proposed by several researchers, for instance in [9]. However, the varying boundary layer may result in a large ϵ for some systems [11]. In [11] the saturation function (18) was replaced with a new continuous function with fixed $\varnothing$ based on system dynamics.

In this paper every chattering will be avoided using the large enough and constant boundary layer thickness together with additional saturation function. Moreover, the steady state error as second problem of the boundary layer will be remedied by an auto adjustable fuzzy equivalent control.

3. The Proposed FSMC

In this section, a novel model-free FSMC strategy is proposed to assurance system's robustness and improved performance. Firstly, the equivalent control is designed by a fuzzy inferential system. Afterward online tuning method is devised for it such that the uncertainty of this dynamic depended part of the FSMC be eliminated, thereby, the steady state error will be canceled completely. Secondly, new switching control is designed by aim of complete elimination of the chattering. It is done by developing conventional switching function with additional saturation function of the sliding variable rate of change. The proposed FSMC controller is constructed as follows (see Figure 2):

$$u_{(k)}=u_{eq\,fuzzy(k)}+u_{s(k)} \tag{20}$$

Where, $u_{s(k)}$ and $u_{eq\,fuzzy(k)}$ are the switching control and fuzzy equivalent control respectively. $u_{(k)}$ is the FSMC output signal as the process input.

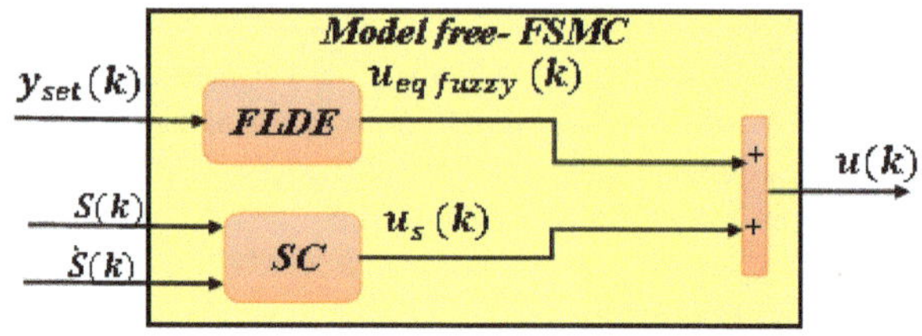

Figure 2. Model free -FSMC controller's elements: Fuzzy Logic Dynamics Estimator (FLDE) and developed Switching Control (SC)

3.1. Design of Equivalent Control of Proposed Model-free FSMC

The SMC whose equivalent part is formulated based on the *approximated model* of the process such as First Order Plus Dead Time (FOPDT) may be robust against inevitable un-modeled uncertainties arises from this modeling process [22], parameters variation and disturbances, but, if the modeling uncertainties vary in large range, the performance of controller will be degraded such that the SMC with conventional alleviated switching function may result in extreme steady state error and chattering on the boundary as mentioned in section 2.2. In this sub-section it is attempted to solve these problems. It is done by estimation of the equivalent part of the SMC using a fuzzy logic inference system based on system's behavior on the steady state condition. Afterward, atuning method is introduced for online adjusting of this fuzzy estimator based on the *difference between output of the SMC and output of its equivalent part* by aim of complete cancellation of this difference (or uncertainty) at steady state condition.

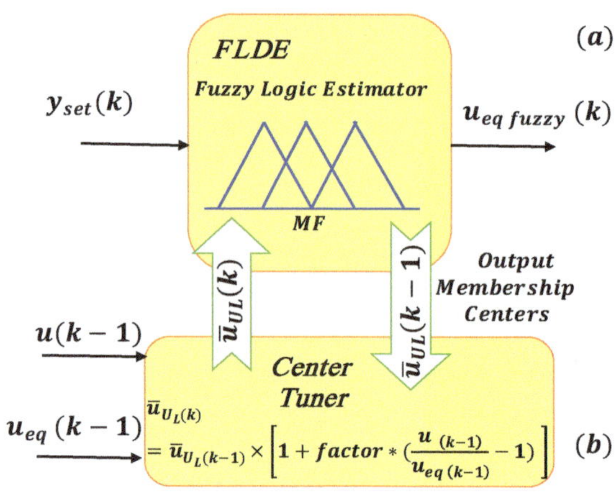

Figure 3. a) FLDE with auto adjustable membership functions centers **b)** Center Tuner for membership functions

3.1.1. The Proposed Methodology

For estimation of the equivalent control u_{eq} a *SISO Fuzzy Logic Dynamic Estimator* (*FLDE*) is designed based on *steady-state I/O* data of underlying plant (u_{ss}, y_{set}) which are used as centers of *O/I* membership functions. These centers are as free parameters which can be updated online by center tuner as depicted in Figure 3.

3.1.2. FLDE Design Procedure

Design of the FLDE is done through two steps; offline predesign step and online tuning step as follows:

Step I) Offline Predesign of pure FLDE

To design the FLDE, *output/input* the offline collected data of the process under control in relevant to the steady-state condition (y_{ss}, u_{ss}) are used. These data are yielded through sensors in the full range of plant operation (which can be obtained from expert knowledge of process engineer). These data are assigned as the primary (pure) input/output centers of the membership functions $(\bar{y}_{A_L}, \bar{u}_{U_L})$ of the pure FLDE. The centers of the output membership function are free to tune online. Fuzzy reference (set-point) value y_{set} and equivalent control u_{eq} are defined with 8 linguistic labels. Triangular Membership Functions (MF) are chosen for the fuzzy variables. One dimensional rule base of the FLDE are given in Table 1, for instance, two typical rules are as follows:

$$\begin{aligned} &IF\ y_{set\,fuzzy}\ is\ small\ THEN\ u_{eq\,fuzzy}\ is\ small.\\ &IF\ y_{set\,fuzzy}\ is\ big\ THEN\ u_{eq\,fuzzy}\ is\ big. \end{aligned} \tag{21}$$

Mamdani product inference engine and Center Average Defuzzifier are used for FLDE estimator. Input and output fuzzy sets are as follows:

Input fuzzy sets:

A_1=Zero; A_2=Very-small; A_3=Small; A_4=Medium; A_5=Big; A_6=Very-Big; A_7=Very Very Big; A_8=Ultra Big

Output fuzzy sets:

U_1=Zero; U_2=Very-small; U_3=Small; U_4=Medium; U_5=Big; U_6=Very Big; U_7=Very Very Big; U_8=Ultra Big

Step II) FLDE's Free Parameters Auto Adjustment

The centers of output membership functions of the FLDE $\bar{u}_{U_L}$ (as free parameters) can be adjusted online based on elimination of difference between output of the SMFC $u_{(k)}$ and output the continuous controller $u_{eq(k)}$. These free parameters are tuned online as follows (see Figure 3):

$$\bar{u}_{U_L(k+1)} = \bar{u}_{U_L(k)} \times \left[1 + factor * reg_{(k)}\right] \tag{22}$$

$$reg_{(k)} = \frac{u_{(k)}}{u_{eq(k)}} - 1 \tag{23}$$

Here, $L = 1:M$ and M is the number of rules.

Also, k is the time-interval at time kT_s and T_s is the sampling period. factor is constant step size. As it is evident, the updating process is not iterative in each sampling period and the centers are updated once per each sampling, thus it is not too time-consuming process and therefore is feasible in practice. While the sliding variable remains at null value, the FLDE membership centers will be invariant but once a deviation is occurred in the sliding variable, the center tuner will attempt to compensate the difference between the FLDE output and the FSMC output automatically, such that the tracking error will be disappeared finally.

Table 1. One dimensional Fuzzy rule base for FLDE. Linguistic variables: B=big, M=medium, S=small, Z=zero, V=very, U=ultra

$y_{setfuzzy}(k)$	Z	VS	S	M	B	VB	VVB	UB
$u_{eqfuzzy}(k)$	Z	VS	S	M	B	VB	VVB	UB

3.2. Design of Developed Switching Control

To design the Switching Control (SC) of the proposed FSMC, in accordance with the Figure 4 an alleviated switching control with sigma saturation function $-k_s \frac{s}{|s|+\varnothing_1}$ is developed by augmenting new sigma saturation function $-k_s \frac{\dot{s}}{|\dot{s}|+\varnothing_2}$ in order to prevent the chattering as follows:

$$u_S = -k_s\left(\frac{s}{|s|+\varnothing_1} + \frac{\dot{s}}{|\dot{s}|+\varnothing_2}\right) \tag{24}$$

Here, k_s are the switching gain and $S, \dot{S}$ are the sliding variable and its rate of change respectively and $|S|, |\dot{S}|$ are their absolute values. $\varnothing_1$, $\varnothing_2$ are constant tuning parameters.

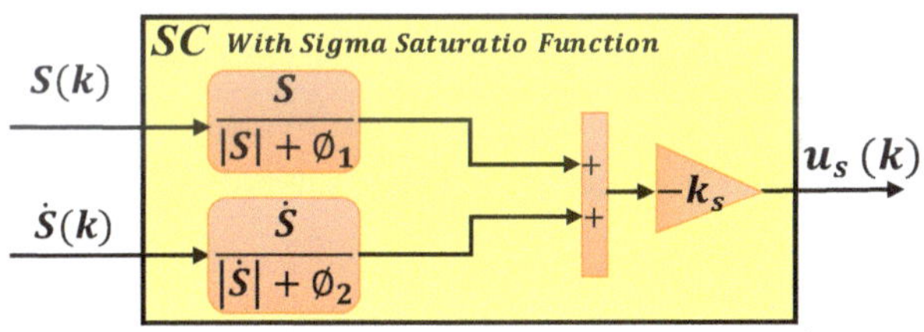

Figure 4. Developed switching control by additional Sigma Function

By choosing the large enough tuning parameter together with the additional sigma function the any chattering and fluctuation (around the sliding surface and on the boundary) which mentioned in sub-section 2.2 will be inhibited completely. It is worth noting that, the source of the chattering are determinant factors in choosing of the $\varnothing_1$, $\varnothing_2$ for example the greater time delay causes the choosing of larger tuning parameters. Also, the steady state error as the second problem of the boundary layer approach specially, if the large thickness is chosen will be eliminated by online tuning of the FLDE as mentioned in in subsection 3.1.

4. Simulation

4.1 Case Study

In this subsection, an isothermal Van de Vusse reactor [24] is chosen as a benchmark for validation of the performance of the proposed model free-FSMC. This high non-linear process is done in the continuous stirred tank reactor (CSTR) in the presence of inverse response behavior, input-delay(dead-time), affected by unmeasured disturbances [25,26]. This reactor is characterized by the following chemical reactions:

$$A \xrightarrow{k1} B \xrightarrow{k2} C \tag{25}$$

$$2A \xrightarrow{k3} D \tag{26}$$

Here, the component A is the reactant (feed of the reactor), B is the main product and C are the undesirable products. k_1, k_2 and k_3 are constants of the reaction rate. The control of the cyclopentenol (B) production from cyclopentadiene (A) by acid-catalyzed electrophilic addition of water in dilute solution is considered [27]. In this process the cyclopentanediol (C) and dicyclopentadiene (D) are as by-products. The reaction rates of A and B (molar rate per unit volume) are as follows:

$$r_A = -k_1 C_A - k_3 C_A^2 \tag{27}$$

$$r_B = k_1 C_A - k_2 C_B \tag{28}$$

The mole balances for components A and B are as follows:

$$\frac{dC_A}{dt} = \frac{F}{V}\left(C_{A,Feed} - C_A\right) - k_1 C_A - k_3 C_A^2 \tag{29}$$

$$\frac{dC_B}{dt} = -\frac{F}{V} C_B + k_1 C_A - k_2 C_B \tag{30}$$

Here, C_A and C_B are the concentrations of the components A, B and F is the molar feed flow rate, consists of pure A. It is supposed that, variation in the feed concentration (disturbance) $C_{A,Feed}$ is unknown. The reactor has constant volume. The density of the solution is assumed to be constant. The $\frac{F}{V}$ is control variable which is controlled by manipulating of dilution rate $\frac{F}{V}$ (see Figure 5). It is worth noting that, this process exhibits a change in sign of gain at peak conversion level of the operating curve. As depicted in Figure 6 it has non-minimum-phase behavior for operation to the left of this peak and minimum phase for operation to the right [14]. Thus, the maximum value of manipulated variable is upper bounded to $\frac{F_{peak}}{V} = 1.3\,\text{min}^{-1}$. The Kinetic parameters of the reaction and process variables are given in Table 2. The MATLAB software was used for simulation and numerical solving the equations of system's dynamics exhibited by Eq. (29), Eq. (30). Also, it is supposed that digital implementation with the sampling time $T_s = 1\,\text{sec}$ and the process time delay are the source of the chattering in this process and the actuator dynamics is assumed to be ideal.

In Accordance with the Figure 6 if the system affected by variation of the feed concentration, the operation curve will be changed, accordingly the FLDE which is designed based on the input/output data of the process under condition $C_{A\,feed} = 10 molL^{-1}$ will have uncertainty.

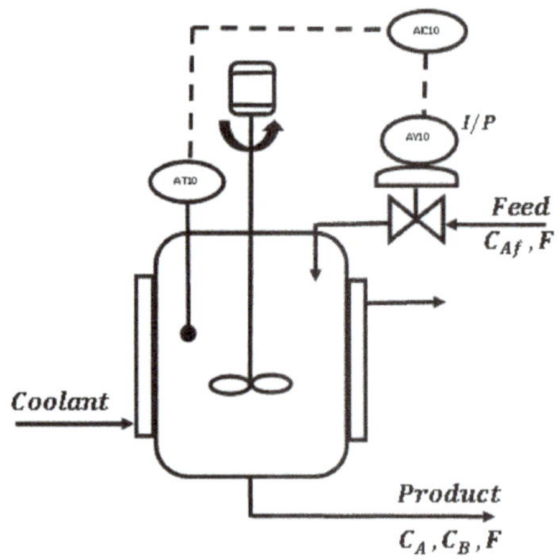

Figure 5. Schematic- isothermal CSTR [28]

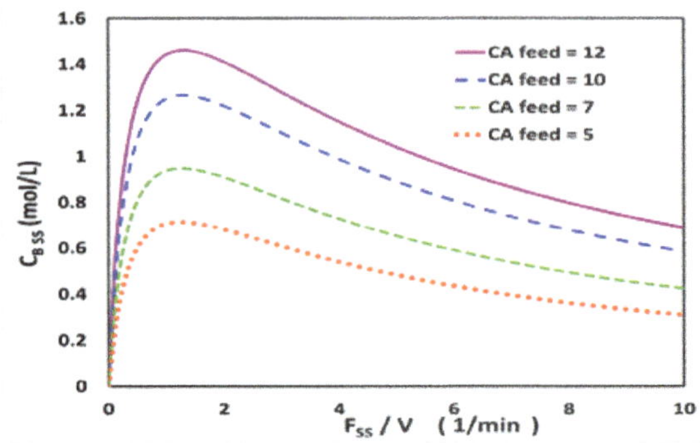

Figure 6. Steady-slate relationships: concentration of product C_{BSS} as a function of the dilution rate $\frac{F_{SS}}{V}$ and $C_{A,Feed}$ [29]

Table 2. Operating parameters of Van De Vusse reactor [14,25]

Parameter	Description	Value
$C_{A,Feed}$	feed concentration	10 molL^{-1}
V	volume of reactor	1.0 L
ρ	mixture density	0.9342 kg L^{-1}
k_1	reaction rate constant	50 h^{-1}
k_2	reaction rate constant	100 h^{-1}
k_3	reaction rate constant	10 Lmol^{-1}h^{-1}

4.2. Numerical Parameters

The sliding variable and the state error is defined as follows:

$$s = e_B, e_B = y - y_{set} = c_B - c_{Bset} \tag{31}$$

Here, C_{Bset} is the set-point concentration.

$$s_{\max} = e_{Bmax} = 1.266 \tag{32}$$

$$\dot{s}_{\max} = \dot{e}_{B\max} = 0.024 \tag{33}$$

From the operating curve depicted in Figure 6, in accordance with the $C_{Afeed}(=10molL^{-1})$, the pick concentration is $C_{Bmax} = 1.266molL^{-1}$.

4.2.1. FLDE's Fuzzy Memory Table

The steady state I/O value of the reactor as the centers of O/I memberships of the FLDE are given in Table 3.1-

Table 3.2. These date were generated by steady state simulation of the process dynamics by MATLB software.

Table 3.1. One dimensional Fuzzy rule base for FLDE

$y_{setfuzzy}(k)$	0	0.2504	0.4341	0.6914
$u_{eqfuzzy}(k)$	0	0.05	0.1	0.2

Table 3.2. One dimensional Fuzzy rule base for FLDE

$y_{setfuzzy}(k)$	1.117	1.1786	1.2345	1.266
$u_{eqfuzzy}(k)$	0.5714	0.7	0.9	1.3

4.2.2 Coefficient Factors of Controllers:

The constants of the proposed model free-FSMC and 3 other controllers which are used for comparison are given as follows. These constant values were yielded from trial and error manually to achieve the best performance:

Controllers:

I) FSMC1: In this controller, equivalent component is pure FLDE (without tuning) and sigma saturation function with fixed tuning parameter is used for its switching control as follows:

$$u = FLDE_{pure} - K_s \frac{s}{|s| + \varnothing_1} \tag{34}$$

To achieve robustness in the face of any large feed concentration fluctuation, the great enough switching gain should be chosen, thus the maximum dilution rate $1.3min^{-1}$ is considered for K_s. The tuning parameter $\varnothing_1$ is computed as 0.40 by trial and error by getting the compromise between tracking error reduction and chattering reduction.

II) FSMC2: This controller is developed of FSMC1 by augmenting additional sigma function as follows:

$$u = FLDE_{pure} - K_s\left(\frac{s}{|s| + \varnothing_1} + \frac{\dot{s}}{|\dot{s}| + \varnothing_2}\right) \tag{35}$$

By supposition $K_s = 1.3$ same as switching gain of the FSMC1. The tuning parameters $\varnothing_1, \varnothing_2$ should be chosen large enough and small enough respectively as $\varnothing_1 = 0.3$ and $\varnothing_2 = 0.036$ to assure prevention of the chattering completely and $\varnothing_1$ should be small enough to achieve better tracking error in comparison with the FSMC1 in the presence of break-like function $\frac{\dot{s}}{|\dot{s}| + \varnothing_2}$.

III) Proposed FSMC: This controller is developed of FSMC2 by auto-adjusting of it equivalent control FLDE as follows:

$$u = FLDE_{autu-tuned} - K_s\left(\frac{s}{|s| + \varnothing_1} + \frac{\dot{s}}{|\dot{s}| + \varnothing_2}\right) \tag{36}$$

The gain K_s is chosen again as 1.3 and $\varnothing_1, \varnothing_2$ and step-size factor are computed as 1.0, 0.0312 and 0.0585 respectively by trial and error by getting the compromise between the achievement of the fast disturbance rejection and prevention of the chattering.

IV) Classical PID Controller:

The tuning parameters of the PID controller are considered as $K_c = 0.7045$; $K_i = 0.9607$; $K_d = 0.0822$ which are computed based on the tuning formulas for Dahlin synthesis [30] to achieve smoother responses than Ziegler-Nichols tuning method. It is done at operating point: $u = 0.2042L\min^{-1}$; $C_B = 0.7$ and $C_A = 1.5714molL^{-1}$.

4.3. Performance Evaluation

The set-point tracking and disturbance rejection performances of the proposed FSMC are evaluated in comparison with 3 other controllers mentioned in previous section. Integral of Absolute Error (IAE), Tracking Error (TE), Settling-Time ST (time taken for the output to reach 95% or 102% of the change in set-point), over shoot percentage and chattering control are used for verification.

$$IAE = \sum_{t=1}^{N} \left| C_B(t) - C_{Bset}(t) \right| .\Delta t \tag{37}$$

4.3.1. Set-point Tracking Performance

To verify the servo tracking performance, the set-point value C_{Bset} is changed step-like from 0.7 $mol.L^{-1}$ to 1.117 $mol.L^{-1}$ from t=0 min to t=15 min. According to Figure 7, and Table 4 it is clearly observed that the proposed method improved transient performance significantly. As it is shown, the settling time were reduced in comparison with the other controllers and less percentage of the overshoot was achieved than SMC1,2. By the SMC2,l ower settling time and percentage of the over shoot were achieved than FSMC1. The PID controller is better than others in term of lower percentage of the overshoot. Moreover, in terms of steady state error, similar performance was achieved by controllers. In accordance with the Figure 7 (b),(c), the proposed method and FSMC2 which were reinforced with the additional alleviated switching term, have less aggressiveness control action to achieve more stable reaching mode near the sliding surface to reduce the chattering thereby which percentage of the overshoot was reduced in comparison with SMC1.

Table 4. Set-point tracking performance of the proposed SMFC in comparison with the other controllers, in relevant to step-like change in Set-point from t=0 min to t=10 min

	PID	FSMC1	FSMC2	Proposed FSMC
IAE	23.67	23.01	15.29	15.22
TE	-1.5E-5	-6.5E-5	1.3E-6	1.6E-13
ST (sec)	116	296	128	84
Overshoot (%)	0.8	27.1	10.4	5.8

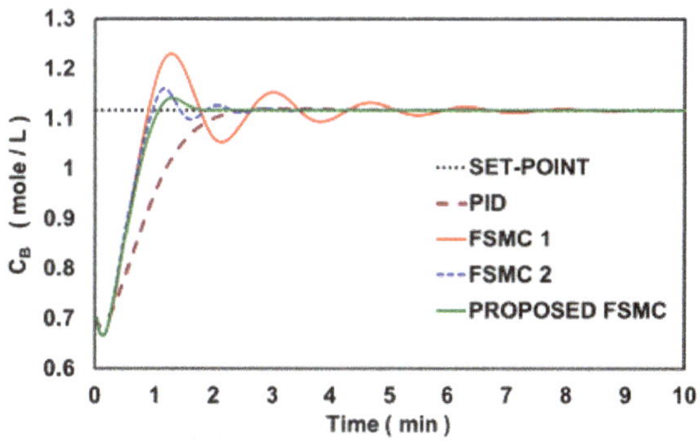

Figure 7(a). Control variable

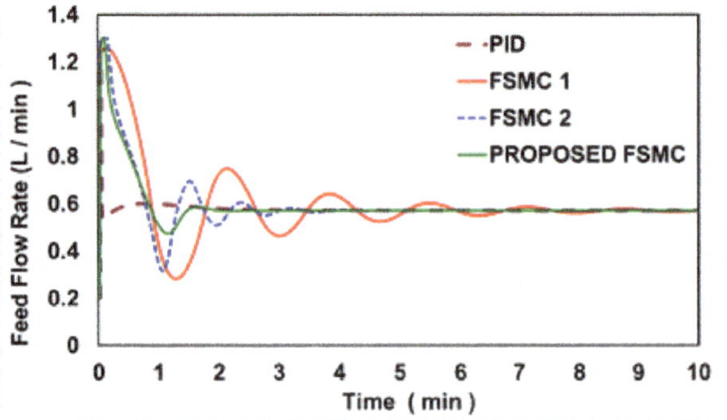

Figure 7 (b). Control action(Feed stream flow rate)

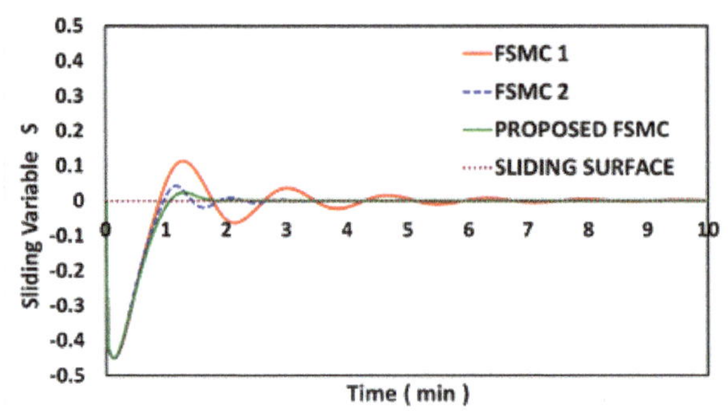

Figure 7 (c). Sliding variable

Figure 7. Set-point-tracking performance

4.3.2 The Regulatory Performance in Load Rejection.

To evaluate the superior robustness of the proposed controller, the feed concentration is changed step-like by +50% and -10% from t=5 min to t=10 min and t=20 min to t=25 min respectively while the set-point value is set in $C_{Bset} = 1.117 molL^{-1}$ as shown in Figure 8. As well depicted in Figure 9 and Table 5, the proposed controller rejected the disturbance successfully in comparison with the other controllers. In Figure 9(a),(c), it is depicted that the proposed FSMC rejected the disturbance completely without any chattering occurrence. The FSMC2 only succeeded in chattering elimination and its performance in rejection of disturbance is not acceptable specially, against large disturbance (50%) against which considerable tracking error (0.0782) was resulted in. The FSMC1 not only didn't succeed in disturbance rejection with tracking error (0.10) similar to FSMC2 but also, only succeeded to eliminate the chattering against small disturbance such that in the face of the large disturbance (50%) it resulted in oscillatory motion of the sliding variable trajectory in the boundary layer, however this oscillation was evanesced and damped finally. The classical PID controller is not sensitive against disturbance and is not robust controller. According to Figure 9(b) the control action of the proposed FSMC was updated successfully until the uncertainty of the FLDE was rejected eventually. It was updated powerfully from 0.571 $L/\min$ to 0.264 $L/\min$ and 0.571 $L/\min$ to 0.810 $L/\min$ against 50% and -10% change in feed concentration respectively. As it is revealed in Figure 9(b),(c) not only the proposed FSMC control action was updated, but also it attempted to break the fast motion of the sliding variable trajectory near the sliding surface to avoid the chattering occurrence similar to FSMC2.

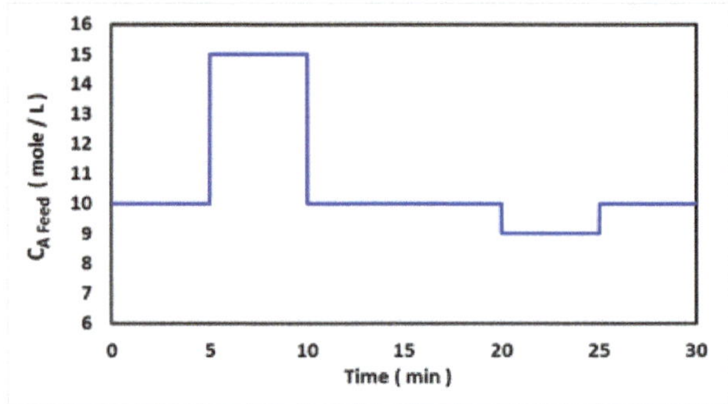

Figure 8. Step-like change in feed concentration (Disturbance)

Table 5. Load rejection performance of the proposed FSMC in comparison with other controllers in relevant to the step-like change in feed concentration from t=5 min to t=10 min

	PID	FSMC1	FSMC2	Proposed FSMC
IAE	19.99	29.76	22.88	4.59
TE	0.0016	0.1022	0.0782	2.3E-5

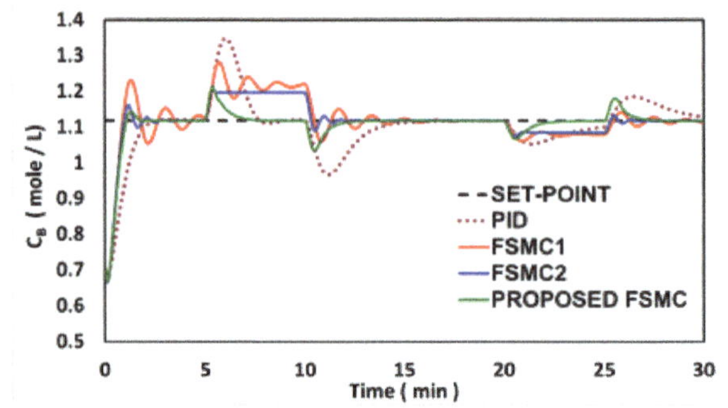

Figure 9 (a). Control variable

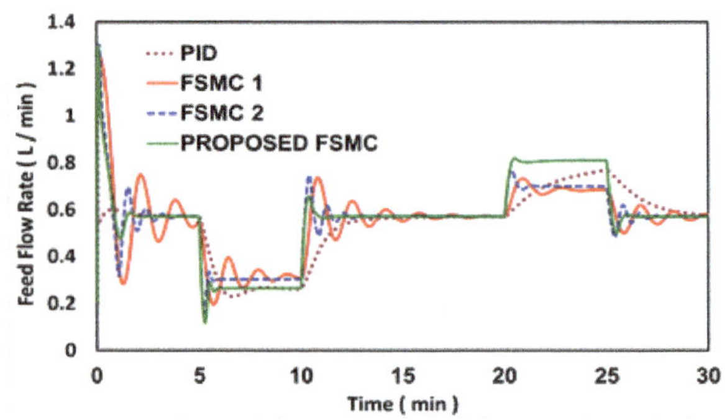

Figure 9 (b). Control action

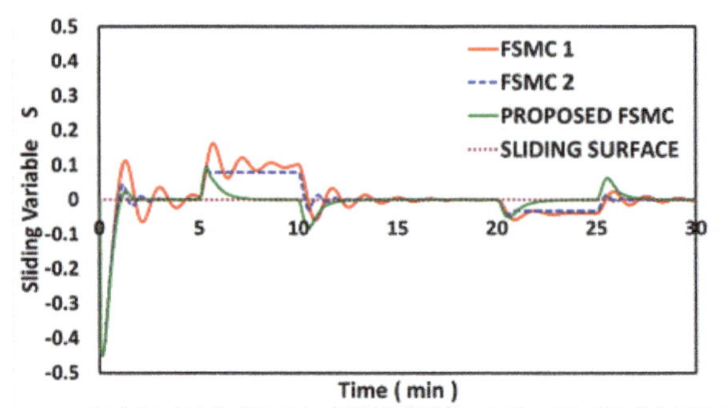

Figure 9 (c). Sliding variable

Figure 9. Disturbance rejection and system regulation performance

It is worth noting that, in accordance with the Figure 10, choosing the small tuning parameter $\varnothing_1 = 0.7$ causes the controller action be speeded up such that disturbance is suppressed faster but at cost of increasing the percentage of the overshoot in servo tracking (5.8% to 25%). Also, larger value of $\varnothing_1$ results in complete inhibiting of the overshoot occurrence (5.8% to 0 %) but at cost of slower response and slower disturbance rejection. These two performance (faster response without any overshoot) is not achieved simultaneously while $\varnothing_1$ is fixed or while sigma functions of switching control in proposed method be combined by constant weighting coefficient. To achieve these performances simultaneously, these sigma function can be integrated with time varying weighting coefficient or tuning parameter $\varnothing_1$ can be allowed to be time varying.

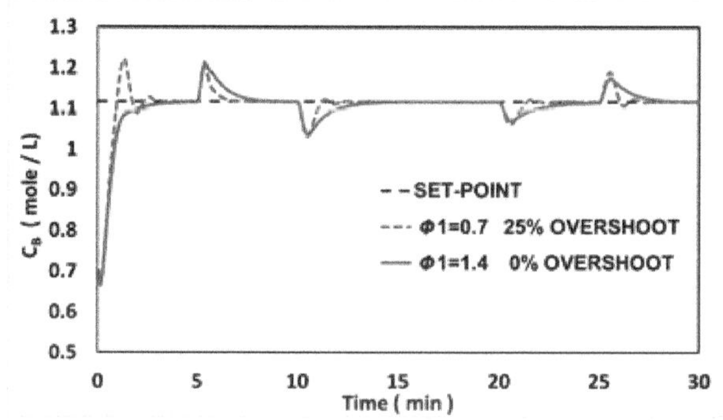

Figure 10. A compromise between overshoot prevention and robustness improvement

5. Conclusions

A novel approach to model-free FSMC was introduced in this paper in order to solve the problems of the pure-SMC can be applied specially in SISO process control. An auto-adjustable fuzzy logic dynamic estimator (FLDE) was proposed in order to improve the robustness in the face of unmeasured disturbances. Also the conventional switching control with large enough boundary layer was reinforced by adding a new saturation function for chattering elimination completely. The proposed method was implemented on an isothermal Van de vusse reactor. The results obviously confirmed the superiority of the new proposed methodology in comparison with two new FSMC and a classical PID control. Robustness was improved effectively specially, in the face of extra disturbances. Also, chattering was removed completely. Moreover, by the proposed method, significant enhancement in transient performance was achieved in terms of shorter settling time with lower overshoot. For further development, the switching control of the proposed method can be developed using fuzzy time varying weighting factor to achieve faster response without any overshoot. Also, we are going to investigate development of the proposed method to apply in MIMO process.

References

[1] K. D. D. Young, V. I. V. I. Utkin, and U. Ozguner, "A Control engineer's guide to sliding mode control," *IEEE Trans. Control Syst. Technol.*, vol. 7, no. 3, pp. 328-342, May 1999.

[2] J.-J. E. Slotine and W. Li, *Applied Nonlinear Control*, vol. 199, no. 1. Prentice-Hall Englewood Cliffs, NJ, 1991.

[3] V. I. Utkin, "Variable structure systems with sliding modes: A Survey," *IEEE Trans. Automat. Contr.*, vol. AC-22, no. 2, pp. 212-222, 1977.

[4] O. M. E. El-Ghezawi, S. A. Billings, and A. S. I. Zinober, "Variable structure systems with and system zeros," in *IEE Proceedings D (Control Theory and Applications)*, vol. 130, no. 1, pp. 1-5, 1983.

[5] B. Bandyopadhyay, F. Deepak, and K.-S. Kim, *Sliding mode control using novel sliding surfaces*, vol. 392. Springer, 2009.

[6] N. Yagiz and Y. Hacioglu, "Fuzzy sliding modes with moving surface for the robust control of a planar robot," *J. Vib. Control*, vol. 11, no. 7, pp. 903-922, Jul. 2005.

[7] F. Piltan, N. Sulaiman, P. Ferdosali, and I. A. Talooki, "Design model free fuzzy sliding mode control: Applied to internal combustion engine," *Int. J. Eng.*, vol. 5, no. 4, pp. 302-312, 2011.

[8] V. Utkin, J. Guldner, and M. Shijun, *Sliding mode control in electro-mechanical systems*, vol. 34. CRC press, 1999.

[9] C.-C. Fuh, "Variable-thickness boundary layers for sliding mode control," *J. Mar. Sci. Technol.*, vol. 16, no. 4, pp. 286-294, 2008.

[10] H.-M. H. Chen, J.-C. J. Renn, and J.-P. J. Su, "Sliding mode control with varying boundary layers for an electro-hydraulic position servo system," *Int. J. Adv. Manuf. Technol.*, vol. 26, no. 1-2, pp. 117-123, 2005.

[11] P. Kachroo and M. Tomizuka, "Chattering reduction and error convergence in the sliding-mode control of a class of nonlinear systems," *IEEE Trans. Automat. Contr.*, vol. 41, no. 7, pp. 1063-1068, Jul. 1996.

[12] L.-C. Yao and J.-S. Chen, "State-dependent boundary layer based on disturbance observer for sliding mode control," *Proc. 2005 CACS Autom. Control Conf.*, 2005.

[13] M. Sulaiman, F. Patakor, and Z. Ibrahim, "New methodology for chattering suppression of sliding mode control for three-phase induction motor drives," *Wseas Trans. Syst. Conntrol*, vol. 9, pp. 1-9, 2014.

[14] A. Shahraz and R. Bozorgmehry Boozarjomehry, "A fuzzy sliding mode control approach for nonlinear chemical processes," *Control Eng. Pract.*, vol. 17, no. 5, pp. 541-550, 2009.

[15] L. A. Zadeh, "Fuzzy sets," *Inf. Control*, vol. 8, no. 3, pp. 338-353, 1965.

[16] G. Chen and Y. H. Joo, "Introduction to fuzzy control systems," *Handb. Appl. Comput. Intell.*, pp. 1-41, 2001.

[17] C.-L. Chen and M.-H. Chang, "Optimal design of fuzzy sliding mode control a comparative study," *Fuzzy Sets Syst.*, vol. 93, no. 1, pp. 37-48, 1998.

[18] W. Guo and D. Liu, "Sliding mode fuzzy dynamic weighting control for a class of nonlinear underactuated systems," *J. Inf. Comput. Sci.*, vol. 13, no. 12, pp. 2661-2668, 2011.

[19] S. Park and S. Rahmdel, "A new fuzzy sliding mode controller with auto-adjustable saturation boundary layers implemented on vehicle suspension," *Int. J. Eng. C Asp.*, vol. 26, no. 12, pp. 1401-1410, 2013.

[20] I. Eksin, M. Guzelkaya, and S. Tokat, "Sliding surface slope adjustment in fuzzy sliding mode controller," in *Proceeding of the 10th Mediterranean Conference on Control and Automation*, pp. 160-168, 2002.

[21] S. N. Dragan Antic, Marko Milojkovic, "Fuzzy sliding mode control with additional fuzzy control component," *Autom. Control Robot.*, vol. 8, no. 1, pp. 25-34, 2009.

[22] O. Camacho, R. Rojas, and W. García-Gabín, "Some long time delay sliding mode control approaches," *ISA Trans.*, vol. 46, no. 1, pp. 95-101, Feb. 2007.

[23] L.-X. Wang, *A Course in Fuzzy Systems and Control*. Prentice-Hall press, USA, pp. 238-248, 1999.

[24] G. J. Van de Vuss, "Plug Flow Type Reactor Versus Tank Reactor," *Control Eng. Sci.*, vol. 19, no. 12, pp. 994-997, 1964.

[25] C.-T. Chen and S.-T. Peng, "A sliding mode control scheme for non-minimum phase non-linear uncertain input-delay chemical processes," *J. Process Control*, vol. 16, no. 1, pp. 37-51, Jan. 2006.

[26] C. Kravaris and P. Daoutidis, "Nonlinear state feedback control of second-order nonminimum-phase nonlinear systems," *Comput. Chem. Eng.*, vol. 14, no. 4, pp. 439-449, 1990.

[27] S. Engell and K. U. Klatt, "Gain scheduling control of a non-minimum-phase CSTR," in *Proceedings of the 2nd European Control Conference*, pp. 2323-2328,1993.

[28] P. Balaguer, V. Alfaro, and O. Arrieta, "Second order inverse response process identification from transient step response," *ISA Trans.*, vol. 50, no. 2, pp. 231-238, 2011.

[29] B. W. Bequette, *Process Dynamics: modeling, analysis, and simulation.* Prentice Hall PTR Upper Saddle River, NJ, pp. 506-516, 1998.

[30] C. A. Smith and A. B. Corripio, *Principles and practice of automatic process control*, vol. 2. Wiley New York, 1985.

Simulation of Three-mass Mechanical System using MATLAB Software

Marek Moravič[1,*], Oskar Ostertag[2], Darina Hroncová[3]

[1]Department of Construction, Automotive and Transport Engineering, Faculty of Mechanical Engineering, Technical University of Košice, Košice, Slovak republic
[2]Department of Applied Mechanics and Mechanical Engineering, Faculty of Mechanical Engineering, Technical University of Košice, Košice, Slovak republic
[3]Department of Mechatronics, Faculty of Mechanical Engineering, Technical University of Košice, Košice, Slovak republic
*Corresponding author: marek.moravic@tuke.sk

Abstract The aim of the paper is the dynamic analysis of selected models of running gear by using MATLAB/Simulink. This paper concerns with numerical approach to a vehicle vertical vibration based on kinematic excitation. Kinematic variables calculation is demonstrated on the quarter three-mass dynamic model with additional shock absorber and spring. There is mathematical model that is numerically solved in MATLAB and is illustrated here together with the simulation in Simulink by block scheme. The results could be used for teaching purposes.

Keywords: *MATLAB/Simulink, signal builder, mathematical model, block*

1. Introduction

Method of releasing is applied to drawing up mathematical model. Kinetic equations are drawn up by Newton's method. The result is kinematic quantities graphically represented and their numerical values. Selecting the way of solution is influenced by concern about mechanical system vibration problems [1]. MATLAB is efficient, interactive setting for scientific and engineering calculations. To be able to solve in Matlab it is necessary to modify system of differential equations of second order to system of differential equations of first order. Detailed calculation process is introduced in work [2].

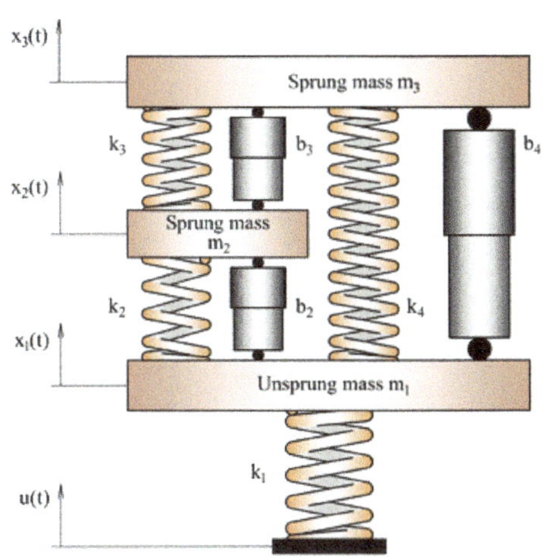

Figure 1. Quarter three-mass model

2. Drawing up Mathematical Model by Method of Releasing

Calculation of kinematic quantities is demonstrated on quarter three-mass dynamic model with additional damper and spring (Figure 1) [2,3]. The system is solved in MATLAB/Simulink.

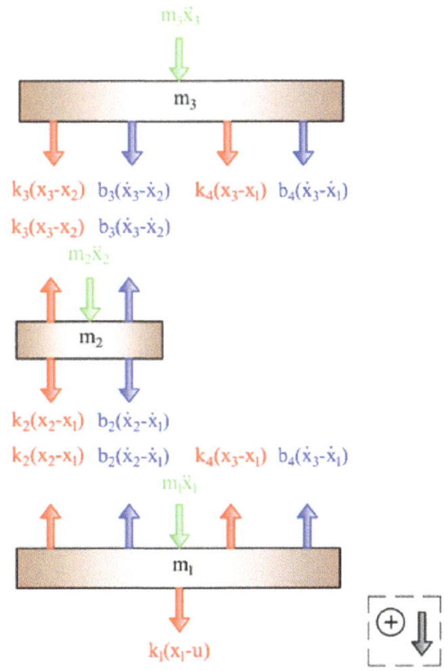

Figure 2. Released quarter model

Running gear is kinetically excited. Kinetic equations of mechanical system are made up by applying of releasing method (Figure 2).

Kinetic equations of damped three-mass mechanical system (Figure 2) are composed by using of releasing method and look like:

$$m_1\ddot{x}_1 - b_2(\dot{x}_2 - \dot{x}_1) - b_4(\dot{x}_3 - \dot{x}_1) + k_1(x_1 - u) - k_2(x_2 - x_1) - k_4(x_3 - x_1) = 0, \tag{1}$$

$$m_2\ddot{x}_2 + b_2(\dot{x}_2 - \dot{x}_1) - b_3(\dot{x}_3 - \dot{x}_2) + k_2(x_2 - x_1) - k_3(x_3 - x_2) = 0, \tag{2}$$

$$m_3\ddot{x}_3 + b_3(\dot{x}_3 - \dot{x}_2) + b_4(\dot{x}_3 - \dot{x}_1) + k_3(x_3 - x_2) + k_4(x_3 - x_1) = 0. \tag{3}$$

We put the substitution:

$$\begin{aligned} x_1 &= x_1, \\ x_2 &= \dot{x}_1, \\ x_3 &= x_2, \\ x_4 &= \dot{x}_2, \\ x_5 &= x_3, \\ x_6 &= \dot{x}_3. \end{aligned} \tag{4}$$

By derivation equations (4) according to time we achieve:

$$\begin{aligned} \dot{x}_1 &= x_2, \\ \dot{x}_2 &= \frac{1}{m_1}\left[b_2(\dot{x}_2 - \dot{x}_1) + b_4(\dot{x}_3 - \dot{x}_1) - k_1(x_1 - u) + k_2(x_2 - x_1) + k_4(x_3 - x_1)\right], \\ \dot{x}_3 &= x_4, \\ \dot{x}_4 &= \frac{1}{m_2}\left[-b_2(\dot{x}_2 - \dot{x}_1) + b_3(\dot{x}_3 - \dot{x}_2) - k_2(x_2 - x_1) + k_3(x_3 - x_2)\right], \\ \dot{x}_5 &= x_6, \\ \dot{x}_6 &= \frac{1}{m_3}\left[-b_3(\dot{x}_3 - \dot{x}_2) - b_4(\dot{x}_3 - \dot{x}_1) - k_3(x_3 - x_2) - k_4(x_3 - x_1)\right]. \end{aligned} \tag{5}$$

By repeated using of equations (4) and modifying we achieve six differential equations of first order:

$$\dot{x}_1 = x_2, \tag{6}$$

$$\dot{x}_2 = \frac{1}{m_1}\left[-(b_2 + b_4)x_2 + b_2 x_4 + b_4 x_6 - (k_1 + k_2 + k_4)x_1 + k_2 x_3 + k_4 x_5 + k_1 u\right], \tag{7}$$

$$\dot{x}_3 = x_4, \tag{8}$$

$$\dot{x}_4 = \frac{1}{m_2}\left[b_2 x_2 - (b_2 + b_3)x_4 + b_3 x_6 + k_2 x_1 - (k_2 + k_3)x_3 + k_3 x_5\right], \tag{9}$$

$$\dot{x}_5 = x_6, \tag{10}$$

$$\dot{x}_6 = \frac{1}{m_3}\left[b_4 x_2 + b_3 x_4 - (b_3 + b_4)x_6 + k_4 x_1 + k_3 x_3 - (k_3 + k_4)x_5\right], \tag{11}$$

3. Solution of Mechanical System Vertical Vibration Using MATLAB/Simulink Program

At first the entry has to be defined – course of obstacle for solution in MATLAB. The function is

$$u(t) = \frac{A}{2}\left(1 - cos\left(2\pi\frac{vt}{L_0}\right)\right), \tag{12}$$

where A is obstacle height, L_0 is its length and v is car speed [4].

The equations (6), (7), (8), (9), (10) and (11) need to be transcribed to m-file of MATLAB (Table 1). The name of file will be saved according to name of function. For numerical solution of differential equations MATLAB applies predefined functions that differ about method of integration. The time course depiction can be gained by standard function ode45 (Figure 3) [5].

Table 1. M-files of MATLAB

fun3M.m	function dx=fun3M(t,x)
	m1=55; m2=400; m3=100;
	k1=230000; k2=30000; k3=50000; k4=0;
	b2=1500; b3=4000; b4=700;
	L0=5; v=15; A=0.03;
	u=(A/2)*(1-cos(2*pi*(v*t/L0))); % obstacle
	dx=[x(2); ...
	(-(b2 + b4)*x(2)+b2*x(4)+b4*x(6)
	-(k1+k2+k4)*x(1)+k2*x(3)+k4*x(5)+k1*u)/m1; ...
	x(4); ...
	(b2*x(2)-(b2+b3)*x(4)+b3*x(6)+k2*x(1)
	-(k2+k3)*x(3)+k3*x(5))/m2; ...
	x(6); ...
	(b4*x(2)+b3*x(4)-(b3+b4)*x(6)+k4*x(1)+k3*x(3)
	-(k3+k4)*x(5))/m3];

course_3H.m	% course depiction – three-mass system
	[t,x]=ode45('fun3M',[0 5],[0; 0; 0; 0; 0; 0]);
	plot(t,x);
	grid on; % turning on and turning off the grid
	xlabel('t [s]');
	ylabel('x1(t) [m], v1(t) [m/s], x2(t) [m], v2(t) [m/s], x3(t) [m], v3(t) [m/s]');
	legend('x1(t)','v1(t)','x2(t)','v2(t)','x3(t)','v3(t)');

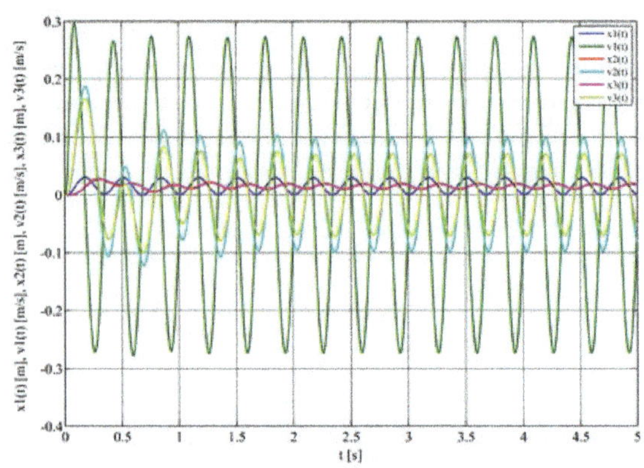

Figure 3. Displacements and velocities depiction

Modeling of mechanical system in Simulink helps us graphically represent the time progressions of kinetic and dynamic quantities. Modifying derived kinetic equations (1), (2) and (3) for three-mass model

$$\ddot{x}_1 = \frac{1}{m_1}\left[-(b_2 + b_4)\dot{x}_1 + b_2\dot{x}_2 + b_4\dot{x}_3 - (k_1 + k_2 + k_4)x_1 + k_2 x_2 + k_4 x_3 + k_1 u\right], \tag{13}$$

$$\ddot{x}_2 = \frac{1}{m_2}\left[b_2\dot{x}_1 - (b_2+b_3)\dot{x}_2 + b_3\dot{x}_3 + k_2x_1 - (k_2+k_3)x_2 + k_3x_3\right] \quad (14)$$

$$\ddot{x}_3 = \frac{1}{m_3}\left[b_4\dot{x}_1 + b_3\dot{x}_2 - (b_3+b_4)\dot{x}_3 + k_4x_1 + k_3x_2 - (k_3+k_4)x_3\right] \quad (15)$$

we make block diagram (Figure 4).

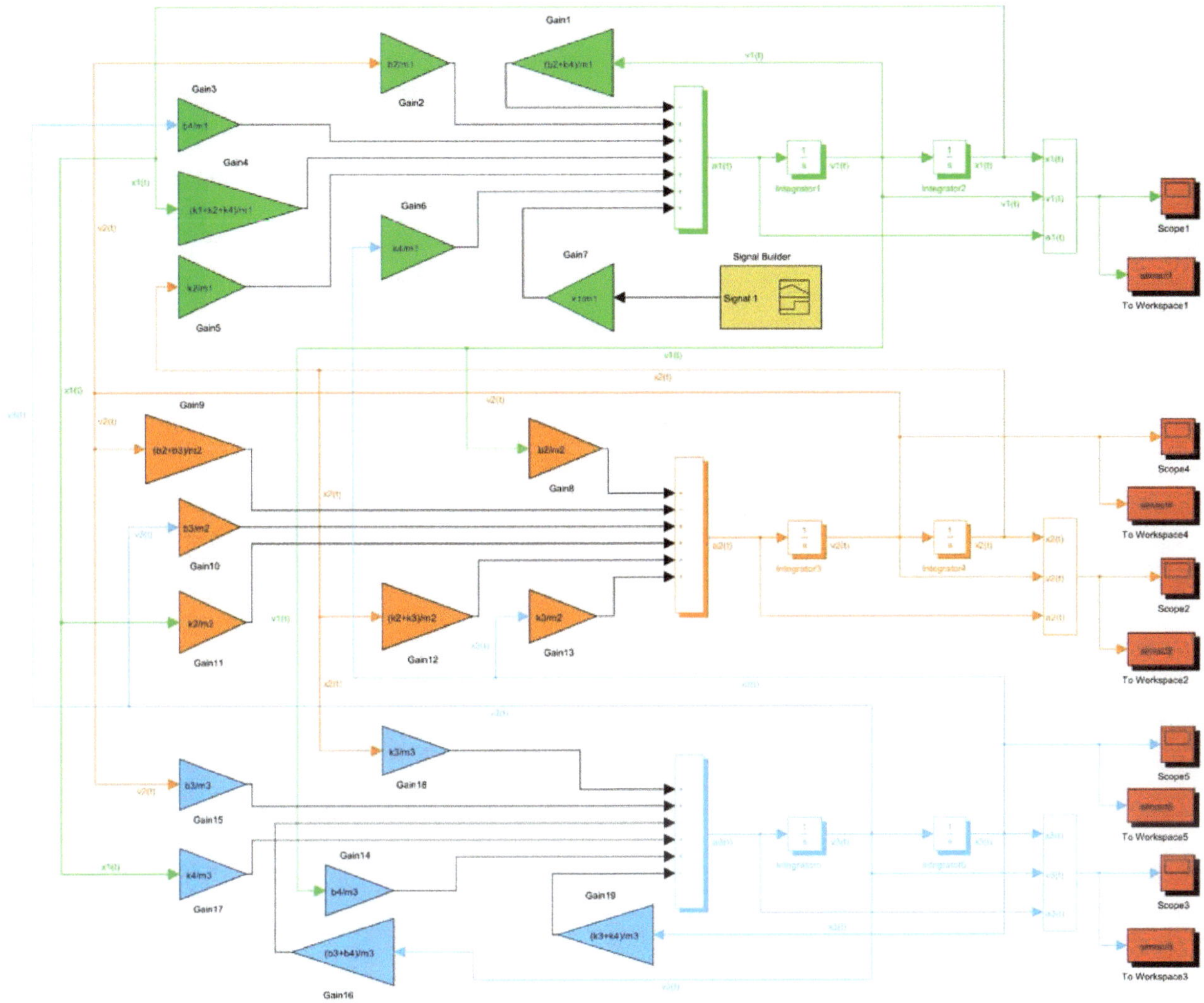

Figure 4. The block diagram of three-mass system in Simulink

Table 2 shows blocks overview from Figure 4, but the library in Simulink contains also other usable blocks for purposes of problem solving.

Table 2. Used blocks in program Simulink

Block	Library	Description
Gain	Math Operations	Multiplication of the input by a constant value
Sum	Math Operations	Sum element
Mux	Signal Routing	Combining into one output
To Workspace	Sinks	Saving the values into Workspace
Scope	Sinks	Diagram representation
Constant	Sources	Entering constant value
Integrator	Continuous	Lowering order
Signal Builder	Sources	Setting up a random excitation

There is a recommendation to change primal value $1e^{-3}$ to $1e^{-15}$ in main menu Simulation, Configuration Parameters before launching a simulation. There is also an option to set time lag of simulation.

The course of obstacle was made up by block Signal Builder. This course needs to be exported File, Export to Workspace and then Start simulation needs to be launched (Figure 5) [3].

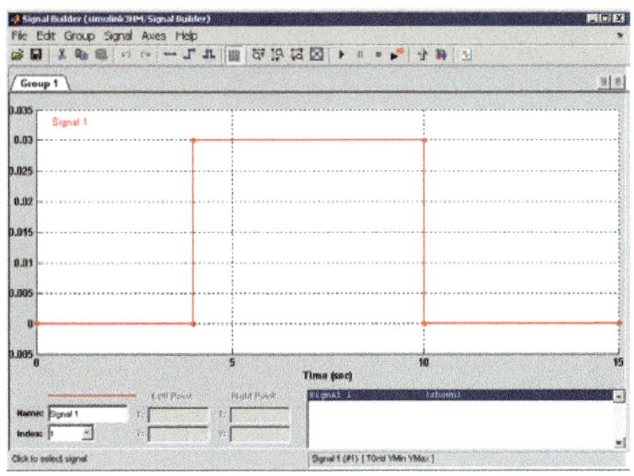

Figure 5. Block Signal Builder

In Figure 6 there is an illustration of defining random excitation in block Signal Builder.

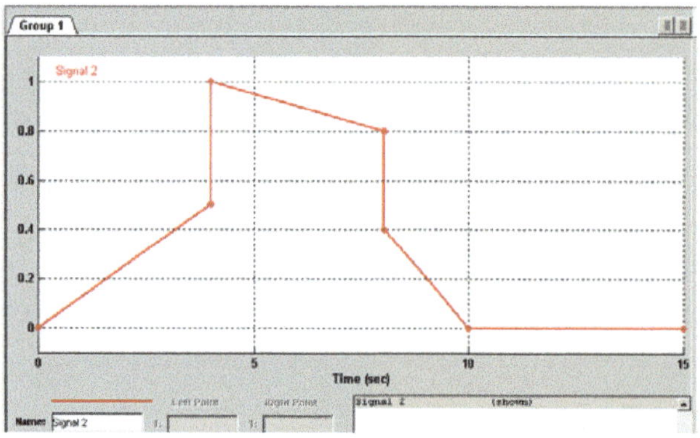

Figure 6. The illustration of defining random excitation in block Signal Builder

Table 3 shows the forms of input data recording to start a simulation in Simulink. These data will be saved into Workspace. It is recommended to use m-file for purposes of faster input editing.

Table 3. The forms of input data recording in Simulink

m-file	% input data - Simulink
	% m - [kg], k - [N/m], b - [N/(m/s)]
	m1=80; m2=300; m3=800;
	k1=230000; k2=40000; k3=55000; k4=3000;
	b2=1700; b3=4000; b4=1000;

Graphical course of quantities is represented by m-file by subplot and plot (Figure 7, Figure 8). Function Plot is entered in Table 4.

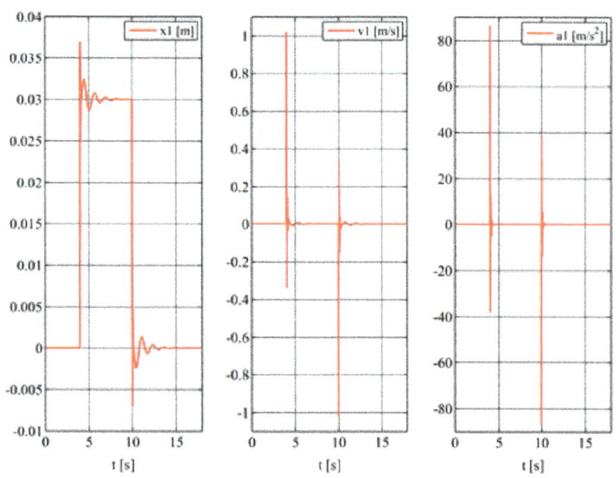

Figure 7. Displacement, velocity and acceleration course of mass 1 next to each other – simout1

Table 4. Function Plot

simulink_simout5_x3.m	% depiction simout5_x3 (mass "3"displacement only) figure(5) set(5,'Name', 'course simout5 in Simulink') plot(simout5.time, simout5.signals.values(:,1:1), 'LineWidth', 2.5) legend('x3=x3(t)') xlabel('t [s]') ylabel('x3 [m]') grid on

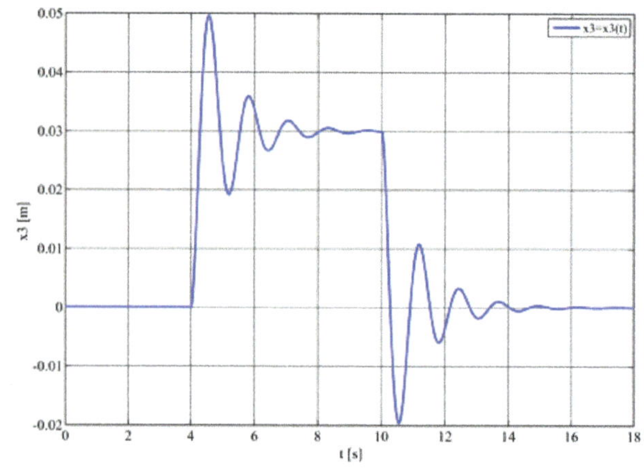

Figure 8. Displacement depiction of mass 3 only – simout5

4. Conclusion

The methodology of three-mass system calculation being solved by deriving of kinetic equations needed for numerical solution in MATLAB is illustrated here together with the simulation in Simulink by block schemes. First-order differential equations need to be assigned for solution in MATLAB and these equations are obtained from modification of second-order differential equations. Making up scheme in Simulink needs to proceed from second-order differential equation that describes system performance.

Acknowledgement

This contribution is a result of the project Slovak Grant Agency – project VEGA No. 1/0937/12 "Development of non-traditional experimental methods for mechanical an mechatronical systems", project KEGA No. 048TUKE-4/2014 "Increasing of knowledge base of students in area of application of embedded systems in mechatronic systems" and project VEGA No. 1/1205/12 "Numerical modeling of mechatronic systems".

References

[1] Hassaan, G., A., "Car Dynamics using Quarter Model and Passive Suspension, Part I: Effect of Suspension Damping and Car Speed," in *International Journal of Computer Techniques*, 1(2). 1-9. 2014.

[2] Moravič, M., *Computer Modelling of the Suspension of the Running Gear* (in Slovak), Diploma thesis, TU of Košice, Faculty of Mechanical Engineering, 2013.

[3] Segľa, Š., Reich, Š., "Optimization of parameters of a semi-active suspension and its comparison with active and passive suspensions", in *Dynamika tuhých a deformovatelných těles 2006*, Ústí nad Labem, 183-190.

[4] Ostertag, O., Fabian, M., Moravič, M., Filičko, J., Boslai, R., "Řešení mechanismu kol s integrovaným elektromotorem, elektronickou brzdou, pružením a zavěšením," in *Cad*, 24(5). 30-33. 2014.

[5] Hroncová, D., "Riešenie mechanických sústav v prostredí MATLAB, Simulink a SimMechanics", in *Dynamika tuhých a deformovatelných těles 2008*, Ústí nad Labem, 69-74.

Logical Circuits and Their Applications

Ľubica Miková*, Michal Kelemen, Alexander Gmiterko, Lukáš Kačmár

Technical University of Košice, Faculty of Mechanical Engineering, Department of Mechatronics, Košice, Slovakia
*Corresponding author: lubica.mikova@tuke.sk

Abstract Logic circuits are inseparable part of almost every modern electronic device. Due to constant raising of requirements of society for technical complexity of these devices, complex integrated circuits are used more frequently. This article deals with mathematical apparatus, which logic circuits work on, and to explain methods of entry and execution of logical functions.

Keywords: *logic cirucits, matlab*

1. Introduction

Logic circuits are part of nearly each electronic device we come in contact with everyday. Cell phone, laptop or plasma television, all these devices contain within complicated, yet compact integrated circuits, which are however made of basic, simple logical circuits.

2. Logics

There are circuits which can be defined by continuous time courses. These are mainly constant, linear and occasionally even nonlinear courses. A good example is sinusoidal that possibly contains a phase shift and is dependent on circuit type. These courses are called *analog signals* [1,2,3].

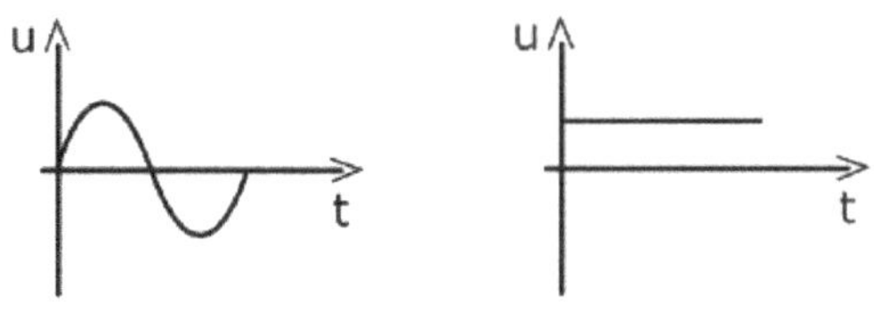

Figure 1. Examples of analog signals [2]

In the process of electronics development there was an improvement of electronic devices, inside which components operating on the principle of numerical methods are nowadays more and more used. Numerical circuits are characterized by discontinuous (step) time change – numerical or digital [2].

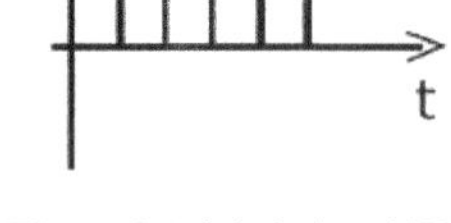

Figure 2. Digital signal [2]

3. Basic Division of Logic Circuits

Logic circuits can be divided according to a few aspects:

According to the logic function they realize

- **combinational** – output variable depends only on input variables in a specific point of time,
- **sequential** – output variable is dependent not only on input variables in a specific point of time, but also on the sequence of input variables in previous points of time [5].

According to their ability to amplify the signal

- **passive**,
- **active [5]**.

3.1. Analysis and Synthesis of Logic Circuits

All logic circuits are as cybernetic systems characterized by behavior defined by logic function, Karnaugh map or truth table and structure, which is expressed by schematics diagram of a logic circuit.

Given the initial impulse and the result of action we know two basic approaches:

- **analysis** – scrutiny of behavior and functioning of designed or already realized circuit,
- **synthesis** – method where, at defined initial conditions of system behavior and requirements on its control, we perform the design of a specific logic circuit that is realizing this behavior [2].

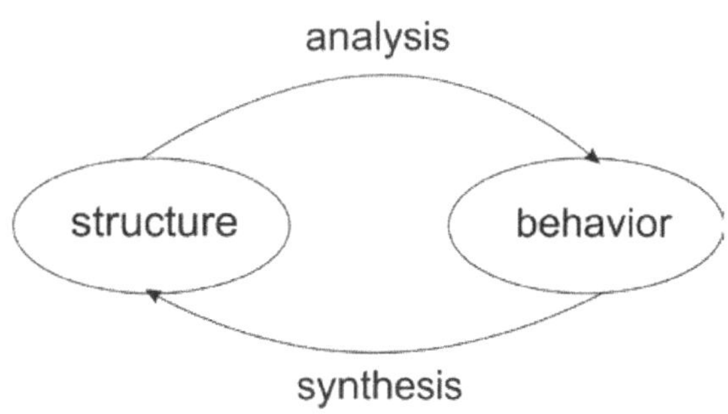

Figure 3. Analysis and synthesis [2]

4. Logical Circuits

4.1. Combinational Logic Circuits

Combinational logic circuit (CLC) has its output state dependent on the immediate state of inputs. Value of output variables is defined only by immediate combination of input variables [4,6].

$$Yi = f(Xj)$$

where i= 1,2,...m, j=1,2,...n.

$Yi=y1,y2,\ldots ym$ – output variables combinational logic circuits

$Xj=x1,x2,\ldots xm$ – input variables combinational logic circuits

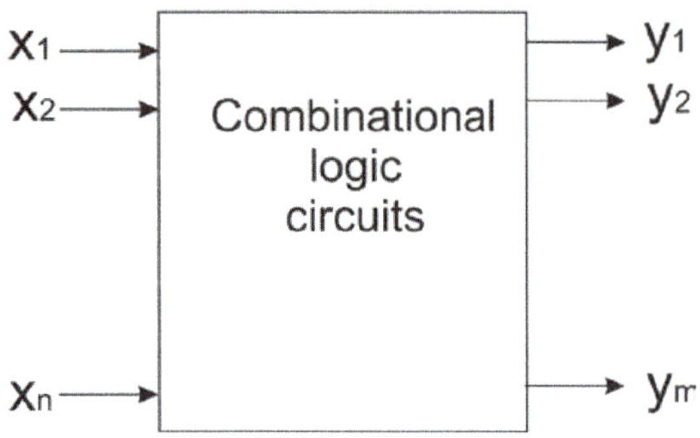

Figure 4. Combinational logic circuits

CLC is realized as a functional whole by connection of logic members or as integrated circuit. Main representers of CLCs are:

- coders and decoders,
- multiplexers and demultiplexers,
- comparators,
- circuits performing arithmetic operations (adders and so on) [5].

4.2. Sequential Logic Circuits

Sequential logic circuit is defined by a circuit where the immediate value of outputs depends on the immediate values of inputs and also on states they were in previously. They contain memory parts for storing previous state of circuit [7].

Function of sequential logic circuits can be described by two equations:

1) Equation for outputs $Yi=(\{Xj\},\{Qpt\})$

2) Equation for following internal state $Qpt+1=(\{Xj\},\{Qpt\})$

$Yi=(y1,y2,\ldots,ym)$ - output set of sequential logic circuits

$Xj=(x1,x2,\ldots,xn)$ - input set of sequential logic circuits

$Qp=(q1,q2,\ldots,qk)$ - internal state set of sequential logic circuits.

So called internal variables $q1,\ q2,\ \ldots,\ qk$ come from outputs of individual flip-flop circuits. Internal state consists of set of all outputs of flip-flop circuits in a specific time t. Index t discerns actual internal state Qpt and index $t+1$ the following state of sequential circuits $Qpt+1$ [7].

Figure 5 illustrates block diagram of sequential logic circuits comprised of combinational and sequential part. The combinational part generates values of outputs and actuating signals of memory circuits. Sequential part contains memory circuits (usually flip-flop circuits), which are controlled by periodical numerical signal with period of T. [7].

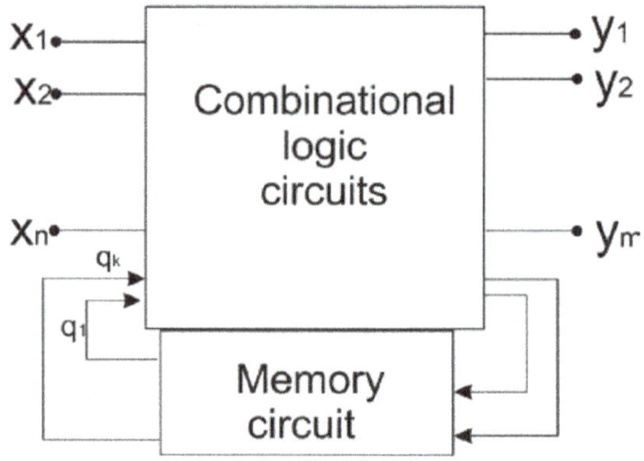

Figure 5. Block diagram of sequential logic circuit

4.3. Examples of Logic Control

One of possible examples of logic control is design of logic circuit for control of water pumping from a well.

Pump will operate if there is insufficient amount of water in the tank inside the house and sufficient amount of water in the well. If there is sufficiency of water in the tank inside the house and insufficiency of water in the well, yellow indicator light will light up. If there is insufficiency of water both in the tank inside the house and in the well, red indicator light will light up.

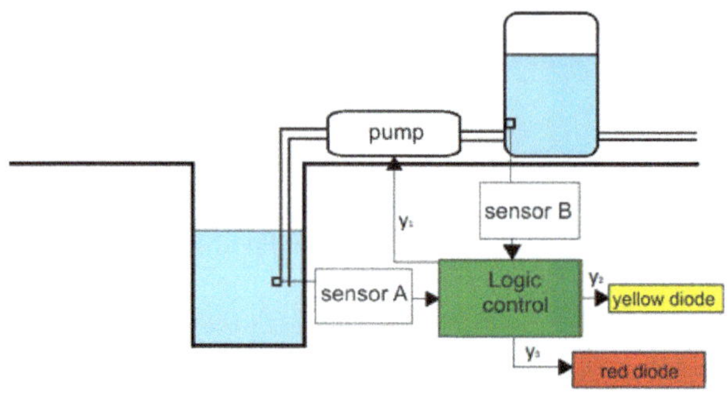

Figure 6. Design of logic circuit for control of water pumping

After compiling a truth table and subsequently a Karnaugh map it is possible to create a design of control of this logic circuit using Matlab/Simulink program.

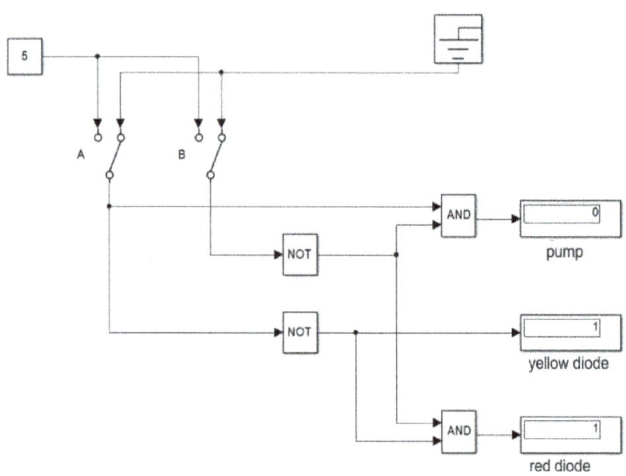

Figure 7. Logic circuit in Matlab/Simulink

Another example can be design of building alarm device. Monitored are two windows on ground floor and entry door. System is designed so that alarm is sounded in the event of infiltration of the object by a stranger.

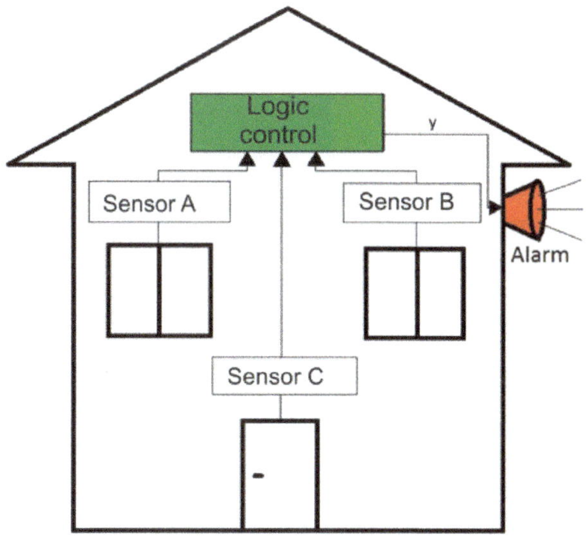

Figure 8. Design of building alarm device

Design of control in Matlab/Simulink program is shown on Figure 9.

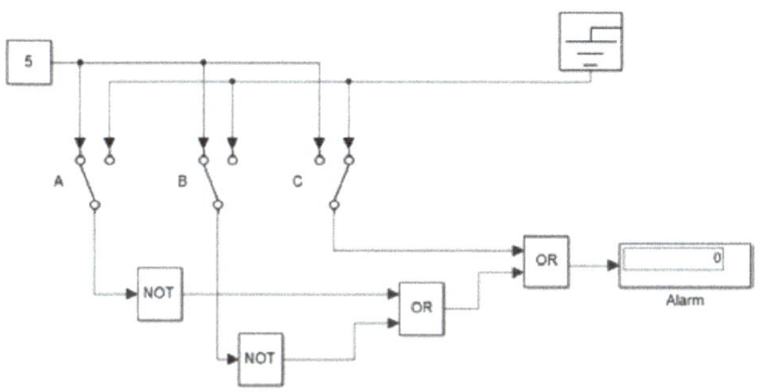

Figure 9. Logic circuit in Matlab/Simulink

Another application of logic control is control of valve of reserve pipeline. Pumps A, B, C and D are connected to pipeline p1 and p2. It is necessary to design a combination control logic circuit for automatic control of valve of the backup pipeline p2, if pump flow rate exceeds the allowed flow rate of pipeline p1 – Q1= 7 l/s, where the pump flow rates are:

- QA = 2 l/s
- QB = 3 l/s
- QC = 3 l/s
- QD = 6 l/s

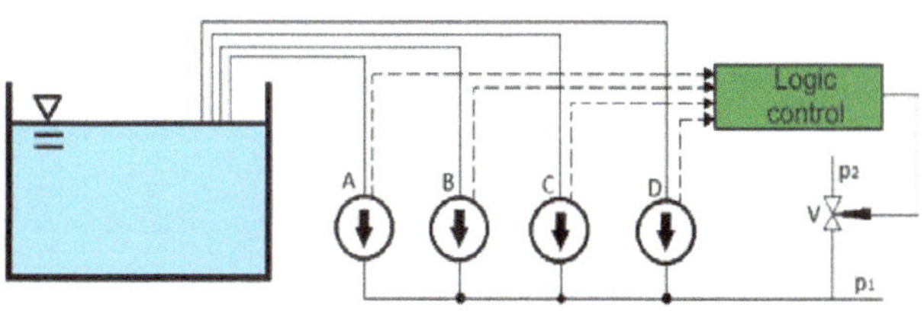

Figure 10. Logic circuit in Matlab/Simulink

Same as in previous examples of logic control, the logic control design diagram is shown on Figure 11.

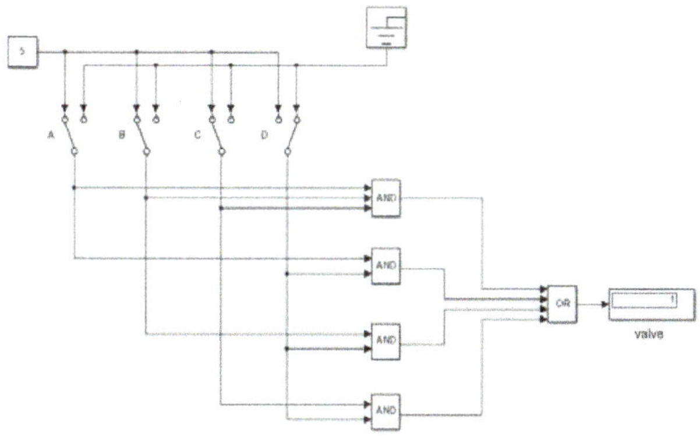

Figure 11. Logic circuit in Matlab/Simulink

5. Conclusion

In the recent years, reversible logic has emerged as a promising technology having its applications in low power CMOS, quantum computing, nano technology and optical computing. Reversible logic circuits provide less power dissipation as well as distinct output assignment for each distinct input [8,9].

The aim of this article was to provide theoretical basis for logic circuits topic and create solution for specific given problems.

Solving of logic control design problems lies in compiling truth tables, which are then used to create Karnaugh maps. Utilizing the rules of handling these maps we put together the resulting logic functions. Logic circuit was created in MATLAB/ Simulink software. Correctness of solution was verified by multiple simulations with changing input parameters and monitoring accuracy of model operation. Simulated model meets all requirements and is easily and intuitively controlled.

Acknowledgement

This contribution is a result of the project Slovak Grant Agency – project VEGA 1/0937/12 “Development of non-traditional experimental methods for mechanical and mechatronical systems”, and project KEGA 048TUKE-4/2014 “Increasing of knowledge base of students in area of application of embedded systems in mechatronic systems”.

References

[1] Náter, E.: Elektronika pre 2. ročník odbor elektrotechnika. 41s. [online] [cit. 17.4.2015]. Dostupné na internete: http://www.spsehalova.sk/ver1/images/norhar/predmetove_komisie/skripta%20_ele_2.pdf.

[2] Elektronika. [online] [cit. 18.4.2015]. Dostupné na internete: http://ostrovskeho.sk/ucivo/data/tvp-2/elektronika-1.cast.pdf.

[3] Diviš, Z., Chmelíková, Z.,Zdrálek, J.: Logické obvody. Ostrava: Vysoká škola baňská – Technická univerzita Ostrava, 1999, 150s., ISBN 80-7078-653-1.

[4] Tóth, Š.: Simulátor logických obvodov. Fakulta riadenia a informatiky, Žilinská univerzita v Žiline, 2008, 59s. [online] [cit. 18.4.2015]. Dostupné na internete: <stefantoth.com/content/Toth-BT.pdf>.

[5] Putiš, M.: Kombinačné logické obvody. Elektrotechnická fakulta, Žilinská univerzita v Žiline, 2008, 65s. [online] [cit. 20.4.2015]. Dostupné na internete: <diplom.utc.sk/wan/2536.pdf>.

[6] Komparátory. [online] [cit. 23.4.2015]. Dostupné na internete: <files.3titt.webnode.sk/200000019-b7014b8f4f/Komparatory.pdf>.

[7] Béres, P.: Simulácia sekvenčných logických obvodov s využitím modulu Stateflow. Elektrotechnická fakulta, Žilinská univerzita v Žiline, 2008, 56s. [online] [cit. 24.4.2015]. Dostupné na internete: <diplom.utc.sk/wan/2505.pdf>.

[8] M. Mahapatro, S. K. Panda, J. Satpathy, M. Saheel: Design of Arithmetic Circuits Using Reversible Logic Gates and Power Dissipation Calculation, 2010 International Symposium on Electronic System Design.

[9] Vasudevan D P, Lala P K and Parkerson J P.: Online Testable Reversible Logic Circuit Design using NAND Blocks, Proceedings of the 19th IEEE International Symposium on Defect and Fault Tolerance in VLSI Systems, 2004.

[10] Ľ. Miková, F. Trebuňa, The application of simulation methods for modeling mechatronic systems, Acta Mechanica Slovaca. Vol. 16, 2012.

Localization and Navigation of Mobile Robot for Outdoor Area with Motion Recognition

Kamil Židek[1,*], Vladislav Maxim[2]

[1]Faculty of Manufacturing Technologies with a seat in Presov, Presov, Slovak Republic
[2]Institute of Automation, Robotics and Mechatronics, FME, Technical University of Kosice, Slovak Republic
*Corresponding author: kamil.zidek@tuke.sk

Abstract This article introduces a prototype of the mobile device intended for monitoring industrial area. The mobile device is designed mainly for outdoor environment. Its main function is manual or automatic remote patrolling. The Paper presents control system and other functions like video recognition and localization sensors. Principles of trajectory navigation by means of GPS system. Monitoring of mobile device position and movements is possible by web interface or desktop solution.

Keywords: *localization, navigation, mobile devices, monitoring, sensors*

1. Introduction

Automatically guided outdoor vehicles are mostly used in transport system. We can easy modify these systems for monitoring or patrolling devices. Common Transport systems intended for indoor part transport are not suitable for outdoor monitoring because there are implemented only sensors for local navigation and carriage for plain surface.

The present state of autonomous transport of inside environment is based on mobile devices following the line on the floor, based on track following magnetic field of conductors inbuilt to floor, or navigation based on reference marks placed, for example, on the factory walls. These solutions are for inside environment fully sufficient, there is relatively low probability of destruction, damage, or pollution of reference marks and lines on the floor.

The situation would change for the described solution when we want to use it in outdoor environments. There is a big probability of dirtying of reference marks and following floor lines. The most suitable solution for outdoor environments is the inductive transport system. There is a problem when the trajectory is too long, there are many junctions. There is a place for alternative solution of navigation for interoperation transport.

This article introduces the solution based on GPS navigation intended for monitoring in outdoor environments. Automatic guard system can be used for night patrolling in production area. All system used in interoperation transport can be used in Guard system. Patrolling system need especially video motion recognition and audio noise level measuring.

Other similar solution of four wheeled mobile robot with car carriage based on GPS navigation are described in these articles [1,2,3].

2. Robot Prototype

The introduced robot construction is a dimension reduced prototype of four-wheeled vehicle. The size of device was modified for experiments in inside environments even though the vehicle is intended only for outdoor environment.

This mobile robot construction has some limitations. For example, the four-wheeled robot is not able to turn in one place, but this limitation has been solved by kinematics estimation. Main structure of the mobile robot was built from aluminum profiles mainly carriage of the robot, other parts as holders etc. are manufactured from acrylic glass. In Figure 1 is shown two prototypes of mobile robot solutions. Final design of device was created in collaboration with Faculty of art, Department of design.

Figure 1. Monitoring robot in outdoor environment

3. Kinematics and Position Estimation

We need some estimation of robot position especially after turning of the robot, because there is used Ackerman type of steering like is described in [4]. This estimation can be used for precising of GPS information. There has been designed the easy system robot position based on speed of its back wheels and the angle of its front wheels. There was abandoned up trust of wheels by turning, front wheels were replaced by one wheel in the axis of the robot. The mathematical model is parametric and it is possible to use it for any four-wheeled mobile carriage with other parameters. Position of the front wheel gives the actual position of the robot. Position of the robot is described by coordinates x, y and by angle of the front wheel α. The principal scheme of solution and software implementation is shown in Fig. 2. The same algorithm is implemented in 3D environment. There can be mapped satellite maps to environment ground.

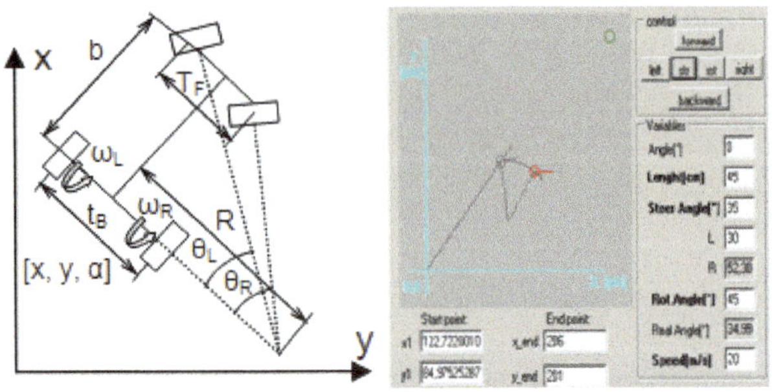

Figure 2. Kinematics scheme for estimation of position in 2D

For a four-wheeled carriage with different front and rear tracks, the mathematical relationship between the speeds of the drive wheels and the steer angles is described in [4] by equation (1), (2).

$$\theta_L = \tan^{-1}\left[\frac{2b(\omega_L - \omega_R)}{\omega_L(t_B + t_F) + \omega_R(t_B - t_F)}\right] \quad (1)$$

$$\theta_R = \tan^{-1}\left[\frac{2b(\omega_L - \omega_R)}{\omega_L(t_B - t_F) + \omega_R(t_B + t_F)}\right]. \quad (2)$$

For a four-wheeled carriage with equal front and rear tracks $t_F = t_B = t$ is described in [5] by equation (3), (4).

$$\theta_L = \tan^{-1}\left[\frac{b}{t}\left(1 - \frac{\omega_R}{\omega_L}\right)\right] \quad (3)$$

$$\theta_L = \tan^{-1}\left[\frac{b}{t}\left(\frac{\omega_L}{\omega_R} - 1\right)\right]. \quad (4)$$

For a vehicle with one steerable wheel at the front, we have the case where $t_F=0$ by equation (5).

$$\theta = \tan^{-1}\left[\frac{2b(\omega_L - \omega_R)}{t(\omega_L + \omega_R)}\right]. \quad (5)$$

4. Control System

The mobile robot control system is based on hierarchical (three level) principle: the low, upper and remote level of control. Low level system generates signals for hardware of the robot (motor, sensors). Upper level of control operates as communication interface for the remote level of control. Upper system is based on Linux OS and handles video, sound and Wi-Fi hardware. The remote level of control has three variants, the desktop application, PDA application or internet interface for web browser based on ASPX technology with AJAX. Every part of the mobile robot control system is described in next chapter. Main principle scheme of control is shown in Figure 3.

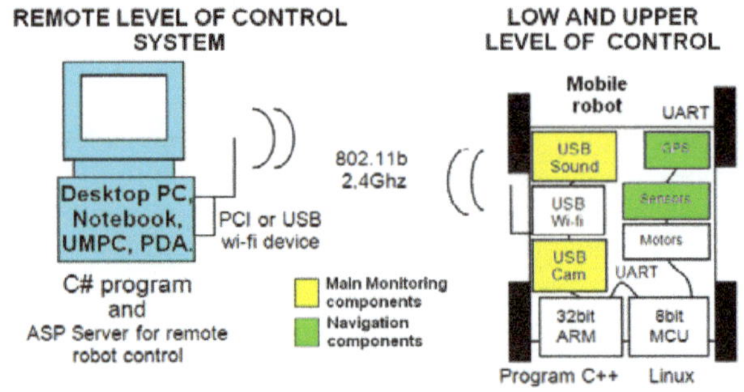

Figure 3. Principle scheme of mobile robot control based on GPS navigation

The low level of control system is designed for these operations: 3x Servo control, 2x DC control, GPS data receiving, monitoring battery status and charging battery. The low level of control system is designated for these operations: 3x Servo control, 1x DC control, GPS data receive, monitoring battery status, charge battery circuit. Battery charge function is designed for 3x Lithium Polymer or Lithium Ion battery pack (11,3V). The board is based on microcontroller ATMEGA128. This microcontroller generates signals for all servomotors and DC motor by internal PWM generator. DC motor drives robot forward and backward. One of servomotors changes direction of the front wheel to the right or to the left. Last two servomotors set position of the camera one for turning the camera left and right, the other for rolling the camera up and down. This board communicates with GPS device with MNEA protocol through the UART communication interface. The second UART communication canal works as the communication interface with the upper level of control by Robot API protocol. We are designed own PCBs for low, upper control system. Both levels of PCB board for control are shown in Figure 4. [6,7]. The control system is designed especially for mobile solution based on Ackerman steering and includes interfaces for all used sensor.

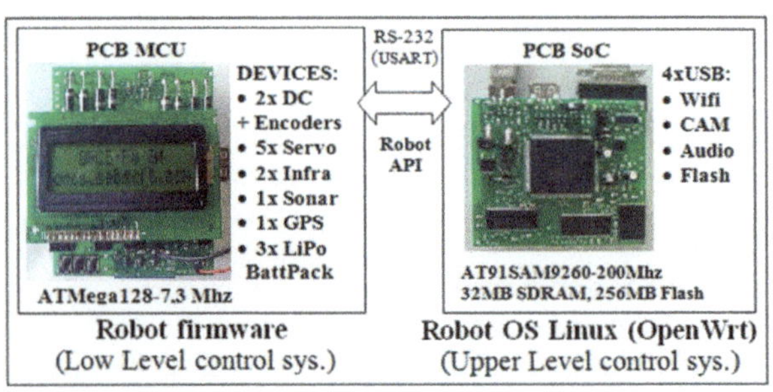

Figure 4. Principle scheme of control system of mobile robot

The low level of control is written in C++ programming language. There is possibility to change the program through the serial port or wirelessly through the implemented boot loader. It is possible to turn off power from camera, servos, GPS...etc. This feature can markedly increase mobile robot working time. The upper level control system is based on PCB board with ARM9 AT91sam9260 SoC. There is installed Linux OS Openwrt console version. Board handles USB webcam, USB audio stick and USB Wi-Fi thru Linux 2.6 device drivers [8]. There is used UART for communication with low level control system. For communication with remote control is used TCP protocol by Wi-Fi connection. Linux application ser2net provides transformation of UART data to TCP packets.

The remote control system provides Robot programming and monitoring interface. This level provides for developers communication protocol, the so called ROBOT API, published on web pages [17]. This feature makes it possible to create fast and effectively the new application based on the introduced hardware solution. We can control robot by Desktop or PDA Application. There is a simple Web interface for Robot monitoring too.

Desktop or PDA application offers connection to the Robot by TCP Client, TCP Server or Serial port. There is described basic functions: you can manual control robot movement by joystick and camera position in two axis by joystick sliders, remote on/off all device, watch robot cam, hear sound and send real-time voice, send simple command to GPS automatic navigation (based on local and global sensors), you can check GPS precision (how many satellites are available and if the GPS has signal).

The web pages for the remote robot control of the mobile robot are based on ASPX in combination with the AJAX technology. The AJAX technology provides maximal speed for web application, especially in page refresh.

The example of web interface for the mobile robot is shown in Figure 5. Only one User will get access to control the mobile robot. Monitoring position of the robot and watching the robot camera is user unlimited. All users are logged to MySQL database and trajectory which they passed. Trajectory is logged as GPS position in 1 second interval [9].

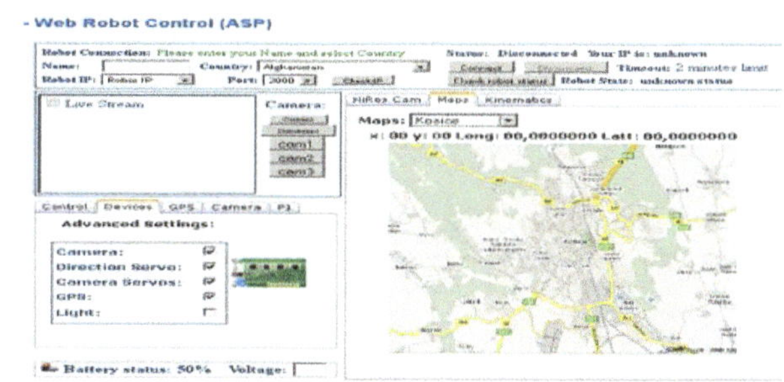

Figure 5. Web application for remote robot control

5. Localization and Navigation

There is described principle of combination global navigation with local sensor object detection.

Navigation of the mobile robot is realized through the acquiring of the position of the robot based on GPS sensor. This sensor sends data in the NMEA protocol. GPS sends several data strings with necessary information about actual position. The introduced solution processes string RMC, which is providing basic data (position, speed, time) [9].

Navigation of the mobile robot is realized by acquiring the position of longitude and latitude transformed to the Cartesian coordinating system and shown on maps of environments. Planned (selected) trajectory of the mobile robot in the Technical University of Kosice is shown in Figure 6. Global navigation working on comparison real robot angle acquired from GPS and required angle counted from checkpoint on planned trajectory.

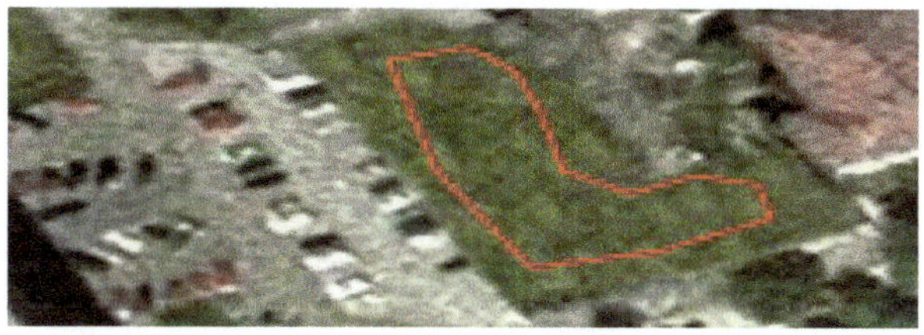

Figure 6. Example of monitoring trajectory in Google Maps

The complex navigation algorithm consists of combination of three separate parts (technologies, theories) combined to one subject:

1. GPS position data.
2. Raster satellite maps.
3. Path search algorithm.

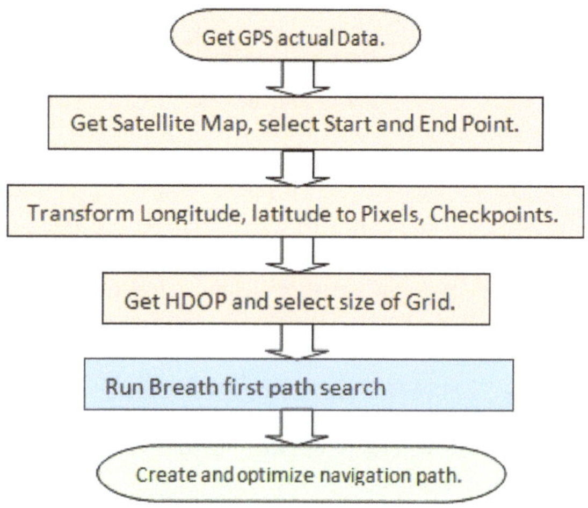

Figure 7. Path planning algorithm principle

There is briefly described navigation algorithm in basic steps. First step is download actual Satellite maps according acquired actual GPS data for start position on maps and selecting manually end position. We must next transfer GPS data to pixels and create vector from start to end point. Next step is measuring distance and create checkpoints for long distances to reduce computing time for breath first algorithm. Actual HDOP acquired from GPS provides input to select size of search algorithm grid. There is executed set of filters for edge detection algorithms to estimate position of obstacles in the map. Then there is started breath first algorithm to find ideal plan path. Last step of algorithm is creating optimized navigation path and segmentation. Simplified diagram of algorithms principle is depicted in Figure 7.

Very simplified actual precision we can get from 2DRMS and HDOP value, estimated by (6). RMS error is directly proportional to HDOP, thus the RMS_Error points would lay on a straight line.

$$\text{RMS\_Error} = \text{HDOP} * \text{2DRMS}. \qquad (6)$$

Figure 8 shows implementation of complex navigation algorithm to path planning.

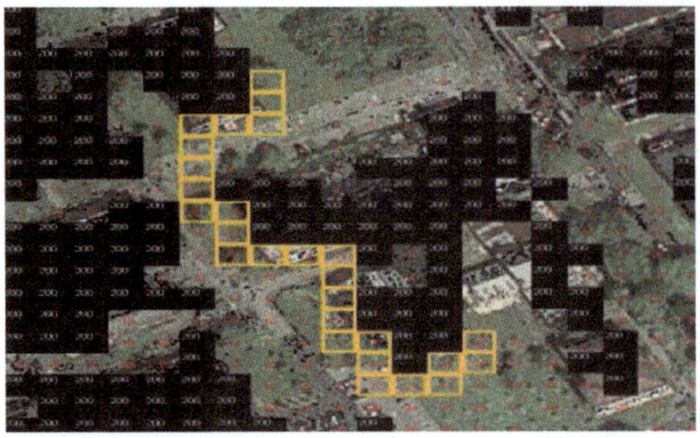

Figure 8. Search algorithm principle in Google maps

The robot can be programmed by teaching, it means that we control robot with a joystick according to the required trajectory, which we control during navigation visually through the camera system, we can check easy collision state, too. We can check the actual position on localization maps. Next kind of control is half automatic. We can enter GPS points which must mobile robot crossed. Last kind of control is entering of control command sequence for the robot without the assistance of GPS system. Last kind of control is not very suitable for outdoor environments and there is a problem in trajectory repeatability. There is a possibility to use it when we have problem with GPS system during very bad weather conditions, when the signal is weak or not available [10].

There is implemented infra sensor in front and back of mobile robot to detect near obstacles. For environment mapping there is implemented ultrasonic sensor with y axis rotation. There is possible to turn in 120° angle range by Servomotor. We can measure distance from 30mm to 3m. Robot must stop movements for environment mapping [11].

Example of mapping area scanned by mobile robot with ultrasonic sensor is shown on the Figure 9. The application calculates an ideal angle of movement between two obstacles.

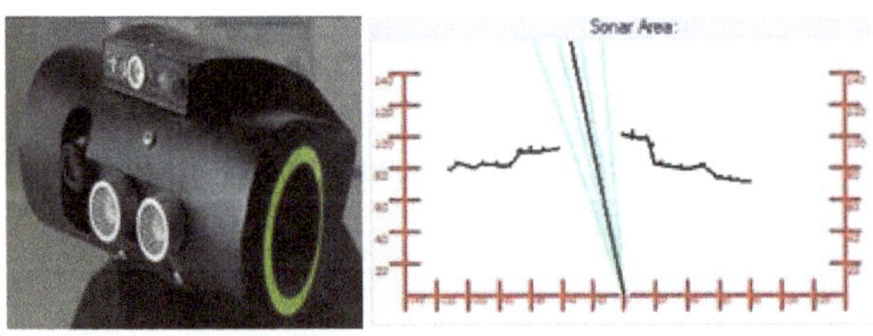

Figure 9. Ultrasonic sensor, simple environment mapping

Extened algorithm for navigation with obstacle avoiding is shown on the Figure 10.

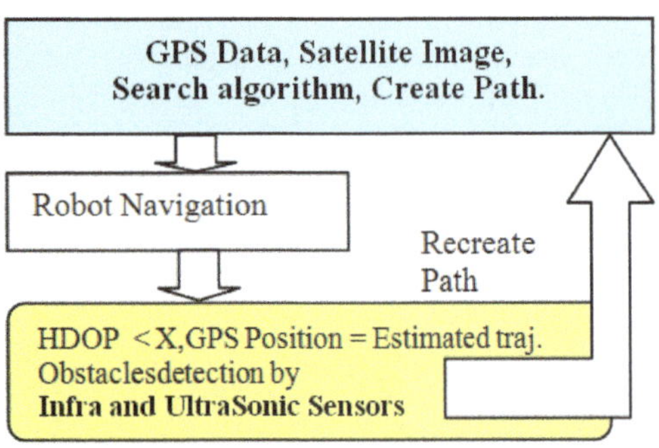

Figure 10. Extended navigation algorithm with sensors

6. Video Recognition System

The camera system consists of the USB camera device connected to Linux OS board. Camera position is controlled by servomotors. The used camera system generates video frames by motion jpeg technology in mjpeg-streamer application. We can get about 30 fps in low resolution 320x240 or 17fps by 640x480. We are acquiring video frames by desktop application C# Open source library. The video camera with resolution 320x240 is fully sufficient for monitoring and implementation of motion recognition. Motion recognition is realized by comparison of two pictures in sequence with variable level of sensitivity. Communication program for audio handling is Gstreamer application. We can get sound to desktop from microphone or send voice to speaker in real time. Alarm signalization is run mp3 sound files from robot by madplay, or generate synthesized voice thru flite program. We used mainly sound system for measuring db level of noise in high sensitive microphone as initial alarm. Mobile robot is monitoring noise during guard trajectories, when there is exceeded selected level of db, Robot stop and searching in vertical and horizontal plane (Servo1, Servo2) motion thru camera and creates screenshot and run alarm. Alarm system principle is showed Figure 11.

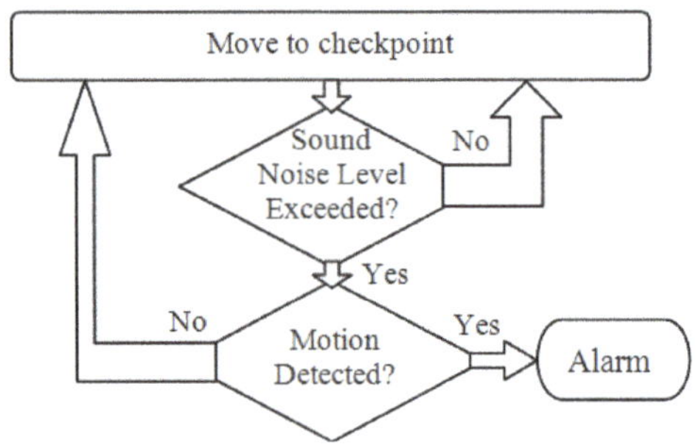

Figure 11. Audio and video system alarm principle

Video and sound hardware implementation is showed in Figure 12 [10].

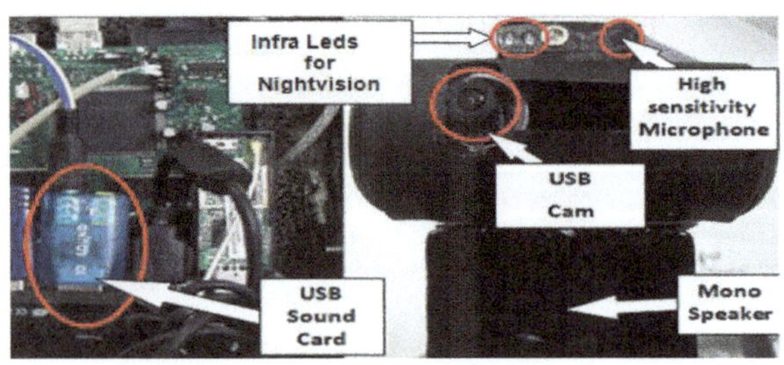

Figure 12. Audio and video system, motion detection

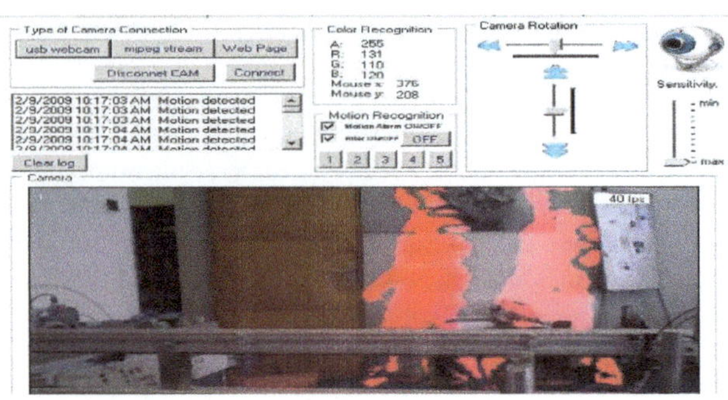

Figure 13. Video system for motion detection

There is possible to change sensitivity of object detection, which helps to filter small object from detection log. Motion detection application is shown on the Figure 13 [11].

7. Example of Implementation to Praxis

There is described example of using the mobile robot in real production conditions as a device for guard system between several halls. This solution is suitable for small companies. There is necessary only one person to setup device, run automatic monitoring or manual remote control. There is possibility to reprogram the trajectory of the mobile robot rapidly and easily without any modification in outdoor environment. The introduced mobile solutions have implemented infra led's for guard and monitor area of production complex in night condition. There is implemented motion recognition for violator detect. The example model situation of using the mobile robot based on GPS navigation is shown in Figure 14 [12].

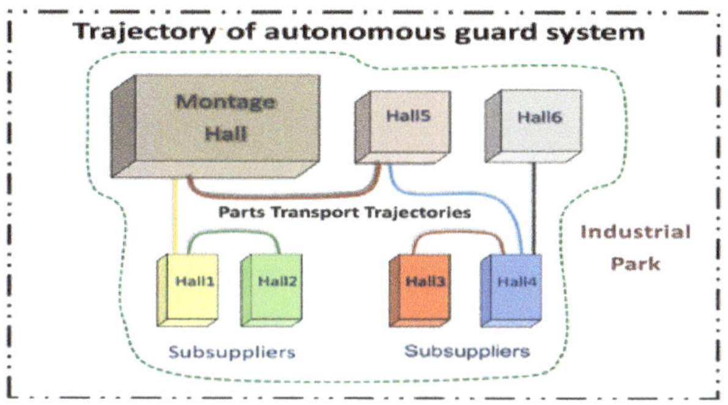

Figure 14. Model situation for monitoring industrial park.

8. Conclusion

The solution introduces mobile robot and its possibility of implementation to guard system based on GPS navigation for outdoor environment. This system does not need any reference marks in outdoor environment. Fast and flexible preprogramming possibility is the next advantage. New feature is Web monitoring of robot state and logging of robot position from GPS. There can be implemented many extension only by changes in software solution. There are two automatic alarm system based on video motion and sound noise level. Next works will be aimed to debugging all system, mainly test reliability of GPS navigation system and implementation of advanced technics of computational intelligence for robot control [13,14].

Acknowledgement

The research was supported by the Project VEGA 1/0911/14 Implementation of wireless technologies into the design of new products and services to protect human health.

References

[1] S. Minamimoto, S. Fujii, H. Yamaguchi, T. Higashino, *Map estimation using GPS-equipped mobile wireless nodes*, Pervasive and Mobile Computing, Volume 6, Issue 6, 2010, pp. 623-641.

[2] R. Chatterjee, F. Matsuno, *Use of single side reflex for autonomous navigation of mobile robots in unknown environments*, Robotics and Autonomous Systems, Volume 35, Issue 2, 2001, pp. 77-96.

[3] N. Ghita, M. Kloetzer, *Trajectory planning for a car-like robot by environment abstraction*, Robotics and Autonomous Systems, Volume 60, Issue 4, 2012, pp. 609-619.

[4] IJ Spark, BC.,Besselink *Zero turn radius vehicle incorporating computer-controlled `intelligent castors'*. In: ASME Computers in Engineering Conference, Minneapolis, USA, 1994. pp. 825-27.

[5] B.C. Besselink, *Computer controlled steering system for vehicles having two independently driven wheels,Comput.* Electron. Agricult., 39 (2003), pp. 209-226.

[6] K.Židek, V.Maxim, *Remote control of mobile robots based on web technologies* , 2008. In: ActaMechanicaSlovaca. Roč. 12, č. 1-A/2008, pp. 161-166.

[7] K.Židek, T.Saloky, V.Maxim, *Implementation of mobile robots to outdoor interoperation transport*, 2008. In: Acta Mechanica Slovaca. Roč. 12, č. 2-A (2008), pp. 761-768.

[8] T. Saloky, J. Piteľ, *Intelligent robots and computer vision*. In.: International Carpathian Control Conference ICCC'2004 – Volume II.,Zakopane, May 25-28, 2004. Krakow: AGH - UST, 2004. pp. 243-246.

[9] HDOP and GPS horizontal position errors, online, [http://users.erols.com/dlwilson/gpshdop.htm].

[10] K.Židek, J.Šeminský, *Wireless audio signal processing for mobile robot system*, 2010. In: Process Control 2010: 9th International, KoutynadDesnou, Czech Republic. - Brno : pp. 187/1-C187/5.

[11] K.Židek, V.Maxim, T.Saloky, *Detekcia pohybu kamerovým systémom*, 2009. In: Strojárstvo extra. Č. 5 (2009), s. 59/1-59/2.

[12] Jurišica, L., Vitko, A., Duchoň, F., Kaštan, D., *Statistical approach to GPS positioning of mobile robot,* Control Engineering and Applied Informatics, volume 12, Issue 2, 2010, pp. 44-51.

[13] A. Hošovský, *"Parametrically Invariant Genetic Fuzzy Adaptive System for a Servomotor"*, Systemnyjanaliz, upravlenije i obrabotkainformaciji, Divnomorskoje, 27. - 29.9.2010, Rostov-na-Donu, DGTU, 2010, pp. 169-173.

[14] J. Boržíková, J. Piteľ, *"Nonlinearity of static characteristics of the antagonistic system"*, Informatizacijatechničeskichsistem i processov, Tom 5, Saratov, 27. 30.5.2008, Saratov: SGTU, 2008, pp. 196-197.

Model of Mobile Robot with Divided Chassis

Ľubica Miková*, Michal Kelemen, Alexander Gmiterko

Technical University of Košice, Faculty of Mechanical Engineering, Department of Mechatronics, Košice, Slovakia
*Corresponding author: lubica.mikova@tuke.sk

Abstract This article is about designing and constructing a four-wheel chassis, which will possess better negotiability of diverse terrain. Analyzed is a used concept of chassis motion control of mechatronic systems on the principle of differential wheel control for the task of active tracking of planned chassis path. Created was a simulation model of chassis mobile system in terms of kinematics, which will be usable in concept of chassis locomotion control on diverse terrain. A model is also required for examining behavior of chassis on diverse terrain, for examining the influence of dimensions on chassis behavior during crossing over roughness of terrain.

Keywords: *mobile robot, simulation*

1. Introduction

One of the main features of mechatronic approach to designing advanced products is complex understanding of the technical object and its computer modeling. The kinematic description of mechanical system issue from an abstraction – simplification of reality.

If we model mechanical and electrical elements as systems with centralized parameters, then it is possible to say, that drawing up the model of mechanical part is clearly the most difficult.

2. Concept of Mobile Robot

Mobile robot poses a solution of four-wheel chassis with autonomous wheel drive 4x4 designed for diverse terrain. It has two degrees of freedom, i.e. linear and rotational motion.

It is possible to use it for various tasks such as inspection of automobile chassis and explosives disposal, pipeline, cavity and difficult to reach location inspection, manipulation with dangerous materials, telecommunication and cable network installation, tactical tasks in fight against terrorists and so on. The device should also be used in educating students from mechatronics and engineering fields of study.

Basis of the device is chassis frame composed of two pieces connected by a passive joint. [14]. Robot chassis has a four-wheel drive that does not lose traction even on diverse terrain surface thanks to the passive joint It enables both parts of frame to randomly tilt depending on terrain difficulty. Hence it is achieved that every wheel in any moment keeps contact with terrain surface [2].

Technical parameters of functional chassis model:

- wheel base: 190 mm
- gauge: 370 mm
- inner diameter: 30 mm
- total mass: 3 kg
- maximal loading capacity: 10 kg
- average motion velocity: 0,3 m/s

Figure 1. Mobile robot with divided chassis

3. Control Concept

The designed chassis uses for its motion differential method of wheel control. Chassis motion should be stable and fluent to avoid slipping between wheels and terrain and to avoid mechanical shock resulting from rapid changes in chassis motion. However during chassis motion unavoidable trajectory deviations between current position and requested trajectory do occur because of path tracking control imperfection using wheels velocity and fault variables from environment (terrain roughness, friction forces changes between wheels and terrain and so forth) [7,8,9].

Trajectory deviations should be corrected online using requested linear and rotational chassis motion velocities by path tracking control [5,11].

In work [7] layout of kinematic control is designed, comprising of three levels of links (dynamic, kinematic and planning). This approach is similar to those, which are applied in robotic manipulators. The designed construction was applied in mobile three-wheel chassis of industrial forklift manipulator.

Wheel mobile robots can be controlled from the aspect of kinematics or dynamics.- Kinematics based control consists of dividing the device into two levels namely kinematic and dynamic loop.

- Dynamics based control considers just one loop of global dynamic control. This has a few disadvantages. Very complex analyses and calculations are needed. This control is very sensitive to parameter uncertainties of model.

Kinematic control is simpler and its global stability is guaranteed if the low-level control link (dynamic loop) is much faster than the middle level of control (kinematic loop). Majority of works solving this kind of problem are based on this assumption.

Input into mobile wheel robot control is tracking reference ref p generated by the high level – planning block. If this reference is dynamically generated (i.e. specified in real time) it is acquired from outer loop, so called planning loop [7].

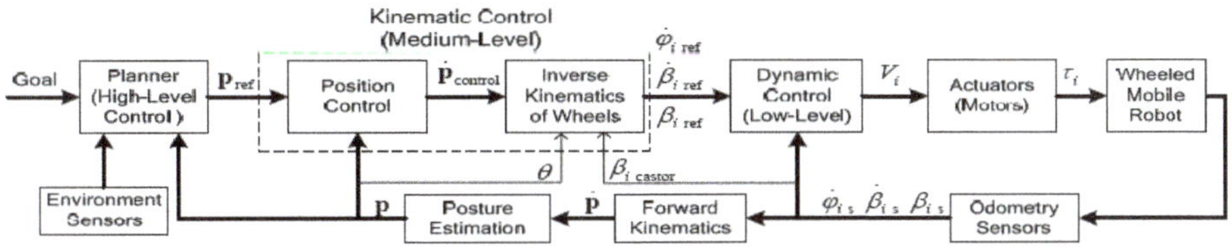

Figure 2. Control scheme of the wheeled mobile robot [7]

4. Kinematic Model

Creation of kinematic model was based on mobile robot, which was designed and constructed in department of applied mechanics, and also on experience

Chassis with four wheels is controlled differentially. Each one of the wheels is independently driven by separate speed servomechanism [10].

Mobile robot always moves in circular trajectory with diameter R (on trajectory projection of which on plane XY is circle). Direct motion occurs when trajectory radius approaches infinity. This radius changes in time depending on desired motion direction [1].

Axes of all wheels have to intersect in one point and that point is an immediate center of rotation [2].

Figure 3. Location of local coordinate systems in linkage

In this case it is difficult to draw up a kinematic model of chassis using common methods, because in the center of gravity there is a joint that enables chassis to better overcome terrain roughness. For description of resulting kinematic variables, such as position and velocity of chassis center of gravity, or positions and velocities of individual wheels, we used transformation matrices [1].

In general cases it is very difficult to directly find matrix of direction cosines and beginning accompaniments. It is however possible to essentially simplify by motion decomposition to a sequence of simple motions composed of so called basic motions [6].

Decompositions such as these are natural component of practically every task layout about body motion. So called matrix kinematics method is effective and general method for solving body motion and basically whole kinematics [4].

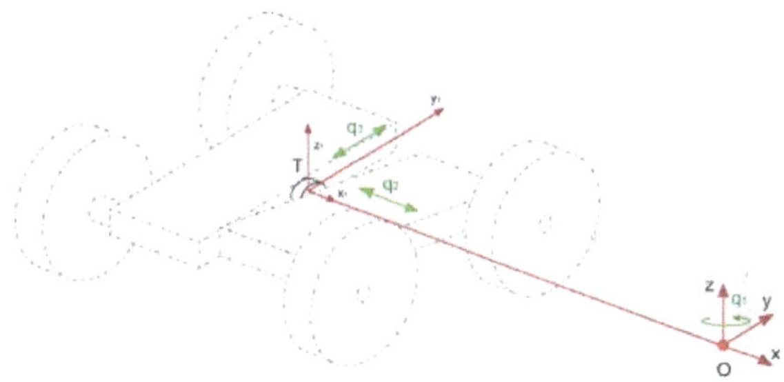

Figure 4. Coordinates q1, q2, q3, q4 implementation

Kinematic model of chassis takes into account real dimensions of GTR2010 robot.

Distance between wheel 1 and geometric center of gravity of chassis in direction of axis x is c = -175 mm and in direction of axis y is a= 95mm.

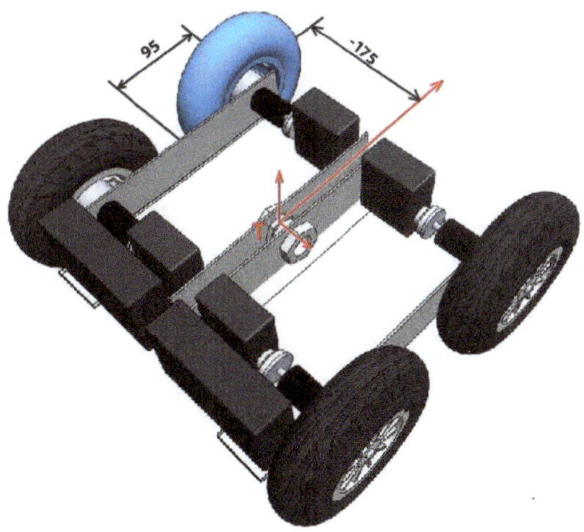

Figure 5. Position of wheel 1 from geometric center of gravity of chassis

Based on constructed transformation matrices it is possible to construct simulation models for defining center of gravity geometric position of chassis and wheels in Matlab/Simulink environment.

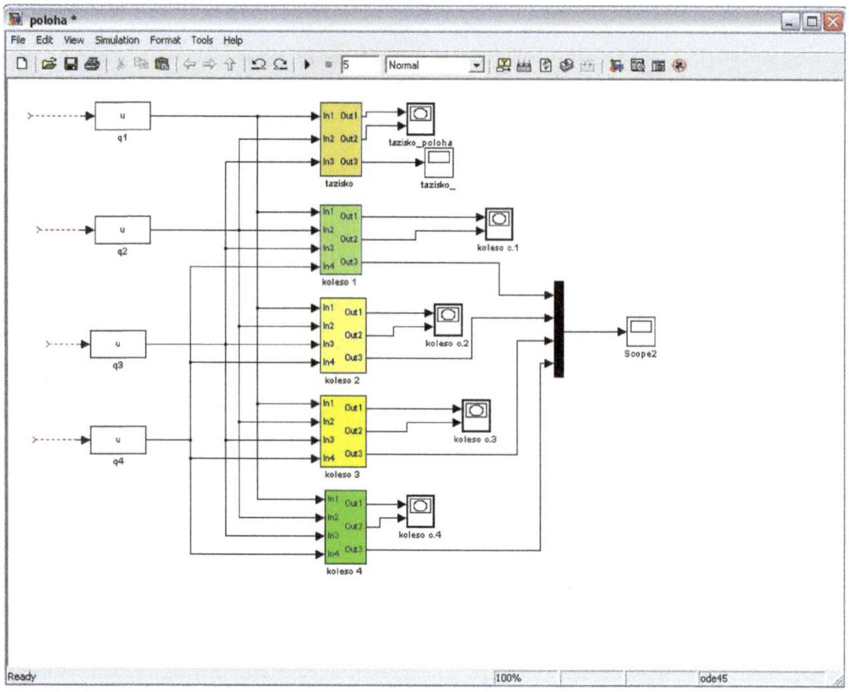

Figure 6. Position Simulation model of position of center of gravity and wheels

Simulation model of position of geometric center of gravity of chassis and individual wheels comprises of several subsystems that have common input. Every subsystem represents position of wheel, or position of center of gravity to beginning O (Instantaneous Center of Rotation).

Layout representing velocity of individual chassis points consists of several subsystems, which represent velocities of wheels 1 - 4 and chassis center of gravity.

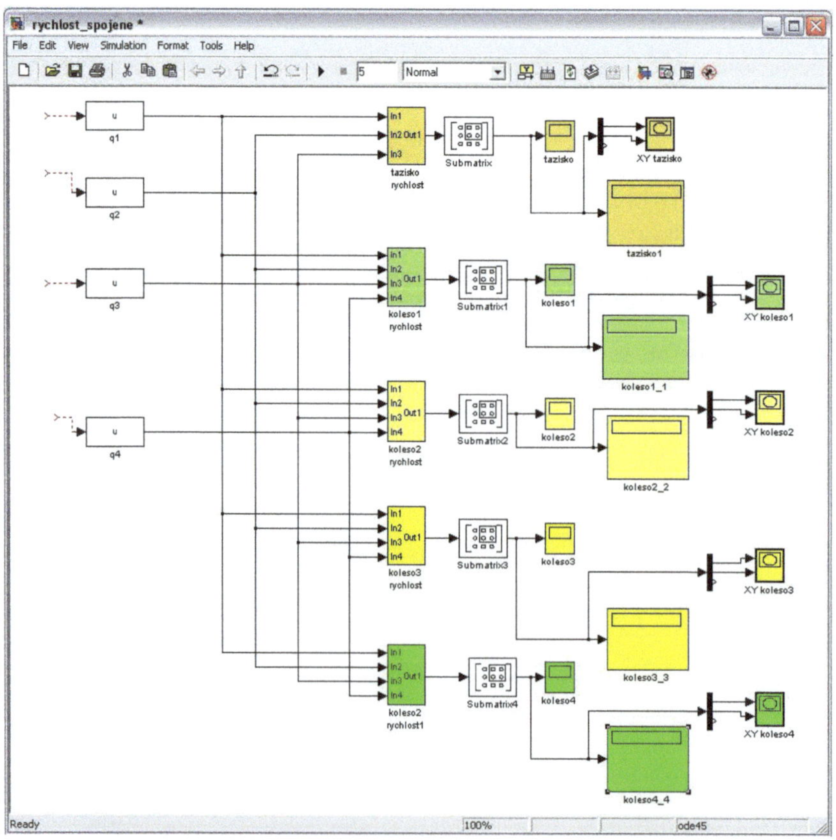

Figure 7. Simulation model of velocity of chassis, center of gravity and wheels

It is possible to use this simulation model to calculate estimated trajectory of center of gravity and wheels of chassis. It is also possible to examine behavior of model in any given combination of chassis dimensions. In connection to this there is also an option of chassis geometry optimization. Furthermore created model provides room for examining influence of terrain roughness on behavior of chassis of mechatronic system.

A significant asset is the option of testing model of such chassis composition of mechatronic system, what enables to find optimal chassis geometry in terms of various aspects of assessment before constructional vehicle solution itself.

5. Conclusion

The concept of control structure using kinematic and dynamic model at the same time provides the benefit of economical utilizing of energy sources. In terms of practical realization it is being considered to implement this structure into functional model, where this structure would in real time ensure chassis locomotion control of mobile mechatronic system. With this application the mobile mechatronic system gains a certain amount of intelligence.

Hence it is possible to use the model for calculating estimated or requested trajectory of chassis center of gravity and wheels. At the same time it is possible to examine model behavior in any given combination of chassis dimensions and relating to this there is an option of chassis geometry optimization in terms of obstacles negotiability on diverse terrain. Furthermore the created model provides room for examination of the influence of terrain roughness on behavior of mechatronic system chassis.

Acknowledgement

This contribution is a result of the project Slovak Grant Agency – project VEGA 1/0937/12 "Development of non-traditional experimental methods for mechanical an mechatronical systems", and project KEGA 048TUKE-4/2014 "Increasing of knowledge base of students in area of application of embedded systems in mechatronic systems".

References

[1] J. SMRČEK, L. KÁRNIK, "Robotics, Industrial robots, Desing – Construction – Solutions," Košice, 2008.

[2] R. GREPL, "Mechatronics-selected problems," Brno 2008.

[3] R. GREPL, "Modeling mechatronics systems in Matlab SimMechanics," Praha, 2007.

[4] R. GREPL, "Kinematics a dynamics mechatronics systems," Brno 2007.

[5] M. NICULESCU, "Experiments in Mobile Robot Control," Mechatronics 2008.

[6] V. BRÁT, "Matrix methods of analysis and synthesis of space committed mechanical systems," Praha, 1981.

[7] L. GRACIA, J. TORNERO, "Kinematic control of wheeled mobile robots. Latin American Applied Research,". Vol. 38, p. 7-16 (2008).

[8] CH. CHIH-FU, H. CHIN-I, F. LI-CHEN, "Nonlinear Control of a Wheeled Mobile Robot with Nonholonomic Constraints," IEEE International Conference on Systems, Man and Cybernetic, p. 5404-5410.

[9] E. IVANJKO,E. KOMSIC, I. PETROVIC, "Simple off-line odometry calibration of differential drive mobile robots," in Proceedings of 16th International Workshop on Robotics in Alpe-Adria-Danube Region, Ljubljana, Slovenia, june 7-9, 2007, p. 164-169.

[10] L. JURIŠICA, A. VITKO, "Mechanics and robotics," AT&P Journal 11/1998, p. 44-45.

[11] E. LUCET, Ch. GRAND, D. SALLÉ, P. BIDAUD, "Dynamic control of the 6WD skid-steering mobile robot RobuROC6 using sliding mode technique," The 2009 IEEE/RSJ International Conference on Intelligent Robots and Systems, October 11-15, 2009 St. Louis, USA.

[12] J. ŠKARUPA, V. MOSTÝN, "Methods and tools of design industrial and service robots," Vienala Košice, 2002.

[13] M. VALÁŠEK, "Kinematics of robotic systems," 2011.

[14] Ľ. MIKOVÁ, M. KELEMEN, T. KELEMENOVÁ, "Four wheels inspection robot with differential wheels control," Acta Mechanica Slovaca, Roc. 12, c.3-B (2008), p. 548-558.

[15] J. HAIYANG, Z. PENG, H. YING, Z. JIANWEI, Z. ZHIZENG, Design and Kinematic Analysis of A Pedicle Screws Surgical Robot, Proceedings of the 2010 IEEE International Conference on Robotics and Biomimetics, December 14-18, 2010, Tianjin, China.

16

Solving the Partial Differential Problems Using Differentiation Term by Term Theorem

Chii-Huei Yu*

Department of Management and Information, Nan Jeon University of Science and Technology, Tainan City, Taiwan
*Corresponding author: chiihuei@nju.edu.tw

Abstract This paper took advantage of the mathematical software Maple as the auxiliary tool to study the partial differential problems of four types of two-variables functions. We can obtain the infinite series forms of any order partial derivatives of these two-variables functions by using differentiation term by term theorem, and hence greatly reduce the difficulty of calculating their higher order partial derivative values. On the other hand, we proposed some examples to do calculation practically. The research methods adopted in this study involved finding solutions through manual calculations and verifying our answers by using Maple.

Keywords: *partial derivatives, infinite series forms, differentiation term by term theorem, Maple*

1. Introduction

As information technology advances, whether computers can become comparable with human brains to perform abstract tasks, such as abstract art similar to the paintings of Picasso and musical compositions similar to those of Beethoven, is a natural question. Currently, this appears unattainable. In addition, whether computers can solve abstract and difficult mathematical problems and develop abstract mathematical theories such as those of mathematicians also appears unfeasible. Nevertheless, in seeking for alternatives, we can study what assistance mathematical software can provide. This study introduces how to conduct mathematical research using the mathematical software Maple. The main reasons of using Maple in this study are its simple instructions and ease of use, which enable beginners to learn the operating techniques in a short period. By employing the powerful computing capabilities of Maple, difficult problems can be easily solved. Even when Maple cannot determine the solution, problem-solving hints can be identified and inferred from the approximate values calculated and solutions to similar problems, as determined by Maple. For this reason, Maple can provide insights into scientific research.

In calculus and engineering mathematics curricula, the study of Laplace equation, wave equation, and other important physical equations are involved the partial differentiation. So the evaluation and numerical calculation of the partial derivatives of multivariable functions are important. On the other hand, calculating the q-th order partial derivative value of a multivariable function at some point, in general, needs to go through two procedures: firstly determining the q-th order partial derivative of this function, and then taking the point into the q-th order partial derivative. These two procedures will make us face with increasingly complex calculations when calculating higher order partial derivative values (i.e. q is large), and hence to obtain the answers by manual calculations is not easy. In this paper, we studied the partial differential problem of the following four types of two-variables functions

$$f(x,y)=\left(\sqrt{\alpha^2x^2+\beta^2y^2}\right)^p\times\frac{\cos\left(p\cot^{-1}\frac{\alpha x}{\beta y}\right)\sinh 2\alpha x-\sin\left(p\cot^{-1}\frac{\alpha x}{\beta y}\right)\sin 2\beta y}{\cosh 2\alpha x+\cos 2\beta y} \tag{1}$$

$$g(x,y)=\left(\sqrt{\alpha^2x^2+\beta^2y^2}\right)^p\times\frac{\cos\left(p\cot^{-1}\frac{\alpha x}{\beta y}\right)\sin 2\beta y+\sin\left(p\cot^{-1}\frac{\alpha x}{\beta y}\right)\sinh 2\alpha x}{\cosh 2\alpha x+\cos 2\beta y} \tag{2}$$

$$u(x,y)=\left(\sqrt{\alpha^2x^2+\beta^2y^2}\right)^p\times\frac{\cos\left(p\cot^{-1}\frac{\alpha x}{\beta y}\right)\sinh 2\alpha x+\sin\left(p\cot^{-1}\frac{\alpha x}{\beta y}\right)\sin 2\beta y}{\cosh 2\alpha x-\cos 2\beta y} \tag{3}$$

$$w(x,y)=\left(\sqrt{\alpha^2x^2+\beta^2y^2}\right)^p\times\frac{-\cos\left(p\cot^{-1}\frac{\alpha x}{\beta y}\right)\sin 2\beta y+\sin\left(p\cot^{-1}\frac{\alpha x}{\beta y}\right)\sinh 2\alpha x}{\cosh 2\alpha x-\cos 2\beta y} \tag{4}$$

where α, β are real numbers, $\beta \neq 0$, and p is an integer. We can obtain the infinite series forms of any order partial derivatives of these four types of two-variables functions using differentiation term by term theorem; these are the major results of this study (i.e., Theorems 1-4), and hence greatly reduce the difficulty of calculating their higher order partial derivative values. The study of related partial differential problems can refer to [1-13]. In addition, we provided some examples to do calculation practically. The research methods adopted in this study involved finding solutions through manual calculations and verifying these solutions by using Maple. This type of research method not only allows the discovery of calculation errors, but also helps modify the original directions of thinking from manual and Maple calculations. Therefore, Maple provides insights and guidance regarding problem-solving methods.

2. Main Results

Firstly, we introduce some notations, formulas and some Maple's commands used in this paper.

2.1. Notations

2.1.1. Let $z = a + ib$ be a complex number, where $i = \sqrt{-1}$, a, $b$are real numbers. We denote a the real part of z by Re(z), and b the imaginary part of z by Im(z).

2.1.2. Suppose m, n are non-negative integers. For the two-variables function $f\ (x,\ y)$, its n-times partial derivative with respect to x, and m-times partial derivative with respect to y, forms a $m+n$-th order partial derivative, and denoted by $\dfrac{\partial^{m+n} f}{\partial y^m \partial x^n}(x,y)$.

2.1.3. Suppose r is any real number, m is any positive integer. Define $(r)_m = r(r-1)\cdots(r-m+1)$, and $(r)_0 = 1$.

2.1.4. evalf ( ); the Maple's command of calculating the approximation.

2.1.5. sum ( ); the command of evaluating the summation.

2.1.6. D [1$3, 2$4] (f) (1, 2); the command of finding the 7-th order partial derivative of $f\ (x,\ y)$ at (1, 2), $\dfrac{\partial^7 f}{\partial y^4 \partial x^3}(1,2)$.

2.1.7. Bernoulli (n); the command of evaluating the n-th bernoulli number.

2.1.8. product (8-j, j = 0..6); the command of determining the product $8 \times 7 \times \cdots \times 2$.

2.2. Formulas

2.2.1. Euler's Formula

$e^{i\theta} = \cos\theta + i\sin\theta$, where θ is any real number.

2.2.2. De Moivre's Formula

$(\cos\theta + i\sin\theta)^p = \cos p\theta + i\sin p\theta$, where p is any integer, and θ is any real number.

2.2.3. ([14])

$\sinh(a+ib) = \sinh a\cos b + i\cosh a\sin b$, where a, b are real numbers.

2.2.4. ([14])

$\cosh(a+ib) = \cosh a\cos b + i\sinh a\sin b$, where a, b are real numbers.

2.2.5. Taylor Series Expression of Complex Hyperbolic Tangent Function ([15])

$\tanh z = \sum_{k=1}^{\infty} \dfrac{2^{2k}(2^{2k}-1)B_{2k}}{(2k)!} z^{2k-1}$, where z is a complex number, $|z| < \dfrac{\pi}{2}$, and B_k are the k-th Bernoulli number.

2.2.6. Taylor Series Expression of Complex Hyperbolic Cotangent Function ([15])

$\coth z = \dfrac{1}{z} + \sum_{k=1}^{\infty} \dfrac{2^{2k} B_{2k}}{(2k)!} z^{2k-1}$, where z is a complex number, $0 < |z| < \pi$.

In the following, we introduce an important theorem used in this study.

2.3. Differentiation Term by Term Theorem ([16])

For all non-negative integers k, if the functions g_k: $(a, b) \to R$ satisfy the following three conditions : (i) there exists a point $x_0 \in (a,b)$ such that $\sum_{k=0}^{\infty} g_k(x_0)$ is convergent, (ii) all functions $g_k(x)$ are differentiable on open interval $(a,\ b)$, (iii) $\sum_{k=0}^{\infty} \dfrac{d}{dx} g_k(x)$ is uniformly convergent on $(a,\ b)$. Then $\sum_{k=0}^{\infty} g_k(x)$ is uniformly convergent and differentiable on $(a,\ b)$. Moreover, its derivative $\dfrac{d}{dx}\sum_{k=0}^{\infty} g_k(x) = \sum_{k=0}^{\infty} \dfrac{d}{dx} g_k(x)$.

Before deriving the major results in this study, we need three lemmas.

2.4. Lemma 1

Suppose a, b are real numbers, $b > 0$, and p is an integer. Then

$$(a+ib)^p$$

$$= \left(\sqrt{a^2+b^2}\right)^p \left[\cos\left(p\cot^{-1}\frac{a}{b}\right) + i\sin\left(p\cot^{-1}\frac{a}{b}\right)\right] \tag{5}$$

$$= \left(\sqrt{a^2+b^2}\right)^p \exp\left(ip\cot^{-1}\frac{a}{b}\right) \tag{6}$$

Proof $(a+ib)^p$

$$=\left[\sqrt{a^2+b^2}\left(\frac{a}{\sqrt{a^2+b^2}}+i\frac{b}{\sqrt{a^2+b^2}}\right)\right]^p$$

$$=\left(\sqrt{a^2+b^2}\right)^p(\cos\theta+i\sin\theta)^p$$

(Where $\theta=\cot^{-1}\frac{a}{b}$)

$$=\left(\sqrt{a^2+b^2}\right)^p\left[\cos\left(p\cot^{-1}\frac{a}{b}\right)+i\sin\left(p\cot^{-1}\frac{a}{b}\right)\right]$$

(By DeMoivre's formula)

$$=\left(\sqrt{a^2+b^2}\right)^p\exp\left(ip\cot^{-1}\frac{a}{b}\right)$$

(By Euler's formula)

2.5. Lemma 2

Suppose a, b are real numbers with $\cosh 2a+\cos 2b\neq 0$. Then

$$\tanh(a+ib)=\frac{\sinh 2a+i\sin 2b}{\cosh 2a+\cos 2b} \tag{7}$$

Proof $\tanh(a+ib)$

$$=\frac{\sinh(a+ib)}{\cosh(a+ib)}$$

$$=\frac{\sinh a\cos b+i\cosh a\sin b}{\cosh a\cos b+i\sinh a\sin b}$$

(By Formulas 2.2.3 and 2.2.4)

$$=\frac{(\sinh a\cos b+i\cosh a\sin b)(\cosh a\cos b-i\sinh a\sin b)}{\cosh^2 a\cos^2 b+\sinh^2 a\sin^2 b}$$

$$=\frac{\sinh a\cosh b+i\sin b\cos b}{\sinh^2 a+\cos^2 b}$$

$$=\frac{\sinh 2a+i\sin 2b}{\cosh 2a+\cos 2b}$$

2.6. Lemma 3

Suppose a, b are real numbers with $\cosh 2a-\cos 2b\neq 0$. Then

$$\coth(a+ib)=\frac{\sinh 2a-i\sin 2b}{\cosh 2a-\cos 2b} \tag{8}$$

Proof $\coth(a+ib)$

$$=\frac{\cosh(a+ib)}{\sinh(a+ib)}$$

$$=\frac{\cosh a\cos b+i\sinh a\sin b}{\sinh a\cos b+i\cosh a\sin b}$$

$$=\frac{(\cosh a\cos b+i\sinh a\sin b)(\sinh a\cos b-i\cosh a\sin b)}{\sinh^2 a\cos^2 b+\cosh^2 a\sin^2 b}$$

$$=\frac{\sinh a\cosh a-i\sin b\cos b}{\cosh^2 a-\cos^2 b}$$

$$=\frac{\sinh 2a-i\sin 2b}{\cosh 2a-\cos 2b}$$

Next, we determine the infinite series forms of any order partial derivatives of the two-variables function (1).

2.7. Theorem 1

Suppose α, β are real numbers, $\beta\neq 0$, p is an integer, and m, n are non-negative integers. If the domain of the two-variables function

$$f(x,y)=\left(\sqrt{\alpha^2x^2+\beta^2y^2}\right)^p\times\frac{\cos\left(p\cot^{-1}\frac{\alpha x}{\beta y}\right)\sinh 2\alpha x-\sin\left(p\cot^{-1}\frac{\alpha x}{\beta y}\right)\sin 2\beta y}{\cosh 2\alpha x+\cos 2\beta y}$$

is $\left\{(x,y)\in R^2\,\middle|\,\beta y>0,\alpha^2x^2+\beta^2y^2<\frac{\pi^2}{4}\right\}$. Then the $m+n$-th order partial derivative of $f(x, y)$,

$$\frac{\partial^{m+n}f}{\partial y^m\partial x^n}(x,y)$$
$$=\alpha^n\beta^m\sum_{k=1}^{\infty}\frac{2^{2k}(2^{2k}-1)B_{2k}(2k+p-1)_{m+n}}{(2k)!}\cdot\left(\sqrt{\alpha^2x^2+\beta^2y^2}\right)^{2k+p-1-m-n}\times\cos\left[(2k+p-1-m-n)\cot^{-1}\frac{\alpha x}{\beta y}+\frac{m\pi}{2}\right] \tag{9}$$

Proof Let $z=\alpha x+i\beta y$, and $\beta y>0$, $\alpha^2x^2+\beta^2y^2<\frac{\pi^2}{4}$, then

$z^p\tanh z$

$$=(\alpha x+i\beta y)^p\tanh(\alpha x+i\beta y)$$
$$=\left(\sqrt{\alpha^2x^2+\beta^2y^2}\right)^p\left[\cos\left(p\cot^{-1}\frac{\alpha x}{\beta y}\right)+i\sin\left(p\cot^{-1}\frac{\alpha x}{\beta y}\right)\right]\times\frac{\sinh 2\alpha x+i\sin 2\beta y}{\cosh 2\alpha x+\cos 2\beta y} \tag{10}$$

(By (5) and (7))

Therefore,

$f(x,y)$

$$=\mathrm{Re}(z^p\tanh z)$$

$$=\mathrm{Re}\left[(\alpha x+i\beta y)^p\sum_{k=1}^{\infty}\frac{2^{2k}(2^{2k}-1)B_{2k}}{(2k)!}(\alpha x+i\beta y)^{2k-1}\right]$$

(By Formula 2.2.5)

$$=\mathrm{Re}\left[\sum_{k=1}^{\infty}\frac{2^{2k}(2^{2k}-1)B_{2k}}{(2k)!}(\alpha x+i\beta y)^{2k+p-1}\right] \tag{11}$$

Using differentiation term by term theorem, differentiating n-times with respect to x, and m-times with respect to y on both sides of (11), we obtain

$$\frac{\partial^{m+n} f}{\partial y^m \partial x^n}(x,y)$$

$$= \alpha^n \beta^m \operatorname{Re}\left[i^m \sum_{k=1}^{\infty} \frac{2^{2k}(2^{2k}-1)B_{2k}(2k+p-1)_{m+n}}{(2k)!} \cdot (\alpha x + i\beta y)^{2k+p-1-m-n} \right]$$

$$= \alpha^n \beta^m \sum_{k=1}^{\infty} \frac{2^{2k}(2^{2k}-1)B_{2k}(2k+p-1)_{m+n}}{(2k)!} \cdot \operatorname{Re}[i^m (\alpha x + i\beta y)^{2k+p-1-m-n}]$$

$$= \alpha^n \beta^m \sum_{k=1}^{\infty} \frac{2^{2k}(2^{2k}-1)B_{2k}(2k+p-1)_{m+n}}{(2k)!} \cdot \left(\sqrt{\alpha^2 x^2 + \beta^2 y^2}\right)^{2k+p-1-m-n} \times \cos\left[(2k+p-1-m-n)\cot^{-1}\frac{\alpha x}{\beta y} + \frac{m\pi}{2}\right]$$

(By (6))

Next, we find the infinite series forms of any order partial derivatives of the two-variables function (2).

2.8. Theorem 2

If the assumptions are the same as Theorem 1. Suppose the domain of the two-variables function

$$g(x,y) = \left(\sqrt{\alpha^2 x^2 + \beta^2 y^2}\right)^p \times \frac{\cos\left(p\cot^{-1}\frac{\alpha x}{\beta y}\right)\sin 2\beta y + \sin\left(p\cot^{-1}\frac{\alpha x}{\beta y}\right)\sinh 2\alpha x}{\cosh 2\alpha x + \cos 2\beta y}$$

is $\left\{(x,y) \in R^2 \middle| \beta y > 0, \alpha^2 x^2 + \beta^2 y^2 < \frac{\pi^2}{4}\right\}$. Then

$$\frac{\partial^{m+n} g}{\partial y^m \partial x^n}(x,y) = \alpha^n \beta^m \sum_{k=1}^{\infty} \frac{2^{2k}(2^{2k}-1)B_{2k}(2k+p-1)_{m+n}}{(2k)!} \cdot \left(\sqrt{\alpha^2 x^2 + \beta^2 y^2}\right)^{2k+p-1-m-n} \times \sin\left[(2k+p-1-m-n)\cot^{-1}\frac{\alpha x}{\beta y} + \frac{m\pi}{2}\right] \tag{12}$$

Proof $g(x,y)$

$= \operatorname{Im}(z^p \tanh z)$

(By (10))

$$= \operatorname{Im}\left[\sum_{k=1}^{\infty} \frac{2^{2k}(2^{2k}-1)B_{2k}}{(2k)!}(\alpha x + i\beta y)^{2k+p-1}\right] \tag{13}$$

Thus, by differentiation term by term theorem, we obtain

$$\frac{\partial^{m+n} g}{\partial y^m \partial x^n}(x,y) = \alpha^n \beta^m \sum_{k=1}^{\infty} \frac{2^{2k}(2^{2k}-1)B_{2k}(2k+p-1)_{m+n}}{(2k)!} \cdot \left(\sqrt{\alpha^2 x^2 + \beta^2 y^2}\right)^{2k+p-1-m-n} \times \sin\left[(2k+p-1-m-n)\cot^{-1}\frac{\alpha x}{\beta y} + \frac{m\pi}{2}\right]$$

In the following, we determine the infinite series forms of any order partial derivatives of the two-variables function (3).

2.9. Theorem 3

If the assumptions are the same as Theorem 1. Suppose the domain of the two-variables function

$$u(x,y) = \left(\sqrt{\alpha^2 x^2 + \beta^2 y^2}\right)^p \times \frac{\cos\left(p\cot^{-1}\frac{\alpha x}{\beta y}\right)\sinh 2\alpha x + \sin\left(p\cot^{-1}\frac{\alpha x}{\beta y}\right)\sin 2\beta y}{\cosh 2\alpha x - \cos 2\beta y}$$

is $\left\{(x,y) \in R^2 \middle| \beta y > 0, \alpha^2 x^2 + \beta^2 y^2 < \pi^2\right\}$. Then

$$\frac{\partial^{m+n} u}{\partial y^m \partial x^n}(x,y) = \alpha^n \beta^m \left(\sqrt{\alpha^2 x^2 + \beta^2 y^2}\right)^{p-1-m-n} (p-1)_{m+n} \cdot \cos\left[(p-1-m-n)\cot^{-1}\frac{\alpha x}{\beta y} + \frac{m\pi}{2}\right] + \alpha^n \beta^m \sum_{k=1}^{\infty} \frac{2^{2k}B_{2k}(2k+p-1)_{m+n}}{(2k)!}\left(\sqrt{\alpha^2 x^2 + \beta^2 y^2}\right)^{2k+p-1-m-n} \times \cos\left[(2k+p-1-m-n)\cot^{-1}\frac{\alpha x}{\beta y} + \frac{m\pi}{2}\right] \tag{14}$$

Proof Let $z = \alpha x + i\beta y$, then

$$z^p \coth z = (\alpha x + i\beta y)^p \coth(\alpha x + i\beta y) = \left(\sqrt{\alpha^2 x^2 + \beta^2 y^2}\right)^p \left[\cos\left(p\cot^{-1}\frac{\alpha x}{\beta y}\right) + i\sin\left(p\cot^{-1}\frac{\alpha x}{\beta y}\right)\right] \times \frac{\sinh 2\alpha x - i\sin 2\beta y}{\cosh 2\alpha x - \cos 2\beta y} \tag{15}$$

(By (5) and (8))

Hence,

$u(x,y)$

$$= \operatorname{Re}(z^p \coth z) \tag{16}$$

$$= \operatorname{Re}\left[(\alpha x + i\beta y)^{p-1} + \sum_{k=1}^{\infty} \frac{2^{2k}B_{2k}}{(2k)!}(\alpha x + i\beta y)^{2k+p-1}\right]$$

(By Formula 2.2.6)
Also, by differentiation term by term theorem, we have

$$\frac{\partial^{m+n}u}{\partial y^m \partial x^n}(x,y)$$

$$=\alpha^n\beta^m\left(\sqrt{\alpha^2x^2+\beta^2y^2}\right)^{p-1-m-n}(p-1)_{m+n}\cdot$$

$$\cos\left[(p-1-m-n)\cot^{-1}\frac{\alpha x}{\beta y}+\frac{m\pi}{2}\right]+$$

$$\alpha^n\beta^m\sum_{k=1}^{\infty}\frac{2^{2k}B_{2k}(2k+p-1)_{m+n}}{(2k)!}\cdot$$

$$\left(\sqrt{\alpha^2x^2+\beta^2y^2}\right)^{2k+p-1-m-n}\times$$

$$\cos\left[(2k+p-1-m-n)\cot^{-1}\frac{\alpha x}{\beta y}+\frac{m\pi}{2}\right]$$

Finally, we obtain the infinite series forms of any order partial derivatives of the two-variables function (4).

2.10. Theorem 4

Let the assumptions be the same as Theorem 1. If the domain of the two-variables function

$$w(x,y)=\left(\sqrt{\alpha^2x^2+\beta^2y^2}\right)^p\times$$

$$\frac{-\cos\left(p\cot^{-1}\frac{\alpha x}{\beta y}\right)\sin 2\beta y+\sin\left(p\cot^{-1}\frac{\alpha x}{\beta y}\right)\sinh 2\alpha x}{\cosh 2\alpha x-\cos 2\beta y}$$

is $\left\{(x,y)\in R^2 \middle| \beta y>0,\alpha^2x^2+\beta^2y^2<\pi^2\right\}$. Then

$$\frac{\partial^{m+n}w}{\partial y^m \partial x^n}(x,y)$$

$$=\alpha^n\beta^m\left(\sqrt{\alpha^2x^2+\beta^2y^2}\right)^{p-1-m-n}(p-1)_{m+n}\cdot$$

$$\sin\left[(p-1-m-n)\cot^{-1}\frac{\alpha x}{\beta y}+\frac{m\pi}{2}\right]+$$

$$\alpha^n\beta^m\sum_{k=1}^{\infty}\frac{2^{2k}B_{2k}(2k+p-1)_{m+n}}{(2k)!}\cdot \tag{17}$$

$$\left(\sqrt{\alpha^2x^2+\beta^2y^2}\right)^{2k+p-1-m-n}\times$$

$$\sin\left[(2k+p-1-m-n)\cot^{-1}\frac{\alpha x}{\beta y}+\frac{m\pi}{2}\right]$$

Proof By (15), we have

$$w(x,y)$$

$$=\mathrm{Im}(z^p\coth z) \tag{18}$$

$$=\mathrm{Im}\left[(\alpha x+i\beta y)^{p-1}+\sum_{k=1}^{\infty}\frac{2^{2k}B_{2k}}{(2k)!}(\alpha x+i\beta y)^{2k+p-1}\right]$$

Using differentiation term by term theorem, we obtain

$$\frac{\partial^{m+n}w}{\partial y^m \partial x^n}(x,y)$$

$$=\alpha^n\beta^m\left(\sqrt{\alpha^2x^2+\beta^2y^2}\right)^{p-1-m-n}(p-1)_{m+n}\cdot$$

$$\sin\left[(p-1-m-n)\cot^{-1}\frac{\alpha x}{\beta y}+\frac{m\pi}{2}\right]+$$

$$\alpha^n\beta^m\sum_{k=1}^{\infty}\frac{2^{2k}B_{2k}(2k+p-1)_{m+n}}{(2k)!}\cdot$$

$$\left(\sqrt{\alpha^2x^2+\beta^2y^2}\right)^{2k+p-1-m-n}\times$$

$$\sin\left[(2k+p-1-m-n)\cot^{-1}\frac{\alpha x}{\beta y}+\frac{m\pi}{2}\right]$$

3. Examples

In the following, for the partial differential problem of the four types of two-variables functions discussed in this study, we proposed four examples and use Theorems 1-4 to obtain the infinite series forms of any order partial derivatives of these functions, and evaluate some of their higher order partial derivative values. On the other hand, we employed Maple to calculate the approximations of these higher order partial derivative values and their solutions for verifying our answers.

3.1. Example 1

Suppose the domain of the two-variables function

$$f(x,y)=\left(\sqrt{25x^2+4y^2}\right)^5\times$$

$$\frac{\cos\left(5\cot^{-1}\frac{5x}{2y}\right)\sinh 10x-\sin\left(5\cot^{-1}\frac{5x}{2y}\right)\sin 4y}{\cosh 10x+\cos 4y} \tag{19}$$

is $\left\{(x,y)\in R^2 \middle| y>0,25x^2+4y^2<\frac{\pi^2}{4}\right\}$. By (9), we obtain any $m+n$-th order partial derivative of $f(x, y)$,

$$\frac{\partial^{m+n}f}{\partial y^m \partial x^n}(x,y)$$

$$=5^n2^m\sum_{k=1}^{\infty}\frac{2^{2k}(2^{2k}-1)B_{2k}(2k+p-1)_{m+n}}{(2k)!}\cdot \tag{20}$$

$$\left(\sqrt{25x^2+4y^2}\right)^{2k+p-1-m-n}\times$$

$$\cos\left[(2k+p-1-m-n)\cot^{-1}\frac{5x}{2y}+\frac{m\pi}{2}\right]$$

Therefore, we can evaluate the 7-th order partial derivative value of $f(x, y)$ at $\left(\frac{1}{5},\frac{1}{2}\right)$,

$$\frac{\partial^7 f}{\partial y^4 \partial x^3}\left(\frac{1}{5},\frac{1}{2}\right)$$
$$=5^3 2^4 \sum_{k=1}^{\infty} \frac{2^{2k}(2^{2k}-1)B_{2k}(2k+4)_7}{(2k)!}(\sqrt{2})^{2k-3}\cos\frac{(2k-3)\pi}{4} \quad (21)$$

Next, we use Maple to verify the correctness of (21).

```
>f:=(x,y)->(sqrt(25*x^2+4*y^2))^5*(cos(5*arccot((5*x)/(2*y)))*sinh(10*x)-sin(5*arccot((5*x)/(2*y)))*sin(4*y))/(cosh(10*x)+cos(4*y));
>evalf(D[1$3,2$4](f)(1/5,1/2),14);
```

$$-2.6908842324474 \cdot 10^7$$

```
>evalf(5^3*2^4*sum(2^(2*k)*(2^(2*k)-1)*bernoulli(2*k)*product(2*k+4-j,j=0..6)/(2*k)!*(sqrt(2))^(2*k-3)*cos((2*k-3)*Pi/4),k=1..infinity),14);
```

$$-2.6908842324474 \cdot 10^7$$

3.2. Example 2

If the domain of the two-variables function

$$g(x,y)=\left(\sqrt{x^2+y^2}\right)^7 \times \frac{\cos\left(7\cot^{-1}\frac{x}{y}\right)\sin 2y+\sin\left(7\cot^{-1}\frac{x}{y}\right)\sinh 2x}{\cosh 2x+\cos 2y} \quad (22)$$

is $\left\{(x,y)\in R^2 \middle| y>0, x^2+y^2<\frac{\pi^2}{4}\right\}$. By (12), we obtain

$$\frac{\partial^{m+n} g}{\partial y^m \partial x^n}(x,y)$$
$$=\sum_{k=1}^{\infty} \frac{2^{2k}(2^{2k}-1)B_{2k}(2k+6)_{m+n}}{(2k)!}\left(\sqrt{x^2+y^2}\right)^{2k+6-m-n} \times \sin\left[(2k+6-m-n)\cot^{-1}\frac{x}{y}+\frac{m\pi}{2}\right] \quad (23)$$

Therefore,

$$\frac{\partial^9 g}{\partial y^3 \partial x^6}\left(\frac{2}{3},\frac{1}{3}\right)$$
$$=-\sum_{k=1}^{\infty} \frac{2^{2k}(2^{2k}-1)B_{2k}(2k+6)_9}{(2k)!}\left(\sqrt{\frac{5}{9}}\right)^{2k-3} \cdot \cos[(2k-3)\cot^{-1}2] \quad (24)$$

```
>g:=(x,y)->(sqrt(x^2+y^2))^7*(cos(7*arccot(x/y))*sin(2*y)+sin(7*arccot(x/y))*sinh(2*x))/(cosh(2*x)+cos(2*y));
>evalf(D[1$6,2$3](g)(2/3,1/3),18);
```

$$-2.68926222712736 \cdot 10^5$$

```
>evalf(-sum(2^(2*k)*(2^(2*k)-1)*bernoulli(2*k)*product(2*k+6-j,j=0..8)/(2*k)!*(sqrt(5/9))^(2*k-3)*cos((2*k-3)*arccot(2)),k=1..infinity),18);
```

$$-2.68926222712726 \cdot 10^5$$

3.3. Example 3

Let the domain of the two-variables function

$$u(x,y)=\left(\sqrt{3x^2+4y^2}\right)^9 \times \frac{\cos\left(9\cot^{-1}\frac{\sqrt{3}x}{2y}\right)\sinh 2\sqrt{3}x+\sin\left(9\cot^{-1}\frac{\sqrt{3}x}{2y}\right)\sin 4y}{\cosh 2\sqrt{3}x-\cos 4y} \quad (25)$$

be $\left\{(x,y)\in R^2 \middle| y>0, 3x^2+4y^2<\pi^2\right\}$. By (14), we have

$$\frac{\partial^{m+n} u}{\partial y^m \partial x^n}(x,y)$$
$$=\sqrt{3}^n 2^m \left(\sqrt{3x^2+4y^2}\right)^{8-m-n}(8)_{m+n} \cdot \cos\left[(8-m-n)\cot^{-1}\frac{\sqrt{3}x}{2y}+\frac{m\pi}{2}\right]+$$
$$\sqrt{3}^n 2^m \sum_{k=1}^{\infty} \frac{2^{2k}B_{2k}(2k+8)_{m+n}}{(2k)!}\left(\sqrt{3x^2+4y^2}\right)^{2k+8-m-n} \times \cos\left[(2k+8-m-n)\cot^{-1}\frac{\sqrt{3}x}{2y}+\frac{m\pi}{2}\right] \quad (26)$$

Thus,

$$\frac{\partial^7 u}{\partial y^4 \partial x^3}\left(\frac{1}{2},\frac{3}{4}\right)$$
$$=2903040+144\sum_{k=1}^{\infty} \frac{2^{2k}B_{2k}(2k+8)_7 3^k}{(2k)!}\cos\frac{(2k+1)\pi}{3} \quad (27)$$

```
>u:=(x,y)->(sqrt(3*x^2+4*y^2))^9*(cos(9*arccot(sqrt(3)*x/(2*y)))*sinh(2*sqrt(3)*x)+sin(9*arccot(sqrt(3)*x/(2*y)))*sin(4*y))/(cosh(2*sqrt(3)*x)-cos(4*y));
>evalf(D[1$3,2$4](u)(1/2,3/4),24);
```

$$3.19133553043736659056376 \cdot 10^7$$

```
>evalf(2903040+144*sum(2^(2*k)*bernoulli(2*k)*product(2*k+8-j,j=0..6)*3^k/(2*k)!*cos((2*k+1)*Pi/3),k=1..infinity),24);
```

$$3.19133553043736659056323 \cdot 10^7$$

3.4. Example 4

Suppose the domain of the two-variables function

$$w(x,y)=\left(\sqrt{9x^2+y^2}\right)^{11} \times \frac{-\cos\left(11\cot^{-1}\frac{3x}{y}\right)\sin 2y+\sin\left(11\cot^{-1}\frac{3x}{y}\right)\sinh 6x}{\cosh 6x-\cos 2y} \quad (28)$$

is $\left\{(x,y)\in R^2 \middle| y>0, 9x^2+y^2<\pi^2\right\}$. By (17), we obtain

$$\frac{\partial^{m+n}w}{\partial y^m \partial x^n}(x,y)$$
$$=3^n\left(\sqrt{9x^2+y^2}\right)^{10-m-n}(10)_{m+n}\cdot$$
$$\sin\left[(10-m-n)\cot^{-1}\frac{3x}{y}+\frac{m\pi}{2}\right]+ \quad (29)$$
$$3^n\sum_{k=1}^{\infty}\frac{2^{2k}B_{2k}(2k+10)_{m+n}}{(2k)!}\left(\sqrt{9x^2+y^2}\right)^{2k+10-m-n}\times$$
$$\sin\left[(2k+10-m-n)\cot^{-1}\frac{3x}{y}+\frac{m\pi}{2}\right]$$

Hence,

$$\frac{\partial^8 w}{\partial y^5 \partial x^3}\left(\frac{2}{3},\frac{4}{3}\right)$$
$$=78\cdot 10!\cdot\cos\left(2\cot^{-1}\frac{3}{2}\right)+$$
$$27\sum_{k=1}^{\infty}\frac{2^{2k}B_{2k}(2k+10)_8}{(2k)!}\left(\frac{52}{9}\right)^{k+1}\cos\left[(2k+2)\cot^{-1}\frac{3}{2}\right] \quad (30)$$

```
>w:=(x,y)->(sqrt(9*x^2+y^2))^11*(-cos(11*arccot(3*x/y))*sin(2*y)+sin(11*arccot(3*x/y))*sinh(6*x))/(cosh(6*x)-cos(2*y));
>evalf(D[1$3,2$5](w)(2/3,4/3),22);
```

$$-5.154590255812792862\cdot 10^8$$

```
>evalf(78*10!*cos(2*arccot(3/2))+27*sum(2^(2*k)*bernoulli(2*k)*product(2*k+10-j,j=0..7)*(52/9)^(k+1)/(2*k)!*cos((2*k+2)*arccot(3/2)),k=1..infinity),22);
```

$$-5.154590255812792852\cdot 10^8$$

4. Conclusion

In this article, we provided a new technique to evaluate any order partial derivatives of four types of two-variables functions. We will use this technique to solve another partial differential problems. On the other hand, the differentiation term by term theorem plays a significant role in the theoretical inferences of this study. In fact, the applications of this theorem are extensive, and can be used to easily solve many difficult problems; we endeavor to conduct further studies on related applications. In addition, Maple also plays a vital assistive role in problem-solving. In the future, we will extend the research topic to other calculus and engineering mathematics problems and solve these problems by using Maple. These results will be used as teaching materials for Maple on education and research to enhance the connotations of calculus and engineering mathematics.

References

[1] A. Griewank and A. Walther, Evaluating derivatives: principles and techniques of algorithmic differentiation, 2nd ed., SIAM, Philadelphia, 2008.

[2] C. H., Bischof, G. Corliss, and A. Griewank, " Structured second and higher-order derivatives through univariate Taylor series," Optimization Methods and Software, Vol. 2, pp. 211-232, 1993.

[3] D. N. Richard, " An efficient method for the numerical evaluation of partial derivatives of arbitrary order, " ACM Transactions on Mathematical Software, Vol. 18, No. 2, pp. 159-173, 1992.

[4] T-W, Ma, " Higher chain formula proved by combinatorics," The Electronic Journal of Combinatorics , Vol. 16, #N21, 2009.

[5] L. E. Fraenkel, " Formulae for high derivatives of composite functions," Mathematical Proceedings of the Cambridge Philosophical Society, Vol. 83, pp. 159-165, 1978.

[6] C.-H., Yu, "Application of Maple: taking the partial differential problem of two-variables functions as an example," Proceedings of 2013 Business Innovation and Development Symposium, Taiwan, B20130113001, 2013.

[7] C. -H. Yu, " Partial derivatives of some types of two-variables functions, " Pure and Applied Mathematics Journal , Vol. 2, No. 2, pp. 56-61, 2013.

[8] C.-H. Yu, " Application of Maple on the partial differential problem of four types of two-variables functions, " Proceedings of the International Conference on Advanced Information Technologies, Taiwan, No. 87, 2013.

[9] C.-H. Yu, "Using Maple to study the partial differential problems," Applied Mechanics and Materials, in press.

[10] C.-H. Yu, " Application of Maple: taking the partial differential problem of some types of two-variables functions as an example, " Proceedings of the International Conference on e-Learning, Taiwan, pp. 337-345, 2013.

[11] C. -H. Yu, " Using Maple to evaluate the partial derivatives of two-variables functions, " International Journal of Computer Science and Mobile Computing , Vol. 2, Issue. 6, pp. 225-232, 2013.

[12] C. -H. Yu, " Evaluating partial derivatives of two-variables functions by using Maple," Proceedings of the 6th IEEE/International Conference on Advanced Infocomm Technology, Taiwan, pp. 23-27, 2013.

[13] C. -H. Yu and B. -H. Chen, "The partial differential problem," Computational Research, Vol.1, No.3, pp. 53-60, 2013.

[14] R. V. Churchill and J. W. Brown, Complex variables and applications, 4th ed., McGraw-Hill, New York, p65, 1984.

[15] Hyperbolic functions, online available from http://en.wikipedia.org/wiki/Hyperbolic_function.

[16] T. M. Apostol, Mathematical analysis, 2nd ed., Addison-Wesley, Boston, p 230, 1975.

17

Controlling Inverted Pendulum Using Performance-Oriented PDC Method

Kamran Vafaee[*], **Behdad Geranmehr**

School of Mechanical Engineering, Iran University of Science and Technology (IUST), Tehran, IRAN
*Corresponding author: vafaee.kamran@gmail.com

Abstract In this paper a performance-oriented Parallel Distributed Compensation (PDC) controller for Stabilizing the Linear Single Inverted Pendulum system is presented as a classical challenging benchmark problem in control engineering. The main idea of the original PDC method is to partition the dynamics of a nonlinear system into a number of linear subsystems, design a number of state feedback gains for each linear subsystem, and finally generate the overall state feedback gain by fuzzy blending of such gains. In Performance-Oriented PDC algorithm the state feedback gains are not considered constant through the linear subsystems, rather, based on some prescribed performance criteria, several feedback gains are associated to every subsystem, and the final gain for every subsystem is obtained by fuzzy blending of such gains. The model-based design methodology for mechatronic systems is a key factor for innovative and operative excellence in the design process. It was shown through this method of designing and simulation studies that application of the Performance-Oriented PDC method to such a challenging problem is robust against the uncertainties, while the control effort is still kept at an acceptable level.

Keywords: *Fuzzy Takagi–Sugeno model, Parallel Distributed Compensation, model-based mechatronic design, Inverted Pendulum*

1. Introduction

Fuzzy logic is known as a powerful and flexible tool for modeling and control of nonlinear systems with uncertainties, especially in systems that are structurally difficult to model due to their inherent natural nonlinearities and other modeling complexities [21].

Model based fuzzy control, with the possibility of guaranteeing the closed loop stability, is an attractive method for control of nonlinear systems. In recent years, many studies are devoted to the stability analysis of continuous time or discrete time model based fuzzy control systems [1,2,3,4]. Among such methods, the method of Takagi–Sugeno [1] has found many applications in modelling of complex nonlinear systems [5,6]. Since T–S fuzzy model is a universal approximator, it can be used to approximate any complex nonlinear system, and can also produce any complex control signal. The concept of sector nonlinearity [7] provided means for exact approximation of nonlinear systems by fuzzy blending of a few locally linear subsystems.

Parallel Distributed Compensation (PDC) is a model-based T–S fuzzy controller whose rule antecedents are equivalent to those of the fuzzy model of the plant [3]. It is a generalization of the state feedback controller to the case of nonlinear systems, using the Takagi–Sugeno fuzzy model [2]. This method is based on partitioning of a nonlinear system dynamics into a number of linear subsystems, for which state feedback gains are designed and blended in a fuzzy sense. Using this technique, the control design procedure will be conceptually simple and natural. The unique advantage of PDC technique is that many of conventional linear control design methods based on both classical and modern theory can be employed in designing the nonlinear T–S fuzzy controllers [22]. One important advantage of using such a method for control design is that the closed-loop stability analysis, using the Lyapunov method, becomes easier to apply. Various stability conditions are proposed for such systems, [3,8], where the existence of a common solution to a set of Lyapunov equations is shown to be sufficient for guaranteeing the closed-loop stability. Some relaxed conditions are also proposed in [9,10,11,12].

Takagi–Sugeno model and parallel distributed compensation are applied to many applications successfully [13,15]. Linear matrix inequality (LMI) technique offers a numerically tractable way for designing a PDC controller, with various design objectives such as stability [2,10,11,16], control [17], H2 control [18] and pole-placement technique [19,20] and others [3].

A Linear Single Inverted Pendulum is comprised of a vertical pendulum pivoted in the end, over a cart moving linearly. The cart can be forced to translate linearly using an electrical motor. The objective is to retain the pendulum around its upper unstable equilibrium position.

Despite the simplicity of the structure, some characteristics such as strong nonlinearity, inherent instability, and the under-actuation of the system, made it a good motivation for researchers to adopt it as a benchmark problem for verification of the effectiveness of new control strategies. Various recent works are reported for control and stabilization of such systems which have been used as benchmarks for control of linear unstable systems, and highly nonlinear dynamical systems [24,25].

This paper, proposes the application of an extension to the conventional PDC method - Performance-Oriented parallel distributed compensation – (proposed in [23]) to the challenging control engineering problem, Linear Single Inverted Pendulum. Unlike the method in [23], it is not used the pole placement method to obtain stabilizing feedback gains, instead LQR control scheme is applied.

This research is organized as follows: the mechatronic design, the Takagi–Sugeno fuzzy model and the PDC design method are shortly reviewed in Section 2 and 3. In section 4 the Performance-Oriented parallel distributed compensation method is presented. Section 5 describes the mathematical modeling of a Linear Single Inverted Pendulum system, which has been manufactured. In section 6, the controller designing procedure is followed. And finally Obtained results from simulating the controlled system, are presented in section 7.

2. Mechatronic Design

Mechatronics can be considered to be an integrative methodology utilizing the technologies of mechanical engineering, electrical engineering, control engineering and computer engineering in order to provide enhanced products, processes and systems [28,29].

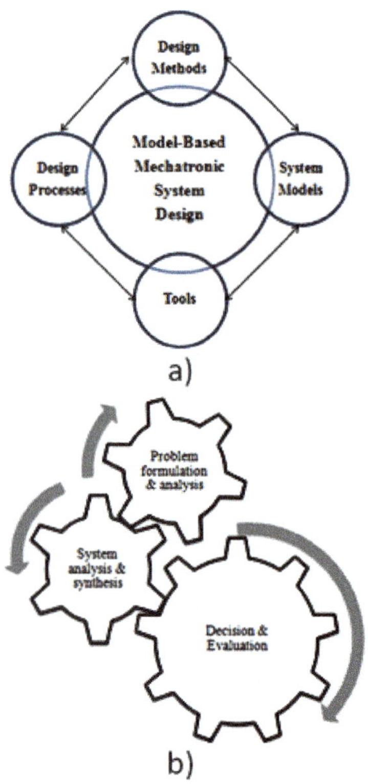

Figure 1. a) Model based design [30,31] and b) Mechatronic design process

A model-based design methodology for mechatronic systems is a key factor for innovative and operative excellence in the design process. Here we present an approach for using hierarchical models in the design process of a mechatronic system, called Linear Single Inverted Pendulum system which is a multi domain system. Demanding mechatronic solutions include an optimized result spanning various domains. The utilization and proper combination of solution principles from different mechatronic domains allow an extended variety and quality of principal solutions. Using traditional design methods this often leads to several iteration steps concerning the whole design process. In the case of a mechatronic system these iterations are especially time consuming as they include intense communication across domain boundaries.

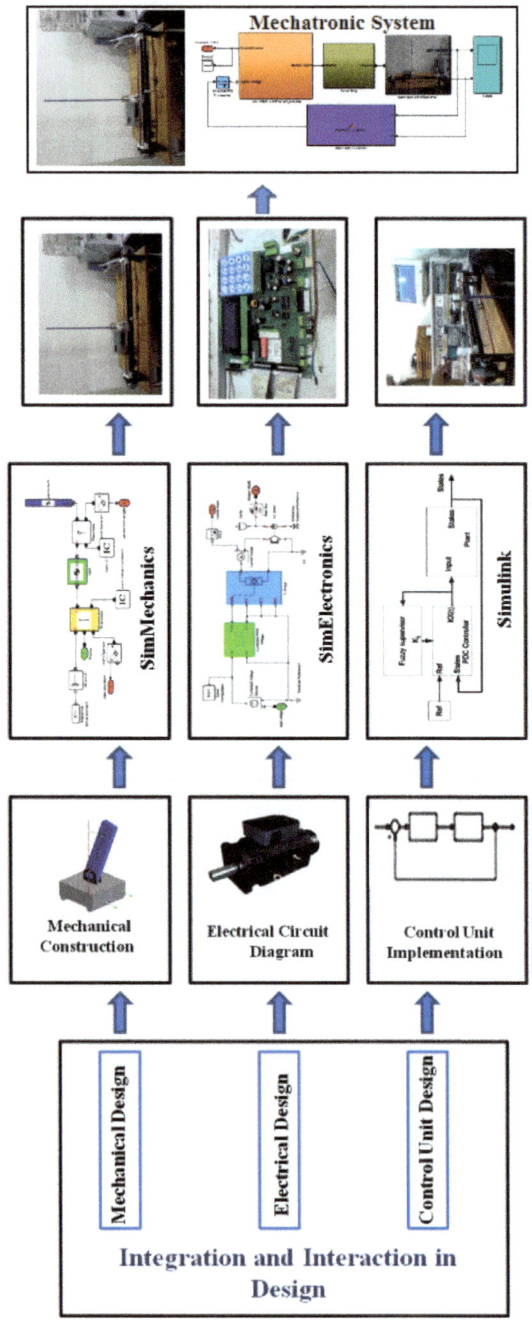

Figure 2. Hierarchical models in the conceptual design

In this study, CATIA is used for modeling and Assembly, ANSYS is used for rigid body dynamics, finite element vibration and strength analyses, and The integration of the design process is achieved with a MATLAB/Simscape (SimMechanics, SimElectronics and SimDriveline) for mechatronic modeling of cart-pole and control simulation (Figure 2 and Figure 3).

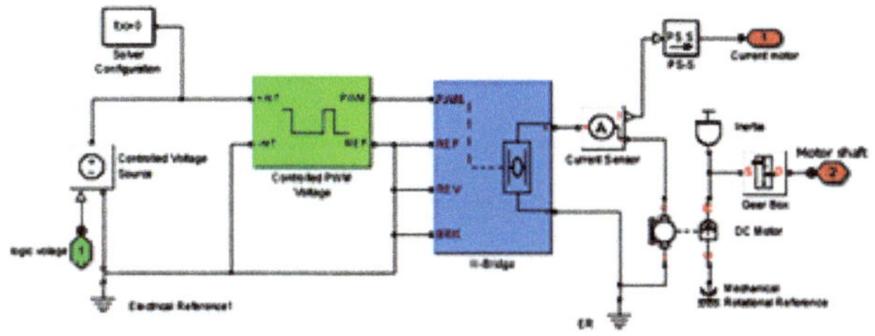

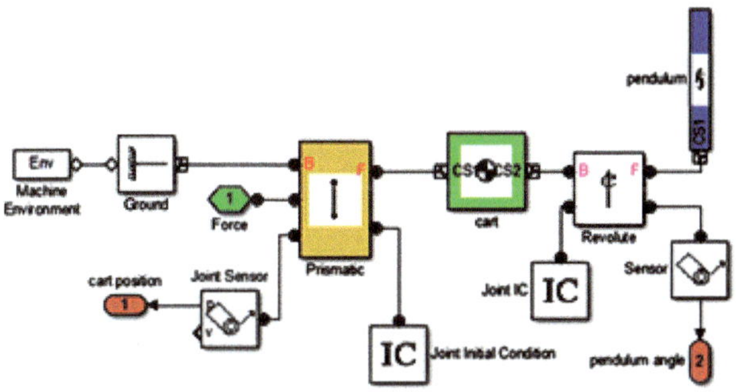

Figure 3. Conceptual design

3. Fuzzy-PDC Control

The history of the PDC method begins with the work of Sugeno and Kang [3]; the name PDC; however, was first used by Wang et al.[4]. The PDC design methodology was improved and the associated stability issues were studied by Tanaka and Sugeno [5]. Wang et al. [4] verified the effectiveness of the PDC by simulating the stabilization of a moving cart inverted pendulum, where the objective of the control was to stabilize the pendulum, without any specific control on the position of the cart.

The fuzzy-parallel distributed compensation (fuzzy-PDC) design approach consists of three steps [2]. The first step is to use the main feature of the Takagi-Sugeno fuzzy model to express the local dynamics of each fuzzy implication rule by a linear model. The overall fuzzy model of the system is achieved by fuzzy "blending" of the linear models. The second step is implementation of the so-called parallel distributed compensation (PDC) scheme on the fuzzy models. And the last step is to check, weather the stability conditions are satisfied by using the linear matrix inequality (LMI) analysis.

Suppose the original nonlinear system satisfies the sector nonlinearity condition [3] i.e., the values of nonlinear terms in the state-space equation remain within a sector of hyper-planes passing through the origin. The fuzzy rule associated with the i th linear subsystem, can then be defined as:

i th rule: If $Z_1(t)$ is M_{i1}, and $Z_2(t)$ is M_{i2}; . . ., and $Z_p(t)$ is M_{ip} then

$$\begin{cases} \dot{x} = A_i * x(t) + B_i * u(t) \\ y(t) = C_i * x(t) \end{cases} \quad fori = 1,2,\ldots,r \tag{1}$$

where, $x(t) \in R^n$ is the state vector, $u(t) \in R^m$ is the input vector, $A_i \in R^{n*n}$, $B_i \in R^{m*n}$, $C_i \in R^{q*n}$, Here, $\{Z_1(t), Z_2(t), \ldots, Z_p(t)\}$ are some nonlinear functions of the state variables obtained from the original nonlinear equation, and $M_{ij}(Z_i)$ are the degree of membership of Z_i in a fuzzy set M_{ij}. Whenever there is no ambiguity, the time argument in $Z(t)$ is dropped. The overall output, using the fuzzy blend of the linear subsystems, will then be as follows:

$$\dot{x}(t) = \frac{\sum_{i=1}^{r} w_i(z)\{A_i x(t) + B_i x(t)\}}{\sum_{i=1}^{r} w_i(z)} = \sum_{i=1}^{r} h_i(t)(A_i x(t) + B_i x(t)) \tag{2}$$

And

$$y(t) = \frac{\sum_{i=1}^{r} w_i(z) * C_i x(t)}{\sum_{i=1}^{r} w_i(z)} = \sum_{i=1}^{r} h_i(t) * C_i x(t) \tag{3}$$

Where

$$w_i(z) = \prod_{j=1}^{l} M_{ij} Z_j \tag{4}$$

And

$$h_i(t) = \frac{w_i(z)}{\sum_{i=1}^{r} w_i(z)} \tag{5}$$

It is also true, for all t, that

$$\begin{cases} \sum_{i=1}^{r} w_i(z) > 0 \\ w_i(z) \geq 0 \end{cases} \quad fori = 1,2,\ldots,r \tag{6}$$

Using the Takagi–Sugeno fuzzy model, a fuzzy combination of the stabilizing state feedback gains, F_i, i=1,2,…,r, associated with every linear subsystem is used as the overall state feedback controller. The general structure of the controller is then as

If $Z_1(t)$ is M_{i1}, and $Z_2(t)$ is M_{i2}; . . ., and $Z_p(t)$ is M_{ip} then

$$u = \frac{\sum_{i=1}^{r} w_i(z) * F_i x(t)}{\sum_{i=1}^{r} w_i(z)} = -\sum_{i=1}^{r} h_i(t) * F_i x(t) \tag{7}$$

The Takagi–Sugeno model and the Parallel Distributed Compensation have the same number of fuzzy rules and use the same membership functions.

4. Performance-Oriented PDC

In the proposed modified PDC, unlike the conventional PDC, state feedback gains, associated with every linear subsystem, are not assumed fixed. Instead, based on some pre- specified performance criteria, several feedback gains are designed and used for every subsystem. The overall gain associated with each of the subsystems, is then determined by a fuzzy blending of such gains, so that a better closed-loop performance can be achieved [23]. The dynamics of the resulting closed-loop control system can be analyzed as follows. Consider the following Takagi–Sugeno model of the plant:

$$\dot{x}(t) = \sum_{i=1}^{r} h_i(t)(A_i x(t) + B_i x(t)) \tag{8}$$

The following structure is proposed for the fuzzy controller rules:

ith rule: If $Z_1(t)$ is M_{i1}, and $Z_2(t)$ is M_{i2} ; . . ., and $Z_p(t)$ is M_{ip}, J(t) is H_i^1; . .., and J(t) is H_i^{qi} then:

$$u_i = \left\{ \sum_{i=1}^{q_i} m_i^n (J(t)) * K_i^n \right\} x(t) \tag{9}$$

where q_i is the number of gain coefficients in the i th subsystem, m_i^n is the relevant membership degree for $J(t)$, K_i^n is the n th state feedback gain associated with the i th subsystem, and H_i^n is the n th membership function for J(t), defined in the i th rule. Here, J(t) is a term depicting a selected performance index; for instance, if one wants to limit the magnitude of the control signal, u(t), then $J(t) = |u(t)|$. For such a selection, the resulting block diagram of the fuzzy blended parallel distributed controller is as shown in Fig 3, where the overall control input generated by the PDC controller is in the form of where, K_i , the overall feedback gain for the i th subsystem, is in the form of

$$u = \sum_{i=1}^{r} h_i(z) * u_i(t) x(t) = -\left\{ \sum_{i=1}^{r} h_i(z) * K_i \right\} x(t) \tag{10}$$

$$K_i = \sum_{i=1}^{q_i} m_i^n (J(t)) * K_i^n \tag{11}$$

Lemma 1. The fuzzy control system (6), with the control strategy (8), is globally, Asymptotically, stable, if there exists a common positive definite matrix, P, such that

$$G_{ii}^{nT} P + P G_{ii}^n < 0 \tag{12}$$

$$\left(\frac{G_{ij}^n + G_{ji}^n}{2} \right)^T P + P \left(\frac{G_{ij}^n + G_{ji}^n}{2} \right) \leq 0 \tag{13}$$

where i < j, $h_i \cap h_j \neq \varphi$, and. The proof is presented step by step in [23].

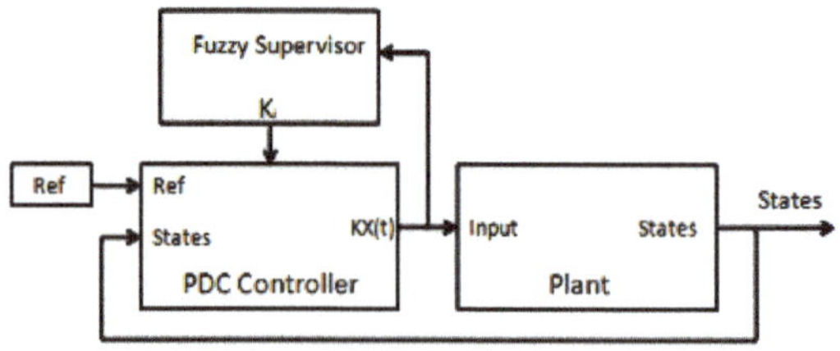

Figure 4. Block diagram of input-bounded controller [23]

5. Single Inverted Pendulum Modeling

A Linear Single Inverted Pendulum is comprised of a vertical pendulum pivoted in the end, over a cart moving linearly. The cart can be forced to translate linearly using an electrical motor. The objective is to retain the pendulum around its upper unstable equilibrium position. Despite the simplicity of the structure, some characteristics such as strong nonlinearity, inherent instability, and the under-actuation of the system, made it a good motivation for researchers to adopt it as a benchmark problem for verification of the effectiveness of new control strategies. Similar behavior of a Linear Single Inverted Pendulum to a number of well-known commercially available systems, e.g., Segway self-balancing human transporter, further motivates such studies.

In order to derive the mathematical model of the Linear Single Inverted Pendulum, some simplifying assumptions are made, e.g., all frictions and the gearbox backlash are neglected. The derived equations are as follows:

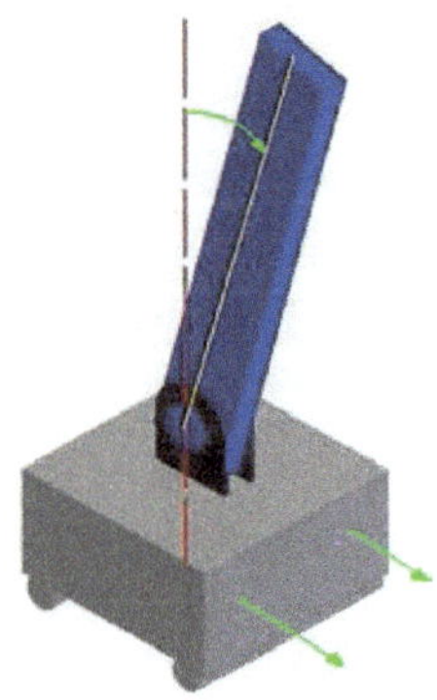

Figure 5. Model for a Linear Single Inverted Pendulum

$$\ddot{x} = \frac{1}{m_c + m_p \sin^2\theta}[f + m_p \sin\theta(l\dot{\theta}^2 + g\cos\theta)] \tag{14}$$

$$\ddot{\theta} = \frac{1}{l(m_c + m_p \sin^2\theta)}[-f\cos\theta - m_p l\dot{\theta}\cos\theta\sin\theta - (m_p + m_c)g\sin\theta] \tag{15}$$

6. Controller Design

In order to overcome the nonlinearities, and the, usually unknown, friction and mechanical backlash of the gearbox conventional heuristic fuzzy control methods can be applied. However, in this research, a Performance-Oriented Parallel Distributed Compensation (PDC) scheme is employed. One of the most important characteristics of the PDC method is its model-based nature, which provides means for a closed-loop stability analysis. Unlike the traditional fuzzy approach in which the control membership functions are determined through a tedious trial and error procedure, the PDC membership functions are derived by a systematic approach using the mathematical model of the system.

In order for the sector nonlinearity condition satisfied, the nonlinear expressions $z_1 = \dfrac{1}{m_c + m_p \sin^2\theta}$, $z_2 = \sin\theta$, $z_3 = \dot{\theta}\sin\theta$ and $z_4 = \cos\theta$ for $\theta(t) \notin \left(-\pi/2, \pi/2\right)$, are expressed as

$$z_1 = \frac{1}{m_c + m_p \sin^2\theta} = \sum_{i=1}^{2} E_i a_i \tag{16}$$

$$z_2 = \sin\theta = (\sum_{i=1}^{2} M_i b_i)\theta \tag{17}$$

$$z_3 = \dot{\theta}\sin\theta = \sum_{i=1}^{2} N_i c_i \tag{18}$$

$$z_4 = \cos\theta = \sum_{i=1}^{2} S_i d_i \tag{19}$$

where $a_1 = \dfrac{10}{12}$, $a_2 = 1$, $b_1 = \dfrac{2}{\pi}$, $b_2 = 1$, $c_1 = -10$, $c_2 = 10$, $d_1 = 0$, $d_2 = 1$. Hence, the membership functions for z are obtained as

$$E_1 = \frac{z_1 - a_2}{a_1 - a_2} \tag{20}$$

$$E_2 = \frac{a_1 - z_1}{a_1 - a_2} \tag{21}$$

$$M_1 = \begin{cases} \dfrac{\sin\theta - \frac{2}{\pi}\theta}{(1-\frac{2}{\pi})\theta} & \theta \neq 0 \\ 1 & \theta = 0 \end{cases} \tag{22}$$

$$M_2 = \begin{cases} \dfrac{\theta - \sin\theta}{(1-\frac{2}{\pi})\theta} & \theta \neq 0 \\ 0 & \theta = 0 \end{cases} \tag{23}$$

$$N_1 = \frac{z_3 - c_2}{c_1 - c_2} \tag{24}$$

$$N_2 = \frac{c_1 - z_3}{c_1 - c_2} \tag{25}$$

$$S_1 = \frac{z_4 - d_2}{d_1 - d_2} \tag{26}$$

$$S_2 = \frac{d_1 - z_4}{d_1 - d_2} \tag{27}$$

Now, the following stabilizing feedback gains are chosen using the LQR method, so that K_1^1 and K_2^1 produce large magnitude inputs for subsystems 1 and 2, respectively, and K_1^2 and K_2^2 induce low magnitude inputs for those subsystems, respectively.

7. Experimental Results

In this section, the performance of the proposed approach is evaluated for the Linear Single Inverted Pendulum problems. In this experiment, simulation results of proposed performance-oriented PDC controller are given for different initial conditions.

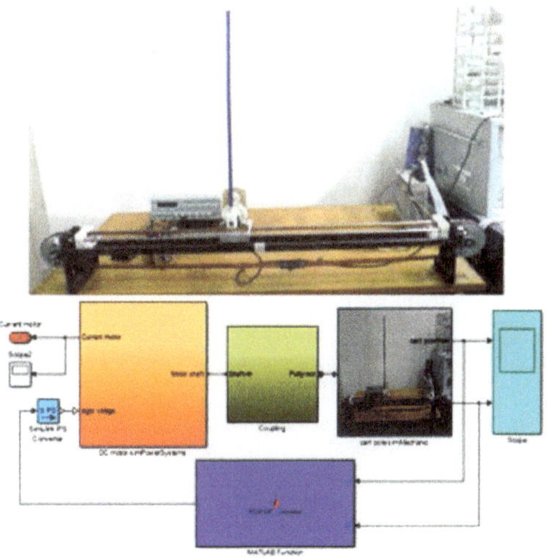

Figure 6. performance- oriented PDC controller for a Linear Single Inverted Pendulum

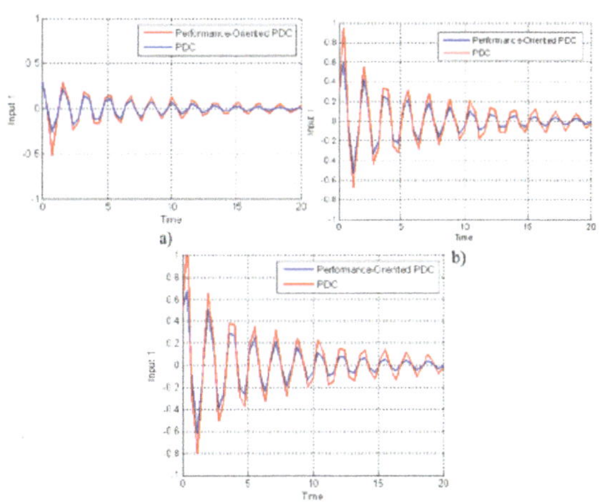

Figure 7. Error comparison of $\theta(t)$ between PDC controller and the Performance-Oriented Controller for Inverted Pendulum system, a) with $\theta(0) = 0.3rad$, b) with $\theta(0) = 0.3rad$ and $\dot{\theta}(0) = 2\,rad/sec$, c) with $\theta(0) = 0.5rad$ and $\dot{\theta}(0) = 2\,rad/sec$

8. Conclusions

In this paper, the Performance-oriented PDC control scheme has been applied to the stabilization control of Linear Single Inverted Pendulum System. The Performance-oriented PDC control scheme realized stabilization of the Inverted Pendulum System with good performance. In comparison with PDC method, it was shown that the Performance-Oriented method is more acceptable. Also It was shown through simulation studies that application of the Performance-oriented PDC method to such a challenging problem is robust against the uncertainties, while the control effort is still kept at an acceptable level.

The framework used in this paper can also be applied to generate nonlinear controllers for other uncertain systems. This plant is designed through a mechatronic process. This way of model-base design leaded the plant to fulfill the control requirements as well as other requirements.

References

[1] T. Takagi, M. Sugeno, Fuzzy identification of systems and its applications to modeling and control, IEEE Trans. Syst. Man Cybern. 15 (1985) 116-132.

[2] H.O. Wang, K. Tanaka, M.F. Griffin, An approach to fuzzy control of nonlinear systems: stability and the design issues, IEEE *Trans. Fuzzy Syst.* 4 (1996) 14-23.

[3] K. Tanaka, H.O. Wang, Fuzzy Control Systems Design and Analysis: A Linear Matrix Inequality Approach, Wiley, New York, 2001.

[4] K. Tanaka, T. Hori, H.O. Wang, A multiple Lyapunov function approach to stabilization of fuzzy control systems, IEEE *Trans. Fuzzy Syst.* 11 (2003) 582-589.

[5] K. Tanaka, M. Sano, A robust stabilization problem of fuzzy control systems and its applications to backing up control of a truck trailer, IEEE *Trans. Fuzzy. Syst.* 2 (1994) 119-134.

[6] J.L.D. Niemann, H.O. Wang, Robust tracking for high-rise/high-speed elevators, in: *Proceedings of American Control Conference*, 1998, pp. 3445-3449.

[7] S. Kawamoto, An approach to stability analysis of second order fuzzy systems, in: First IEEE *International Conference on Fuzzy Systems*, 1992, pp. 1427-1434.

[8] C.S. Ting, Stability analysis and design of Takagi-Sugeno fuzzy systems, Inf. *Sci.* 176 (2006) 2817-2845.

[9] E. Kim, New approaches to relaxed quadratic stability conditions of fuzzy control systems, IEEE Trans. Fuzzy *Syst.* 8 (2000) 523-534.

[10] B.C. Ding, H.X. Sun, P. Yang, Further study on LMI-based relaxed nonquadratic stabilization conditions for nonlinear systems in the Takagi-Sugeno's form, Automatica 42 (2006) 503-508.

[11] C.H. Fang, Y.S. Liu, S.W. Kau, L. Hong, C.H. Lee, A new LMI-based approach to relaxed quadratic stabilization of T-S fuzzy control systems, IEEE *Trans. Fuzzy Syst.* 14 (2006) 386-397.

[12] K. Tanaka, T. Ikeda, H.O. Wang, Fuzzy regulator and fuzzy observer: relaxed stability conditions and LMI- based designs, IEEE *Trans Fuzzy Syst.* 6 (1998) 250-265.

[13] M. Sugeno, G.T. Kang, Fuzzy modeling and control of multilayer incinerator, Fuzzy *Sets Syst.*, 1986, pp. 329-346.

[14] S.K. Hong, R. Langari, Robust fuzzy control of a magnetic bearing system subject to harmonic disturbances, IEEE *Trans. Cont. Syst. Tech.* 8 (2000) 366-371.

[15] P.P. Bonissone, V. Badami, K.H. Chaing, P.S. Khedkar, K.W. Marcelle, M.J. Schutten, Industrial applications of fuzzy logic at general electric, Proc. IEEE 83 (1995) 450-465.

[16] K. Tanaka, M. Sugeno, Stability analysis and design of fuzzy control systems, Fuzzy *Sets Syst.* 45 (1992) 135-156.

[17] K.R. Lee, E.T. Jeung, H.B. Park, Robust fuzzy H1 control for uncertain nonlinear systems via state feedback: an LMI approach, Int. J. Fuzzy *Sets Syst.* 120 (2001) 123-134.

[18] Y.C. Lin, J.C. Lo, Robust H2 fuzzy control via dynamic output feedback for discrete-time systems, in: IEEE *Conference on Fuzzy Systems*, 2003, pp. 1384-1388.

[19] J. Joh, R. Langari, E. Jeung, W. Chiuig, A new design method for continuous Takagi-Sugeno fuzzy controller with pole-placement constraints: an LMI approach, in: IEEE *Conference on Systems, Man, Cybernet*, 1997, pp. 2969-2974.

[20] G. Kang, W. Lee, Design of TSK fuzzy controller based on TSK fuzzy model using pole-placement, in: IEEE *Conf. on Fuzzy Systems*, 1998, pp. 246-251.

[21] H.X. Li, H.B. Gatland, A new methodology for designing a fuzzy logic controller, IEEE *Tran. Syst., Man, Cybernet.* 25 (3) (1995) 505-512.

[22] X. Ban, X. Gao, X. Huang, H. Yin, Stability analysis of the simplest Takagi-Sugeno fuzzy control system using Popov criterion, Internat. *J. Innovative Computing, Inform. Control* 3 (5) (2007) 1087-1096.

[23] M. Seidi, A.H.D. Markazi, Performance-oriented parallel distributed compensation, *J. Franklin Inst.* (2010).

[24] Åström K.J, Furuta K. Swinging up a pendulum by energy control. *J Automatica* 2000; 36: 287-95.

[25] Wei Q, Dyawansa W.P, Levine W.S. Nonlinear controller for an inverted pendulum having restricted travel. *J Automatica* 1995; 31: 841-850.

[26] Wang H.O, Tanaka K, Griffin M.F. Parallel Distributed Compensation of Nonlinear Systems by Takagi-Sugeno Fuzzy Model. In: Proc. FUZZ IEEE/IFES' 95 1995; 531-8.

[27] G.F. Franklin, J.D. Powell, A. Emami-Naeini Feedback Control Of Dynamic Systems, 2002, *Prentice Hall, New Jersey,* USA

[28] De Silva CW.: Mechatronics - an integrated approach. *CRC Press Boca Raton, London, New York, Washington* DC (2005).

[29] Werner Dieterle, 2005. Mechatronic systems: Automotive applications and modern design methodologies. *J of Annual Reviews in Control* .

[30] Hehenberger P. (2012): Advances in Model-Based Mechatronic Design", *Trauner Verlag.*

[31] Vajna S., Weber Chr., Bley H., Zeman K., Hehenberger P. (2009): "CAx für Ingenieure - Eine praxisbezogene Einführung" (CAx for Engineers) , 2009, 550 pages, 264 figures, ISBN: 978-3-540-36038-4, *Springer Berlin Heidelberg.*

18

Automation for Monitoring Elderly Americans

Alice J. Lin[1,*], Charles B. Chen[2]

[1]Marshall University, Huntington,West Virginia, USA
[2]*West Virginia*University, Morgantown,West Virginia, USA
*Corresponding author: lina@marshall.edu

Abstract Moreelderly Americans are living alone than ever before. Since elderly Americans are exposed to more risks of falling as they get older,reliably detecting falls for elderly Americans hasbecome an important field of research.Currently, there are somedevices that can assist the elderly, but they are not real-time, accessible, or particularly effective. We designeda novel automatic system for monitoring healthy independent living. The system will contain the devices for fall detection, surrounding environment monitoring, as well as measuring a person's blood pressure, pulse, and oxygen saturation in real time. With this automatic monitoring system, a person's state is not only controlled by that individual; rather, everything is automated so that even if a person falls unconscious or becomes extremely injured, it will still take the necessary steps to call for assistance. The system we proposed is aimed towards both healthy individuals as well as those with disabilities and chronic conditions.

Keywords: *automatic system, monitor, healthcare*

1. Introduction

The aging American population is already bringing pressure on healthcare and social services. As the number of elderly individuals continues to rise, the situation is likely to get worse. The elderly population (people 65 years or older) numbered 40.4 million in 2010, which was a 15.3% increase since 2000. The number of Americans aged 45-64 who will reach 65 over the next two decades increased by 31% during this decade. Over one in eight Americans is an elderly American. Persons reaching age 65 have an average life expectancy of an additional 18.8 years.The population 65 and over is projected to increase to 55 million in 2020 (a 36% increase for that decade). By 2030, there will be about 72.1 million older persons, over twice thenumber in 2000 [1].

Falls occur frequently in the elderly population and significantly impair their quality of life. It is estimated that more than one in three elderly individualsliving at home fall at least once a year. About 10% of falls result in serious injury (e.g. fractures). Falls are a leading cause of injury-related hospital admissionsand deaths for older people [2]. The risk of falling also rises with increasing age.Falls also lead to decreased mobility, fear of falling, and death [3,4,5].

Treatment of the injuries and complications associated with falls costs the U.S. over thirty billion dollars annually. Sixty percent of falls occur at home and the possibility of not having any assistance in case of unconsciousness or extreme injury are primary reasons why many otherwise healthy individuals are forced to leave the comfort and privacy of their own home to live in an assisted-care environment. Furthermore, a fall can have a psychological impact even if the senior is not physically injured. After a fall,many seniors become so afraid of falling again that they limit their activities. This in turn decreases their fitness, mobility, and balance and leads to decreased social interactions, reduced satisfaction with life and higher likelihood of depression. This fear cycle then increases the risk of another fall [6].

2. Background

Accelerometers with low-cost and low-power features make a wearable and reliable fall detection system possible. Multiple sensors with accelerometers placed at various locations in the body are used for real-time human movement detection [7,8,9]. Many systems [3,10,11,12,13] employ triaxial accelerometers to detect fall according to the acceleration of body motion and posture angle. To achieve better accuracy, later systems [14,15,16] detect fall using accelerometers with barometric pressure sensors, image processors, and gyroscopes.

Information Technology for Assisted Living at Home (ITALH) is a project using new technology to help older citizens live more comfortably [17,18]. The ITALH includes two items: the IVY project concerns detecting falls at home or in office environments, and the SensorNet project concerns developing an integrated, safe, and wireless sensor to monitor the user. However, those previous systems have several restrictions [19]: the methods are device-centric, not user-centric; the devices are expensive and complicated; and the information received by the doctor is insufficient to make an accurate diagnosis in a timely fashion. In most of the systems, the final decisions are based on the data collected from the sensors and the user cannot express his ideas on his own initiative. He can only passively accept the decision. In addition, some of the previous systems use acoustic or vibration sensors and image processing software. Most of them are very costly and not universally accessible. Ordinary users cannot control them at their own will. Few systems send an SMS message as a simple alarm. However, the text message is not enough to describe a patient's symptoms, so caregivers cannot have an accurate assessment of the situation.

Home healthcare sentinel system (HONEY) [19] is a home-based fall detection system. It uses a triaxial accelerometer to trigger the detection and deploy speech recognition systemsand images to reduce the false positives. In that system, if the triaxial accelerometer does not detect the fall, the alert will not be sent out. If a person falls slowly, the fall will not be detected.

For real-time reconstruction of 3D scenes from video, Narayanan et al. [20] computed depth maps using multi-baseline stereo and merged them to produce viewpoint-based visible surface models. Holes due to occlusion are filled from nearby depth maps.Koch et al. [21] presented a volumetric approach for fusion as part of an uncalibrated 3D modeling system.Sato et al. [22] proposed a volumetric method based on voting. Each depth estimate votes not only for the most likely surface but also for the presence of free space between the camera and the surface.Werner and Zisserman [23] proposed an approach to reconstruct buildings. It uses sparse points and line correspondences to discover the ground and facade planes.Vaishet al.presentsa method which adopts techniques from classical stereo reconstruction, matching corresponding pixels in all images of the light field using essentially robust patch-based block matching [24]. Cornelis et al. [25] presented a system for near real-time city modeling that employs a simple model for the geometry of the world.

3. Proposed Approach

Due to the limitations of the above technologies, we design a system that monitors the potential risks to healthy,elderly citizens and notifies us of emergency situations in real time. Figure 1 showsthe framework of the system.By wearing Watching-Over-Me (WOM), a person will be monitored not only in his or her home but also in places where the person spends plenty of time.

The system will contain the devices for fall detection, environmental monitoring,as well as measuring a person's lifecharacteristic markers.

3.1. Fall Detection

A triaxial accelerometer is integrated into thefall sensor, and the fall sensor sends early warning information if the trigger conditions are met with handling the three axes' sample values. Many smartphones have the triaxialaccelerometer. We can use these smartphones, such as the HTC G3 Hero smartphone, as fall sensors. In addition, the Bluetooth module and a high performance processor on the G3 satisfy the fall sensor's requirements. .

It is known and verified that a sensor based on a triaxial accelerometer can distinguish the body movements more precisely when it is fixed on the patient's waist [26]. The triaxial accelerometer will put out three acceleration values on the x-, y- and z-axis at every sampling point. When the body is stationary, the total acceleration of the body is vertical down. When the body is moving, the acceleration changes along with the movement intensity. Fall sensorsare based on the assumption that a fall is usually associated with a magnitude impact. An estimation of the degree of body movement intensity can be obtained from the signal vector magnitude (SVM). Define SVM by the relation:

$$SVM = \sqrt{x_i^2 + y_i^2 + z_i^2}$$

where x_i is the i-th sample value of the x-axis signal (similarly for y_i and z_i). Therefore, comparing the SVM to a preset SVM threshold allows detecting the associated fall. Similarly, when the body falls, the space relationship between body and ground also changes significantly. In order to determine the space posture of the body, Tilt Angle (TA) is defined as the angle between positive z-axis and SVM by the relation:

$$\mathrm{TA} = \arccos(\mathrm{z}/\mathrm{SVM})$$

where z is the sample value of the z-axis signal. TA refers to the relative tilt of the body in space.

We also need an angle distinction between the upright postures of sitting and standing, as well as lying in various conditions. Karantonis's works [9] provide the range of TA corresponding to the different body postures: if thepatient's TA is from 0 to 20°, it is classified as standing, and values from 20 to 60° indicates a sitting posture; if TA is between 60 and 120°, it is regarded as a lying posture. In most cases, a fall starts from a standing posture, and directly ends with lying on the floor. However, no fall would be predicted if the user falls in such a way that he or she was not parallel with the ground. This is important in various cases during a fall. A user will try to grasp a wall, chair, or other objects and end up slumping next to the object, such as sitting on a chair, rather than lying on the floor.Therefore, a sitting posture following a magnitude SVM is regarded as a fall.

3.2. Surrounding Environment Monitoring

WOM (see Figure 1) can be attached toan elder person's side at all times.One set of sensors are image sensors –tiny cameras, which are mounted on the elder person's clothes or in a hat. With today's camera hardware it has become possible to capture real-time images of the surroundings. Our method is based on detecting surrounding environmental information. Our advanced algorithm will make the system learn from environmental data, and then construct the 3D scene.

There has also been a considerable amount of work involving 3D reconstruction from aerial images [27,28]. The system collects video streams and generates automatic, real-time 3D reconstruction from videos of scenes. The core algorithms operate on the frames of a single video-camera as it moves in space. The reconstructions are based on frames captured at different times by the same camera under the assumption that the scenes remain static [29]. Compared to laser scanning, an attractive property of passive, image-based stereo techniques is the ability to create a 3D representation solely from photographs and to easily capture the scene from different viewpoints [30].

3.3. Health Data Monitoring

Another set ofsensors monitors the individual's vitals, such as blood glucose, blood pressure, pulse, oxygen saturation and sweat pH.

Additionally, new glucose-monitoring devices [31] can measure blood glucose levels in a non-invasive fashion. The wearable glucose sensor, which measures glucose concentration and hydration levels, consists of a laser device that produces a wavefront of light to illuminate the skin and nearby arteries on the wrist, and a camera that monitors changes in the light as it scatters off the skin. The rehydration levels are measured because the strength of the signals produced are affected by muscle weakness, one of the major signs indicating mild to moderate dehydration. It uses the speckle effect to directly monitor the glucose concentration in the bloodstream, as well as the wearer's relative hydration level.

Pulse Tracker [32] could be used to monitor pulse non-invasively with a sensor that measuresheart rate without being affected by movement. It uses the 'speckle effect', where interference patterns are produced on images when laser light reflects from an uneven surface or scatters from an opaque material. The scatterpatterns of light change when the reflected material is moving (such as blood through blood vessels). Speckle effect could be used to measure flow pulsations even when the light source used to create the speckle pattern was moving.

Wrist blood pressure monitorsproduce the most accurate reading whether you are at home or outdoors [33]. Wrist blood pressure monitorsare sensitive to wrist position with its motion-sensor technology, and automatically keeps a history of the individual's measurements. Italso gives the wearer the option of sharing the data with his or her primary care provider. Measuring methods are oscillometricand automatic inflation [33]. There is also an automated blood pressure and pulse detector on the wrist [34].

Oxygen Saturation Monitoring [35] is based on a synergy of photoplethysmography (PPG) and blood flow sensors. It can be built into the bottom of a wristwatch to measure oxygen saturation and pulse at the wrist. This technology exploits the synergy of a set of noninvasive optical sensors. . The PPG sensor is designed such that it collectsthe same amount of physiological information as conventional fingertip probes. Another side of the solution is using a complementary blood flow sensor. This sensor usesa coherent light scattering technique, the "dynamic speckle", that allows continuous measurements of blood rheological parameters.

The use of wearable sensors to monitor various health related biometric parameters during daily activities has attracted increasing interest recently. A number of methods currently exist for the measurement of sweat pH. The most popular of these is the pH-probe, which can be placed in contact with the skin in order to measure sweat pH [36].

Sweat pH monitoring device [37] is an ideal wearable sensor for measuring the pH of sweat since it does not contain any electronic part for fluidic handling or pH detection and because it can be directly incorporated into wristbands, which are in continuous contact with the skin. In addition, due to the micro-fluidic structure, fresh sweat is continuously passing through the sensing area providing the capability to perform continuous real-time analysis. It is able to provide immediate feedback regarding sweat composition. Sweat analysis is attractive for monitoring purposes as it can provide physiological information directly relevant to the health and performance of the wearer without the need for an invasive sampling approach.

A wearable electronic-free micro-fluidic device for the continuous monitoring of pH in sweat during exercise is described by Curto et al. [38]. The sensing capability is based on ionic liquid hydrogels, containing pH sensitive dyes capable of reporting pH activity in the range of3 to 10.

Salvo et al. [39] developed a prototype for wearable sweat rate sensors integrated onto a textile. It can be worn directly on the body and continuously monitors sweat rate. They rely on the measurement of the water-vapor pressure gradient near the skin in unventilated chambers: the steeper the gradient, the higher the water evaporation rate from the skin. To assemble the sweat-rate sensor, a pocket was created on two fabric nets. The first humidity sensor was at a distance of 0.2 cm from the skin, while the second was 1 cm from the skin. Since sensitivity is inversely proportional to the distance between sensors, a compromise was made in order to have good sensitivity and a difference in the humidity values greater than inaccuracy in the single humidity measurements. The mesh of the net was large enough to offer negligible resistance to the diffusion of the water-vapor flow.

3.4. The Processing

Our system firstacquires real-time data from the surrounding environment and the health information from the person wearing WOM. The devices periodically compare the real-time data to the acquired data. When an abnormal event occurs, it will make a decision and alertthe pre-determined parties (ambulance, caregiver, family members, etc.).

The system will base its ability to detect progressive declines in physical and cognitive abilities by determining the abnormal scene. To achieve this, the system compares recent input data of the scene with the data of the routine scene. For example, suppose a person falls down or is staggering. The system integrates the environmental input with the health data to make the decision if the person is in critical condition. If the system detects this kind of situation it automatically calls 911. It also triggers an alert system in the WOM. If there are people around, they will also be alerted.

The monitoring unit transforms the data into a real-time 3D scene and 3D character. The data is continuously fed and the character and surrounding environment are updated all the time. The person's activities can be viewed on mobile devices. The person wearing the WOM can control who can see his or her activities.

4. Conclusion

We have presented a novel design for healthy independent living. It is reliable, safe, and simple. It is easy to use and has intuitive user interfaces with consideration for a user's disability or impairment. Thedesign provides feedback in meaningful forms, whether auditory, visual, or tactile. Most importantly, oursystem for healthy independent living engages, empowers, and motivates the individuals with respect to his or her abilities.

The advantage of our system is that we are able to detect the person's dynamic state. Our method is based on surrounding environmental input, fall detection, and health data as auxiliary information. As long as the person is wearing WOM, the entire system is automated. Even if the person suddenly falls unconscious, the device will still take the appropriate actions. Thus, this technology has multiple benefits and can be targeted for both disabled and healthy individuals alike.

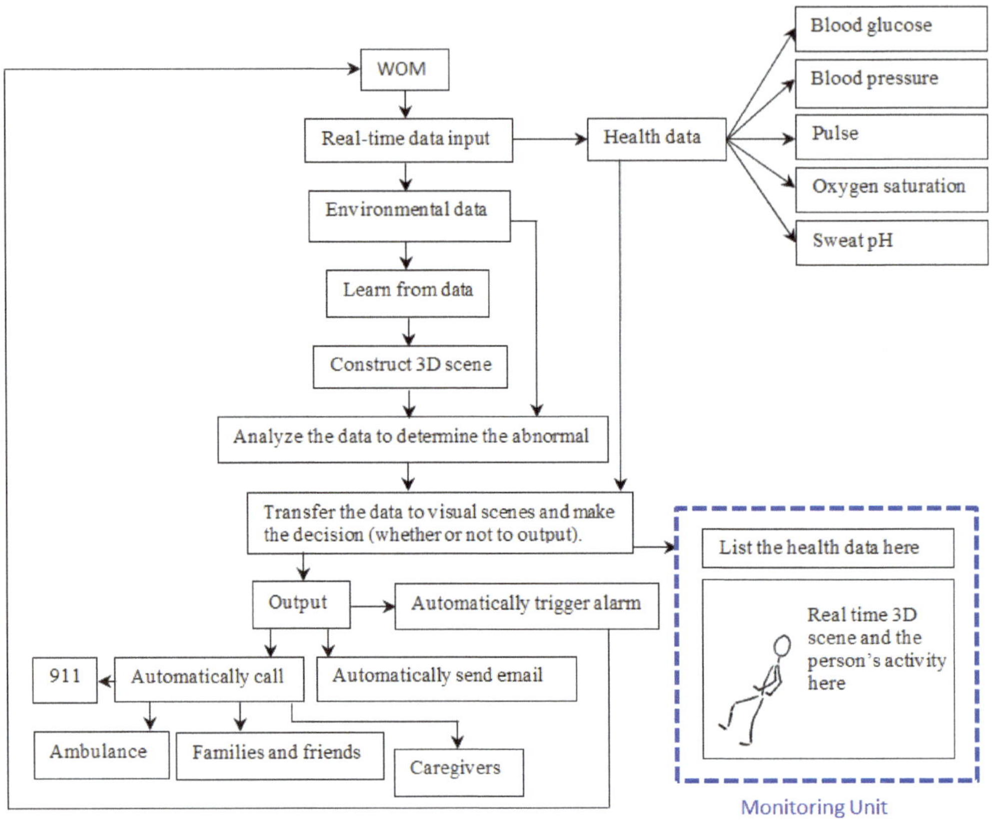

Figure 1. The framework of the system

References

[1] U.S. Department of Health and Human Services, "A Profile of Older Americans: 2011- Administration on Aging," 2011.

[2] D. o. V. A. Australian Government, "HomeFront guide to preventing falls for older people," pp. 1-59, 2006.

[3] K. Doughty, R. Lewis, and A. McIntosh, "The design of a practical and reliable fall detector for community and institutional telecare," *J.Telemed. Telecare,* vol. 6, no. Suppl. 1, pp. S150-154, 2000.

[4] A. Ozcan *et al.*, "The relationship between risk factors for falling and quality of life in older adults," *BMC Public Health,* vol. 5, pp. 90, 2005.

[5] R. W. Sattin, D. A. L. Huber, et al., "The incidence of fall injury events among the elderly in a defined population," *Am. J. pidemilo.,* vol. 131, pp. 1028-1037, 1990.

[6] B. C. M. o. H. P. O. o. t. Provincial Health Officer, "Prevention of Falls and Injuries Among the Elderly: A Special Report From the Office of the Provincial Health Officer," 2004.

[7] J. Fahrenberg *et al.*, "Assessment of posture and motion by multichannel piezoresistive accelerometer recordings," *Psychophysiol.,* vol. 34, pp. 607-612, 1997.

[8] F. Foerster, and J. Fahrenberg, "Motion pattern and posture: Correctly assessed by calibrated accelerometers," *Behav. Res. Meth. Instrum. Comput.,* vol. 32, pp. 450-457, 2000.

[9] D. M. Karantonis *et al.*, "Implementation of a real-time human movement classifier using a triaxial accelerometer for ambulatory monitoring. , vol. 10, no. 1,," *IEEE Transaction on Information Technology in Biomedicine,* vol. 10, no. 1, 2006.

[10] R. Sallen *et al.*, "Low power tri-axial ambulatory falls monitor," *In Proc. 10th Int. Conf. on Biomedical Engineering*, 2000.

[11] G. Williams et al., "A smart fall and activity monitor for telecare applications," Proc. 20th Annual Int. Conf. of the IEEE Engineering in Medicine and Biology Society, 1998.

[12] M. Mathie, J. Basilakis, and B. G. Celler, "A system for monitoring posture and physical activity using accelerometers," *23rd Annual Int. Conf. of the IEEE Engineering in Medicine and Biology Society*, 2001.

[13] G. Brown, "An accelerometer based fall detector: Development, Experimentation and Analysis," 2005.

[14] Q. Li *et al.*, "Accurate, fast fall detection using gyroscopes and accelerometer-derived posture information," *2009 Sixth International Workshop on Wearable and Implantable Body Sensor Networks*, pp. 138-143, 2009.

[15] F. Bianchi *et al.*, "Barometric pressure and triaxial accelerometer-based falls event detection," *IEEE Transaction on Neural Systems and Rehabilitation Engineering,* vol. 18, no. 6, 2010.

[16] A. Williams, D. Ganesan, and A. Hanson, "Aging in place: Fall detection and localization in a distributed smart camera network," *In Proceedings of the 15th International Conference on Multimedia*, pp. 892-901, 2007.

[17] M. Eklund, "Executive Summary: Information Technology for Assisted Living at Home (ITALH)," 2006.

[18] M. Eklund, T. Hansen, and S. Sastry, "Information technology for assisted living at home: building a wireless infrastructure for

assisted living," *The 27th Annual International Conference of the EMBS*, pp. 3931-3934, 2005.

[19] Q. Zhang, L. Ren, and W. Shi, "HONEY: A Multimodality Fall Detection and Telecare System," *Telemed J E Health,* vol. 19, no. 5, pp. 415-429, 2013.

[20] P. J. Narayanan, P. W. Rander, and T. Kanade, "Constructing virtual worlds using dense stereo," *Computer Vision, 1998. Sixth International Conference*, pp. 3-10 1998.

[21] R. Koch, M. Pollefeys, and L. V. Gool, "Robust calibration and 3d geometric modeling from large collections of uncalibrated images," *DAGM*, pp. 413-420, 1999.

[22] T. Sato *et al.*, "Dense 3-d reconstruction of an outdoor scene by hundreds-baseline stereo using a hand-held video camera.," *Journal of Computer Vision,* vol. 47, no. 1-3, pp. 119-129, 2002.

[23] T. Werner, and A. Zisserman, "New Techniques for Automated Architectural Reconstruction from Photographs," *Proceedings of the 7th European Conference on Computer* pp. 541-555, 2002.

[24] V. Vaish *et al.*, "Reconstructing Occluded Surfaces Using Synthetic Apertures: Stereo, Focus and Robust Measures," *Proceedings of the 2006 IEEE Computer Society Conference on Computer Vision and Pattern Recognition,* vol. 2, pp. 2331-2338 2006.

[25] N. Cornelis, K. Cornelis, and L. V. Gool, "Fast compact city modeling for navigation pre-visualization," *Computer Vision and Pattern Recognition, 2006 IEEE Computer Society Conference,* vol. 2, pp. 1339-1344 2006.

[26] J. Chen et al., "Wearable sensors for reliable fall detection," In Proceedings of the 2005 IEEE Engineering in Medicine and Biology 27th Annual Conference, pp. 3551-3553, 2005.

[27] A. Fischer *et al.*, "Extracting buildings from aerial images using hierarchical aggregation in 2d and 3d," *Computer Vision and Image Understanding,* vol. 72, no. 2, pp. 185-203, 1998.

[28] Z. Zhu, A. R. Hanson, and E. M. Riseman, "Generalized parallel-perspective stereo mosaics from airborne video," *IEEE Trans. on Pattern Analysis and Machine Intelligence,* vol. 26, no. 2, pp. 226-237, 2004.

[29] M. Pollefeys *et al.*, "Detailed Real-Time Urban 3D Reconstruction from Video," *International Journal of Computer Vision,* vol. 78, no. 2-3, pp. 143-167, 2008.

[30] C. Kim *et al.*, "Scene Reconstruction from High Spatio-Angular Resolution Light Fields," *Proceedings of ACM SIGGRAPH (Anaheim, USA, July 21-25, 2013), ACM Transactions on Graphics,*, vol. 32, no. 4, pp. 73:1-73:12, 2013.

[31] N. Ozana *et al.*, "Improved noncontact optical sensor for detection of glucose concentration and indication of dehydration level," *Biomedical Optics Express,* vol. 5, no. 6, pp. 1926-1940 2014.

[32] M. Nemati *et al.*, "Dynamic light scattering from pulsatile flow in the presence of induced motion artifacts," *Biomedical Optics Express,* vol. 5, no. 7, pp. 2145-2156 2014.

[33] "Wireless Blood Pressure Wrist Monitor," Retrieved 2014, from http://www.ihealthlabs.com/blood-pressure-monitors/wireless-blood-pressure-wrist-monitor/.

[34] "Wrist blood pressure monitors," Retrieved 2014,from http://www.beurer.com/web/us/products/bloodpressure/wrist.php.

[35] "Oxygen Saturation Monitoring," Retrieved 2014, from http://www.medgadget.com/2013/11/oxitone.html.

[36] D. Morris *et al.*, "Bio-sensing Textile Based Patch with Integrated Optical Detection System for Sweat Monitoring," *Sensors and Actuators B: Chemical,* vol. 139, no. 1, pp. 231-236, 2009.

[37] V. F. Curto *et al.*, "Real-time sweat pH monitoring based on a wearable chemical barcode micro-fluidic platform incorporating ionic liquids," *Sensors and Actuators B: Chemical*, no. 171-172 pp. 1327-1334, 2012.

[38] V. F. Curto et al., "Wearable micro-fluidic pH sweat sensing device based on colorimetric imaging techniques," 15th International Conference on Miniaturized Systems for Chemistry and Life Sciences, October 2-6, Seattle, Washington, 2011.

[39] P. Salvo *et al.*, "A Wearable Sensor for Measuring Sweat Rate," *IEEE Sensors Journal,* vol. 10, no. 10, pp. 1557-1558, 2010.

Design of Nonlinear CSTR Control System using Active Disturbance Rejection Control Optimized by Asexual Reproduction Optimization

Navid Yazdanparast[1,*], Mehdi Shahbazian[1], Masoud Aghajani[2], Saeed Pour Abed[3]

[1]Department of Instrumentation and Automation Engineering, Petroleum University of Technology, Ahwaz, Iran
[2]Department of Chemical Engineering, Petroleum University of Technology, Ahwaz, Iran
[3]Senior expert in Instrumentation Unit, Sarkhoon&Qeshm Gas treating Company, Bandar Abbas, Iran
*Corresponding author: Navid.yazdan@gmail.com

Abstract This paper proposes an optimal Active Disturbance Rejection Control (ADRC) based on using Asexual Reproduction Optimization (ARO) to control the temperature of a nonlinear CSTR. The parameters of non-isothermal continuous stirred tank reactor (CSTR) are varying with time caused by fouling and the deactivation and regeneration of the catalyst. Furthermore, in the exothermal region, dynamic behavior of this reactor is unstable. Therefore, designing an efficient controller in this complicated situation is difficult and challenging. ADRC is used as a robust method to control the temperature of CSTR in the situation that the CSTR parameters are varying with time. The parameters of ADRC are difficult to adjust and if these parameters tuned properly, it performs more efficiently in setpoint tracking and disturbance rejection. In this paper Controller design is represented as an optimization problem. The parameters of ADRC are tuned by ARO and then by Particle Swarm Optimization (PSO). The performance of ADRC tuned by ARO (ADRC-ARO) is compared with the performance of ADRC tuned by PSO (ADRC-PSO) and PID controller. The simulation results that the proposed ADRC-ARO method reveals robustness and better performance in both setpoint tracking and disturbance rejection with faster response time and less settling time.

Keywords: *Active Disturbance Rejection Control, Asexual Reproduction Optimization, Particle Swarm Optimization, Non-isothermal continuous stirred tank reactor, time varying parameters, robustness*

1. Introduction

One of the most important and fundamental process in chemical industries is continuous stirred tank reactor (CSTR). CSTR is a highly nonlinear system and its parameters affect its complex dynamic severely. Due to fouling and deactivation and regeneration of the catalyst the parameters of CSTR are varying with time. In addition, in the exothermal operating region of CSTR the dynamic behavior of this reactor is unstable. So design a proper controller for such CSTR systems is rather complicated and requires more efforts. Various methods have been developed to control the CSTR system [1-6]. In [1] Power-shaping control is proposed to control the CSTR. In [4,5] a design method based on using artificial bee colony (ABC) algorithms and a robust predictive control approach are proposed respectively to control an unstable nonlinear CSTR.

In this paper, we focus on a novel optimal Active Disturbance Rejection Control (ADRC) strategy to control the temperature of CSTR system. ADRC paradigm was proposed by Professor Han [7,8,9]. The key idea of the ADRC is to estimate the disturbances and the unknown parts of the model (denoted as total disturbances) and compensate them by suitable input signal in real time [9]. In ADRC approach, Extended State Observer (ESO) is employed to estimate the total disturbances of the system as an extra state. So the most important part of the ADRC is Extended State Observer (ESO). In comparison with other approaches ESO has exhibited advantages, because it requires little knowledge of the plant. Enormous successful applications of ADRC prove the capability of this method for control design [10,11,12,13].

In this work, we proposed a design for the 2nd-order ADRC to control the temperature of a non-isothermal CSTR with time varying parameters. There are numerous parameters typically used in the 2nd-order ADRC controller, and some of them influence the control performance more than others. So it is very important, how to optimize the parameters, as shown in [14-17]. In this paper, an objective function is proposed to optimize parameters of ADRC. Then Asexual Reproduction

Optimization (ARO) and Particle Swarm Optimization (PSO) are employed to solve the optimization problem. Then the performance of ADRC tuned by ARO (ADRC-ARO) is compared with the performance of ADRC tuned by PSO (ADRC-PSO) in controlling the temperature of CSTR in both constant and time varying parameters.

The rest of this paper is organized as follows. In section 2 ADRC design is introduced. First an objective function is proposed and then ARO and PSO method are introduced to minimize this objective function. In section 3 the detailed CSTR system description with constant and time varying parameter CSTR are presented and the controllers are applied to the CSTR. And the illustrative results are provided. Finally the conclusion is presented in section 4.

2. Methodology Description

2.1. Active Disturbance Rejection Control

ADRC is a robust control method that rejects disturbances, ADRC focuses more around the expansion of the model with an additional and imaginary state variable, indicating everything that are not included in the mathematical equation of the system. ESO (Extended State Observer), the brain of the ADRC, estimates internal and external disturbances and unmodeled part of the system as an extra state. This extra state is canceled by a proper control signal before it influences the system as shown in Figure 1.

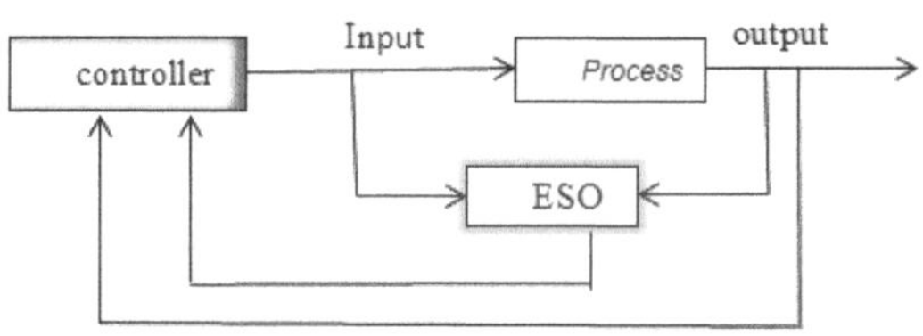

Figure 1. ADRC block diagram

The adaptive capability and robustness of ADRC make it a powerful technique for control in cases where the full information of the system is not accessible

An nth order system is chosen as an instance, which could be represented by the following state space equations:

$$\begin{cases} \dot{x}_1 = x_2 \\ \dot{x}_2 = x_3 \\ \vdots \\ \dot{x}_n = f(x_1, x_2, \ldots, x_n, w, t) + bu(t) \\ y = x_1 \end{cases} \tag{1}$$

Where y is the output, b is a constant, u is the input, f is the total disturbance containing internal disturbance and external disturbance, w is the external disturbance and t is time. This model is canonical, and many physical nonlinear systems can be found in this shape.

ESO (Extended State Observer), the most important part of the ADRC, estimates the total disturbance of the plant and cancels it online before it affects the system. The central idea in ESO is to use a completed state space model of $f(x_1, x_2, \ldots, x_n, w, t)$.

The ESO dynamics are proposed as following equation:

$$\begin{cases} \dot{z}_1 = z_2 - \beta_1 E_1 \\ \dot{z}_2 = z_3 - \beta_2 E_1 \\ \vdots \\ \dot{z}_n = z_{n+1} + \hat{b}u - \beta_n E_1 \\ \dot{z}_{n+1} = -\beta_{n+1} E_1 \end{cases} \tag{2}$$

In equation (2):

$$z_1 = \hat{x}_1, z_2 = \hat{x}_2, \ldots z_n = \hat{x}_n, z_{n+1} = \hat{f} \tag{3}$$

State variable z_{n+1} is the estimate of f and β_i are observer gains. Variable E_1 is defined as follows:

$$E_1 = z_1 - y \tag{4}$$

β_i Coefficients are chosen so that the matrix A_o is negative definite:

$$A_o = \begin{pmatrix} -\beta_1 & 1 & 0 & \cdots & 0 \\ -\beta_2 & 0 & 1 & \cdots & 0 \\ \vdots & 0 & 0 & \ddots & 1 \\ -B_{n+1} & 0 & 0 & 0 & 0 \end{pmatrix} \tag{5}$$

For a second order system that stated by following equation we have:

$$\ddot{y} = f(y, \dot{y}, w, t) + bu \tag{6}$$

Observer structure is:

$$\begin{pmatrix} \dot{z}_1 \\ \dot{z}_2 \\ \dot{z}_3 \end{pmatrix} = \begin{pmatrix} -\beta_{01} & 1 & 0 \\ -\beta_{02} & 0 & 1 \\ -\beta_{03} & 0 & 0 \end{pmatrix} \begin{pmatrix} z_1 \\ z_2 \\ z_3 \end{pmatrix} + \begin{pmatrix} 0 & \beta_{01} \\ b & \beta_{02} \\ 0 & \beta_{03} \end{pmatrix} \begin{pmatrix} u \\ y \end{pmatrix} \tag{7}$$

$$z_1 = \hat{y}, z_2 = \dot{\hat{y}}, z_3 = \hat{f} \tag{8}$$

$\hat{f}$ is the estimation of the total disturbance. The control signal then can be built according to the following equation:

$$u = \frac{u_0 - z_3}{b} \tag{9}$$

Reduces (6) to a simple double-integral plant:

$$\ddot{y} = u_0 \tag{10}$$

This can be easily controlled. The difference between this and all of the previous methods is that no analytical model of f is assumed here. The only thing required is the knowledge of the order of the plant model and the approximate value of b in (6) [8]. For the control signal u_0 is in the form of:

$$u_0 = \ddot{y}_d - k_d(z_2 - \dot{r}) - k_p(z_1 - r), k_p, k_d > 0 \tag{11}$$

Where r is the setpoint. The controller parameters, k_d and k_p, will be tuned by proposed methods in the next sections. Equation 11 shows that instead of using speed of output signal an estimate of it, z_2, is used. So ADRC has no need to speed of the output signal and this is an advantage of ADRC algorithms.

It is proven in [18] ADRC is stable if the characteristic polynomial for the state feedback design (11) $s^2 + k_d s + k_p$ is Hurwitz. So when the parameters of k_d and k_p are positive ADRC is stable.

The gains observer:

$$[\beta_{o1}, \beta_{o2}, \beta_{o3}]^T \tag{12}$$

To simplify the tuning process, we consider the parameters as:

$$L = [3\omega_o, 3\omega_o^2, \omega_o^3]^T \tag{13}$$

ω_o *is* observer bandwidth.

A number of studies have attempted to tune parameters of the ADRC [14,15,16,17]. In this paper ARO and PSO, as two optimization methods, are used to determine the best value of b and k_d and k_p to achieve a proper response in a control problem. In the next section a proper objective function will be proposed.

2.2. Objective Function for Optimal ADRC

There are several limitations for overshoot and control signal. Therefore, in this paper integral of the squared derivative of error is added to The Integral of the Squared Error (ISE) to guarantee the smoothness of the response and the control signal and minimize the overshoot. So the proposed objective function is:

$$J = \int_0^\infty |e|^2 .dt + \lambda . \int_0^\infty |de/dt|^2 \tag{14}$$

Where λ is the smoothness weight coefficient of the response that varies between 0 and 1 and it is adjustable. If we need a smooth response without overshoot, we use λ near 1. If we just focus on fast response and don't mind about overshoot it should be near 0. In next sections two optimization methods are used to minimize this objective function.

2.3. Asexual Reproduction Optimization

Asexual Reproduction Optimization (ARO) is an individual and a model free optimization technique. It is proposed in 2010. ARO inspired from the budding procedure of asexual reproduction. ARO principally is an evolutionary algorithm. [19,20]

Each individual produces a childthat in ARO called bud. Then, they participate in a competition with each other and more deserving person is selected based on a performance directory that has been defined as an objective function optimization problem.

A binary string indicates each individual which is a decision variable vector, $X = (x_1, x_2, \ldots, x_n); \rightarrow X \in R_n$. Each variable is characterized as the chromosome which prepared by a number of bits, called genes. Thus, for a chromosome with a length of L the first bit is the sign of the individual. The second and third bit of L shows the integer part, L_1, and decimal part, L_2, of the chromosome, respectively. So $L = L1 + L2 + 1$ and the length of each individual in the n-the dimensional space is $Li = n \times L$ *[19,20]*.

In ARO each solution in the search space is an organism in its environment that resources are limited and the most deserving organism can stay alive. To start the algorithm, an individual is accidentally initiated in the space *S*. The individual reproduces an offspring labeled bud by a particular operator known as reproduction mechanism. According to a performance index or a fitness function the parent and its offspring compete to stay alive. If the bud wins the competition, its parent will be rejected. Therefore, the bud is replaced with its parent and it becomes the new parent. If the parent wins the bud will be thrown away. When the stopping criteria are satisfied the algorithm is stopped. [19,20].

2.4. Particle Swarm Optimization

Particle Swarm Optimization (PSO) algorithm was introduced for the first time in 1995. PSO is a population-based evolutionary optimization algorithm. It is inspired from social behavior of bird and fish. In this method, a swarm has a set of population members that called particle. A position and velocity are defined for each particle. The position indicates a nominee solution in the solution space. The velocity transfers it from one position to another in the multidimensional solution space. A fitness function is used to calculate the particle value of each position, until the convergence criteria are satisfied [21].

3. Case Study: Nonlinear Non-isothermal Continuous Stirred Tank Reactor, Implementation and Results

3.1. Case Study Descriptions

In this section, first ADRC that is tuned by ARO will be applied to a non-isothermal CSTR with constant parameters and its performance will be compared to ADRC tuned by PSO and both of them will be compared with PID control. Then, all of them will be applied to an exothermic CSTR with time varying parameters including activation energy and heat transfer coefficient.

Consider a simple liquid-phase, irreversible chemical reaction where chemical species A reacts to form species B. The reaction can be written as $A \rightarrow B$. We assume that the rate of reaction is first-order with respect to component A:

$$r = kC_A \tag{15}$$

Where r is the rate of reaction of *A* per unit volume, k is the reaction rate constant, and C_A is the molar concentration of species *A*. For single-phase reactions, the rate constant is typically a strong function of reaction temperature given by the Arrhenius relation[6]:

$$k = k_0 \exp(-E/RT) \tag{16}$$

Where k_0 is the frequency factor, *R* is the gas constant, and E is the activation energy.

The diagram of the CSTR is shown in Figure 2. The inlet flow consists of pure species *A* with molar concentration C_A. Cooling coils, fix the reaction mixture at the desired operating temperature by removing heat that is produced in the reaction.

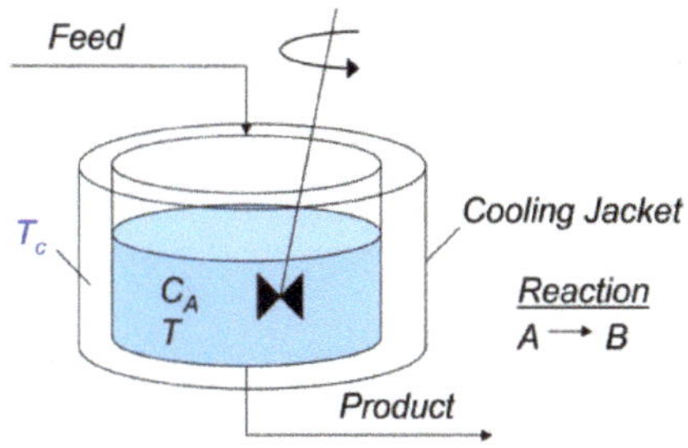

Figure 2. Non-isothermal CSTR

The process model consists of two nonlinear ordinary differential equations are given [6]:

$$\frac{dT}{dt}=\frac{q}{V}(T_f-T) + \frac{-\Delta H}{\rho cp}k_0C_A\exp(-\frac{E}{RT})+\frac{UA}{V\rho cp}(T_c-T) \quad (17)$$

$$\frac{dC_A}{dt}=\frac{q}{V}\left(C_{Af}-C_A\right)-k_0C_A\exp\exp\left(-\frac{E}{RT}\right) \quad (18)$$

Where T_C, coolant temperature, is the manipulated variable; and T, reactor temperature, is the controlled variable. To produce a desire component concentration, coolant flow rate influences the reactor temperature and reaction rate. Here it is assumed that the coolant flow rate is fixed. Due to physical limitations and reactor safety it is necessary to control the temperature accurately. Other parameters are given in Table 1. [6]

The **MATLAB** software is employed for simulation and numerical solving the equations of the system's dynamics described by Eq. (17), Eq. (18). Also, it is considered the continuous implementation with the sampling time $T = 0.001\sec$. The values of equilibrium points are given in Table 1.

Table 1. Parameters of non-isothermal CSTR

Parameter	Value	Parameter	Value
Feed concentrationC_{Af}	1mol/lit	Feed flow rateq	100 lit/min
Feed temperature T_f	350K	Activation energy $\frac{E}{R}$	8750 k
Mixture densityρ	1000kg/m^3	*Rate constant* k_0	7.2×10^{10} min^{-1}
Reactor volume V	100lit	Overall heat transfer coefficient UA	5×10^4 j/min.k
Heat capacity C_p	0.239 j/g.k	Coolant temperature T_css	300 k
Hear of reaction -ΔH	5×10^4 j/min.k	Concentration of A C_Ass	0.5 mol/lit
Reactor temperature Tss	350 k		

3.2. Case Study 1: Temperature Control of Non-isothermal CSTR with Constant Parameters

The open-loop step responses to step changes in Tc is shown in Figure 3. It is seen that the process is strongly nonlinear and highly unstable.

In order to control the temperature of this oscillating plant, ADRC-ARO, ADRC-PSO and PID approaches are applied. In Figure 4 the setpoint tracking behaviors of ADRC-ARO and ADRC-PSO and PID are compared. For setpoint changes, as shown in Table 2, ADRC-ARO shows a faster response than the ADRC-PSO control strategies. PID shows faster response than ADRC-ARO but it delivers a very big overshoot and longer settling time in setpoint changes.

Figure 5 shows the disturbance rejection performance of ADRC-ARO and ADRC-PSO controller strategies. Because the performance of PID was so poor in disturbance rejection it is preferred not to compare it with other methods. For -30K change in inlet feed temperature at t=100, as disturbance, ADRC-ARO exhibits a faster response toward setpoint when compared to ADRC-PSO.

In comparison with ADRC-PSO and PID, ADRC-ARO shows better performance with much faster response to the setpoint.

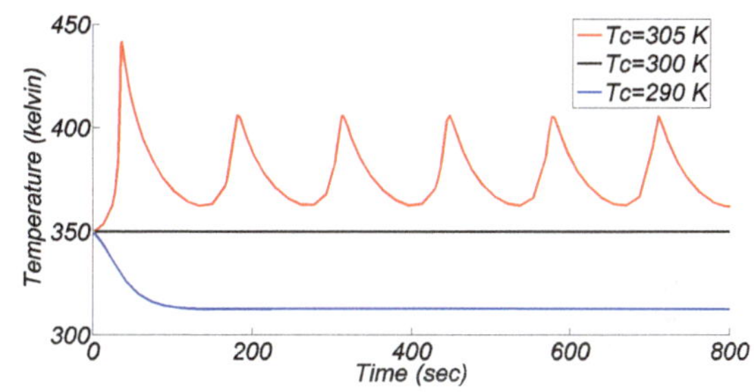

Figure 3. Reactor temperature variation with step changes in cooling water temperature Tc from i) 300 K to 305 K and ii) 300 K to 290 K.

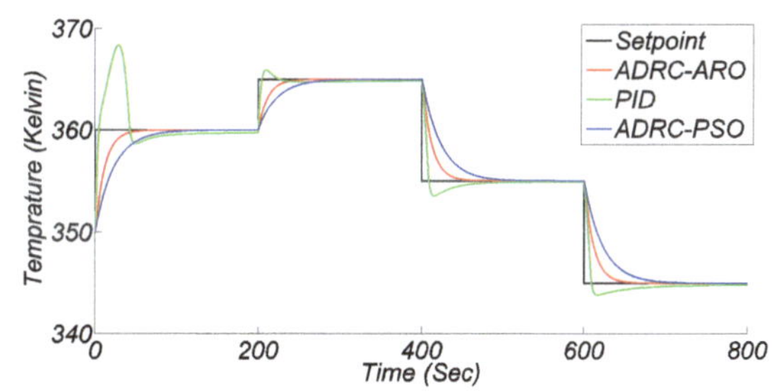

Figure 4. Setpoint tracking of CSTR with constant parameters

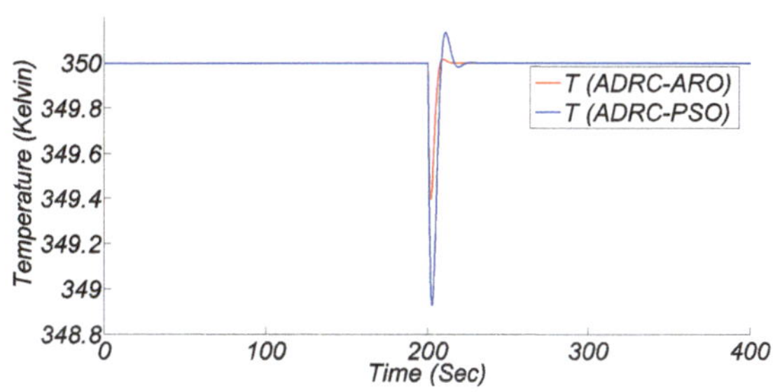

Figure 5. The disturbance rejection performance for -30 K step change in the inlet feed temperature (Tf) of the CSTR at t=100sec

Table 2. Set point tracking performance of controllers

Controller	Settling time (2%) from 350K to 360K	Rise time
ADRC-ARO	**49.4 Sec**	**10.3 Sec**
ADRC-PSO	88.1 Sec	18.9 Sec

3.3. Case Study 2: Temperature Control of the Non-isothermal CSTR with Time Varying Parameters

In this section, the previously ADRC-ARO and ADRC-PSO and PID control strategies will be applied to the non-isothermal CSTR with time varying parameters. Two case studies will be examined in which the CSTR shows strong nonlinearities. First, it is supposed that the heat transfer coefficient and so the overall heat transfer coefficient is varying with time due to fouling, then the Time varying activation energy caused by deactivation and regeneration of the catalyst are considered to show the robustness capability of ADRC approach.

3.3.1. Time Varying Heat Transfer Coefficient

Fouling is one of the most serious issues in heat transfer equipment. In the period of process operation, when a material is deposited on a heat transfer surface, Fouling happens. In practical applications, heat transfer surfaces usually become polluted with deposits and this makes extra resistance to the stream of heat. There are two usual cases fouling film over a period of time. One is asymptotic fouling. In this case, the resistances to heat transfer increases fast when the operation starts and becomes asymptotic to a steady state value at the end. The other one is linear fouling, where the fouling resistance increases linearly during the entire process operation. In this paper, we consider just second case that the fouling film increase linearly over the period of process operation. Therefore, the overall heat transfer coefficient is given by following equation [22]:

$$U_d = \Phi_h(t)U,\ \Phi_h(t) = 1 - 0.01t,\ \Phi_c(t) = 1 \qquad (19)$$

$\Phi_c(t)$ and $\Phi_h(t)$ are time-varying parameters which indicate the catalyst deactivation / regeneration and heat transfer fouling [22].

The open-loop step responses to step changes in T_c is shown in Figure 6. It is seen that the process is strongly nonlinear and non-stationary.

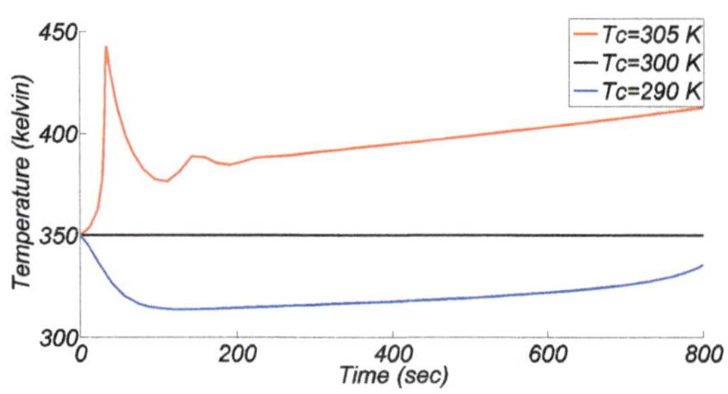

Figure 6. The open - loop response of the CSTR with the time varying heat transfer coefficient to step changes in cooling water temperature T_c from i) 300 K to 305 K and ii) 300 K to 290 K

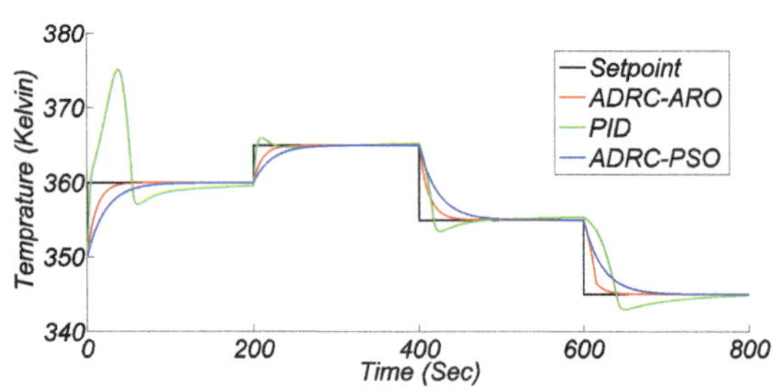

Figure 7. Setpoint tracking performance of CSTR with time varying heat transfer coefficient

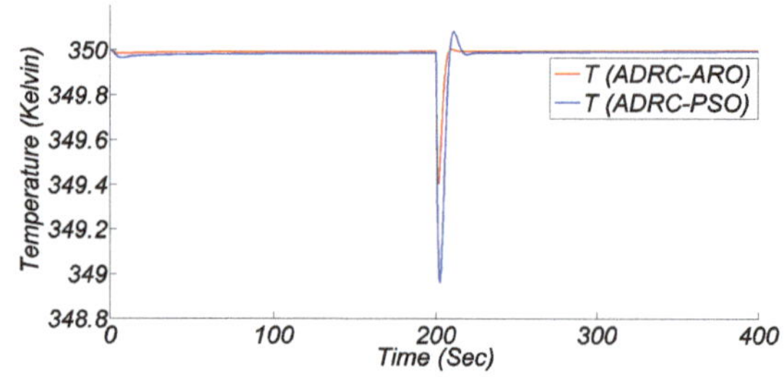

Figure 8. The disturbance rejection performance for -30 K step change in the inlet feed temperature (Tf) of the CSTR with time-varying heat transfer coefficient at t=100sec.

Figure 7 shows that the setpoint tracking performance of ADRC-ARO and ADRC-PSO are only slightly affected by the time varying heat transfer coefficient. PID controller, however, severely affected by fouling. In setpoint tracking ADRC-ARO has faster response with less settling time than ADRC- PSO. In disturbance rejection in Figure 8, ADRC-ARO exhibits faster response than ADRC-PSO. So ADRC-ARO for both setpoint tracking and disturbance rejection acts efficiently.

3.3.2. Time Varying Activation Energy

The temperature dependence of the rate expression is usually represented by the rate constant through the Arrhenius in (16).

Although the activation energy is not affected by temperature in the moderate temperature range, there are some exceptions. Other factors known to affect the activation energy include pressure and catalyst in the reaction. In the presence of a catalyst, an important factor which affects the rate of reaction is the deactivation of the catalyst by poisoning. There is no exact theoretical expression for the deactivation process. However, some empirical expressions have been reported. In this case study, the general equation for k is given by [23]:

$$k = k_0 \exp\exp\left(-\frac{E}{RT}\right).\phi_c(t) \qquad (20)$$

$$\phi_c(t) = \exp(-\alpha t) \qquad (21)$$

3.3.2.1. Catalyst Deactivation ($\alpha > 0$)

In the presence of a catalyst, the reaction rate is greatly influenced by the catalyst condition. During the period of reaction, because of poisoning, the catalyst may be deactivated. The catalyst deactivation is usually unpredictable, so an empirical correlation is used just for the simulation purpose.

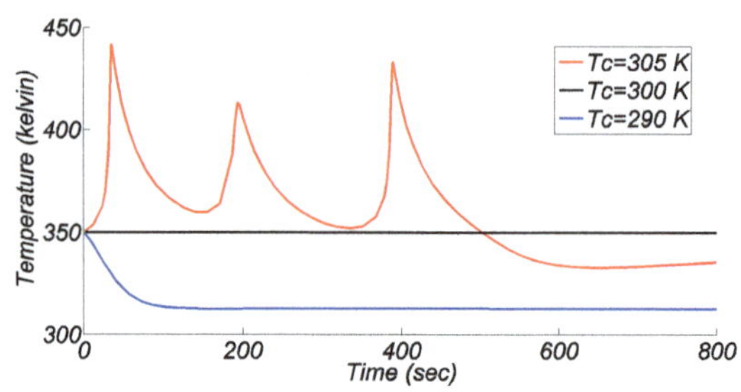

Figure 9. The open-loop response of the CSTR with time varying activation energy due to catalyst deactivation coefficient to step changes in cooling water temperature T_c from i) 300 K to 305 K and ii) 300 K to 290 K.

The open-loop responses to step changes in Tc is shown in Figure 9.

The setpoint tracking and disturbance rejection of different controllers are shown in Figure 10 and Figure 11. We can see that the performance of the ADRC-ARO is much better than ADRC-PSO and PID in setpoint tracking. In Figure 11 the temperature of the inlet feed (Tf) is decreased by -30k and shows that both ADRC-ARO and ADRC-PSO methods reject it in less than 30 second. Obviously the performance of ADRC-ARO in setpoint tracking and disturbance rejection is better than ADRC-PSO.

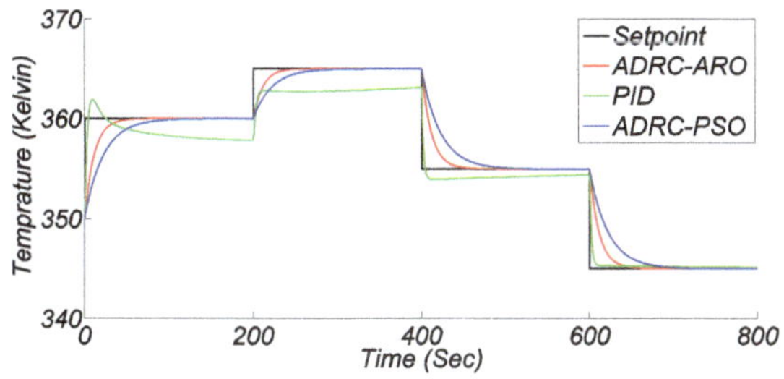

Figure 10. Setpoint tracking performance of CSTR with time varying activation energy due to catalyst deactivation ($\alpha > 0$)

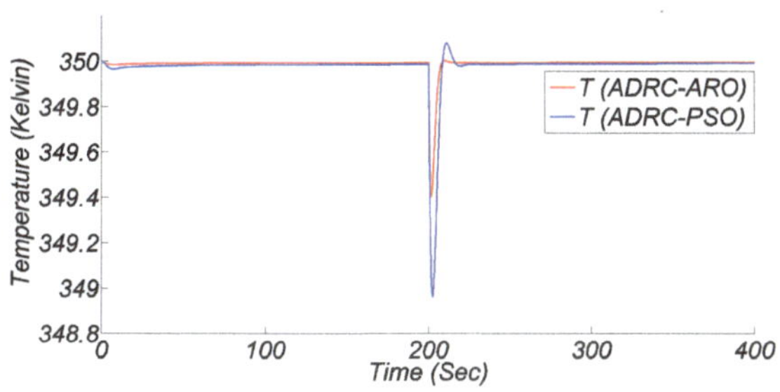

Figure 11. The disturbance rejection performance for -30 K step change in the inlet feed temperature (Tf) of the CSTR with time-varying activation energy due to catalyst deactivation at t=100sec.

3.3.2.2. Catalyst Regeneration $(\alpha < 0)$

Catalyst regeneration is a last case that is considered in this work. During this process, the catalyst is reactivated and the reaction rate can get to be very large. Catalyst regeneration could be an extremely serious situation. A great amount of heat is produced; however, it cannot be removed in time. This may result in the reactor to lose temperature control and equipment damage. Once in a while a blast may cause extreme harm. The open-loop simulation shown in Figure 12 exhibits this point.

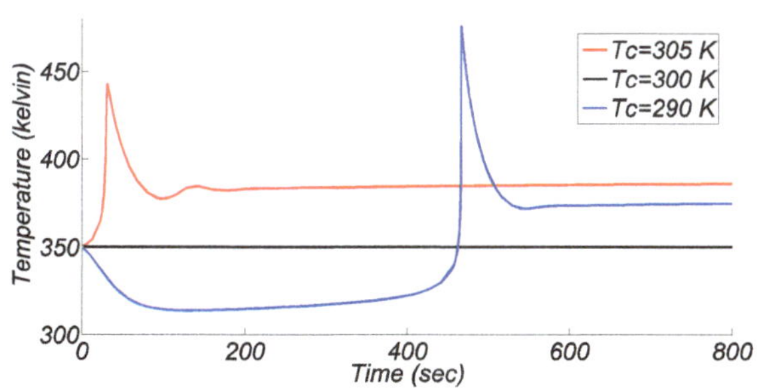

Figure 12. The open-loop response of the CSTR with time varying activation energy due to catalyst regeneration to step changes in cooling water temperature Tc from i) 300 K to 305 K and ii) 300 K to 290 K.

In Figure 13 and Figure 14 the performance of controllers in setpoint tracking and disturbance rejection are demonstrated. Figure 13 shows that the performance of the ADRC-ARO is much better than ADRC-PSO and ADRC-PSO is better than PID because it has less settling time and overshoot. In Figure 13 Similar to constant parameter and time varying heat transfer coefficient, PID method starts with a big overshoot and is affected more from regeneration. So unlike ADRC-ARO and ADRC-PSO, performance of PID is seriously influenced by this change which is so dangerous and can damage to the reactor. Overall the performance of ADRC-ARO is more superior than other controllers mentioned in this paper.

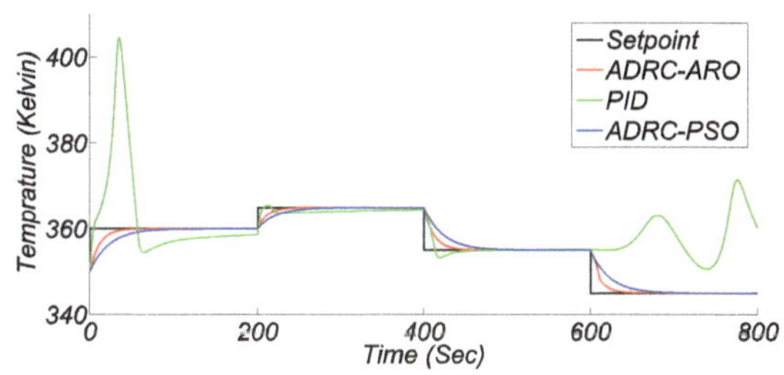

Figure 13. Setpoint tracking performance of CSTR with time varying activation energy due to catalyst regeneration

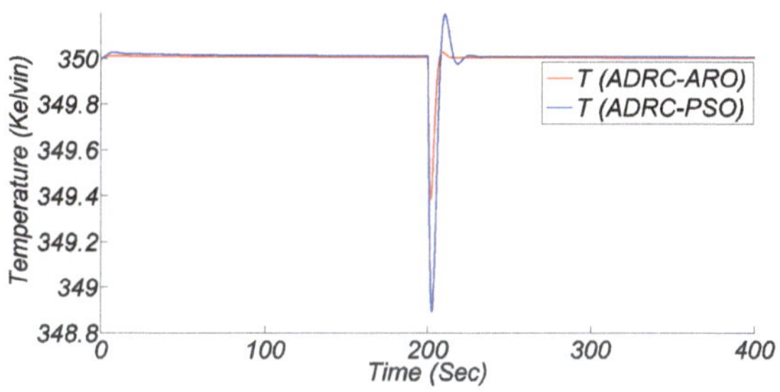

Figure 14. The disturbance rejection performance for -30K change in the inlet feed temperature (Tf) of the CSTR with time-varying activation energy due to catalyst regeneration at t=100sec.

4. Conclusion

In this paper, an optimal active disturbance rejection controller, ADRC-ARO, was proposed to control the temperature of the CSTR. The ADRC-ARO is applied to a non-isothermal CSTR. First, the performance of this approach was studied in CSTR with constant parameters and then with time varying parameters including heat transfer coefficient due to fouling and activation energy caused by deactivation and regeneration of the catalyst. The performance of ADRC-ARO was compared with ADRC-PSO approach. ADRC-ARO revealed Improvement specially in setpoint tracking with faster response time and less settling time in all case studies in this paper. Then both approaches have been compared with PID control. Obviously, ADRC-ARO and ADRC-PSO performed better than PID, this comparison applied just because we have seen the extreme scope of the nonlinearity by PID control. Finally, it has been proven that ADRC-ARO has an excellent performance in controlling the non-isothermal CSTR in the situation when the PID controller failed. It has been demonstrated that the ADRC-ARO controller is robust enough in an extensive range of operating conditions. Generally, if the parameters of ADRC tuned properly, it performs more efficiently in setpoint tracking and rejecting external and un-modeled disturbances.

Acknowledgement

The authors would like to thank the technical and financial supports provided by the National Iranian Gas Company–Sarkhoon & Qeshm Gas treating Company (NIGC-SQGC).

References

[1] A. Favache and D. Dochain, "Power-shaping control of reaction systems: The CSTR case," *Automatica,* vol. 46, pp. 1877-1883, 2010.

[2] F. Martinsen, L. T. Biegler, and B. A. Foss, "A new optimization algorithm with application to nonlinear MPC," *Journal of Process Control,* vol. 14, pp. 853-865, 2004.

[3] C.-T. Chen and S.-T. Peng, "Intelligent process control using neural fuzzy techniques," *Journal of process control,* vol. 9, pp. 493-503, 1999.

[4] W.-D. Chang, "Nonlinear CSTR control system design using an artificial bee colony algorithm," *Simulation Modelling Practice and Theory,* vol. 31, pp. 1-9, 2013.

[5] V. Ghaffari, S. V. Naghavi, and A. Safavi, "Robust model predictive control of a class of uncertain nonlinear systems with application to typical CSTR problems," *Journal of Process Control,* vol. 23, pp. 493-499, 2013.

[6] D. E. Seborg, D. A. Mellichamp, T. F. Edgar, and F. J. Doyle III, *Process dynamics and control*: John Wiley & Sons, 2010.

[7] H. Jingqing, "Auto-disturbances-rejection Controller and Its Applications [J]," *Control and Decision,* vol. 1, 1998.

[8] J. Han, "From PID to active disturbance rejection control," *Industrial Electronics, IEEE transactions on,* vol. 56, pp. 900-906, 2009.

[9] J. Han, "Active disturbance rejection control technique-the technique for estimating and compensating the uncertainties," *National Defense Industry Press, Beijing,* pp. 197-270, 2008.

[10] B. Sun and Z. Gao, "A DSP-based active disturbance rejection control design for a 1-kW H-bridge DC-DC power converter," *Industrial Electronics, IEEE Transactions on,* vol. 52, pp. 1271-1277, 2005.

[11] D. Wu, X. Wang, T. Zhao, and W. Lv, "Application of active disturbance rejection to tracking control of a fast tool servo system," in *Control Applications, 2005. CCA 2005. Proceedings of 2005 IEEE Conference on*, 2005, pp. 547-552.

[12] Q. Zheng and Z. Gao, "An energy saving, factory-validated disturbance decoupling control design for extrusion processes," in *Intelligent Control and Automation (WCICA), 2012 10th World Congress on*, 2012, pp. 2891-2896.

[13] C.-E. Huang, D. Li, and Y. Xue, "Active disturbance rejection control for the ALSTOM gasifier benchmark problem," *Control Engineering Practice,* vol. 21, pp. 556-564, 2013.

[14] H.-S. Li and X.-F. Zhu, "On parameters tuning and optimization of active disturbance rejection controller," *Control Engineering of China,* vol. 11, pp. 419-423, 2004.

[15] Y.-l. SHI, C.-z. HOU, and H.-b. SU, "Auto-disturbance-rejection controller design based on particle swarm optimization algorithm," *Journal of System Simulation,* vol. 2, p. 040, 2008.

[16] Z.-j. KANG, Q.-w. WANG, L. Huang, B. LIU, and L. XU, "Design of Auto-disturbance-rejection Controller of Rectifier in HVDC System Based on the Improved Genetic Algorithm [J]," *High Voltage Engineering,* vol. 8, p. 017, 2008.

[17] Z. Gao, "Scaling and bandwidth-parameterization based controller tuning," in *Proceedings of the American Control Conference*, 2006, pp. 4989-4996.

[18] Z. Gao, "Active disturbance rejection control: a paradigm shift in feedback control system design," in *American Control Conference, 2006*, 2006, p. 7 pp.

[19] T. Mansouri, A. Farasat, M. B. Menhaj, and M. Reza Sadeghi Moghadam, "ARO: A new model free optimization algorithm for real time applications inspired by the asexual reproduction," *Expert Systems with Applications,* vol. 38, pp. 4866-4874, 2011.

[20] A. Farasat, M. B. Menhaj, T. Mansouri, and M. R. S. Moghadam, "ARO: A new model-free optimization algorithm inspired from asexual reproduction," *Applied Soft Computing,* vol. 10, pp. 1284-1292, 2010.

[21] Y. Shi and R. Eberhart, "A modified particle swarm optimizer," in *Evolutionary Computation Proceedings, 1998. IEEE World Congress on Computational Intelligence., The 1998 IEEE International Conference on*, 1998, pp. 69-73.

[22] J. B. Rawlings and J. G. Ekerdt, *Chemical reactor analysis and design fundamentals*: Nob Hill Pub, Llc, 2002.

[23] R. Perry and C. Chilton, "1973, Chemical Engineers Handbook," ed: McGraw–Hill, New York.

ANFIS Based Identification and Control of Distillation Process

M. Shahbazian, H. Jazayerirad, M. Ebnali*

Department of Instrumentation and Automation Engineering, Ahwaz Faculty of Petroleum Engineering, Ahwaz, Iran
*Corresponding author: mehdiebnali@gmail.com

Abstract Typical production objectives in distillation process require the delivery of products whose compositions meets certain specifications. The distillation control system, therefore, must hold products composition as near the set points as possible in the faces of upsets. Since product quality cannot be measured easily and economically online, the control of product quality is often achieved by maintaining a suitable tray temperature near its set point. Tray temperature control method, however, is not a proper option for a multi-component distillation column, because the tray temperature does not correspond exactly to the product composition. To overcome this problem, secondary measurements can be used to infer the product quality and adjust the values of the manipulated variables. In this paper we have used a novel adaptive network fuzzy inference system (ANFIS) based inferential control approach for distillation process. ANFIS is used for identifying the distillation process and building two composition estimators to estimate the compositions of the bottom and sidestream products. The developed estimators are tested and results show that the predictions made by ANFIS structure are in good agreement with the results of simulation by ASPEN HYSYS process simulation package. In addition inferential control by implementation of ANFIS based online composition estimators is more superior to traditional tray temperature control method because of less integral time absolute error.

Keywords: *distillation column, process identification, adaptive network fuzzy inference system, inferential control*

1. Introduction

Distillation of multi-component mixtures is one of the most common separation operations in the chemical industry. The distillation control system must hold product composition as near the set point as possible in the faces of upsets. The control is difficult because the product quality cannot be measured easily and economically online. This is because the instrumentation is either very expensive and / or measurement lags and sampling delays make impossible to design an effective control system.

A solution to this problem is the use of tray temperature control method. The temperature control is based on the assumption that the product composition can satisfy its specification when an appropriate tray temperature is kept constant at its set point. If tray temperatures are to be used, the issue is selecting the best tray or trays on which temperature is held constant. This problem has been discussed in the distillation literature for over a half century. The pioneering books by Buckley [3], Rademaker et al. [14] and Shinskey [16] present a number of alternative procedures for making this selection. Hundreds of papers have appeared that use temperature control with each author using one particular method for control tray selection. A recent discussion of one approach to this problem is given in Marlin [10].

Tray temperature control method is a proper option for binary distillation columns at constant pressure, but in the case where feed composition or feed flow rate changes in a multi-component distillation tower, it is quiet difficult to keep product quality at its set point by using temperature control, because the tray temperature does not correspond exactly to the product composition. In addition pressure changes also cause temperature variation. To solve this problem, secondary measurements in conjunction with a mathematical/empirical model of process can be used to estimate the product quality.

The distillation Process can be modeled using input-output data from experimental tests. Regression and neural network modeling technique are commonly used for this purpose. Morris et al. [11] examined the contribution that various network methodologies can make to the process modeling and control toolbox. Feed forward networks with sigmoidal activation functions, radial bases function networks and auto associative networks were reviewed and studied using data from industrial processes. The concept of dynamic networks was introduced with an example of nonlinear predictive control.

MacMurray et al. [9] described the modeling of packed distillation column with artificial neural network (ANN) and provide an example of complex modeling. The change in the sign of the gain was observed under various operating conditions.

Kung et al. [6] proposed algebraic projection analysis and provide an analytical solution for optimal hidden units size and learning rate of the back propagation neural networks. Baratti et al [1] used neural networks to control distillation column.

Singh et al. [17] by using artificial neural network estimated composition of distillate in distillation column. The structure of neural network has been made by input and output data, which inputs are temperatures and outputs are composition of distillate. Artificial neural networks were applied to predict and estimate vapor–liquid equilibrium data for ternary systems saturated with salt by Nguyen et al. [13].

Tamura et al. [18] have discussed the capabilities of a neural network with a finite number of hidden units and shown with the support of mathematical proof that a four layered feed forward network is superior to three layered feed forward network in terms of the number of parameters needed for the training data.

Murata et al. [12] have investigated the problem of determining the optimal number of parameters in neural network from statistical point of view. The proposed new information criterion therein measures the relative merits of two models having the same structure but different number of parameters and concludes whether more number of neurons should be added to the network or not.

Kano et al. [4] presented a control scheme to control the product composition in a multi-component distillation column. The distillate and bottom compositions are estimated from online measured process variables. The inferential models for estimation product compositions are constructed using dynamic partial least squares (PLS) regression, on the base of simulated time series data. From the detailed dynamic simulation results, it is found that the cascade control system based on a proposed dynamic (PLS) model works much better than the usual tray temperature control system.

Kano et al. [5] proposed a new inferential control scheme termed as "Predictive Inferential Control". In predictive inferential control system, future compositions predicted from online measured process variables are controlled instead of the estimates of current compositions. The key concept is to realize the feedback control with a feed forward effect by the use of inherent nature of a distillation column.

Sbarbaro et al [15] presented the traditional approach to include multi-dimensional information into conventional control systems and proposed a new structure based on pattern recognition. The artificial neural networks and finite state machines as a frame work for designing the control system is used.

Taking advantage of technique developed by Kolmogorov, Kurkova [7] provided a direct proof of the universal approximation capabilities of perceptron type network with two hidden layers.

Lippmann [8] demonstrated the computational power of different neural net models and the effectiveness of simple error correction training procedures. Single and multilayer perceptrons, which can be used for pattern classification, are described as well as Kohonen's feature map algorithm, which can be used for clustering or as a vector quantizer.

To our knowledge, there is no ANFIS based inferential control approach for distillation process. The main aim of this paper is to present a novel identification and control study concerning an industrial distillation column using adaptive network fuzzy inference systems. ANFIS structure is used for building two nonlinear models of the distillation column. The investigated column operates the separation of benzene from the outlet stream of a reforming unit. The benzene concentration in both the bottom and sidestream has to be controlled.

The research work was carried out in the following steps: (i) steady state simulation of the process; (ii) dynamic simulation of process; (iii) identification of process; (iv) inferential control of process based on empirical model of the system and (v) comparison between the proposed inferential control and the traditional tray temperature control strategy.

2. Methods

2.1. Steady State Simulation of the Process

A schematic representation of the examined column is given in Figure 1. The column has 21 valve trays: feed enters at tray 4 and bottom, sidestream and top products are removed at trays 1, 16, 21 respectively. Composition of the feed stream is illustrated in table 1. The column was simulated by means of ASPEN HYSYS process simulation package and steady state operating conditions of column are reported in table 2.

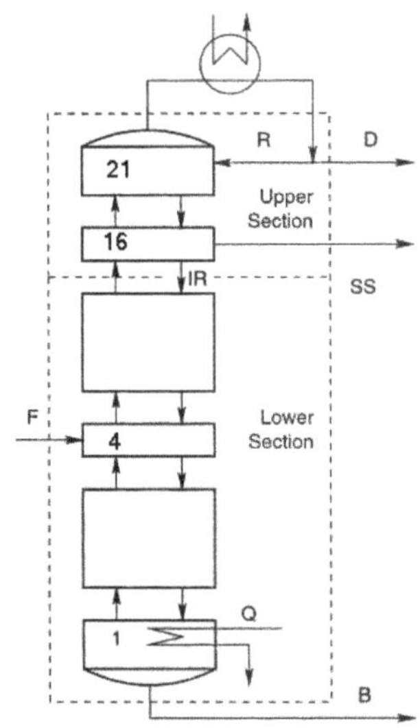

Figure 1. Schematic layout of the column

Table 1. Composition of column feed

Component Name	Concentration (wt %)
n-butane	0.30
i-pentane	5.60
n-pentane	4.07
2,2 dimethyl butane	1.20
2 methyl pentane	6.79
n-hexane	4.32
2,4 dimethyl pentane	0.61
Benzene	5.44
2 methyl hexane	5.46
n-eptane	2.09
Toluene	23.04
2 methyl eptane	1.75
Ethylbenzene	3.82
o-xylene	20.57
1,3,5 trimethyl benzene	14.94

The column feed stream is the outlet stream from the reforming unit. The purpose of this column is to reduce the benzene concentration in the reformed bottom product gasoline, which is sent to the blending unit.

Table 2. Steady state operating conditions of the column

Stream	Property	Unit	Value
Feed	Flow rate	Kg/hr	67500
	temperature	^{0}C	126
	Benzene concentration	Wt %	5.44
Distillate	Flow rate	Kg/hr	5210
	Benzene concentration	Wt %	0.5
	Drum temperature	^{0}C	59.6
Sidestream	Flow rate	Kg/hr	14480
	Benzene concentration	Wt %	21
Bottom	Flow rate	Kg/hr	47810
	Benzene concentration	Wt %	1.2
	Reboiler duty	Kj/hr	2.32×10^7
	Reflux ratio	-	11.8

2.2. Dynamic Simulation of the Process

The dynamic behavior of the column was simulated by ASPEN HYSYS DYNAMIC simulation package. Tray liquid hydraulic was predicted by Glitsch procedure and the equilibrium constants were calculated by SRK relation. The simulation model takes into account the distillation tower as well as all the auxiliary apparatus (reboiler, condenser, reflux drum, measuring devices and control elements).

As illustrated in Figure 2, for design of distillation column control system, it is proposed that the sidestream flow rate is manipulated to control the tray 17 temperature in order to keep the benzene concentration in the side stream constant, the reflux drum liquid level is controlled by manipulating the distillate flow rate, reboiler liquid level is controlled by manipulating bottom product flow rate and finally tray 2 temperature is controlled by manipulating the reboiler duty in order to keep the bottom product benzene concentration constant.

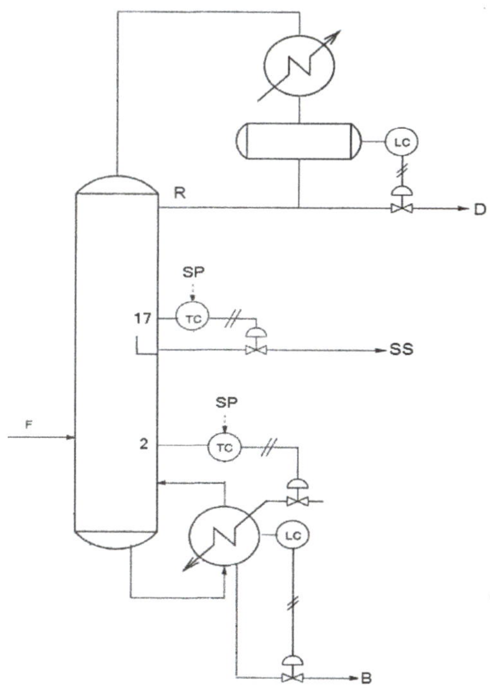

Figure 2. Traditional tray temperature control technique applied to distillation column

2.3. Process Identification

2.3.1. Input signal design and data collection

Simulated data for building inferential estimators are obtained under the following conditions. A connection is established between MATLAB and HYSYS software. The pseudo random binary signals of bounded and varying amplitude (within ±5% of the steady state values) are introduced by MATLAB as feed temperature and total feed flow rate changes during dynamic simulation of the column by HYSYS simulator. For realizing slow composition changes, each signal is filtered by a first order lag model. Total simulation time is 8 hours and process data are sampled every 10 seconds by MATLAB.

Simulated data for validating inferential models are obtained under almost the same conditions as described above. Total simulation time is 4 hours.

If the input-output data described above are obtained when the temperature controllers are in the automatic mode, the data do not include the operational conditions when the controlled temperatures undergo large changes. However, when inferential composition control is applied instead of temperature control, the trays temperature fluctuates greatly. Thus, when an inferential model is used for composition control, the accuracy of estimation may deteriorate due to large changes of tray temperature. In order to improve the accuracy, the inferential model must be built using appropriate data, which include large fluctuation of the temperature. For this purpose, the proportional gain of bottom temperature controller is changed between ±50% of base controller gain in the simulation in order to change reboiler duty. Figure 3 represents the input signals applied to the process for collecting identification data.

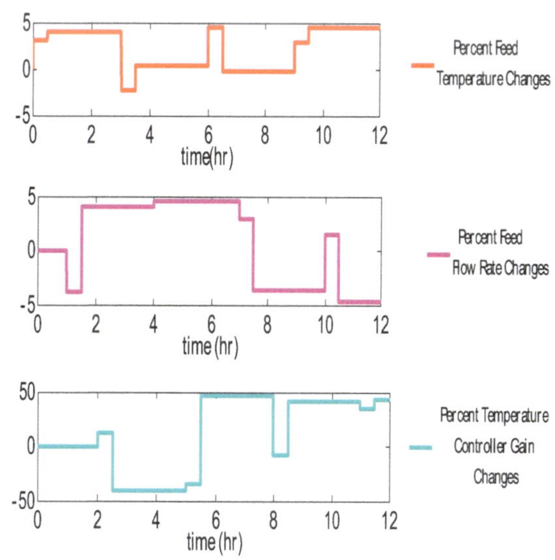

Figure 3. Input signals applied to the process for collecting of identification data

2.3.2. Data Preprocessing

Collected data have different magnitudes depending on the units adopted. This can cause larger magnitude variables to be dominant over smaller ones during the training process. Data scaling is therefore needed. Min-max normalization method is one of the common scaling methods. Min-max normalization is given by:

$$a' = \frac{a - \min_a}{\max_a - \min_a}\left(\max_{a'} - \min_{a'}\right) + \min_{a'} \quad (1)$$

Where
a is unscaled variable;
a' is scaled variable
$\min_a$ is minimum value of unscaled variable
$\max_a$ is maximum value of unscaled variable
$\min_{a'}$ is minimum value of scaled variable
$\max_{a'}$ is maximum value of scaled variable
In this research, collected data are scaled to [0,1] interval. Since collected data are the results of simulation and have no inconsistent data, so other preprocessing operation such as outlier detection, missing value replacement and data de-noising are not required.

2.3.3. Model Structure Selection

One modification of a neural network structure is to replace some or all components of neuron by fuzzy logic operations. Conventional neural networks are used to approximate functions from numerical input-output data. Fuzzy-neural networks are a more general computational structure with which function approximation can be extended to linguistic data.

To illustrate the use of neural networks for fuzzy inference, consider a fuzzy rule base consisting of only two Sugeno-Takagi rules:

$$R_1 : \textit{If } x_1 \textit{ is } A_1 \textit{ and } x_2 \textit{ is } B_1 \textit{ then } y = f_1(x) \quad (2)$$

$$R_2 : \textit{If } x_1 \textit{ is } A_2 \textit{ and } x_2 \textit{ is } B_2 \textit{ then } y = f_2(x) \quad (3)$$

Where A_i and B_i are fuzzy sets and

$$f_1(x) = Z_{11}x_1 + Z_{12}x_2 + Z_{13} \quad (4)$$

$$f_2(x) = Z_{21}x_1 + Z_{22}x_2 + Z_{23} \quad (5)$$

When numerical input $x=(x_1,x_2)$ is presented, the inference mechanism will produce the numerical output

$$y^* = \frac{A_1(x_1)B_1(x_2)f_1(x) + A_2(x_1)B_2(x_2)f_2(x)}{A_1(x_1)B_1(x_2) + A_2(x_1)B_2(x_2)} \quad (6)$$

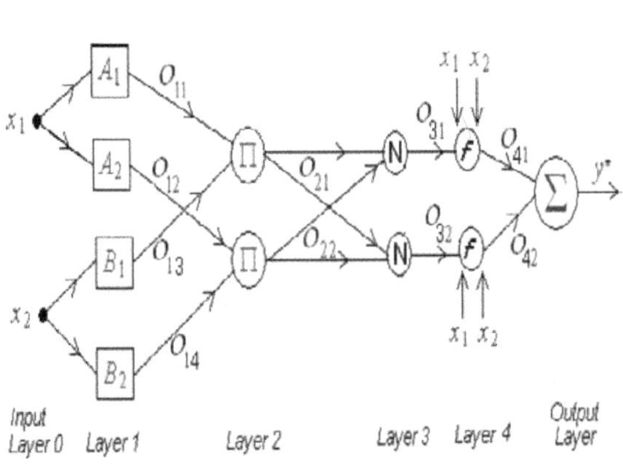

Figure 4. First order Sugeno fuzzy model with two rules

A fuzzy neural network for implementing the above formula is shown in Figure 4. The observed input $x=(x_1,x_2)$ is presented to layer 1 by input layer 0. The output of layer 1 is:

$$(O_{11}, O_{12}, O_{13}, O_{14}) = (A_1(x_1), A_2(x_1), B_1(x_2), B_2(x_2)) \quad (7)$$

Layer 2 consists of fuzzy neurons with an aggregation operator being some T-norm. If product T-norm is used the output of layer 2 is:

$$(O_{21}, O_{22}) = (A_1(x_1)B_1(x_2), A_2(x_1)B_2(x_2)) \quad (8)$$

Layer 3 is normalizer. The output of layer 3 is:

$$(O_{31}, O_{32}) = \left(\frac{O_{21}}{O_{21}+O_{22}}, \frac{O_{22}}{O_{21}+O_{22}}\right) \quad (9)$$

The fuzzy neurons in layer 4 output the values

$$(O_{41}, O_{42}) = (O_{31}f_1, O_{32}f_2) \quad (10)$$

Finally the output layer calculates the estimated output by summing:

$$y^* = O_{41} + O_{42} \quad (11)$$

Of course, this neural network type for representing the inference procedure for a rule base of two rules can be extended in an obvious way to an arbitrary number of rules.

In the ANFIS structure, the parameters of the premise and consequence play the role of weights in neural network systems. The ANFIS learning algorithm consists of adjusting the mentioned parameters from sample data $\left((x_1^k, x_2^k), y^k\right), k = 1, \ldots, N.$

2.3.4. Validation of Model

The estimated model can be evaluated on the basis of mean squared error of prediction (MSEP), which is calculated by applying the models to the validation data.

$$MSEP = \frac{1}{N}\sum_{n=1}^{N}(b(n) - \hat{b}(n))^2 \quad (12)$$

Where b is a measurement of the product composition, $\hat{b}$ is its estimate and N is the number of measurements.

3. Results and Discussion

3.1. Inferential Models

The output variables to be estimated are the benzene concentration in bottom and sidestream products. The inputs to the inferential estimators are various combinations of 21 tray, condenser and reboiler temperatures and feed, distillate, bottom and sidestream flow rates. Table 3 illustrates the properties of fuzzy inference systems and MSEP of validation data for the following combinations of input variables to the inference systems (cases A1 to A10 for estimation of bottom product benzene concentration and cases B1 to B10 for estimation of sidestream benzene concentration).

A1. *T1, T2, T3*
A2. *T1, T2, T3, T4*
A3. *T1, T2, T3, T4, RebT*
A4. *T1, T2, T3, RebT, FF*
A5. *T1, T4, T7, T10, T13, T16, T19*
A6. *T3, T4, RebT, BF, FF*
A7. *T1, T2, T3, RebT,RDT,FF*
A8. *RebT, RDT, FF, BF, SSF*
A9. *T1, T16, T21, RebT, RDT*
A10. *T1, T2, RebT, RDT, FF, BF*

B1. *T15, T16, T17*
B2. *T14, T15, T16, T17, T18*
B3. *T16, T17, RDT, RebT*
B4. *T16, T17, SSF, FF*
B5. *T1, T16, T21, RDT, RebT*
B6. *RDT, RebT, FF, SSF, BF*
B7. *T1, T15, T16, T17, T21*
B8. *T1, T4, T7, T10, T13, T16, T19*
B9. *T16, T17, T18, RDT, RebT, FF, SSF*
B10. *T16, T17, T18, RDT, FF, SSF*

Table 3. Fuzzy inference system properties and MSEP for some combination of inputs to ANFIS structure

Case	Inference system properties			MSEP for bottom product benzene concentration ($x10^{-3}$)
	Number of fuzzy rules	Number of linear parameters	Number of nonlinear parameters	
A1	2	8	12	640.14
A2	2	10	16	71.21
A3	2	12	20	72.78
A4	2	12	20	50.17
A5	2	16	28	6.60
A6	2	12	20	80.54
A7	2	14	24	29.32
A8	2	12	20	0.81
A9	2	12	20	31.05
A10	2	14	24	0.21
Case	Inference system properties			MSEP for sidestream benzene concentration ($x10^{-3}$)
	Number of fuzzy rules	Number of linear parameters	Number of nonlinear parameters	
B1	2	8	12	364.44
B2	2	12	20	6.32
B3	2	10	16	32.20
B4	2	10	16	89.51
B5	2	12	20	6.47
B6	2	12	20	21.84
B7	2	12	20	24.80
B8	2	16	28	4.88
B9	2	16	28	1.65
B10	2	14	24	0.34

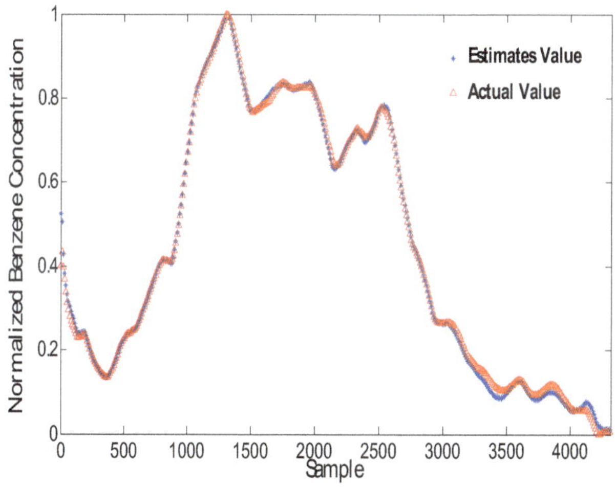

Figure 5. Estimated and actual value for bottom product benzene concentration

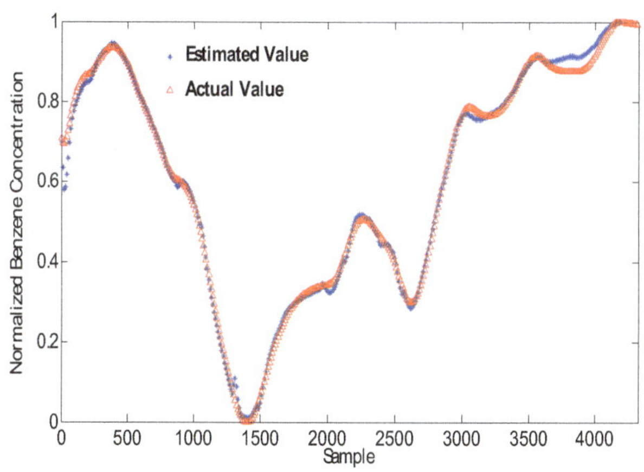

Figure 6. Estimated and actual value for sidestream benzene concentration

Examination of various combinations of inputs show that combinations recommended in cases A10 and B10 are the most efficient for estimation of Benzene concentration in the bottom product and sidestream, respectively, based on MSEP. Figure 5 and Figure 6 visualize the estimated values versus the actual ones for both training (samples number from1 to 2880) and validation data (samples number from 2881 to 4320) of mentioned cases.

3.2. Control Strategy and Results

On the basis of the control analysis reported in section 2.2, the inferential control strategy of the column shown in Figure 7 was adopted. A cascade control scheme has been assumed for each control variable, with benzene concentration on the primary (master) loop and tray temperature on the secondary (slave) one. The efficiency of the control strategy was examined on the basis of the responses of the process to step change of the benzene composition of the feed stream (±10%), feed temperature (±5%) and feed total mass flow rate (±5%). The evaluation of control efficiency was done by calculating the integral time absolute error (ITAE) (see table 4) over a horizon of 3 hours after the implementation of disturbance.

The response of the system to step change in feed stream benzene concentration (+10%), when the benzene concentration of only one product stream is controlled by the cascade inferential scheme or the benzene concentration of two product streams are controlled by the cascade inferential scheme (full inferential) and also keeping constant the relevant tray temperature were compared in Figure 8 and Figure 9. It can be observed that:

The traditional tray temperature control scheme lead to the significant offset for both the bottom and sidestream.

The response of full inferential scheme and single variable inferential scheme are very close to each other.

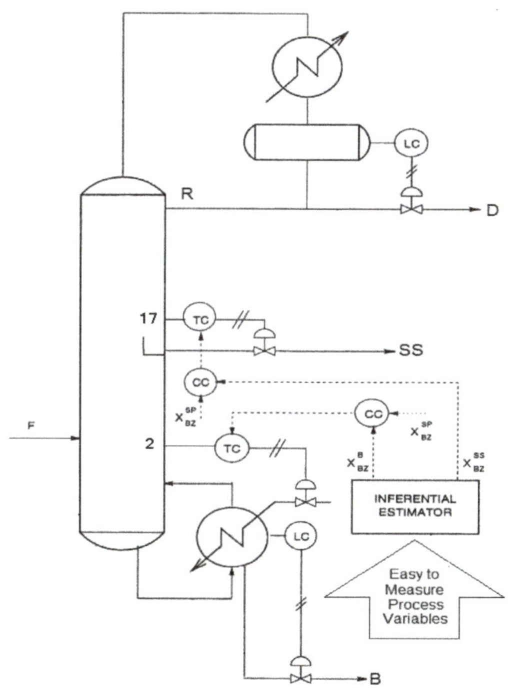

Figure 7. Layout of column inferential control configuration

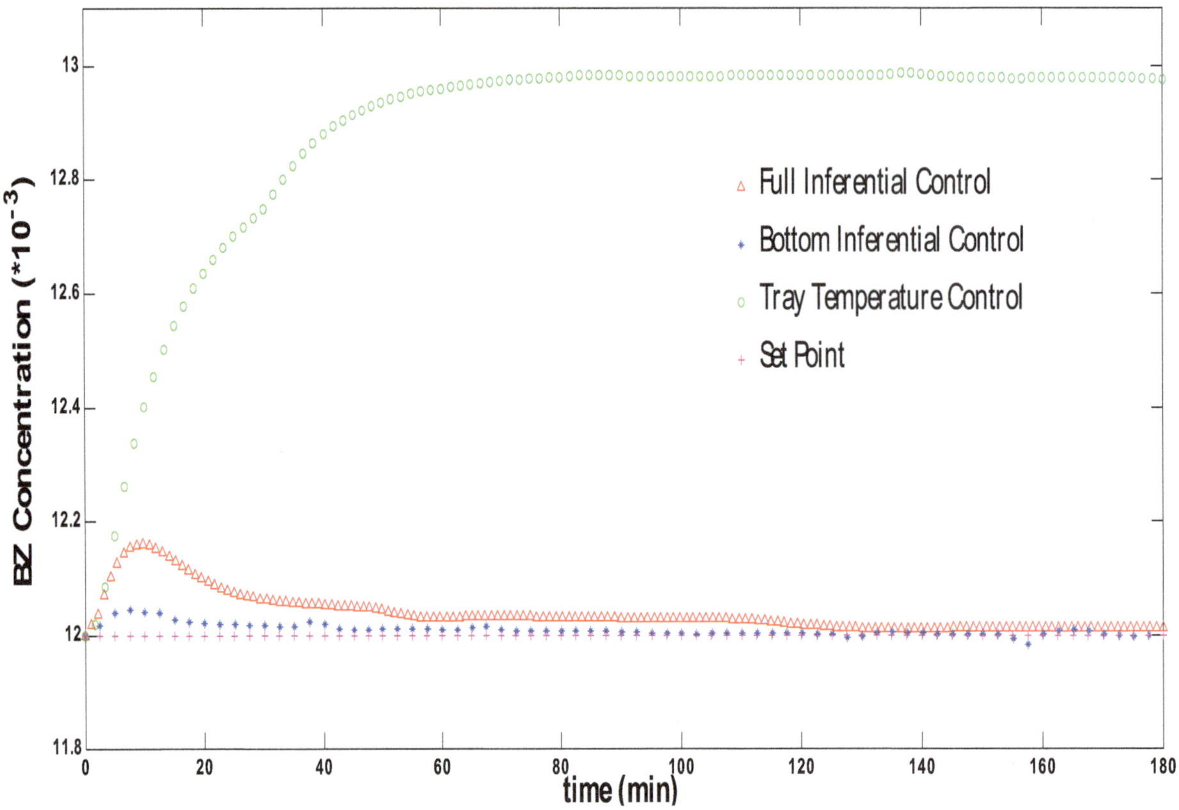

Figure 8. Comparison between the performance of the traditional tray temperature, full inferential and single variable inferential bottom product controller in response to step change in feed stream benzene concentration

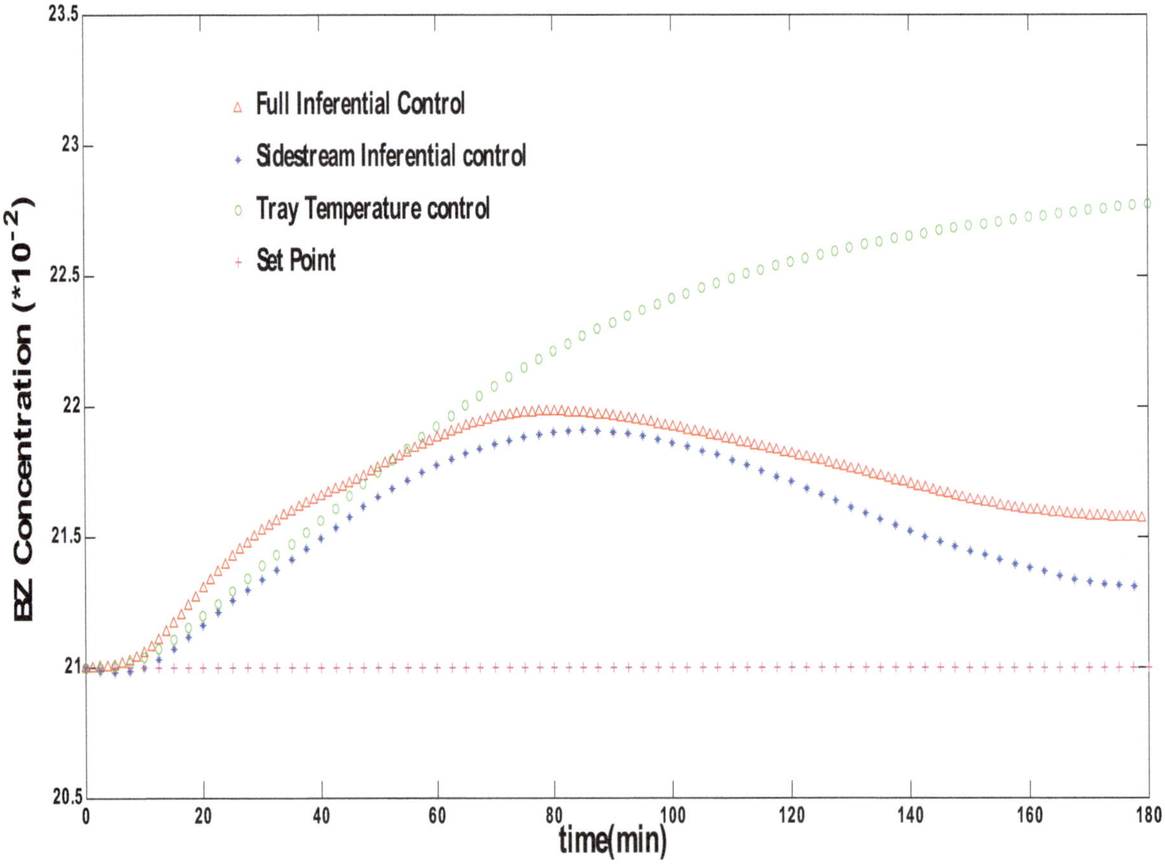

Figure 9. Comparison between the performance of the traditional tray temperature, full inferential and single variable sidestream composition controller in response to step change in feed stream benzene concentration (+10%)

Table 4. ITAE for benzene concentration for different control schemes and different disturbances

Disturbance	Stream	Control Strategy	ITAE
Step change of feed stream benzene concentration (+10%)	Sidestream	Tray Temperature	10931.17
		Single Variable Inferential	3362.50
		Full Inferential	4187.40
	Bottom	Tray Temperature	701.65
		Single Variable Inferential	18.38
		Full Inferential	18.56
Step change of feed stream benzene concentration (-10%)	Sidestream	Tray Temperature	9806.32
		Single Variable Inferential	3265.21
		Full Inferential	3984.41
	Bottom	Tray Temperature	564.70
		Single Variable Inferential	18.04
		Full Inferential	20.32
Step change of feed stream temperature (+5%)	Sidestream	Tray Temperature	432.15
		Single Variable Inferential	26.46
		Full Inferential	32.46
	Bottom	Tray Temperature	256.00
		Single Variable Inferential	8.62
		Full Inferential	11.45
Step change of feed stream temperature (-5%)	Sidestream	Tray Temperature	801.34
		Single Variable Inferential	154.06
		Full Inferential	198.54
	Bottom	Tray Temperature	311.10
		Single Variable Inferential	24.32
		Full Inferential	30.80
Step change of feed stream total flow rate (+5%)	Sidestream	Tray Temperature	1721.70
		Single Variable Inferential	466.25
		Full Inferential	497.68
	Bottom	Tray Temperature	1010.30
		Single Variable Inferential	56.14
		Full Inferential	120.00
Step change of feed stream total flow rate (-5%)	Sidestream	Tray Temperature	3051.76
		Single Variable Inferential	854.08
		Full Inferential	924.25
	Bottom	Tray Temperature	1925.12
		Single Variable Inferential	124.31
		Full Inferential	208.31

At the end of transient state a small offset of sidestream benzene concentration can be observed when both the single variable and full inferential strategies are applied. This is due to the inability of sidestream composition estimator to estimate as accurate as bottom composition estimator (see Figure 5 and Figure 6). The value of offset is greater when full inferential scheme is used. This is the consequence of interaction between the individual control loops (coupling problem).

4. Conclusion

Distillation column is one of the equipments is used in chemical industry. Products of this column must have specific composition concentration. The devices for measuring the product composition are costly and have high delay time. Inferential control by using of secondary measurements in conjunction with an empirical model of process can be the replacement of direct composition control method.

In this research, a column with 21 trays, condenser and reboiler is studied. 4320 data is generated by implementation of proper disturbances to the process. 2880 data is used for training of ANFIS based composition estimators and 1440 data is used for validation of the models. The MSEP criterion is used to identify the optimal combination of input variables to inference system.

It is concluded that the use of ANFIS based composition estimators leads to the predictions that are in good agreement with the results of simulation by ASPEN HYSYS process simulation package, also inferential control of distillation column by implementation of ANFIS based online composition estimators is more superior to traditional tray temperature control method because of less integral time absolute error.

Nomenclature

B: Bottom Stream
BF: Bottom Stream Flow Rate (Kg/hr)
BZ: Benzene
CC: Composition Controller
D: Distillate Stream
F: Feed Stream
FF: Feed Stream Flow Rate (Kg/hr)
IR: Internal Reflux Stream
LC: Level Controller
N: Number Of Samples
Q: Reboiler Duty (Kj/hr)
R: External Reflux Stream
RDT: Reflux Drum Temperature (°C)
RebT: Reboiler Temperature (°C)
SS: Side Stream
SSF: Sidestream Flow Rate (Kg/hr)
SP: Set Point
TC: Temperature Controller
Ti: Tray *i* Temperature (°C)
Wt: Weight (Kg)
x: Linguistic Variable
X: Concentration In Weight Percent
Z: Constant coefficient

References

[1] Baratti,R., Corti, S. and Servida, A., "A Feed forward Control Strategy for Distillation Columns," *Artificial Intelligent Engineering*, 11. 405-412.1997.

[2] Bettoni, A., Bravi, M. and Chianese, A. "Inferential Control of Sidestream Distillation Columns," *Computer and Chemical Engineering*, 23. 1737-1744. 2000.

[3] Buckley, P.S, *Techniques of Process Control*, Wiley, 1964.

[4] Kano, M., Showchaiya, N., Hasebe, S. and Hashimoto, I., "Inferential Control System of Distillation Compositions Using Dynamic Partial Least Square Regression," *Journal of Process Control*, 10. 157-166. 2000.

[5] Kano, M., Showchaiya, N., Hasebe, S. and Hashimoto, I., "Inferential Control of Distillation Compositions: Selection of Model and Control Configuration," *Control Engineering Practice*, 11. 927-933. 2003.

[6] Kung, S.Y and Hwang, J.N, *An Algebraic Projection Analysis for Optimal Hidden Units Size and Learning Rates in Back Propagation Learning*, Princeton University, Department of Electrical Engineering, Princeton, NJ 08544.

[7] Kurkova, V., "Kolmogorov's Theorem and Multi Layer Neural Networks," *Neural Networks*, 5. 501-506. 1992.

[8] Lippmann, R.P, *Neural Nets for Computing*, Lincola Laboratory, M.I.T, Lexington, MA 02173, U.S.A.

[9] MacMurray, J.C. and Himmelblau, D.M., "Modeling and Control of a Packed Distillation Column Using Artificial Neural Networks," *Computer and Chemical Engineering*, 19.1088. 1995

[10] Marlin, T.E, *Process Control: Designing Processes and Control Systems for Dynamic Performance*, McGraw-Hill, 2000.

[11] Morris, A.J., Montague, G. A. and Willis, M.J., "Artificial Neural Networks: Studies in Process Modeling and Control," *Transaction of Institute of Chemical Engineering*, 72 (A). 3-19. 1994.

[12] Murata, N., Yoshizawa, S. and Amari, S., "Network Information Criterion Determining the Number of Hidden Units for an Artificial Neural Network Model," *IEEE Transaction of Neural Networks*, 5. 865- 872.1994.

[13] Nguyen, V.D., Tan, R.R., Brondial, Y. and Fuchino, T., "Prediction of Vapor-Liquid Equilibrium Data for Ternary Systems Using Artificial Neural Networks," Fluid Phase Equilibra, 254. 188-197. 2007.

[14] Rademaker, O., Rijnsdorp, J.E. and Maarleveld, A., Dynamics *and Control of Continuous Distillation Units*, Elsevier, 1975.

[15] Sbarbaro, D., Espinoza, P. and Araneda, J., "A Pattern Based Strategy for Using Multidimensional Sensors in Process Control," *Computer and Chemical Engineering*, 27. 1943. 2003.

[16] Shinskey, F.G, *Distillation Control*, McGraw-Hill, 1977.

[17] Singh, V., Gupta, I. and Gupta, H.O. "ANN Based Estimator for Distillation Inferential Control," *Chemical Engineering Process*, 44. 785-795. 2005.

[18] Tamura, S., Tateishi, M., "Capabilities of a Four Layered Feed Forward Neural Network: Four Layers Versus Three," *IEEE Transaction of Neural Networks*, 8. 251-255.1997.

In the Issue of Complex Plane Motions of the Robot

Lozhkin A.G.[1,*], Abramov I.V.[2], Nikitin Yu.R.[2], Arkhipov I.O.[1], Sultanov R.O.[1]

[1]Departments "Software", M.T. Kalashnikov Izhevsk State Technical University, Izhevsk, Russia
[2]Mechatronic Systems department, M.T. Kalashnikov Izhevsk State Technical University, Izhevsk, Russia
*Corresponding author: lag@istu.ru

Abstract Two methods (classical and direct) for calculating an elliptical path for the kinematic mechanisms is considered briefly. A new method is proposed for such calculations. It is simpler, convenient for further calculations and can be applied to arbitrary curves in some cases. This method can serve as a basis for building simple robots with complex trajectories.

Keywords: *ellipse, calculation of trajectory, kinematic mechanisms, plane differentiable curves*

1. Introduction

Many standards such as the life cycle of a product, company management, etc. assert that the right decisions to be applied in the early stages, which guarantees places without high costs in the following stages. Until now, the first author of the article used the new methods of valuation solely for the design or CAD or CAM [6,8,9] but the quality of product design starts much earlier in the development phase of the kinematic mechanisms by team of authors. One of the major tasks of this phase is to calculate the trajectory of motion of the mechanism. Rational calculation of the trajectory of motion actuators automatic machines is important for efficient production. Calculate the trajectories of the kinematic mechanisms studied and necessary mathematical theory in papers [1,2,5,7,10,12,13]. The calculations of the motion trajectory of the kinematic mechanisms are often yields a system of parametric equations of a circle or ellipse

$$\begin{cases} x = a\cos t + h\sin t \\ y = g\cos t + b\sin t \end{cases} \quad (1)$$

where $a,b,c,d,t \in \mathbf{R}$.

2. Transformation of the Circle

It is proposed to consider the parametric system as a transformation of the circle. Canonical equation of the transformed circle can be converted to determine the inverse transformation of the matrix $T = \begin{pmatrix} a & h \\ g & b \end{pmatrix}$. Matrix T^{-1} is equal to $T^{-1} = \begin{pmatrix} \frac{b}{ab-gh} & \frac{-h}{ab-gh} \\ \frac{-g}{ab-gh} & \frac{a}{ab-gh} \end{pmatrix}$ [Efimov 1967].

Consequently the canonical equation of a circle is as follows:

$$\left(\frac{bx}{ab-gh} - \frac{hy}{ab-gh}\right)^2 + \left(\frac{ay}{ab-gh} - \frac{gx}{ab-gh}\right)^2 - 1^2 = 0$$

$$\frac{b^2x^2}{(ab-gh)^2} - \frac{2bhxy}{(ab-gh)^2} + \frac{h^2y^2}{(ab-gh)^2} + \frac{a^2y^2}{(ab-gh)^2} - \frac{2agxy}{(ab-gh)^2} + \frac{g^2x^2}{(ab-gh)^2} - 1^2 = 0,$$

$$\frac{b^2+g^2}{(ab-gh)^2}x^2 - 2\frac{(bh+ag)}{(ab-gh)^2}xy + \frac{a^2+h^2}{(ab-gh)^2}y^2 - 1^2 = 0. \quad (2)$$

It is known from analytic geometry [Efimov 2009], how to get the angle of rotation of the quadratic form of the canonical equation of the second order line. If the general equation written in the form $Ax^2 + 2Bxy + Cy^2 + Dx + Ey + F = 0$, the angle of rotation can be found is $\tan 2\alpha = \frac{2B}{A-C}$.

We obtain from (1) the values of the coefficients: $A = \frac{b^2+g^2}{(ab-gh)^2}$; $B = -\frac{(bh+ag)}{(ab-gh)^2}$; $C = \frac{a^2+h^2}{(ab-gh)^2}$; $D = 0$; $E = 0$; $f = -1^2$. The angle of rotation of the resulting curve will be equal to

$$\tan 2\alpha = \frac{-\frac{2(bh+ag)}{(ab-gh)^2}}{\frac{b^2+g^2}{(ab-gh)^2} - \frac{a^2+h^2}{(ab-gh)^2}},$$

$$\tan 2\alpha = \frac{2(bh+ag)}{(a^2+h^2)-(b^2+g^2)}. \quad (3)$$

We solve the equation (1) from the characteristic equation $Pz = \lambda z$ [3], where $z = \begin{pmatrix} x \\ y \end{pmatrix}$. It is obtain solutions.

$$\lambda_{1,2} = \frac{A+C \pm \sqrt{(A+C)^2 - 4(AC-B^2)}}{2},$$

$$\lambda_{1\,2} = \frac{\frac{b^2+g^2}{(ab-gh)^2} + \frac{a^2+h^2}{(ab-gh)^2} \pm \sqrt{\left(\frac{b^2+g^2}{(ab-gh)^2} + \frac{a^2+h^2}{(ab-gh)^2}\right)^2 - 4\frac{b^2+g^2}{(ab-gh)^2}\frac{a^2+h^2}{(ab-gh)^2} - \frac{(bh+ag)^2}{(ab-gh)^4})}}{2}$$

$$\lambda_{1,2} = \frac{b^2+g^2+a^2+h^2 \pm \sqrt{(b^2+g^2+a^2+h^2)^2 - 4((b^2+g^2)(a^2+h^2) - (bh+ag)^2)}}{2(ab-gh)^2}.$$

The solution

$$\begin{cases} x = \lambda_1 \cos(t+\alpha) \\ y = \lambda_2 \sin(t+\alpha) \end{cases} \quad (4)$$

is obtained, but it is difficult to apply in the coefficients for the other calculations. Engineer must choose one right decisions of the two $\lambda_{1,2}$ additionally.

It is proposed to use another method [7,8,9]. The method is based on the permutation symmetry and other symmetries [8].

3. Direct Analytical Method of Arbitrary Linear Transformations

We calculate the angle β which is symmetrical own corner α for system (1)

$$\tan 2\beta = \frac{2(gb+ha)}{a^2-h^2-b^2+g^2} \quad (5)$$

in the first step. The angle is determined from the two equations

$$\tan\alpha_1 = \frac{a\tan\beta - h}{b - g\tan\beta} \quad (6)$$

and

$$\tan\alpha_2 = \frac{b\tan\beta + g}{a + h\tan\beta}. \quad (7)$$

Angles are equal if the calculation and transformation is correct. The coefficients $\lambda = \begin{pmatrix} c \\ d \end{pmatrix}$ are equal

$$c_1 = \frac{a\cos\beta + h\sin\beta}{\cos\alpha}, \quad (8)$$

$$c_2 = \frac{b\sin\beta + g\cos\beta}{\sin\alpha}, \quad (9)$$

$$d_1 = \frac{a\sin\beta - h\cos\beta}{\sin\alpha}, \quad (10)$$

$$d_2 = \frac{b\cos\beta - g\sin\beta}{\cos\alpha}. \quad (11)$$

The initial system will be:

$$\begin{cases} x = c_1\cos(t+\alpha) \\ y = d_1\sin(t+\alpha) \end{cases} \quad (12)$$

and

$$\begin{cases} x = c_2\cos(t+\alpha) \\ y = d_2\sin(t+\alpha) \end{cases}. \quad (13)$$

Both solutions are valid and the choice is not necessary. General solution for any curve could not get unfortunately. If the result of the calculations obtained systems $\begin{cases} x = mf_x - nf_y \\ y = mf_x + nf_y \end{cases}$, $\begin{cases} x = kmf_x - nf_y \\ y = -knf_x + mf_y \end{cases}$, $\begin{cases} x = mf_x + knf_y \\ y = -nf_x + kmf_y \end{cases}$, where f_x and f_y arbitrary trigonometrically functions, the solution can be found similarly. For example, the system $\begin{cases} x = \cos t - 2\sin 2t \\ y = \cos t + 2\sin 2t \end{cases}$ can be reduced to the form $\begin{cases} x = c\cos(t+\alpha) \\ y = d\sin 2(t+\alpha) \end{cases}$.

4. Theoretical Tests

Submitted theorems are very simple and allow the reader to the proof without reference to sources in a foreign language. Let's consider the sufficiency of all the theorems only. Theorem 3 is published the first.

Theorem 1. Angles α_1 (6) and α_2 (7) is equivalence.

Proof.

Let's equating the formula (6) and (7) in value $\tan\alpha$: $\frac{b\tan\beta + g}{a + h\tan\beta} = \frac{a\tan\beta - h}{b - g\tan\beta}$, $(b\tan\beta + g)(b - g\tan\beta) = (a\tan\beta - h)(a + h\tan\beta)$ or $v_1 = v_2$. Let's show $v_1 = (b\tan\beta + g)(b - g\tan\beta)$: $v_1 = b^2\tan\beta + gb - gb\tan^2\beta - g^2\tan\beta$, $v_1 = (b^2 - g^2)\tan\beta + gb(1 - \tan^2\beta)$. The value $v_2 = (a\tan\beta - h)(a + h\tan\beta)$ is determinate: $v_2 = a^2\tan\beta - ha + ah\tan^2\beta - h^2\tan\beta$, $v_2 = (a^2 - h^2)\tan\beta - ha(1 - \tan^2\beta)$. Results equate to $(a^2 - h^2)\tan\beta - ha(1 - \tan^2\beta) = (b^2 - g^2)\tan\beta + gb(1 - \tan^2\beta)$, $(a^2 - h^2)\tan\beta - (b^2 - g^2)\tan\beta = gb(1 - \tan^2\beta) + ha(1 - \tan^2\beta)$,

$(a^2-h^2-b^2+g^2)\tan\beta=(gb+ha)(1-\tan^2\beta)$,

$\frac{2\tan\beta}{1-\tan^2\beta}=\frac{2(gb+ha)}{a^2-h^2-b^2+g^2}$,

$\frac{\tan\beta}{1-\tan^2\beta}=\frac{gb+ha}{a^2-h^2-b^2+g^2}$,

$\tan 2\beta=\frac{2(gb+ha)}{a^2-h^2-b^2+g^2}$. The resulting formula coincides with the formula (5) and hence the angle α is the same in both formulas.

Remark. Theorem is intended to linearly independent transformations only.

Theorem 2. Parameters $c_{1,2}$ when $b\sin\beta+g\cos\beta\neq 0$ and $d_{1,2}$ when $b\cos\beta-g\sin\beta\neq 0$ are equivalence.

Proof.

1. Let's assume the opposite for coefficient c. Let's $c_1\neq c_2$, $c_1,c_2\neq 0$ than $c_1-c_2=\Delta$ from (8) and (9):

$\frac{b\sin\beta+g\cos\beta}{\sin\alpha}=\frac{a\cos\beta+h\sin\beta}{\cos\alpha}+\Delta$,

$\frac{1}{\sin\alpha}=\frac{a\cos\beta+h\sin\beta}{\cos\alpha(b\sin\beta+g\cos\beta)}+\frac{\Delta}{b\sin\beta+g\cos\beta}$,

$\frac{\cos\alpha}{\sin\alpha}=\frac{a\cos\beta+h\sin\beta}{b\sin\beta+g\cos\beta}+\frac{\Delta\cos\alpha}{b\sin\beta+g\cos\beta}$. We have from (7) and theorem 1 $\tan\alpha=\tan\alpha+\frac{\Delta\cos\alpha}{b\sin\beta+g\cos\beta}$.

Equality holds when $\cos\alpha=0$ or $\Delta=0$. Since $\cos\alpha$ in general not equal 0 then $\Delta=0$ and parameters are equivalence.

2. Let's assume the opposite for coefficient d. Let's $d_1\neq d_2$, $d_1,d_2\neq 0$ than $d_1-d_2=\Delta$ from (10) and (11):

$\frac{b\cos\beta-g\sin\beta}{\cos\alpha}=\frac{a\sin\beta-h\cos\beta}{\sin\alpha}+\Delta$,

$\frac{1}{\cos\alpha}=\frac{a\sin\beta-h\cos\beta}{\sin\alpha(b\cos\beta-g\sin\beta)}+\frac{\Delta}{b\cos\beta-g\sin\beta}$,

$\frac{\sin\alpha}{\cos\alpha}=\frac{a\sin\beta-h\cos\beta}{b\cos\beta-g\sin\beta}+\frac{\Delta\sin\alpha}{b\cos\beta-g\sin\beta}$. We have from (6) and theorem 1 $\tan\alpha=\tan\alpha+\frac{\Delta\sin\alpha}{b\cos\beta-g\sin\beta}$.

Equality holds when $\sin\alpha=0$ or $\Delta=0$. Since $\sin\alpha$ in general not equal 0 then $\Delta=0$ and parameters are equivalence.

Theorem 3. The own angle in classical method and direct analytical method is equivalence.

Proof.

Let's equating the formula (6) and (7) in value $\tan\beta$. Let's deduce the angle $\tan\beta$ of the formula (7):

$\tan\alpha_1=\frac{a\tan\beta-h}{b-g\tan\beta}$, $\tan\alpha_1(b-g\tan\beta)=a\tan\beta-h$,

$\tan\alpha_1 b-g\tan\alpha_1\tan\beta=a\tan\beta-h$,

$\tan\alpha_1 b+h=a\tan\beta+g\tan\alpha_1\tan\beta$, $\frac{\tan\alpha_1 b+h}{a+g\tan\alpha_1}=\tan\beta$.

Let's deduce the angle $\tan\beta$ of the formula (6):

$\tan\alpha_2(a+h\tan\beta)=b\tan\beta+g$,

$\tan\alpha_2 a+h\tan\alpha_2\tan\beta=b\tan\beta+g$

$\tan\alpha_2 a-g=b\tan\beta-h\tan\alpha_2\tan\beta$, $\frac{\tan\alpha_2 a-g}{b-h\tan\alpha_2}$ =$\tan\beta$.

Results equate to $\frac{\tan\alpha_1 b+h}{a+g\tan\alpha_1}=\frac{\tan\alpha_2 a-g}{b-h\tan\alpha_2}$,

$(\tan\alpha_1 b+h)(b-h\tan\alpha_2)=(\tan\alpha_2 a-g)(a+g\tan\alpha_1)$.

We have from theorem 1 $\alpha_1=\alpha_2$ that $(b\tan\alpha+h)(b-h\tan\alpha)=(a\tan\alpha-g)(a+g\tan\alpha)$ or $v_1=v_2$. Let's consider $v_1=(b\tan\alpha+h)(b-h\tan\alpha)$:

$v_1=b^2\tan\alpha+bh-bh\tan^2\alpha-h^2\tan\alpha$,

$v_1=(b^2-h^2)\tan\alpha+bh(1-\tan^2\alpha)$. Value is

$v_2=(a\tan\alpha-g)(a+g\tan\alpha)$

$v_2=a^2\tan\alpha-ga+ag\tan^2\alpha-g^2\tan\alpha$,

$v_2=(a^2-g^2)\tan\alpha-ag(1-\tan^2\alpha)$. Results equate to

$(a^2-g^2)\tan\alpha-ag(1-\tan^2\alpha)=(b^2-h^2)\tan\alpha+bh(1-\tan^2\alpha)$,

$(a^2-g^2)\tan\alpha-(b^2-h^2)\tan\alpha=ag(1-\tan^2\alpha)+bh(1-\tan^2\alpha)$,

$(a^2-g^2-b^2+h^2)\tan\alpha=(ag+bh)(1-\tan^2\alpha)$,

$\frac{\tan\alpha}{1-\tan^2\alpha}=\frac{ag+bh}{a^2+h^2-g^2-b^2}$, $\tan 2\alpha=\frac{2(ag+bh)}{a^2+h^2-g^2-b^2}$.

We find (2).

The proof of the transformation parameters can be obtained similarly to Theorem 2, but it is cumbersome for publication in the journal.

Let coefficients of matrix are $a=b=k$, $h=g=0$.

Angle β is equivalence $\frac{2(gb+ha)}{a^2-h^2-b^2+g^2}=\tan 2\beta$,

$\frac{2(0\times k+0\times k)}{k^2-0^2-k^2+0^2}=\tan 2\beta$, $\frac{0}{0}=\tan 2\beta$. Since the angle is not define. Let coefficients h and g are equality value ε , where $\varepsilon\to 0$, than $\lim_{\varepsilon\to 0}\frac{4\times\varepsilon\times k}{k^2-\varepsilon^2-k^2+\varepsilon^2}=\tan 2\beta$,

$\lim_{\varepsilon\to 0}\frac{4\times\varepsilon\times k}{0}=\tan 2\beta$, $\infty=\tan 2\beta$ и $\beta=\pi/4$. Angle α is $\tan\alpha_1=\frac{a\tan\beta-h}{b-g\tan\beta}$, $\alpha_1=\frac{\pi}{4}$ or

$\tan\alpha_2=\frac{b\tan\beta+g}{a+h\tan\beta}$, $\alpha_2=\frac{\pi}{4}$. Angle α is turned out the same for both formulas. Coefficients c and d are

$c_1=\frac{a\cos\beta+h\sin\beta}{\cos\alpha}=\frac{k\cos(\pi/4)+0\sin(-\pi/4)}{\cos(\pi/4)}=k$,

$c_2=\frac{b\sin\beta+g\cos\beta}{\sin\alpha}=\frac{k\sin(\pi/4)+0\cos(\pi/4)}{\sin(\pi/4)}=k$,

$d_1=\frac{a\sin\beta-h\cos\beta}{\sin\alpha}=\frac{k\sin(\pi/4)-0\cos(\pi/4)}{\sin(\pi/4)}=k$,

$d_2=\frac{b\cos\beta-g\cos\beta}{\cos\alpha}=\frac{k\cos(\pi/4)-0\sin(\pi/4)}{\cos(\pi/4)}=k$.

Thus the formulas for converting performed in the vicinity of the transformation $\begin{pmatrix} x' \\ y' \end{pmatrix} = \begin{pmatrix} k & 0 \\ 0 & k \end{pmatrix} \begin{pmatrix} x \\ y \end{pmatrix}$, but do not exist for accurate conversion.

Let coefficients of matrix are $a = b = \cos\varphi$, $h = -\sin\varphi$, $g = \sin\varphi$. Angle β is equivalence $\frac{2(gb + ha)}{a^2 - h^2 - b^2 + g^2} = \tan 2\beta$, $\frac{2(\cos\varphi\sin\varphi - \cos\varphi\sin\varphi)}{\cos^2\varphi - \sin^2\varphi - \cos^2\varphi + \sin^2\varphi} = \frac{0}{0}$. Since the angle is not define also. Let coefficients of matrix are $a = b = \cos\varphi + \varepsilon$, $h = -\sin\varphi - \varepsilon$, $d = \sin\varphi + \varepsilon$, where $\varepsilon \to 0$. Angle β will seek, as $b^2 = a^2$: $\lim\limits_{\varepsilon\to 0} \frac{2((\cos\varphi + \varepsilon)(\sin\varphi + \varepsilon) - (\cos\varphi + \varepsilon)(\sin\varphi + \varepsilon))}{-\sin^2\varphi + 2\varepsilon\sin\varphi - \varepsilon^2 + \sin^2\varphi + 2\varepsilon\sin\varphi + \varepsilon^2} = \tan 2\beta$, $\lim\limits_{\varepsilon\to 0} \frac{0}{4\varepsilon\sin\varphi} = \tan 2\beta$, $0 = \beta$. Angle α is $\tan\alpha_1 = \frac{a\tan\beta - h}{b - g\tan\beta}$, $\tan\alpha_1 = \frac{\cos\varphi\tan 0 + \sin\varphi}{\cos\varphi - \sin\varphi\tan 0} = \tan\varphi$, $\alpha_1 = \varphi$ or $\tan\alpha_2 = \frac{b\tan\beta + g}{a + h\tan\beta}$, $\tan\alpha_2 = \frac{\cos\varphi\tan 0 + \sin\varphi}{\cos\varphi + \tan 0(-\sin\varphi)} = \tan\varphi$, $\alpha_2 = \varphi$. Angle α is turned out the same for both formulas. Coefficients c and d :

$$c_1 = \frac{a\cos\beta + h\sin\beta}{\cos\alpha} = \frac{\cos\varphi\cos 0 + \sin 0(-\sin\varphi)}{\cos\varphi} = 1 ,$$

$$c_2 = \frac{b\sin\beta + g\cos\beta}{\sin\alpha} = \frac{0\cos\varphi + 1\sin\varphi}{\sin\varphi} = 1 ,$$

$$d_1 = \frac{a\sin\beta - h\cos\beta}{\sin\alpha} = \frac{\cos\varphi\sin 0 + \sin\varphi\cos 0}{\sin\varphi} = 1 ,$$

$$d_2 = \frac{b\cos\beta - g\sin\beta}{\cos\alpha} = \frac{\cos\varphi\cos 0 - \sin\varphi\sin 0}{\cos\varphi} = 1 .$$

Findings are consistent with the conclusion of the previous conversion.

Method can not be used to rotate $\begin{pmatrix} \cos\phi & -\sin\phi \\ \sin\phi & \cos\phi \end{pmatrix}$ and scaling (homothety) $\begin{pmatrix} k & 0 \\ 0 & k \end{pmatrix}$ and some other transformations (see theorem 2) [Lozhkin 2012].

5. Geometrical Modeling

Example of geometric modeling the method is presented to calculate the trajectory of the astroid in language AutoLisp. Astroid (hypocycloid modulo 4) is described by two identical systems of parametric equations $\begin{cases} x = \frac{3}{4}R\cos\frac{t}{4} + \frac{1}{4}R\cos\frac{3t}{4} \\ y = \frac{3}{4}R\sin\frac{t}{4} - \frac{1}{4}R\sin\frac{3t}{4} \end{cases}$ and $\begin{cases} x = R\cos^3\frac{t}{4} \\ y = R\sin^3\frac{t}{4} \end{cases}$ [Gibson 2001].

The program algorithm is quite simple:

1. The requested transformation parameters;
2. We output the desired astroid will transform every part of the line of it. Output is black;
3. Forward transformation parameters;
4. Output a new astroid with parameters obtained in yellow.

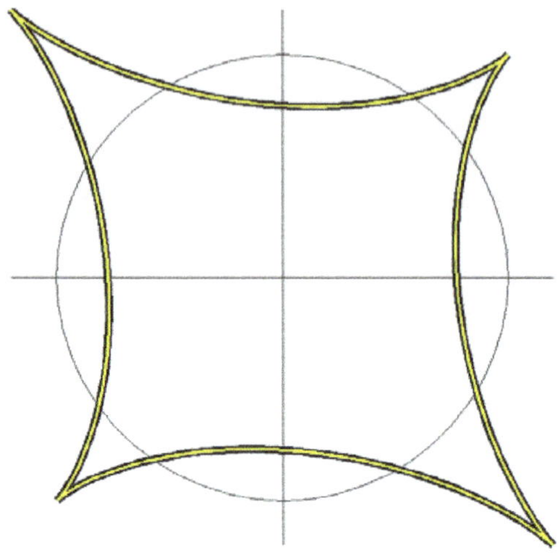

Figure 1. Normal transformation an astroid

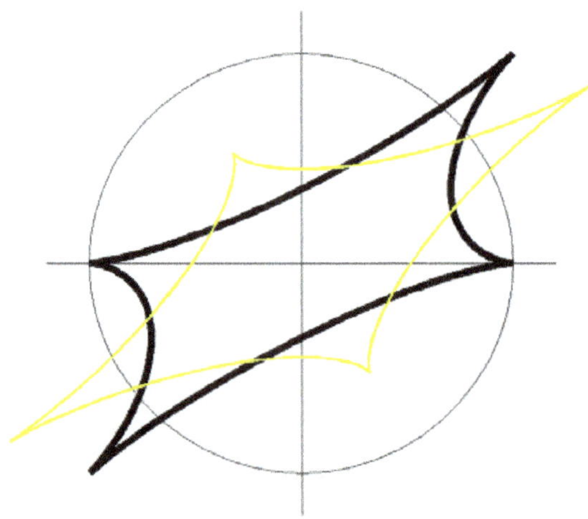

Figure 2. "Bad" transformation.

The result of the programs for conversion $\begin{pmatrix} 1 & 1.2 \\ 1 & -1.2 \end{pmatrix}$ shown in Figure 1. The method works. Parameters are calculated correctly. The result of the programs for conversion $\begin{pmatrix} 1 & 1 \\ 0 & 1 \end{pmatrix}$ is shown in Figure 2. Method does not work. The experiment was carried out in AutoCAD 2009.

Two rules to observe when moving algorithm to other languages:

1. Calculate the arc tangent function of two arguments;
2. Remember in which quadrant is a calculation.

It should be noted that the theoretical and modeling researches of Jordan curves is not full.

6. Experiment

Experiments were carried out on the basis of the developed stand with dimensions 1300x1400x1500 mm to check the adequacy of the developed mathematical model. Photo booth is shown in Figure 3.

Figure 3. Photo experimental stand

The design of the stand was assembled from aluminum profiles, linear modules and fasteners RK Rose + Krieger. Aluminium profiles RK Rose + Krieger Blocan shown in Figure 4.

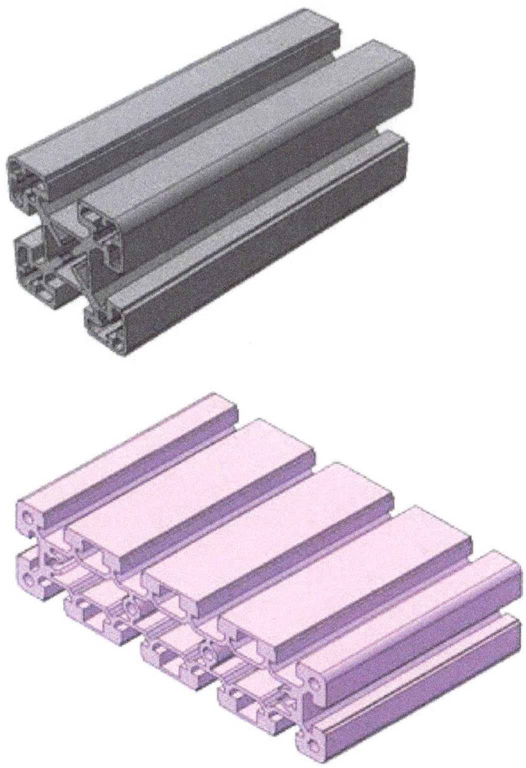

Figure 4. Aluminum profiles RK Rose+Krieger Blocan (3D model)

Aluminum profiles are profiled rods with longitudinal T-slots. The surface profile is finishing. Linear Module RK Rose + Krieger Blocan have aluminum construction, possess high hardness, large carrying capacity. Assembly is protected from dust and dirt. Linear guides have been installed on a wide aluminum profile and secured the appropriate bracket, Figure 5.

Figure 5. Model roller linear guides

Two profiles with guides were bonded together using an aluminum sheet on the top side. Four movable slider on the guide are connected to the same sheet of aluminum, 3D model shown in Figure 6.

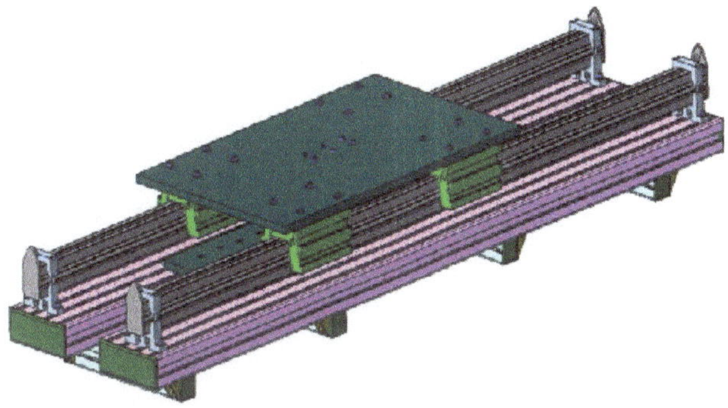

Figure 6. Model design guide stand

Experiments have shown a difference between the theoretical and actual trajectory less than 5%.

Simple and cheap robot to move along the path of the executive body of the ellipse and the circle may develop on the basis of the proposed mathematical model. Photo pneumatic robot is shown in Figure 7.

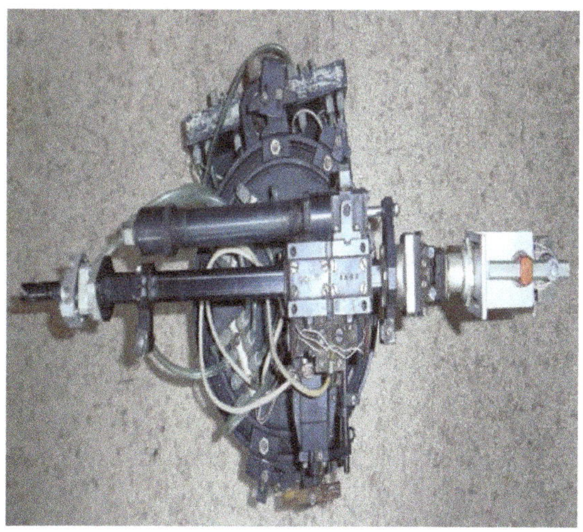

Figure 7. Photograph of the pneumatic robot

This robot can be carved with milling cutter details in the form of an ellipse and a circle made of different materials, such as glass, metal, ceramics and wood. The robot can with laser burn in different materials is not necessarily circular holes.

7. Plane Differentiable Curves

The authors studied the expression of symmetry Euler not only for conic sections but also more complex forms. Four groups of linear $\mathbf{H}_1 = \begin{pmatrix} m & -n \\ m & n \end{pmatrix}$, $\mathbf{H}_2 = \begin{pmatrix} -m & n \\ m & n \end{pmatrix}$, $\mathbf{H}_3 = \begin{pmatrix} km & n \\ -kn & m \end{pmatrix}$, $\mathbf{H}_4 = \begin{pmatrix} m & kn \\ -n & km \end{pmatrix}$, where $k, m, n \in \mathbf{R}$, can be used DAM for any Jordan curve, but the method is not applicable for rotation and dilation for any curve, because they not change the discrete structure of Euclidean plane.

The proposed theory allows the calculation of complex mechatronic systems. Let there be a simple kinematical mechanism built on the basis of three gears: central, moving inside the central wheel and moving outside of the central wheel (Figure 8). Let a point on the first wheel moves according to the law $\begin{cases} x = r\cos t \\ y = r\sin t \end{cases}$. Let the motion of the second and third wheels are defined by the systems of parametric equations $\begin{cases} x = r\cos^3 \dfrac{t}{4} \\ y = r\sin^3 \dfrac{t}{4} \end{cases}$ (asteroid), $\begin{cases} x = 3r\cos t - r\cos 3t \\ y = 3r\sin t - r\sin 3t \end{cases}$ (epicycloids).

It is necessary to design an elliptical gear mechanism to the trajectory of the first wheel to a system of parametric equations $\begin{cases} x = -1.2r\cos t + r\sin t \\ y = 1.2r\cos t + r\sin t \end{cases}$ from second group, where $m = 1.2$ and $n = 1$. The method of calculation can be applied to all wheels. Geometric modeling results are shown in Figure 9.

Solution of the characteristic equation $\mathbf{T}\vec{v} = \lambda\vec{v}'$ can be different. Vector $\vec{v}$ does not change after the transformation of the ellipse. Vector belongs to the set $\vec{v}' \in \left\{ \begin{pmatrix} x \\ y \end{pmatrix}, \begin{pmatrix} y \\ x \end{pmatrix}, \begin{pmatrix} -y \\ -x \end{pmatrix} \right\}$ after the transformation of the planar differentiable curve. Parameters symmetric transformations find possible and to simplify the system of parametric equations for the complex movements as a result. Real coefficients can be used in systems of parametric equations. Thus the analytical solution for calculating opened for the complex trajectories of the mechatronic system. Precision of designing is defined within a computer data storage and precision of machining equipment.

Symmetries has an important influence on the trajectory of this is obvious. Let anticipated changes in trajectory when applying an arbitrary linearly independent transformation. The system of parametric equations is divided into two or four portions of movement: $\begin{cases} x = c_1 f_x(t) \\ y = d_1 f_y(t) \end{cases}$, $\begin{cases} x = c_1 f_y(t) \\ y = d_1 f_x(t) \end{cases}$, $\begin{cases} x = -c_1 f_x(t) \\ y = -d_1 f_y(t) \end{cases}$, $\begin{cases} x = -c_1 f_y(t) \\ y = -d_1 f_x(t) \end{cases}$. This hypothesis is tested in research now.

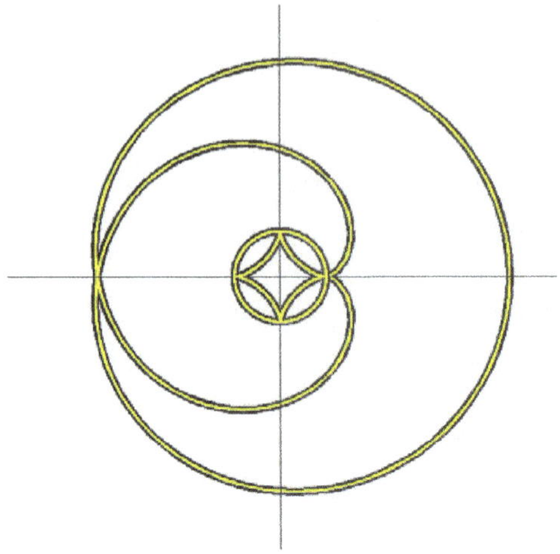

Figure 8. Simple kinematical mechanism

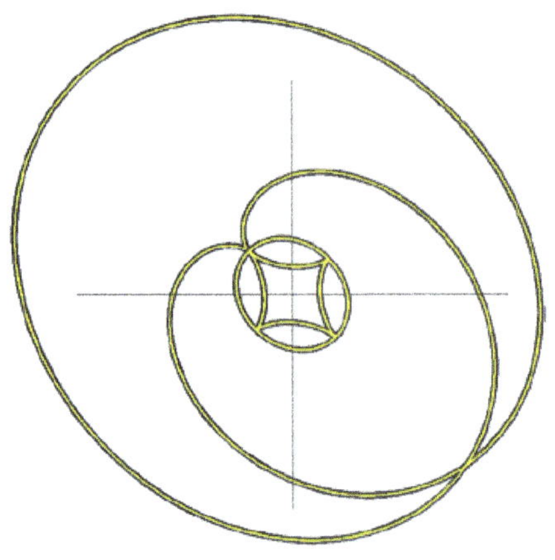

Figure 9. Complex kinematical mechanism

8. Summary

If the trajectory of the point of the kinematic mechanism is represented by the parametric equations that can be used a new effective method. Parametric system is used to convert the equations of motion in a circle. The method allows to find the exact characteristics of the mechanism are based on the kinematic trajectory of its motion. The method described by simple algorithms more uncomplicated compared with the classical method.

Acknowledgement

Supported by Minobrnauki of Russian Federation, Grant GZ/TVG 14(01.10).

References

[1] Biagiotti, L. and Melchiorri, C. Trajectory Planning for Automatic Machines and Robots. Springer, Science & Business Media, 2008.

Hirsh, H., Coen, M.H., Mozer, M.C., Hasha, R. and Flanagan, J.L, "Room service, AI-style," *IEEE intelligent systems*, 14 (2). 8-19. Jul.2002.

[2] Božek, P. and Chmelikova, G. Implementation of robot offline programming. In *Annals of DAAAM and Proceedings of DAAAM Symposium*. pp. 0331-0332.

[3] Efimov N.V. Quadratic forms and matrices. Moscow, Science, 2009. (In Russian).

[4] Virgala, I., Frankovský, P., Kenderová, M.: Friction Effect Analysis of a DC Motor. In: American Journal of Mechanical Engineering. Vol. 1, no. 1 (2013), p. 1-5.

[5] Gibson, C.G. Elementary geometry of differentiable curves: an undergraduate introduction. Cambridge, University press, 2001.

[6] Frankovský, P., Hroncová, D., Virgala, I.: Kinematics analysis of the six member mechanism in MSC Adamsview, In: Acta Simulatio. Roč. 1, č. 1 (2015), s. 17-22.

[7] Lozhkin, A.G. Applied planemetry with singular transformations. Ekaterinburg, IE UrO RAS, 2009. (In Russian).

[8] Lozhkin, A. About definition of singular transformation by N.V. Efimov, Geometry, topology, algebra and number theory, applications. *Abst. of International conference dedicate to the 120th anniversary of B.N. Delone*, Steklov MI RAS, Lomonosov MSU, Moscow, August 2010, pp. 160-162.

[9] Lozhkin, A. and Dyukina, N. Automorphisms: from mirror to symmetry of knowledge. Lozhkin A.G., Izhevsk-Glazov, 2011. (In Russian).

[10] Lozhkin, A. and Dyukina, N. Structurization of analytical geometry on the base of symmetries. Saarbruken, LAP, 2012. (In Russian).

[11] Nikitin, Y.R. and Abramov I.V. CNC machines diagnostics, *Proceedings 13-th International Simposium on Mechatronics*, Trencinske Teplice, June 2010, pp. 89-91.

[12] Trefilov, S.A. and Kalinkin, A.A. and Nikitin, Y.R. High-maneuverable transport robot, *University Review*, 2007, Vol. 1, No 2, Alexander Dubcek University of Trencin, Slovakia, Izhevsk State Technical University, Russia, pp. 34-39.

[13] Malindžáková, M., Straka, M., Rosová, A., Kanuchová, M., Trebuňa, P.: Modeling the process for incineration of municipal waste, Przemysl chemiczny, 2015, 94, No. 8, 1000-1004.

Machines for In-Pipe Inspection

Alexander Gmiterko, Ivan Virgala, Lubica Miková, Peter Frankovský, Tatiana Kelemenová, Michal Kelemen*

Department of Mechatronics, Technical University of Kosice, Faculty of Mechanical Engineering, Kosice, Slovak Republic
*Corresponding author: michal.kelemen@tuke.sk

Abstract The paper deals with in-pipe machines based on directional friction and inertial stepping principle. Both type are developed for inner pipe with diameter 11 mm. The main purpose is inspection of inner pipe wall as prevention of cracks.

Keywords: *In-pipe, machine, locomotion, insert, template*

1. Introduction

In-pipe micromachines are able to move in the pipe to inspect or to repair the pipe or other special tasks. Pipe is as confined space i.e. this is constraint for micromachine dimensions, degrees of freedom etc. There are a lot of interdisciplinary problems in design and realisation of in-pipe micromachine. It is difficult to choose suitable actuators, sensors, and power supply etc.

In term of biological analogy, it is possible to divide ways of locomotion into two main groups (Figure 1) [1-20]:

- artificial locomotion,
- biological inspired locomotion.

In term of physical principle, we can divide both these groups into several basic locomotion ways (see Figure 1). It is important to say that it is not final and changeless dividing.

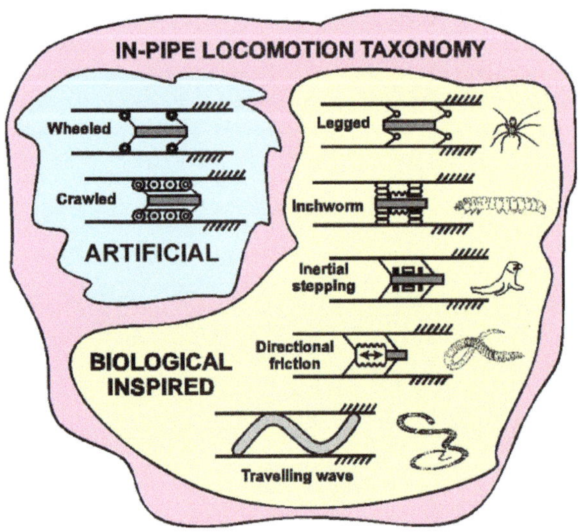

Figure 1. Ways of in-pipe machine locomotion

Wheeled locomotion principle: Wheels are standard synthetic components, which are very often used for locomotion purposes. However inner pipe wall is not perfect. There are a lot of obstacles, which have various characters. When inner pipe wall is dirty, wheels tend to slipping.

Crawled locomotion principle: Tracks is also unfinished rotating element. It is most adaptable to surface then wheels but it is not so often used as wheels.

Legged locomotion principle: Legs are components, which have inspiration in biology (e.g. spider, cockroach etc.). Many biological organisms are able to locomote via legs and reach high speed. These organisms overcome difficult obstacles and there are fascinating for designers.

Inchworm-like principle: The inchworm strategy comes from biological example. The inchworm is capable of manoeuvring in extremely small spaces, it can do so in arbitrary orientations to gravity and can withstand substantial external forces attempting to diverge it from its intended course. It can do these things because its mobility system is governed by a simple rule: "Never let go of what you're holding until you're holding something else!“.

Inertial stepping locomotion principle: The principle based on fact that a part of device (inertial mass) oscillates with suitable frequency. Backward tendency of motion is damped.

Worm-like locomotion principle: The locomotion uses difference characteristic of friction between device and pipe wall. Forward friction force is less then backward friction force. It causes that device locomotes in forward direction.

Travelling wave locomotion principle: Device has articulated body and generates travelling wave from head to tail [4,5,6,7].

2. In-pipe Machine Based on Directional Friction

Subject of the article is in-pipe micromachine which locomotes via worm-like principle based on directional friction. Motive forces must be generated internally and then the system will use friction and constraints and shape changes to move. With humans, the frictional forces are isotropic, i.e., sliding a foot forward cause the other foot to slide backwards. That is why we must lift one foot to reposition it while the other foot remains static. But snakes and worms remain in contact with the ground and can have anisotropic frictional forces because of their scales [21].

When sliding forward the frictional forces are minimal, but when a body segment slides backward, the scales dig in and the frictional force becomes very large. Worm can be substituted with one spring, two masses (Figure 2). When the spring is expanded (l increases) scale B slides over the ground and scale A grips the ground. Then the spring is contracted and scale A slides and scale B digs in and grips.

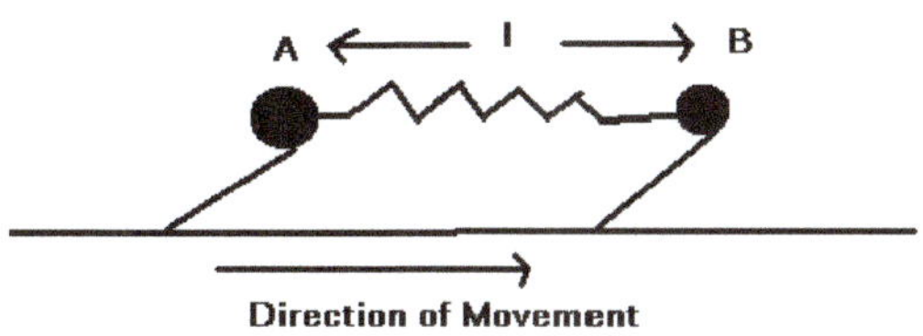

Figure 2. Worm-like in-pipe locomotion principle – directional friction principle

So changing (oscillating) the spring length will result in forward motion. A real worm consists of many segments like the above (Figure 2). To prevent having only one scale gripping, the worm sends a wave of compression from its head to its tail. In a real worm, the wave is a square wave. However, with the simple spring-mass model, a square wave creates shape distortions and a sine wave works better describes the locomotion [21].

The in-pipe micromachine (Figure 3) locomotes via worm-like principle described above. It consists of a piezoactuator (1), which converts electrical energy form to mechanical energy form. Just piezoactuator transformation cannot provide in-pipe locomotion. So there is a need of other mechanical parts – bristle plate (2, 3) and bristles (4). These mechanical parts add another function to piezoactuator. Energy transformation can be controlled with any microcomputer. Finally, locomotion can be obtained with this synergic integration.

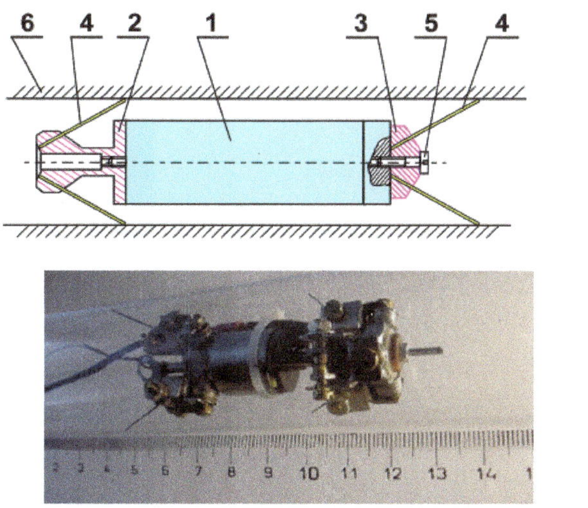

Figure 3. Worm-like in-pipe machine based on directional friction principle. 1 – piezoactuator, linear motor; 2 – head bristle plate; 3 – back bristle plate; 4 – bristles; 5 – screw; 6 – pipe

Mechanical parts provide force coupling with inner pipe wall and diagonally attached bristles, which cause anisotropic character of friction force between bristle tip and pipe wall.

The control system, which generates suitably modulated voltage, provide actuator controlling. There is a need of organs and parts optimisation for effective utilisation of energy transformation. Hence, mechatronics design approach is necessary. It means that actuator is supported with mechanical part. Another possible way of improving is to add another function to bristles. So, implementation of another actuator gives to bristles new function. Another actuator is able to drive bristle parameters, and these bristles become an intelligent "smart bristles". Bristle properties cannot be improved, moreover, by classic design approach, so there is a place for intelligence integration. Finally, the integration gives better properties to system, which becomes competitive in this class of micromachine.

Bristle is simple one leaf spring with constant cross-section. Bristles are attached as cantilevers to a bristle plate diagonally. A characteristic property is bigger deformability, which is typical for springs. High deformability is ensured with using of material with high compliance otherwise with using of material with high stiffness which is suitable shaped.

Bristles are unloaded if micromachine is out of the pipe. Bristles are designed with parameters which secure that bristle tip span are bigger than inner pipe diameter. So, after in-pipe micromachine application into pipe, bristles are deformed. Let's assume that deformations of pipe are negligible. Deformations of bristles depend on geometrical deviations of inner pipe diameter and roundness deviation. It means that, bristle tip span has to equal with inner pipe diameter.

If the in-pipe micromachine doesn't locomote, there is only normal force, which is applied to inner pipe wall. The normal force depends on bristle deflection, bristle stiffness and assembling bristle angle. When excitation force will be applied and it will be continuously increased, if micromachine starts locomotion, value of the excitation force will be equal to adhesive part of friction force Fto.

Bristles are also used as part for creating force coupling with inner pipe wall in [17,18,19,20,21]. Diagonally bristle attaching is suitable because of anisotropic character of friction between bristle tip and pipe wall. The friction force in forward direction is less than friction force in backward direction. The difference of these friction forces causes the forward locomotion of in-pipe micromachine. Work [17] doesn't deal with other details about this phenomenon and doesn't specify conditions, which is necessary for the phenomena. Bristles are substituted with fins and blades in [17,18,19,20,21].

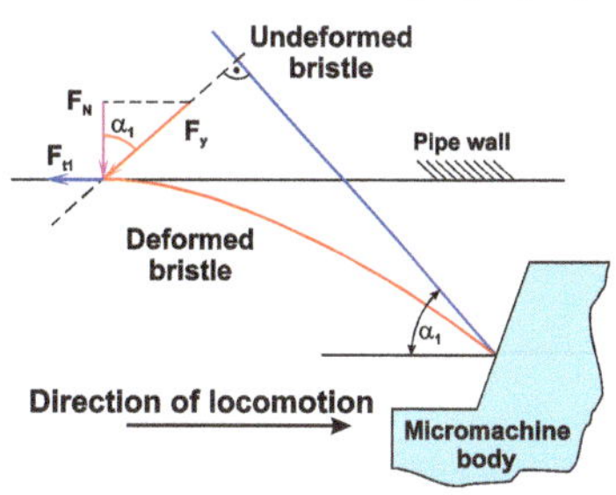

Figure 4. Forces on bristle tip in forward locomotion

Friction force between inner pipe wall (Figure 4) and bristle tip can be described with equation:

$$F_{t1} = \frac{F_{N1} \cdot f}{1 + f \cdot tg\alpha_1} \quad (1)$$

Analogical, it is also possible to derive friction force for moving in backward (Figure 5) direction and it is given as:

$$F_{t2} = \frac{F_{N2} \cdot f}{1 - f \cdot tg\alpha_1} \quad (2)$$

where:

F_{t1} – friction force between bristle tip and pipe wall in forward direction

F_{t2} – friction force between bristle tip and pipe wall in backward direction

f – adhesive coefficient of friction

α_1 – bristle angle to angle of body

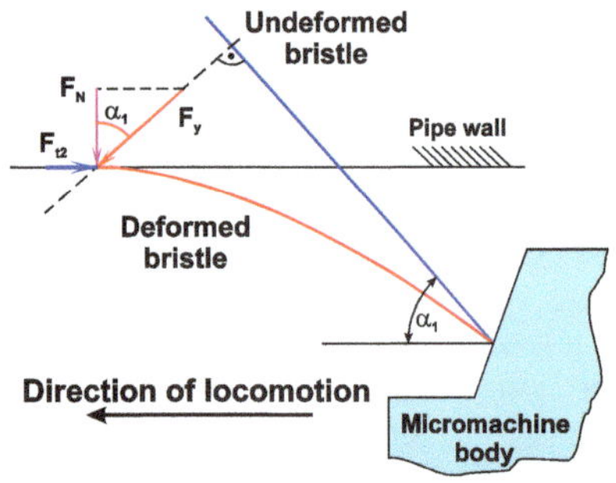

Figure 5. Forces on bristle tip in backward locomotion

If we will assume that normal force will not change, we can write equations (1) and (2) into one common equation:

$$F_{t1,2} = \frac{F_N \cdot f}{1 \pm f \cdot tg\alpha_1} \quad (3)$$

As we can see from equation (3), friction force in forward direction is still less than friction force in backward direction.

If the value of coefficient of friction f will converge to cotgα_1, then friction force in backward direction will converge to infinity, so micromachine will be automatically locked.

The self-locking mechanism causes forward locomotion of most worms and snakes. It is very difficult to influence of friction coefficient value, so we have to find different way of bristle optimisation. The criterion of optimisation is the maximal difference between forward and backward friction force.

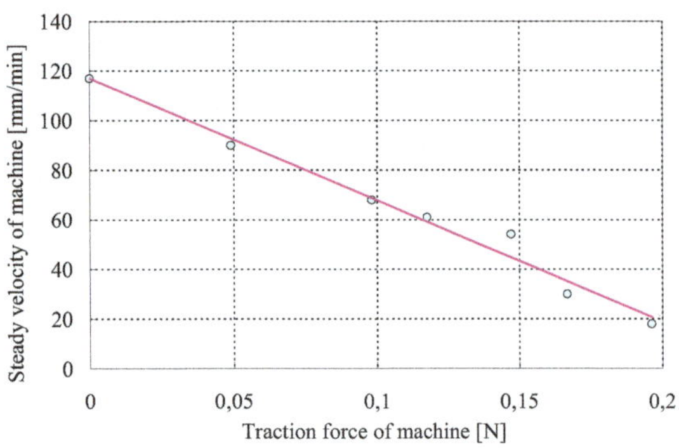

Figure 6. Machine velocity versus traction force of machine

Figure 6 shows the machine velocity dependence on traction force of in-pipe machine.

3. In-pipe Machine Based on Inertial Stepping Principle

Figure 7 shows the outlook photograph of the developed micromachine. Dimensions are 10 mm in diameter, 45 mm in length, 10 g in weight.

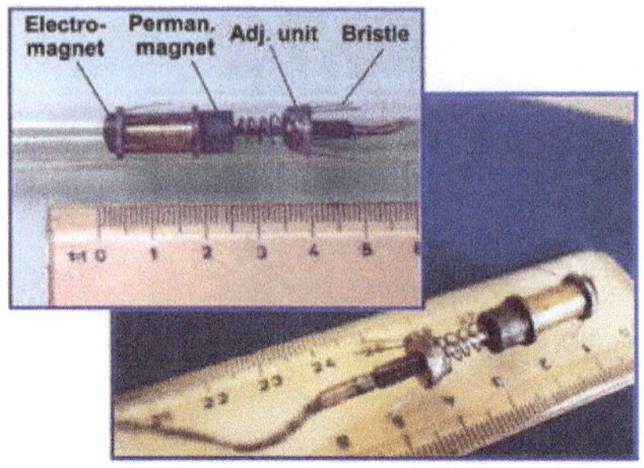

Figure 7. In-pipe machine based on inertial stepping principle

The micromachine is composed of these units: a electromagnet, a adjusting unit, a permanent magnet, a guide rod, a damping spring and bristles. The bristles serve as a clamping element which contacts the pipe wall to hold the weight of machine both in horizontal pipe and in the vertical pipe [22].

Figure 8 shows the mobile principle of the micromachine. When the repulsive force produced by a magnetic field of the electromagnet effects on the permanent magnet, the magnet moves from the electromagnet. During the contact of magnet with the damping spring, the repulsive force is finished. As result, the magnet returns to electromagnet. In the moment of impact of the magnet with electromagnet, the micromachine moves forward, because impact force exceeds the maximum static friction force of the micromachine bristle tip with pipe wall. Consequently, by repeating this cycle the micromachine moves forward. In addition, the spring function of the bristle enables to move in-pipe whose diameter is little of variety.

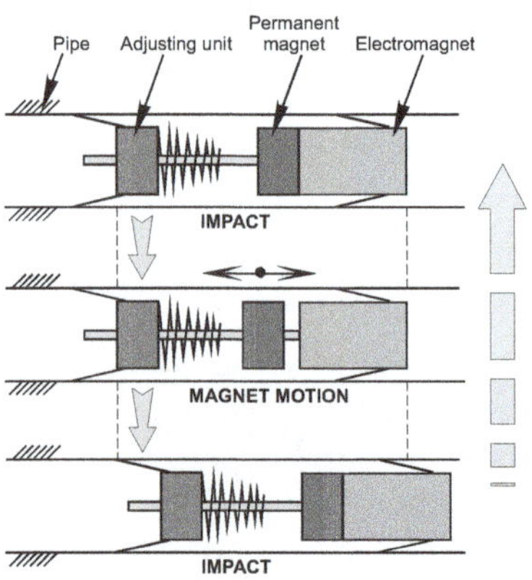

Figure 8. Simplified motion cycle of in-pipe machine based on inertial stepping principle

The speed of the micromachine depends on the distance electromagnet from the spring and the repulsive force duration. The maximum speed was 20 mm/s in the horizontal direction and 15 mm/s in the vertical direction in the glass pipe, at 10-12 ms repulsive force duration and at 1,5 – 2 mm of the distance the spring from the magnet (in while of turned off electromagnet). Figure 9 shows speed versus load characteristic.

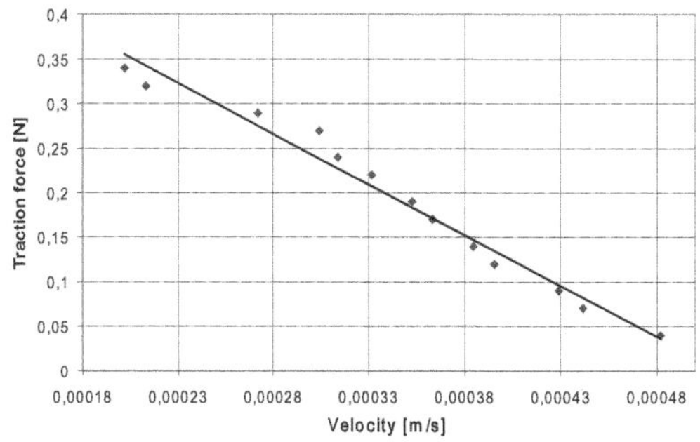

Figure 9. Speed versus load characteristic

4. Conclusion

In-pipe machines are able to locomote inside pipes and they are used mainly for inspection tasks of pipe systems like steam generators, heat exchangers, pipeline for gas, oil, water etc. Inside pipe are technologic remainders in elbows, fittings, reductions. Also inner diameter are not constant, cross-section of pipe has big deviation of roundness. Roundness on inner pipe wall varies and pipe wall are very dirty and very often pipe wall is covered with sediments. Pipe is constrained space and all these factors are affecting the design of the in-pipe machines [23,24,25,26,27,28].

Acknowledgement

The work has been accomplished under the research projects No. VEGA 1/1205/12 and KEGA 048TUKE-4/2014 financed by the Slovak Ministry of Education".

References

[1] Suzumori, K., Miyagawa, T., Kimura, M., and Hasegawa Y., (1999) "Micro Inspection Robot for 1-in Pipes", *IEEE/ASME Transactions on Mechatronics*, Vol. 4, No. 3, September 1999, 286-292.

[2] Hirose, S., Ohno, H., Mitsui, T., and Suyama, K., (1999) "Design of In-Pipe Inspection Vehicles for φ25, φ50, φ150 Pipes," *Proceedings of the 1999 IEEE International Conference on Robotics and Automation*, Detroit, Michigan, May 1999, pp2309-2314.

[3] Roslin, N. S. et al. (2012) "A Review: Hybrid Locomotion of In-pipe Inspection Robot", *Procedia Engineering* 41 (2012) 1456-1462.

[4] Kakogawa A, Ma S. "Mobility of an in-pipe robot with screw drive mechanism inside curved pipes". In: *Proc. IEEE international conference on robotics and biomimetics*. Tianjin (China); 2010. p. 1530-5.

[5] Zhang, Y., Yan, G. (2007) "In-pipe inspection robot with active pipe-diameter adaptability and automatic tractive force adjusting", *Mechanism and Machine Theory* 42 (2007) 1618-1631.

[6] Jun, Ch., Tao, Ch., Zongquan, D., (2011) "Design method of Modular Units for Articulated in-Pipe Robot Inspecting System", 2011 *IEEE Second International Conference on Digital Manufacturing & Automation*, 389-392.

[7] Lee, D., Park, J., Hyun, D., Yook, G., Yang, H., (2012) "Novel mechanisms and simple locomotion strategies for an in-pipe robot that can inspect various pipe types", *Mechanism and Machine Theory* 56 (2012) 52-68.

[8] Lim, J., Park, H., Moon S., and Kim B., (2007), "Pneumatic Robot Based on Inchworm Motion for Small Diameter Pipe Inspection", *Proceedings of the 2007 IEEE International Conference on Robotics and Biomimetics*, December 15-18, 2007, Sanya, China, 330-335.

[9] Kuwada, A., Tsujino, K., Suzumori, K., Kanda, T., (2006), "Intelligent Actuators Realizing Snake-like Small Robot for Pipe Inspection", *Proc. of International Symposium on Micro-NanoMechatronics and Human Science*, 2006, Nagoya, 1-6.

[10] Liu, W. et al. (2010) "An in-pipe wireless swimming microrobot driven by giant magnetostrictive thin film", *Sensors and Actuators A* 160 (2010) 101-108/ Wei Liu, Xinghua Jia, Fuji Wang, Zhenyuan Jia.

[11] Hayashi, I., Iwatsuki, N." and Iwashina, S., (1995) "The Running Characteristics of a Screw-Principle Microrobot in a Small Bent Pipe", *Proc. of Sixth International Symposium on Micro Machine and Human Science 1995*, 225-228.

[12] Jong-Hoon, K., Sharma, G. ; Iyengar, S.S. (2010) "FAMPER: A fully autonomous mobile robot for pipeline exploration". *2010 IEEE International Conference on Industrial Technology (ICIT)*, Vina del Mar, 14-17 March 2010, p. 517-523.

[13] Bertetto AM, Ruggiu M., "In-pipe inch-worm pneumatic flexible robot". *In: Proc. IEEE/ASME international conference on advanced intelligent mechatronics*, vol. 2. Italy; 2001. p. 1226-31.

[14] Qiao, J, Shang, J, Goldenberg, A., "Development of inchworm in-pipe robot based on self-locking mechanism". *IEEE/ASME Transactions on Mechatronics*, Digital Object Identifier 10, 1109/TMECH; 2012s.

[15] Takahashi, M., Hayashi, I., Iwatsuki, N., Suzumori, K., and Ohki, N., "The development of an in-pipe microrobot applying the motion of an earthworm," in *Proc. IEEE 5th Int. Symp. Micro Machine and Human Sciences*, 1994, pp. 35-40.

[16] Neubauer, W., "Locomotion with articulated legs in pipes or dusts," *Robot. Autonomous Syst.*, vol. 11, nos. 3-4, pp. 163-169, 1993.

[17] Idogaki, T., "Characteristics of piezoelectric locomotive mechanism for an in-pipe micro inspection machine". *Proc. of MHS'95*, p.193-198. Nagoya, Japan.

[18] Aoschima, S., Tsujimuri, T., Yabuta, T., (1989) "Design and analysis of a midget mobile robot using piezo vibration for mobility in a thin tube". *Proc. of the International Conference on Advanced Mechatronics*, Tokyo, 1989, p. 659-663.

[19] Fukuda, T. et al., "Giant magnetostrictive alloy (GMA) applications to micro mobile robot as a micro actuator without power supply cables," in *Proc. IEEE Int. Workshop Micro Electro Mechanical Systems (MEMS)*, Jan. 1991, pp. 210-215.

[20] Degani, A., Feng, S., Choset, H., and Mason, M. T., "Minimalistic, Dynamic, Tube Climbing Robot", *Proc. of 2010 IEEE Int. Conf. on Robotics and Automation*, Anchorage Convention District, May 3-8, 2010, Anchorage, Alaska, USA. (2010) 1100-1101.

[21] Ostertag, O., Ostertagová, E., Kelemen, M., Kelemenová, T., Buša, J., and Virgala, I., (2014) "Miniature Mobile Bristled In-Pipe Machine", *Int J Adv Robot Syst*, 2014, 11:189.

[22] Yum, Y.J., Hwang, H.S., Kelemen, M., Maxim, V., and Frankovský, P. (2014) "In-pipe micromachine locomotion via the inertial stepping principle", *Journal of Mechanical Science and Technology* 28 (8) (2014), 3237-3247.

[23] Vitko, A., Jurišica, L., Kľúčik, M., Duchoň, F., (2008) "Context Based Intelligent Behaviour of Mechatronic Systems", *Acta Mechanica Slovaca*. Vol. 12, No. 3-B. pp. 907-916.

[24] Koniar, D., Hargaš, L., Hrianka, M., "The application of DICOM 7th standard in LabView", *Proc. of Biom. Eng.*, Kladno (2007).

[25] Dekan, M., Duchoň, F., Jurišica, L., Vitko, A., Babinec, A., "iRobot Create Used in Education", *Journal of Mechanics Engineering and Automation*. - Vol. 3, Iss. 4, 2013, pages 197-202, (2013).

[26] Kelemenova, T. et al. (2012) "Bristled In-pipe Machine Inside Pipe With Geometric Deviations", Procedia Engineering, Elsevier, Volume 48, 2012, Pages 287-294. 852, 282-287.

[27] Kelemen M., Virgala I., Miková Ľ., Frankovský P., Experimental Identification of Linear Actuator Properties, Acta Mechanica Slovaca. Volume 19, Issue 1, Pages 42-47.

[28] Hargaš, L., Hrianka, M., Koniar, D., and Izák, P., "Quality Assessment SMT Technology by Virtual Instrumentation", *Applied Electronics 2007*, Pilsen, 5. – 6. 9. 2007, (2007).

23

Embedded Sensors in Monitoring of Human Daily Activities

Dusan Simsik[1,*], Alena Galajdova[1], Robert Rakay[1], Daniela Onofrejova[2]

[1]Department of Automation, Control and Human-Machines Interactions, Mechanical Engineering Faculty, Technical University of Kosice, Slovakia
[2]Department of Industrial Engineering and Management, Mechanical Engineering Faculty, Technical University of Kosice, Slovakia
*Corresponding author: dusan.simsik@tuke.sk

Abstract The aim of the paper is to describe some results from experiments with embedded sensors using inertial sensors as wearable sensors. Authors were involved in last years in the development of ICT services for monitoring of elderly persons daily activities monitoring, and their home environment status monitoring and control as well.Main goal of the current research interest is to add embedded sensors to mentioned services for monitoring of the critical physiological parameters to get more information about users' behavior and critical situations.

Keywords: *wearable sensors, social service, inertial sensor, experiments*

1. Introduction

Wearable sensors are continuously emerging over last decade and they are presenting important innovation tool for services associated with healthcare. There are several typical applications of wearable devices such as fall detection [1,2], distance walked and indoor positioning [3,4] or physiological parameters, e.g. blood pressure or ECG [5]. Ageing population [6] in developed countries opens an important topic, how to enable suitable and economically feasible health and social care using new unobtrusive technology for new type of social/health care services.

In our laboratory, we are trying different technology, preferably a wireless technology to assist in services for elderly and other vulnerable persons too. The key parameters for long-term monitoring of user's states based on measurements provided by wearable sensors were proposed to innovate existing laboratory system equipped with stationary sensors and actuators. For selected parameters, the framework for processing and evaluation of measurements was developed and its usability was examined in living lab – lab that simulates household's conditions. Emphasis was on the robustness of the solution as well as on its applicability in various conditions. The attention was paid to complementary data procession from several sensors with intention of minimization of divert distortive effects or with intention of undesirable sensors' properties.

Our research approach follows the rules for an effective work aiming in fast implementation of the applied research into real life. Therefore, we concentrated on existing technology on the market and related services [6,7].

Current situation on the market with embedded technology and services offers several partial solutions based on the implementation of new functions into smart watches, mobile phones or tablets. There are quite complex solutions between them like PAMSys, PLUXbiosignal, Simmer, Sensoplex, etc. (Karchnak 2015) [8].

However, there is still a space for the development of more complex services with affordable prices for the group of vulnerable persons.

2. MEMS Sensor Development and Trials

Our research group developed a sensor based on MEMS technology originally intended for tracking of a remote controlled car model. Later, we have been looking for other applications like assessment of a person's mobility and her/his range of motion [6,7,8]. As we had already used optical methods for human motion analysis in our laboratory, we decided to use MEMS sensor for a persons' motion abilities evaluation because we expected more simple measurement and data evaluation too. The further step after measurements of biomechanical and kinesiological parameters was to include such sensor into technology for monitoring of a person's physical activities and states.

2.1. Design of the Sensor

A small circuit board was developed aiming in minimizing of sensor size (PCB). Our functional sample was double-sided board PCB, and after adding other units our prototype got a cover too as shown in Figure 1.

Figure 1. IMU prototype

As sensor of acceleration, we used 3-axial accelerometer MMA7431 from FreeScale and gyroscope ITG3200 from InvenSense for angular velocity sensing. As control unit, we used microcontroller ATMEGA32L.

Communication of prototype with PC via USB is provided by FT232RL chip. Device communicates with PC via USB that is serving also as power supply. For implementation of wireless technology, UART pins RX and TX are ready. [9]

Implemented accelerometer MMA7431 is analog device, so MCU has to convert data to digital form utilizing its 10bit ADC. Gyroscope ITG3200 has 16-bit ADC on-chip, so the digital output can be send to MCU using I2C. As mentioned above, UART is employed for communication – FT232RL for USB or XBee, etc. Data is coded into a communication chain that is designed according to the following form:

$, x, y, z, T, 255, XX, YY, ZZ, 13, 10

Where:

$ - is the beginning of the chain,

x, y, z – are incoming raw acceleration data

T - is output of temperature sensor on gyroscope

255 – is separator character

XX, YY, ZZ – is gyroscope data

13, 10 – is end of the chain

Each of incoming sensor data are encoded by two bytes – high and low. These data need to be compiled and converted into one number and processed after that. Testing of device was performed using USB for communication as well as using wireless technology.

2.1.1.Monitoring of critical situations - monitoring scenario

A very important issue for a successful monitoring is an optimal selection of a scenario. Variety of wearable sensors and their functions enable different innovative solutions adjusted to specific type of persons and their environments. Monitoring of daily activities is now commonly integrated with functionalities oriented on long time monitoring of physiological parameters served for the health care aims. Clients then may stay at home and in the same time their basic functions are under supervision. In such way, wearable sensors offer important information about health conditions, activities of clients, and their behavior.

Analysis of the long time collected data enables identification and even prediction of a critical health situation. For example, changes in gait speed, and hip extension may indicate a negative changes in health conditions related to a person's mobility. Such data may play an important role in the health and social care systems, especially for vulnerable persons, elderly including.

The main goal for implementing of wearable sensors is to enable identification of dangerous behavior, or health status of vulnerable persons without creating an obstructive home environment. In the same time such service will improve safety, self-confidence, and well-being of users. Such complex goal may be reached by the integration of wearable sensors into the intelligent environments, where stationary sensors network exists, and control system provides supervisory tasks and actions.

2.2. Complex Supervising System

The general structure of a complex supervising system consists of:

1. Stationary sensors of environment.
2. Wearable sensors for motion activities and physiological parameters.
3. Communication interfaces.
4. Techniques for data processing and evaluation.
5. Models of user's behavior and actual behavioral analysis to discover significant changes.

Monitoring of users living in a supported environment is a subject of many research projects. Many of them are focusing on the design of complex supervising system. However, many of them stress a specific role of such system like saving energy [10,11,12], recognition of individual users, and monitoring of users health status [13,14,15,16,17]. For example, the system may identify possible Parkinson's disease in behavioral changes of a person like slowing medium walking speed, and changes in hip extension [18]. Above mentioned researchers apply different methods for data analysis. Basically, the system may work in the following structure shown in Figure 2.

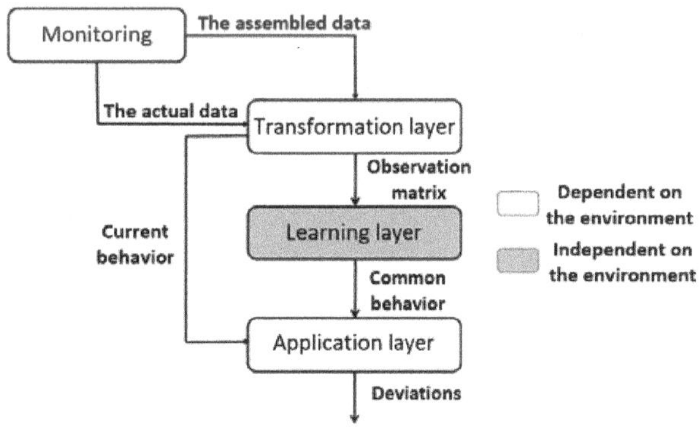

Figure 2. System architecture for detection of changes in user's behavior (adapted from [15])

2.2.1.Wearable sensors role in the monitoring scenario

Stationary sensors play an important role in monitoring of the environmental data like temperature, lighting, energy consumption, but also for identification of daily activities of users. Embedded wearable sensors add more personal information dealing with health status or critical situations during physical activities – fall, etc. Therefore both kinds of sensors are important for identification of critical situations, and behavioral changes too.

Wearable sensors usually offer the following data:

- Numbers of steps per day n(i).
- Medium walk speed v(i).

- Medium step lengths l(i)
- Energy consumption e(i)
- Percentage of time spent in static and dynamic activities a(i).
- Total time of dynamic activities t(i).
- Extension in hip joint r(i).
- Numbers of activations of the panic button p(i).
- Part of the room space where user spends majority of time during day q(i).

Wearable sensors must be warned all the day. In the case of a critical accidental situation, a signal is generated and sent to supervising system (social centre with 24 hours service, or a member of family, etc.). Potential dangerous situations and necessary actions are described in the next chapter.

2.2.2.Classification of Critical Situations

Definition of dangerous and potentially dangerous situations depends on many factors. Analysis of human motion activities offers prediction of many dangerous health critical states. In Table 1, we summarized the most frequent acute statuses and raised actions:

Table 1. Classification of dangerous, potentially dangerous situations

Acute state, potentially dangerous	Action invoked
Lower number of steps comparing to the long-time average value	Inspection from a health centre and/or a carer, identification of a reason, motivation to increase physical activity
Larger angle in hip extension	Inspection from a health centre or a carer, identification of a reason, treatment/rehabilitation
Lowering walk speed	Examination by a medical doctor, assessment of a fall risk
Greater variability in the step length	Examination by a medical doctor, and neurological specialist
Lowering a ratio of time spent in dynamic activities	Inspection from a health centre and/or a carer, identification of a reason, motivation to increase physical activity
Permanently raising time of lying in the bed	Inspection from a health centre and/or a carer, identification of a reason, motivation to increase physical activity
Fall detection	Complementary inspection, sending an alarm message, inspection by a carer, when accident repeats examination by a medical doctor, and proper treatment/rehabilitation, removing barriers at home

There is a relation between less walking intensity - lowering number of steps, length of walk, and walk speed- and fall risk; variation in step length and in the same time increasing of hip extension may indicate the Parkinson disease. Increasing time spent in the bed may indicate a depression, or great mobility problems, etc. These are reasons to monitor above mentioned parameters, evaluate them in long time period, and provide actions when necessary.

3. Experimental Testing

Fall detection plays an important role in vulnerable person's life. Fall is the main reason of many injuries, even deaths, especially in elderly population. Fast reaction and contacting of caregivers or family can save life of injured person. There are many different ways how to recognize the fall, however utilizing MEMS sensors has following advantages:

- device is small and reliable,
- device can work remotely,
- capable of employing the smart algorithms,
- low cost device.

3.1. Fall Detection Device

Development of reliable and robust algorithm of fall detection is crucial. One of the possible ways of fall detection is comparison of absolute value of acceleration to the threshold and after passing this threshold follows computation of the tilt. For such sensing, one triaxial accelerometer is sufficient. Accelerometer senses gravitational vectors in each axis. Amplitude of measured gravity changes its value as sinus α between sensing axis and horizontal plane (Figure 3).

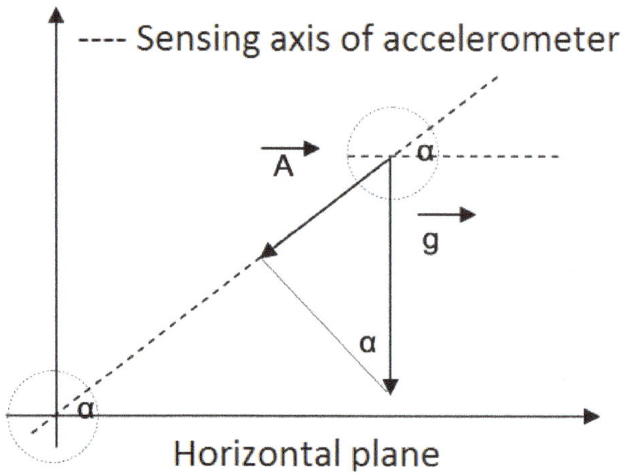

Figure 3. Sensing axis and horizontal plane

Our device contains 3-axial accelerometer and gyroscope. Accelerometer provides information about higher acceleration values that can indicate fall. But similar information gives also a jump, or stair climbing. Gyroscope provides us with angular velocity and we can derive tilt value from that too. By combining these data – acceleration, angular velocity and tilt, we obtain complex motion information. Setting the appropriate thresholds for each parameter is of utmost priority.

According to the possibility of obtaining information about acceleration, angular velocity and tilt, we proposed simple fall algorithm (Figure 4). Absolute value of acceleration with frequency of 20Hz is monitored, and if it passes threshold $|A_T|$ then the following 30 samples are monitored for passing two other thresholds – angular velocity threshold $|\omega_T|$ and tilt threshold $|\theta_T|$. If all conditions are true, then a fall has occurred with a high probability.

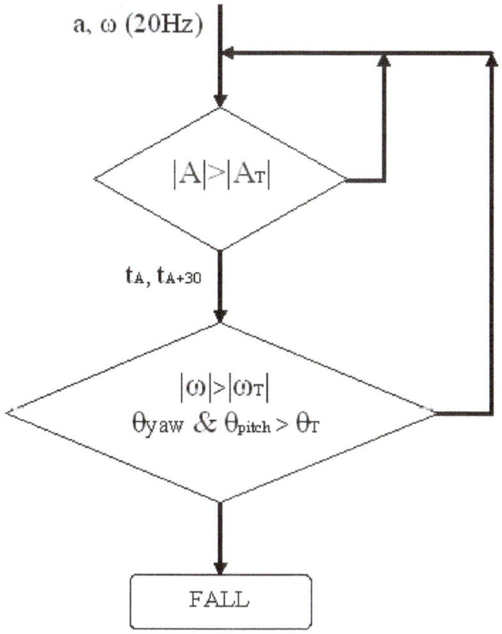

Figure 4. Fall detection algorithm

To verify proposed algorithm, we performed experiments. IMU was placed on the belt of the proband. During experiment, data was sent to PC with frequency of 20Hz. Person was walking on the floor, with little disturbing elements such as small stairs, and after more than 25 seconds asimulated fall occurred. Our threshold parameter settings were following: $|A_T|$=2,5g, $|\omega_T|$=300°/s, $|\theta_T|$=45°. These thresholds were set by empirical testing. Graphical interpretations of experimental data are shown in Figure 5.

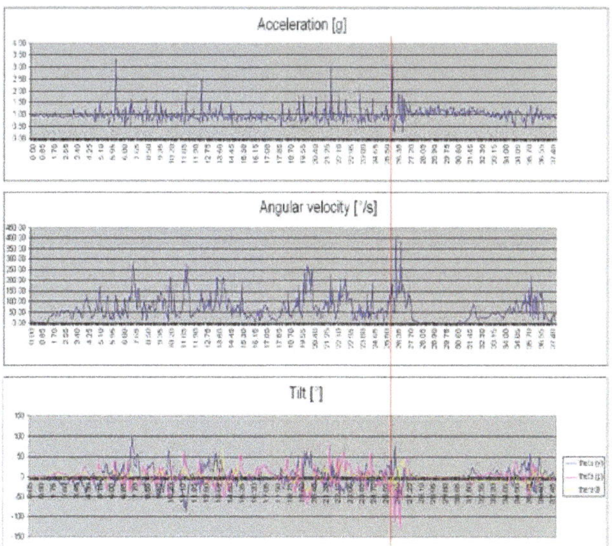

Figure 5. Visualized data from experiment

The device has to be reliable, it is essential to detect true alarm and neglect false ones. We decided to assess reliability of proposed algorithm as a ratio of false alarms to all alarms, as in (1):

$$r = 1 - \frac{A_F}{A_A} \qquad (1)$$

where r – reliability, A_F – number of false alarms, A_A – number of all alarms

Results were satisfactory – reliability was r=0,91. We assessed feasibility of acceleration as determinant of fall detection using (1), but results were not satisfying at all (r < 0,6). It means that acceleration itself does not provide enough information for fall detection. Consequently, using of appropriate sensor combinations and data fusion as angular velocity and tilt are essential. Acceleration data as a trigger for further assessment of other movement elements (e.g. angular velocity) seem to be acceptable and suitable solution.

Proposed algorithm was created in C++ and implemented into PC application for fall detection. In our future work we plan to apply various wireless technologies, such as XBee for communication of fall detector with control system. System will run on Raspberry Pi due to its low cost and appropriate properties.

Acknowledgement

This work has been supported by the Slovak Grant Agency VEGA contract Nb. 1/0911/14 "Implementation of wireless technologies into the design of new products and services to protect human health".

References

[1] Huang, C.N., Chan, C.T.,A ZigBee-Based Location-Aware Fall Detection System for Improving Elderly Telecare, International Journal of Environmental Research and Public Health, vol. 11, 2014, 4233-4248.

[2] Bianchi, V., Grossi, F., De Munari, I., Ciampolini, P., MuSA: a multisensor wearable device for AAL. Proceedings of the Federated Conference on Computer Science and Information Systems, 2011, 375-380.

[3] Bishop, E., Li, Q., Walking speed estimation using shank-mounted accelerometers. IEEE International Conference on Robotics and Automation. 2010, 5096-5101.

[4] Lee, Ch. Y., Lee, J.J., Estimation of Walking Behavior Using Accelerometers in Gait Rehabilitation. International Journal of Human-friendly Welfare Robotic Systems, vol. 3, no. 2, 2002, 32-35.

[5] Patel, S. et al., A review of wearable sensors and systems with application in rehabilitation. Journal of NeuroEngineering and Rehabilitation, 9, 21, 2012, 1-17.

[6] Biros, O., Karchnak, J., Simsik, D., Multiagent approach in home care based on open architecture. 2013. In: American Journal of Mechanical Engineering. Vol. 1, no.7, 2013, p. 217-220.

[7] Karchnak, J., Simsik, D., Jobbagy, B., Galajdova, A., Onofrejova, D., MEMS sensors in evaluation of human biomechanical parameters. In: Procedia Engineering: Modelling of Mechanical and Mechatronic Systems MMaMS 2014: 25th-27th November 2014, High Tatras, Slovakia. Netherland: Elsevier, 2014 Vol. 96, 2014, p. 209-214. Available: http://www.sciencedirect.com/science/article/pii/S1877705814032044#...[Accessed 27 oct 2015].

[8] Karchnak, J., Simsik, D., Jobbagy, B., Onofrejova, D., Feasibility Evaluation of Wearable Sensors for Homecare Systems. In: Acta Mechanica Slovaca. Nb. 2, 2015, 1-8. Available: http://www.actamechanica.sk/e-publications/2015/volume-19-issue-no-2...[Accessed 21sept 2015].

[9] Simsik, D., Karchnak, J., Jobbagy, B.,Galajdova, A.,Design of inertial module for rehabilitation device.In: SAMI 2013: IEEE 11th International Symposium on Applied Machine Intelligence and Informatics proceedings: January 31 - February 2, 2013, Herľany, Slovakia. Print: Budapest: IEEE, 2013, 33-36.

[10] Vavilov, D., Melezhik, A., Platonov, I., Smart Home User's Behavior Prediction. In: IEEE Third International Conference on Consumer Electronics (ICCE). 2013, s. 261-264.

[11] Bao, K., Allerding, F., Schmeck, H., User Behavior Prediction for Energy Management in Smart Homes. In: Eighth International Conference on Fuzzy Systems and Knowledge Discovery (FSKD). 2011, s. 1335-1339.

[12] Chen, C., Cook, D.J., Behavior-based Home Energy Prediction. In: Proceedings of the 2012 Eighth International Conference on Intelligent Environments. IEEE, s. 57-63.

[13] Fatima, I. et all., A Unified Framework for Activity Recognition-Based Behavior Analysis and Action Prediction in Smart Homes. Sensors (Basel, Switzerland) 13.2 .2013, 2682-2699.

[14] Abdelsalam, H., Cook, D.J., Schmalz, M., Smart Home-Based Health Platform for Behavioral Monitoring and Alteration of Diabetes Patients. Journal of diabetes science and technology (Online), 2009, 141-148.

[15] Aztiria, A., Farhadi, G., Aghajan, H., User Behavior Shift Detection in Ambient Assisted Living Environments. Ed. Antonio Jara et al. JMIR mHealth and uHealth, 2013.

[16] Aztiria, A., Augusto, J.C., Izaguirre, A., Autonomous Learning of User's Preferences improved through User Feedback. In: Proceedings of BMI. 2008, 72-86.

[17] Kadouche, R., Chikhaoui, B., Abdulrazak, B., User's behavior study for smart houses occupant prediction. In: Annales des Télécommunications, vol. 65, 2010, 539-543.

[18] Whittle, M.V., Gait Analysis: An Introduction. Butterworth-Heinemann, 3. edition, p. 232.

Optimal Controller for Single Area Load Frequency Control via LQR and Legendre Wavelet Function

Hatem Elaydi[1,*], Mohammed Wadi[2]

[1]Electrical Engineering Department, Islamic University of Gaza, Gaza, Palestine
[2]Electrical Engineering Department, Yildiz Technical University, Istanbul, Turkey
*Corresponding author: helaydi@iugaza.edu.ps

Abstract Load frequency control (LFC) is an important tool to insure the stability and reliability of power systems. Nowadays power demand is increasing continuously and the biggest challenge is to provide uninterrupted electrical energy to consumers under changing conditions. For satisfactory operation, the frequency of power system should be kept near constant value or within standard value; in addition to that an uninterrupted power should continue to flow despite the variations of load in different areas. Many techniques have been proposed to guarantee the uninterrupted flow of power and the constant value of frequency and to overcome any deviations. In this study, optimal control based on Legendre wavelet function and linear quadratic regulator (LQR) techniques are applied to single area power system in order to approximate the gain of controller to overcome the fluctuations of frequency in electrical power systems. This paper models a single area power system with optimal control strategies based on LQR and Legendre wavelet transform and presents simulation results using Simulink/MATLAB. The paper also compares the performance of the LQR and Legendre wavelet transform controllers. A numerical example is solved in order to demonstrate the effectiveness of the proposed technique.

Keywords: *Load Frequency Control (LFC), Automatic Generation Control (AGC), Legendre Wavelet Function, Linear Quadratic Regulator (LQR), Single Area Power System*

1. Introduction

During the last three decades, power generation systems have become more complex and largely interconnected and they required automatic generation control (AGC) in order to keep the systems operating safely [1,2,3,4]. Automatic generation control is also referred to as load frequency control (LFC). The frequency of a power system is affected by changes in real power, while the voltage magnitude is dependent on changes in reactive power. Thus, active and reactive powers can be controlled separately to meet different criteria. In an isolated power system, the main objective of an LFC is to maintain a balance of a power system by maintaining the frequency within an acceptable tolerance around a nominal system frequency. The reliability, security and stability are three required properties in order to balance any power system efficiently. Reliability is defined as the ability of the system to adequately supply a near continuous electric services with few interruptions over an extended period of time, security is defined as the resistance of a power system to disturbances to supply consumers with power without discontinuity and indicates the level of robustness of the power system to a disturbance at any instant in time, while stability is defined as the ability of the power system to return to an equilibrium operating state by keeping frequency and voltage to be within acceptable standard levels [5].

Typically, the power output of the generators and the difference between the instant loads change the nominal frequency. If the amount of production is more/less than demanded power, the speed of generators increases/decreases; hence, frequency increases/decreases. When deviation occurs, frequency is brought to a desired level by controlling the speed of generators. Thus, generated power and frequency in the system should be adjusted according to the consumption; this process is called LFC model.

In an isolated power system, frequency control is considered a challenging issue due to low system inertia, nonlinear effects of physical components such as governor dead zone and generation rate constraints. Furthermore, any mismatch between generation and demand deviates the system frequency from scheduled values. High frequency deviation may lead to system collapse. In order to maintain constant nominal frequency, an accurate and fast acting controller such as optimal controller is required [6]. This study proposes an optimal control strategy based on Legendre wavelet function to control load frequency. The optimal control is based on LQR control and utilizes the direct method to parameterize the cost function and constraints using Legendre wavelet functions. The optimal control strategy is applied to a single area power system with a load frequency control model in order to bring the steady-state frequency error to zero.

This paper is organized such as: section 2 talks about load frequency control of a single area power systems, section 3 covers Optimal control using Legendre functions, section 4 shows the simulated results, section 5 concludes this paper.

2. Load Frequency Control Model of Single Area Power Systems

In large electrical power system, nominal frequency depends significantly on the balance of produced and consumed active power. When active power imbalance occurs in any part of the system, it results in changes in the frequency of the entire system. Industrial loads connected to electrical power system are very sensitive to the quality of electrical energy, mainly the frequency component. Thus, the steady-state frequency error in the system must stay within acceptable values in order to keep the balance [7]. Figure 1 shows the relationship between system frequency and load which is inversely proportional. Possible increase in the load reduces the nominal frequency of the system. This alternation in frequency is sensed by a regulator in the primary control loop; subsequently, the rotational speed of the turbine is increased leading to an increase in the produced power. Load frequency control is accomplished using two stages: primary control and secondary control. Figure 2 illustrates the main components of LFC for single area power system which can be represented in a matrix form as shown in equation (1) [8,9].

$$\begin{bmatrix} \Delta\dot{P}_g \\ \Delta\dot{P}_t \\ \Delta\dot{\omega} \end{bmatrix} = \begin{bmatrix} \frac{-1}{T_g} & 0 & \frac{-1}{RT_g} \\ \frac{1}{T_t} & \frac{-1}{T_t} & 0 \\ 0 & \frac{1}{2H} & \frac{-D}{2H} \end{bmatrix} \begin{bmatrix} \Delta P_g \\ \Delta P_t \\ \Delta\omega \end{bmatrix} + \begin{bmatrix} 0 \\ 0 \\ \frac{-1}{2H} \end{bmatrix} \Delta P_L + \begin{bmatrix} \frac{1}{T_g} \\ 0 \\ 0 \end{bmatrix} \Delta P_{ref} \quad (1)$$

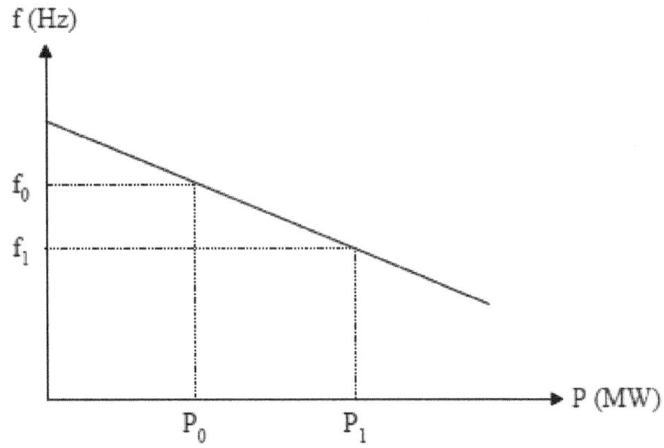

Figure 1. Variation of Load frequency Characteristic [10]

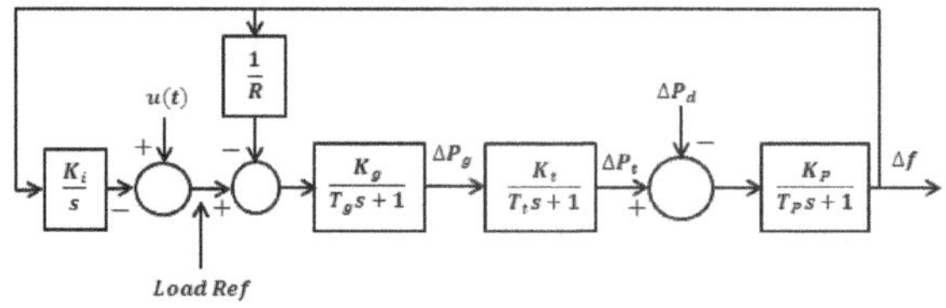

Figure 2. LFC for Single Area Power Control System

3. Linear Quadratic Optimal Problem

Optimal control is an important science that deals with nonlinear optimal control problem (OCP) and the main objective of optimal control is to find an optimal controller that can be applied to a nonlinear system as well as to a linear and time varying system to minimize a certain cost function within the system's physical constraints. In other words, the main task of an optimal controller is to transfer the system from its initial state to the final state with minimum cost function [11]. Generally, optimal control can be classified as direct and indirect methods. Indirect methods are based on converting OCP into two-point boundary value problem (TPBVP), then solving the last one by Euler Lagrange technique or Hamilton-Jacobi-Bellman equation (HJBE). Direct methods can be implemented using discretization or parameterization methods. Mathematically, the optimal control problem can be considered as finding the optimal controller $u^*(t) = -Kx(t)$ that minimize a performance index such

$$J = \int_0^{t_f} \left(X^T Qx + u^T Ru \right) dt \quad (2)$$

Subject to linear constraints and initial condition

$$\dot{x} = Ax(t) + Bu(t),\ x(0) = x_0 \quad (3)$$

Where $t \in [0, t_f]$, $x, x_0 \in R^n, u \in R^m$, A,B are n×n and n×m constant matrices respectively. Q is an n×n positive semidefinite matrix and R is an m×m positive definite matrix.

3.1. Optimal Control via Legendre Wavelet Function

One of the most famous indirect optimal control techniques that are used recently to approximate linear time invariant (LTI), linear time varying (LTV) and nonlinear systems are wavelet functions [12]. Legendre wavelet function has been used to approximate different systems within the subspace of trained initial condition and is considered a powerful function to give accurate approximation. Legendre wavelet function is defined such as [13].

$$\Phi_{nm}(t) = \begin{cases} \sqrt{m + \frac{1}{2}}\, 2^{K/2} P_m\left(2^K t - 2n + 1\right) for\ \frac{2n-2}{2^K} \le t \le \frac{2n}{2^K} \\ 0 \qquad \qquad otherwise \end{cases}$$

Where P_m is the Legendre polynomial of order m; n refers to the section number, $n=1,2,\ldots 2^{K-1}$; K is the scaling parameter and can assume any positive integer and $t\in[0,1]$. Legendre polynomial can be defined as in [16] as follows:

$$P_m(x)=\frac{1}{2^m m!}\frac{d^m}{dx^m}\left(x^2-1\right)^m \tag{5}$$

State and control variables can be approximated using Legendre scaling function as follows:

$$x_i(t)=\sum_{n=1}^{2^{K-1}}\sum_{m=0}^{M} a_{nm}^i\phi_{nm}(t)\; i=1,2,\ldots,s \tag{6}$$

$$u_i(t)=\sum_{n=1}^{2^{K-1}}\sum_{m=0}^{M} b_{nm}^i\phi_{nm}(t)\; i=1,2,\ldots,r \tag{7}$$

3.2. Quadratic Programming Problem

Here, the quadratic form of optimal problem is constructed and may be solved easily using Matlab [13].

$$\min_{z} z^T Hz \tag{8}$$

Subject to equality constraints

$$Fz=h \tag{9}$$

where

$$z=\begin{bmatrix} a \\ b \end{bmatrix}, H=\begin{bmatrix} Q\otimes I_N & O_{Ns\times Nr} \\ O_{Nr\times Ns} & R\otimes I_N \end{bmatrix},$$

$$F=\begin{bmatrix} \left(A\otimes P^T\right)-I_{Ns} & \left(B\otimes P^T\right) \\ \left(I_s\otimes\Phi'\right) & O_{\left(2^{K-1}-1\right)s\times Nr} \end{bmatrix}$$

$$h=\begin{bmatrix} -\xi_o\delta \\ O_{\left(2^{K-1}-1\right)s\times Nr} \end{bmatrix}$$

The optimal vector z* can be also calculated by the standard quadratic programming method [13] as follows:

$$z^*=H^{-1}F^T\left(FH^{-1}F^T\right)^{-1}h \tag{10}$$

3.3. Optimal Control via LQR

The linear quadratic regulator (LQR) is used to design optimal controllers for linear systems with quadratic performance index. LQR can be realized by finding optimal closed-loop state feedback control. In this article, LQR is applied to determine an appropriate state feedback gain along with the minimum variance method to design an LFC for single area power system. LQR can achieve good performance with minimum cost and can be easily designed using MATLAB. On the other hand, LQR has many limitations such as: it requires full state feedback, its function is only to guarantee stability, and it uses quadratic indices. This paper proposes a Legendre wavelet function to design optimal controller and compares its performance with an LQR controller.

4. Simulation Results

A comparison between an optimal controller based on Legendre wavelet function and an LQR control is made with the purpose of controlling LFC. The comparison is made in terms of performance with respect to frequency deviation (Δf), settling time (t_s) and steady state error (SSE). The parameters of the numerical example which is solved using Simulink/Matlab and programming subscripts are shown in Table 1. In addition, the solution of an example consists of three scenarios: the first one contains no controller, the second solution uses an LQR control, and the final scenario uses an optimal controller based on Legendre wavelet function.

Table 1. Parameters of single area power station [11]

Phase	Value
Turbine Time Constant $(\tau_T)Sec$	0.5
Governor Time Constant $(\tau_g)Sec$	0.2
Inertia Constant (H)	5
Speed Regulation (R)	0.05
Frequency Sens. Load Coeff. (D)	0.8
Nominal Frequency $(f)Hz$	60
Base Power $(S_{Base})MVA$	1000

The state space system and performance index of above example are respectively as follow [11]:

$$A=\begin{bmatrix} -5 & 0 & -100 \\ 2 & -2 & 0 \\ 0 & 0.1 & -0.08 \end{bmatrix} B=\begin{bmatrix} 0 \\ 0 \\ -0.1 \end{bmatrix}$$
$$Q=\begin{bmatrix} 20 & 0 & 0 \\ 0 & 10 & 0 \\ 0 & 0 & 5 \end{bmatrix} R=0.15 \tag{11}$$

$$J=\int_0^\infty\left(20x_1^2+10x_2^2+5x_3^2+0.15u^2\right)dt \tag{12}$$

4.1. First Scenario

In the first scenario, a block diagram of LFC is constructed using Simulink and solved without using any controller with the power increasing by 200 MW (0.2 p.u). Figure 3 and Figure 4 show the Simulink block diagram of LFC and the frequency deviation respectively; while Table 2 summaries the obtained results.

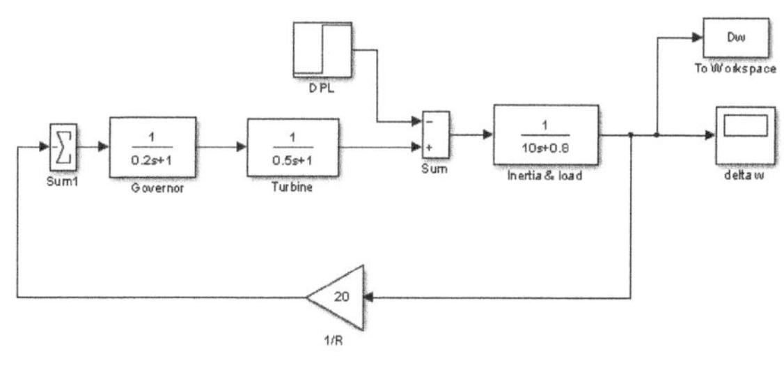

Figure 3. Simulink block diagram of LFC without controller

Table 2. Results of first scenario for LFC

Phase	Value
Frequency Deviation $(\Delta f), p.u$	0.015
Frequency Deviation $(\Delta f)Hz$	0.90
Settling Time $(t_s), Sec$	7
Steady State Error (SSE)	1×10^{-2}

Table 3. Results of second scenario for LFC using LQR

Phase	Value
Frequency Deviation $(\Delta f), p.u$	0.0007
Frequency Deviation $(\Delta f)Hz$	0.042
Settling Time $(t_s), Sec$	0.6
Steady State Error (SSE)	7×10^{-4}

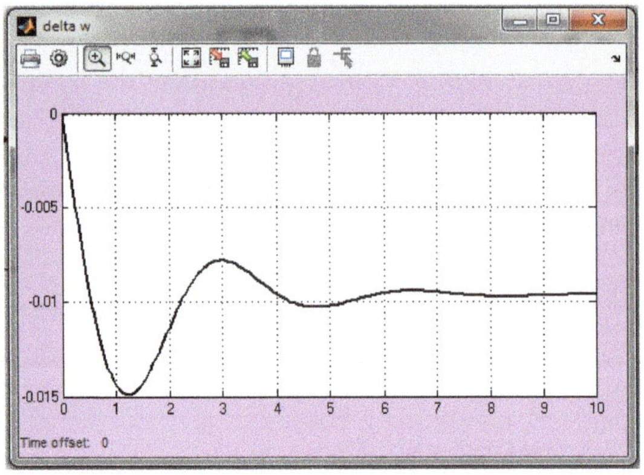

Figure 4. Frequency deviation of first scenario for LFC without controller

4.2. Second Scenario

In the second scenario, an LQR optimal controller is designed to approximate the state feedback gain vector. The LQR controller is designed using M-file and Simulink. The block diagram and the frequency deviation response are shown in Figures 5 and 6 respectively. Table 3 summarizes the obtained results. However, the optimal gain vector k=[6.41 1.10 -112.60] is obtained at end of simulation.

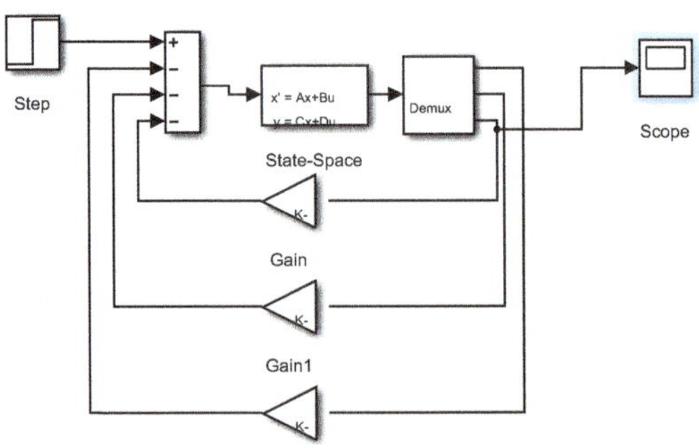

Figure 5. LQR block diagram

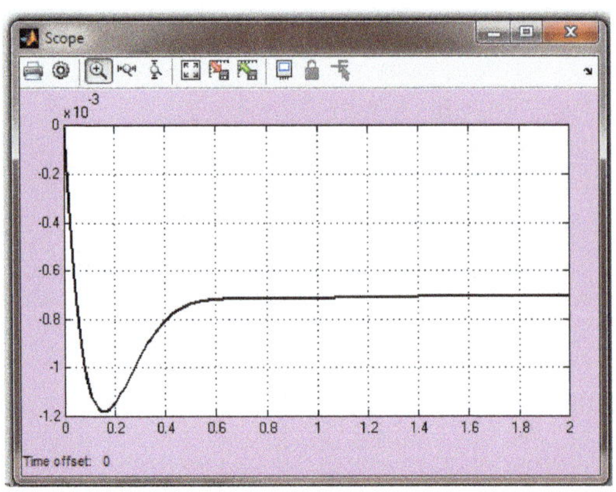

Figure 6. Frequency deviation of 2nd scenario for LFC with LQR

4.3. Third Scenario

A complete reference should contain the name(s) of the author(s) and/or editor(s), the title of the article, the name of the book or conference proceedings where appropriate, and bibliographic information about the article such as the name of the publisher, the city of publication, and the page numbers. The basic concept is that the reference should be sufficiently complete so that the reader could readily find the reference and can judge the authority and objectivity of the reference.

In the third scenario, an optimal controller based on Legendre wavelet function is obtained using the approximation techniques to design the feedback gain vector K by converting the problem into a quadratic form that can be easily solved using Matlab. Furthermore, approximation via Legendre function is based on two factors: the order of Legendre polynomial (M) and the positive integer scaling parameter (K) which increases the accuracy of approximation proportionally. In this paper, the problem is solved for M=4, K=2 and M=6, K=2. Figures 7 and 8 illustrate the response of LFC for M=4 and M=6 respectively; while Table 4 summaries the obtained results.

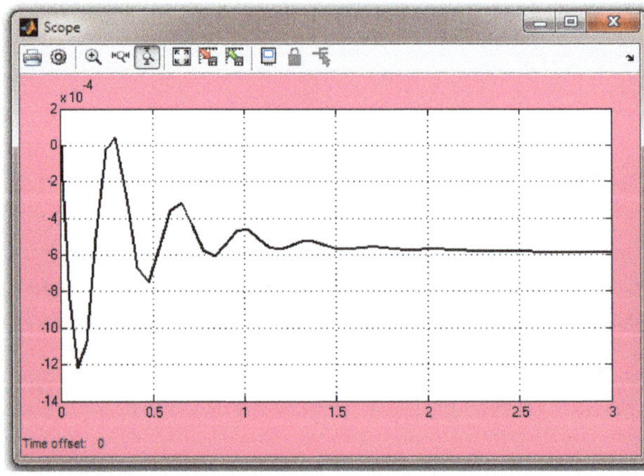

Figure 7. Frequency deviation of third scenario for LFC via Legendre function at M=4, K=2

Table 4. Results of third scenario for LFC using Legendre function

Phase	M=4, K=2	M=6, K=2
Frequency Deviation $(\Delta f), p.u$	0.0006	0.00001
Frequency Deviation $(\Delta f)Hz$	0.0360	0.0006
Gain Vector (K)	[5.96 2.88 - 15.43]	[-66.15 118 1016.2]
Settling Time $(t_s), Sec$	1.5	0.6
Steady State Error (SSE)	6×10^{-4}	1×10^{-5}

As seen from Table 5, we can conclude that the optimal controller via Legendre wavelet function provides better performance and response than LQR. Both controllers offer the same settling time (t_s), but the Legendre wavelet function controller gives better performance in terms of the frequency deviation and steady state error (SSE) by fifteen times less than LQR.

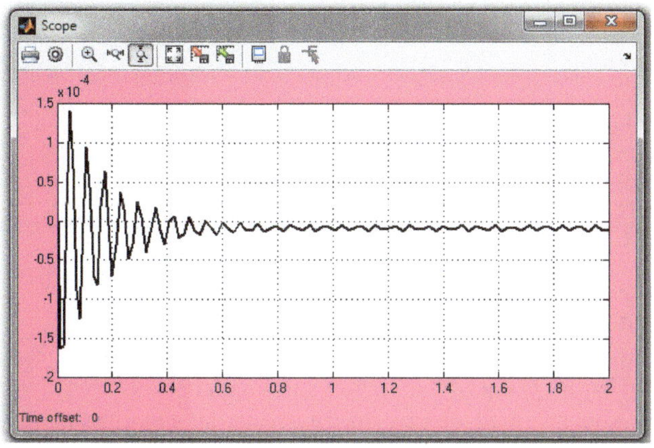

Figure 8. Frequency deviation of third scenario for LFC via Legendre function at M=6, K=2

Table 5. Comparison for LFC between LQR and Legendre function

Phase	LQR	Legendre Function
Frequency Deviation $(\Delta f), p.u$	0.0007	0.00001
Frequency Deviation $(\Delta f) Hz$	0.042	0.0006
Gain Vector (K)	[6.41 1.10 -112.60]	[-66.15 118 1016.2]
Settling Time $(t_s), Sec$	0.6	0.6
Steady State Error (SSE)	7×10^{-4}	1×10^{-5}

5. Conclusion

In this paper, a new optimal controller via Legendre wavelet function was proposed. A comparison was made between an LQR optimal controller and an optimal controller based on Legendre wavelet function in terms of performance aspect in single areas power systems. Simulation results showed that the optimal controller based on Legendre wavelet function approximation method outperformed the LQR controller in terms of less frequency deviation and steady state error; while both had the same settling time. A numerical example demonstrated the effectiveness of the proposed optimal control via Legendre wavelets function over LQR control.

References

[1] Benjamin NN, Chan WC., "Variable structure control of electric power generation," *IEEE Transaction on Power Apparatus & Systems*, PAS-101 (2). 376-380, 1982.

[2] Pan CT, Liaw CM., "An adaptive controller for power system load-frequency control," *IEEE Transactions on Power Systems*, 4 (1). 122-128, 1989.

[3] Kothari ML, Nanda J, Kothari DP, Das D., "Discrete-mode automatic generation control of a two-area reheat thermal system with new area control error," *IEEE Transaction on Power Systems*, 4 (2). 730-738, 1989.

[4] Wang Y, Zhou Y R, Wen C., "New robust adaptive load-frequency control with system parametric uncertainties," *IEEE Transactions on Generation, Transmission & Distribution*, 141 (3). 184-190, 1994.

[5] Gillian R. L., "Frequency control on an island power system with evolving plant mix," PhD, National University of Ireland, Dublin, Ireland, 2005.

[6] Usman A, Divakar B., "Simulation study of load frequency control of single and two area systems," In *IEEE 2012 Global Humanitarian Technology Conference*, 214-219, Oct 2012.

[7] Demiroren A, Yeşil, E., "Automatic generation control with fuzzy logic controllers in the power system including SMES units," *Electrical Power & Energy Systems*, 26. 291-305, 2004.

[8] George G, Jeong W L., "Analysis of load frequency control performance assessment criteria," *IEEE Transaction on Power Systems*, 16 (3). 520-525, 2001.

[9] Adnan B, Ebenezer JA., "Load frequency control with fuzzy logic controller considering nonlinearities and boiler dynamics," ACSE, 18 (3).15-20, 2009.

[10] Gozde, H, Taplamacıoglu, MC, Kocaarslan, I and Senol, MA., "Particle swarm optimization based pi-controller design to load-frequency control of a two area reheat thermal power system," *Journal of Thermal Sciences & Technologies*, 30 (1). 13-21, 2010.

[11] Saadat H., *Power System Analysis, 3rd ed. New York*, NY, USA: McGraw-Hill, 2002.

[12] Jaddu H., "Optimal control of time-varying linear systems using wavelets," PhD, Japan Advanced Institute of Science and Technology, Ishikawa, Japan, 2006.

[13] Wadi M., "Solving quadratic optimal control problems using Legendre scaling function and iterative technique," MSc, Islamic University of Gaza, Gaza, Palestine, 2011.

25

Diagnostics of Product Defects by Clustering and Machine Learning Classification Algorithm

Kamil Židek[1,*], Vladislav Maxim[2]

[1]Faculty of Manufacturing Technologies with a seat in Presov, Presov, Slovak Republic
[2]Institute of Automation, Robotics and Mechatronics, FME, Technical University of Kosice, Slovak Republic
*Corresponding author: kamil.zidek@tuke.sk

Abstract The article deals with usability of clustering and machine learning classification algorithm for search systematic surface errors. The main idea is to propose a methodology for the automated identification, diagnostics and localization of systematic errors in mass production. The introduced methodology consists of three levels: image processing for error parameterization, clustering for creating of errors classes (teach data) and prediction of new samples by machine learning algorithm. We conducted experiments with density based clustering algorithm DBSCAN. For classification we use multilayer perceptron MLP/ANN.

Keywords: *inspection, clustering, machine learning, image processing*

1. Introduction

Vision systems can be used in automated production processes for inspection, guidance, identification, measurement, tracking and counting, in many diverse industries. Vision systems may effectively replace human inspection in demanding cases such as nuclear industry, chemical industry, etc. In most cases, industrial automation systems are designed to inspect only known objects at fixed positions, characterized defects of faulty items and take actions for reporting and correcting these faults and replacing or removing defective parts from the production line [1].

At present, the development of computational performance embedded systems allows parallel image processing with advanced search for systematic errors by using clustering and machine learning classification algorithms. This data can then be used to identify the causes of these errors as well as continuous monitoring of product quality. This article describes automated methodology suitable for systematic error identification in mass production. We conducted experiment with clustering and classification algorithm. The first section describes advantages of selected clustering and classification algorithm. Main sections deal with description of methodology for automated systematic error diagnostics.

Implementation section contains brief description of prototype hardware and software solution.

2. Clustering and Classification Algorithm

In machine vision where it is supposed mass production we cannot use repeatedly clustering algorithm continuously for each new product error due to the large amount of data and the limitations of computing hardware.The solution is looking for algorithms combination from clustering and machine learning with fast prediction during mass production process. Clustering in our case tries to group a set of errors and find whether there is some relationship between its parameters. In general, the classification use acquired set of group classes for teaching and creates classification model and then can predict very fast to which class a new error belongs to. Clustering may then be repetitively performed with longer delays.

Algorithms can be divided to many groups for example hierarchical FLANN [2], density based DBSCAN and statistical K-means [3]. The most suitable is DBSCAN clustering algorithm because is it not necessary define number of clustering group. It is supposed that clustering algorithms provide finer distribution to classes like manually fixed supervised conditional setup of parameters [4,5].

Density-Based Spatial Clustering of Applications with Noise DBSCAN can find non-linearly separable clusters. Unlike other clustering algorithms that require many parameters, such as the number of clusters in the set, to be known and defined before computation, the DBSCAN algorithm has only two input parameters: the minimum size of a cluster and the maximum distance between points in a cluster. The algorithm operates by cycling through all points in the data set and calculating the number of neighbors each point has, which is defined as the number of other points that are within the minimum distance of the original point. Any data point that has fewer neighbors than the minimum cluster size parameter is declared to be a noise point that is not associated with any cluster. [6].

For example this dataset Figure 1 cannot be adequately clustered with k-means or Gaussian Mixture EM clustering.

Generalized DBSCAN is GDBSCAN and OPTICS, the fastest is hierarchical modification HDBSCAN [5,6].

Classification algorithm can be divided too to many groups of principles for example: statistical method, artificial intelligence, decision trees, boosted algorithms … etc [7-15]. In the next section are described chosen method based on ANN/MLP.

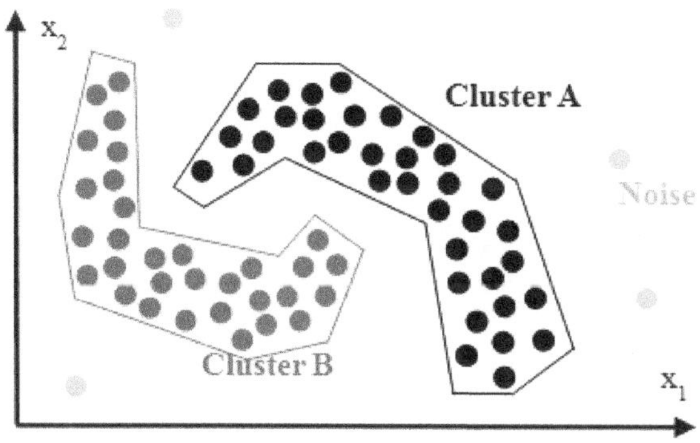

Figure 1. Dataset suitable for DBSCAN algorithm

Neural Network ANN/MLP. To classify the data into classes can also be advantageously used artificial neural network. Feedforward artificial neural network, or more specifically, a multi-layer Perceptron (MLP), the most commonly used type of neural network. MLP form consists of an input layer, an output layer, and one or more hidden layers. MLP each layer comprises one or more neurons associated with directionally neurons of the previous layer and also the next ones. The Figure 2 shows a three-layer Perceptron with three inputs, two outputs and hidden layer with three neurons [16].

All neurons in the MLP are similar and each has the stored value in the form of scales. The results of the neurons are transformed with the activation function f, which is usually the same but may be different for different neurons. The most commonly used activation function is symmetric sigmoid. As the previous algorithms, it is necessary learning using samples which must be placed at the entry to the exit pertaining to the output layer.

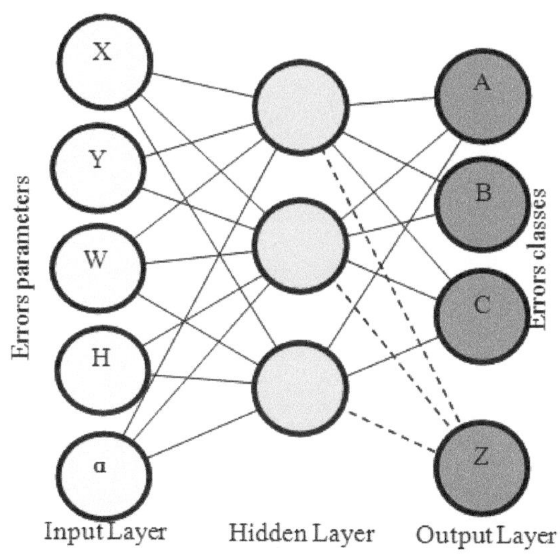

Figure 2. Example of multi-layer perceptron for classification

The main disadvantage is learning process of artificial neural networks which is extensively time consuming, especially with compare to each of presented machine learning classification algorithms.

3. Methodology of Error Clustering, Classification

Stopping production may be made primarily on the basis of a large number of errors in a row or on a time interval which we can show the error rate of production. This method is now used when stopped production on the basis of bad and good products, but the problem is that it tells us nothing about the character of the errors if it is a random or systematic error. If we cannot determine the character of the error is difficult to detect and remove the cause of their formation.

The clustering of large dataset is time consuming for desktop PC. It is not possible and necessary to run clustering for any new error detection in recognition loop. The solution can be combination of clustering method with classification teaching and prediction. The clustering provides slow basic assignment of similar errors to classes. Classification creates prediction model from training dataset provided by clustering. Trained classifier provides very fast assignment of new error to adequate class by prediction.

The main principle of surface error similarity diagnostics during operation we can divide to three steps:

- Image processing with contour detection – extraction of surface error with parameters,
- Search for similar errors by clustering algorithms, preparing of teach data,
- Create of classification model based on teaching and prediction of class assignment for new error.

Image processing algorithm for parameterization of detected errors is out of scope of this article and is in detail described in earlier article about error diagnostics [17].

We assume with these parameters of errors stored partially depended on three layers Figure 3.

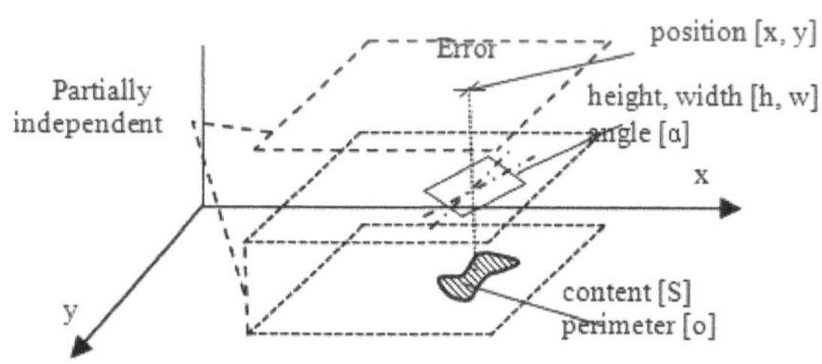

Figure 3. Extraction of parameters to partially dependent layers

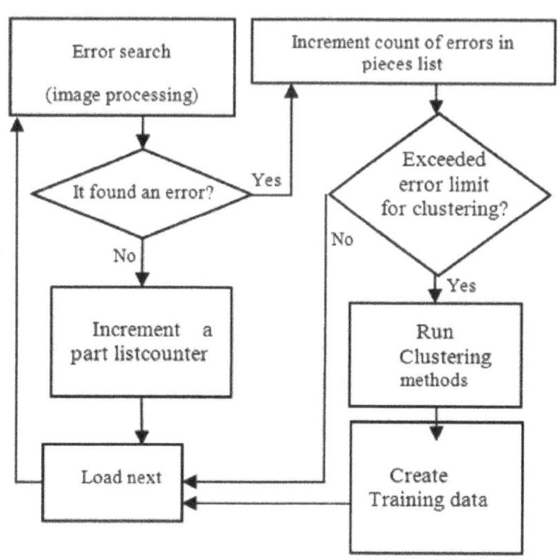

Figure 4. Principle algorithm of errors recognition

Second level of clustering algorithm threshold principle is shown on the Figure 4.

Principle for classification is universal for any proposed algorithm and is shown on the Figure 5.

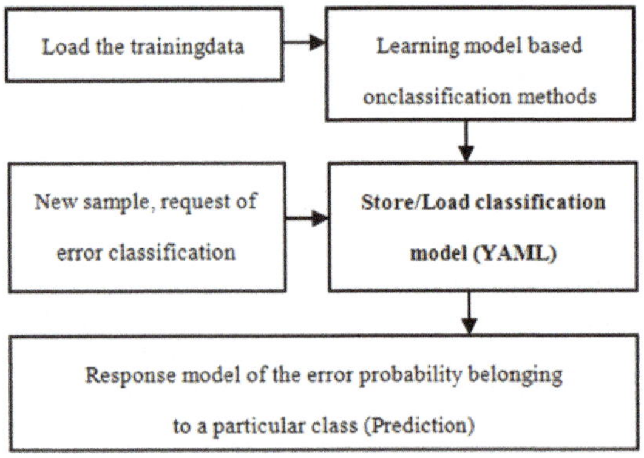

Figure 5. Classification of errors by teaching and prediction

Uses this methodology can help detect systematic errors from its location, size, perimeter, within components. The time variation of these parameters can predict degradation speed of errors (size increase or decrease).

4. Hardware and Software Solution of Vision System

The algorithms for clustering and classification were tested on two different platforms desktop PC (x86) and embedded (ARM) core: Intel Xeon x64 platform with (Ubuntu OS), and embedded board (SBC) with ARM architecture OlinuxinoA13(Fedora) and Raspberry PI (Raspberrian). Prototype stand with two different embedded systems and different camera sensors Firewire (1), USB camera (2) and CSI camera (3) is shown on the Figure 6.

Desktop PC uses Firewire camera, Olinuxino the USB web camera and Raspberry Pi CSI. The real-time vision task requires the industrial camera systems with Firewire for PC and high speed CSI interface for embedded systems.

For the vision recognition task exists many completed free or commercial libraries. For our task was selected OpenCV library (C++ version), because is free, open source, very fast, platform independent and includes current cutting-edge algorithms from image processing and machine learning (clustering, classification, prediction … etc.). The next reason was native support for algorithm acceleration with CUDA, OpenCL and multithreading TBB.

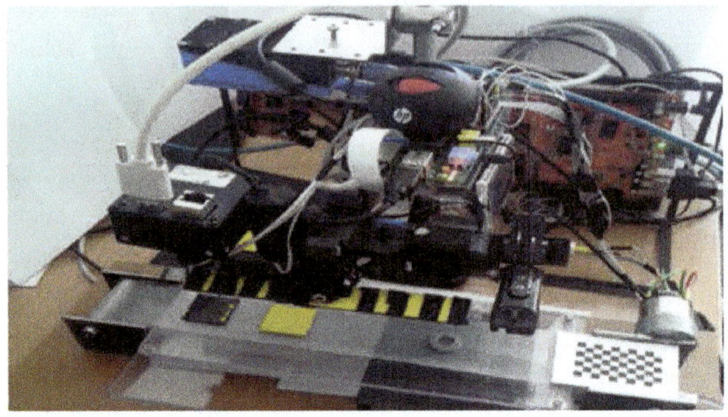

Figure 6. Experimental vision and image processing stand

Software solution of the vision system consists from two parts:

- Image processing and machine learning based on fast C++ Opencv library,
- Graphical user interface based on web technologies with AJAX support.

GUI uses PHP scripts as bridge between C++ and HTML5/jQuery web visualized pages. Web applications are platform independent that was a basic point to use these technologies. The next advantages are remote access and saving power of processing device because the performance is used by browser on client side.

Graphical web GUI interface for control diagnostics process, conditional and advanced clustering is shown on the Figure 7. We assume that error data from recognition process can be available simultaneous. This necessitates a network database to store data. Recognition results are stored in integrated SQlite3 database and can be accessed remote by web GUI. This set of error data in certain defined interval creates basic training set for clustering.

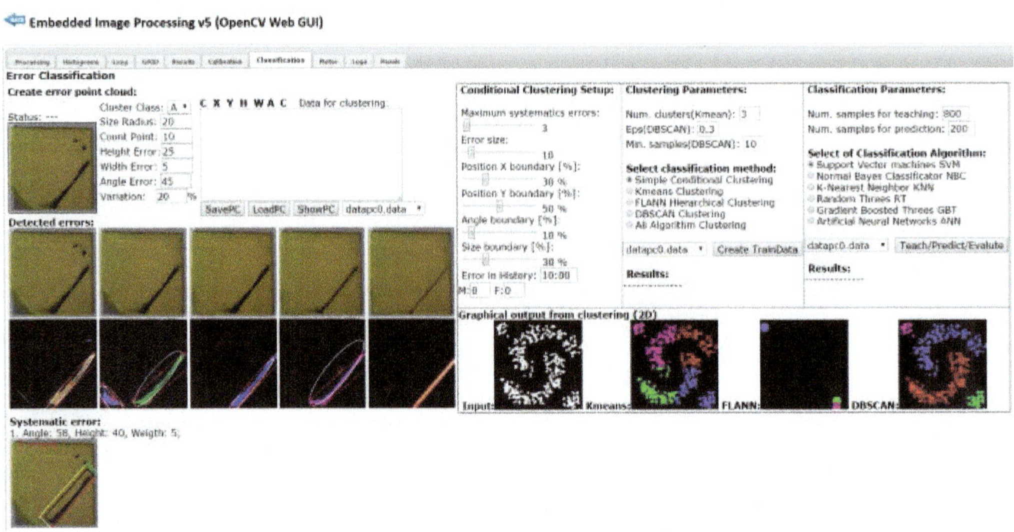

Figure 7. GUI for error parameter extraction by image processing

5. Experiments with Clustering/ Classification Algorithms

We conducted the experiment with metal square parts 35x35 mm painted by polyurethane yellow paint. Possible errors were classified to the four basic classes (A, B, C, D). These errors can arise as result of rough handling, manipulation, drying and poor surface pretreatments before painting. Examples of surface errors samples used in experimental verification are shown on the Figure 8.

Figure 8. Samples used for experiments.

Result of clustering in DBSAN in comparison to next two algorithms is shown on the Figure 9.

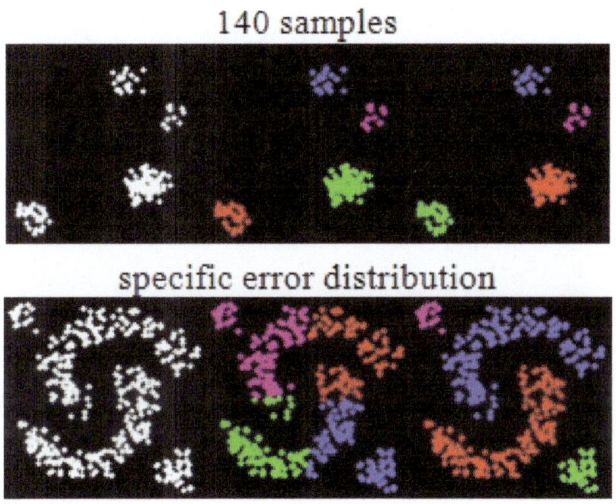

Figure 9. Picture from left: Clear data, Kmeans, DBSCAN

For comparison of result we compare reliability of MLP/ANN teaching with other classification algorithms. Graphical representation is shown on the Figure 10.

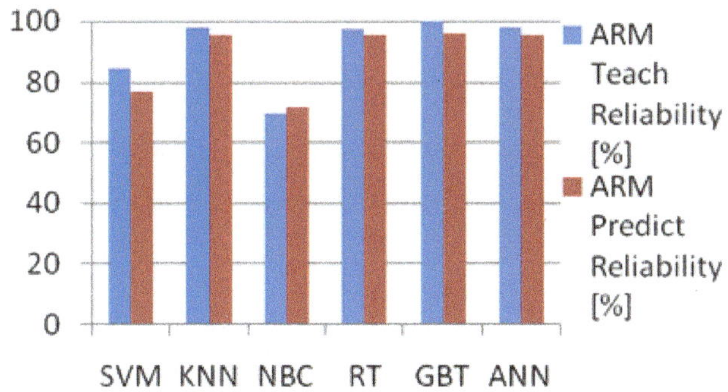

Figure 10. Teaching and prediction reliability for 1000 samples

The experiments show that the most reliable algorithm is Gradient Boosted Threes [GBT] with reliability 99,9 / 95,8 % and neural networks [ANN] 97,8 / 95,3 %. In terms of the duration of the learning model was K-Nearest Neighbor [KNN] the fastest one, achieving sufficient accuracy subsequent classification of new patterns of 97,8 / 95,2 %. In real task, the classification learning must be repeated in certain defined intervals, it requires minimum time of new samples teaching and the minimum time in class assignments with adequate reliability.

6. Conclusions

The article describes methodology of systematic error diagnostics by clustering and classification of advanced machine learning methods and algorithms. Designed software recognizes surface error, creates database of errors and classify them to error classes.

Density based clustering (DBSCAN) is ideal solution for specific class distribution, but with significantly increasing delay for large datasets. We compare reliability of DBSCAN with K-Means algorithm.

For classification we use MLP/ANN algorithm. We compare reliability with these classification algorithms: Support Vector Machines (SVM), Random Threes, Gradient Boosted Threes, K-Nearest Neighbor, Normal Bayes Classifier. ANN algorithm can be more precise after some optimization. Data results from recognition are stored in database and can be used for statistics of production process efficiency. The result of the experiment can be a different for another set of training and test data. The next works will be aimed to experiments with other advanced machine learning algorithms.

Acknowledgement

The research was supported by the Project VEGA 1/0911/14 Implementation of wireless technologies into the design of new products and services to protect human health.

References

[1] E. R. Davies, Computer & Machine Vision, *Theory Algorithms Practicalities*, Elsevier, p. 934.

[2] S. Har-Peled, B. Sadri, How fast is the k-means method?, *Proceedings of the sixteenth annual ACM-SIAM symposium on Discrete algorithms*, pages, Philadelphia, PA, USA, 877–885, (2005).

[3] D. Arthur, S. Vassilvitskii,K-means++: the advantages of careful seeding, *Proceedings of the eighteenth annual ACM-SIAM symposium on Discrete algorithms*, (2007).

[4] W. H. E. Day, H. Edelsbrunner, Efficient algorithms for agglomerative hierarchical clustering methods, *Journal of Classification*, 1,7-24, (1984).

[5] M. Muja, D. G. Lowe, Fast Approximate Nearest Neighbors with Automatic Algorithm Configuration, *in International Conference on Computer Vision Theory and Applications*, VISAPP'09, (2009).

[6] M. Ester, H.-P. Kriegel, J. Sander, X. Xu E. Simoudis, J. Han, U. Fayyad, A density-based algorithm for discovering clusters in large spatial databases with noise, *Proceedings of the 2nd International Conference on Knowledge Discovery and Data Mining*, vol. 1AAAI Press (1996), pp. 226-231.

[7] C. C. Chang, C.-J. Lin., LIBSVM, a library for support vector machines, *ACM Transactions on Intelligent Systems and Technology*, 2:27: 1-27:27, (2011).

[8] O. Kadri, L. H. Mouss, M. D. Mouss, Fault diagnosis of rotary kiln using SVM and binary ACO, *The Journal of Mechanical Science and Technology*, vol. 26, no. 2, pp.601-608, (2012).

[9] K. Fukunaga, Introduction to Statistical Pattern Recognition., New York: Academic Press, (1990).

[10] R.K. Agrawal, R. Bala, Incremental Bayesian classification for multivariate normal distribution data, Volume 29, Issue 13, pp.1873-1876, (2008).

[11] D. Coomans, D.L. Massart, Alternative k-nearest neighbour rules in supervised pattern recognition, k-Nearest neighbour classification by using alternative voting rules, , AnalyticaChimica, Acta 136, pp. 15-27, (1982).

[12] B. Yao, F. Li,P. Kumar, K-Nearest Neighbor Queries and KNN-Joins in Large Relational Databases (Almost) for Free, *Data Engineering (ICDE), 2010 IEEE 26th International Conference*, pp. 4-15, (2010).

[13] L. Breiman, A. Cutler, Random Forests, available at: <www.stat.berkeley.edu/~breiman/RandomForests/cc_graphics.htm>.

[14] Bo-Suk Yang, Random forests classifier for machine fault diagnosis, *The Journal of Mechanical Science and Technology*, vol. 22, no. 9, pp.1716-1725, (2008).

[15] J. Friedman, Greedy Function Aproximation, A Gradient Boosting Machine, Feb. 24, 1999, available at: <docs.salford-systems.com/GreedyFuncApproxSS.pdf>.

[16] Y. LeCun, L. Bottou, G.B. Orr and K.R. Muller, Efficient backprop, in Neural Networks, Tricks of the Trade, Springer Lecture Notes in Computer Sciences 1524, pp.5-50.

[17] K. Židek, E, Rigasová, Diagnostics of Products by Vision System, Applied Mechanics and Materials, Trans Tech Publications, Switzerland, , Vol. 308, 33-38, (2013).

26

Improving Manufacturing Processes Using Simulation Methods

Sławomir Kłos[1], Justyna Patalas-Maliszewska[1], Peter Trebuna[2,*]

[1]University of Zielona Góra, Faculty of Mechanical Engineering, Licealna 9, 65-417 Zielona Góra, Poland
[2]Technical University of Kosice, Faculty of Mechanical Engineering
*Corresponding author: peter.trebuna@tuke.sk

Abstract Computer simulation is a very important method for studying the efficiency of manufacturing systems. This paper presents the results of simulation research about how buffer space allocated in a flow line and operation times influence the throughput of a manufacturing system. The production line in the study consists of four stages and is based on a real machining manufacturing system of a small production enterprise. Using Tecnomatix Plant Simulation software, a simulation model of the system was created and set of experiments was planned. Simulation experiments were prepared for different capacities of intermediate buffers located between manufacturing resources (CNC machines) and operation times as input parameters, and the throughput per hour and average life span of products as the output parameter. On the basis of the experiments, the impact of the allocation of intermediate buffer capacities on production efficiency is analysed.

Keywords: *computer simulation, production line, buffer allocation, throughput*

1. Introduction

Computer simulation is a powerful method for designing and analysing manufacturing processes in industry. Today, within a relatively short period of time, it is possible to design or redesign a manufacturing system to implement new production processes using simulation software. On the basis of the layout of a plant and a description of the technology in use, manufacturing resources can be allocated and the performance of the system can be estimated. A simulation model enables us to analyze various alternatives of the manufacturing system configuration and gauge the influence of the different parameters on the efficiency of different manufacturing processes. Advanced tools of computer simulation enable the analysis of the general performance of manufacturing systems, scheduling methods, support of facility layouts, automated material handling, automated guided vehicle systems design, etc. By doing this, we can reduce manufacturing costs and improve the system's throughput. Typically, however, the building of a simulation model of a manufacturing system is time-consuming and only enables the solution of individual problems. Therefore, it is often the case that a generalization of the results of simulation research is difficult. A very important advantage of a computer simulation is the possibility to create a model of a manufacturing system whose behaviour could be very close to the real system. Modelling the processes of a manufacturing system and production processes is time-consuming and needs expert knowledge which increases the costs of the simulation method. To reduce the modelling time, often some simplifications of a manufacturing system are introduced which do not significantly affect the behaviour of the system. Especially for companies which implement multi-assortment, repetitive production; it is very important to design and organize the production system effectively. Each new project requires the design of a technological process. To analyze the efficiency of a new production process, a model should be created that includes operation and setup times, the number of workers, the capacity of intermediate storage buffers, the availability of machines and tools, etc.

1.1. Literature overview

Simulation has been successfully implemented in a lot of research related to manufacturing system design and operation. Computer simulation offers very effective tools for visualizing, understating, and analyzing the dynamics of manufacturing systems [1]. Due to its complexity and importance, the buffer allocation problem has been studied widely and numerous publications are available in the literature [2]. Stanley and Kim presented results of simulation experiments made for buffer allocations in closed serial-production lines [3]. For the line, a single buffer space is the room and the associated material handling equipment that is needed to store a single job that is a work-in-process, and buffer allocation is the specific placement of a limited number of buffer spaces in a production line. The authors demonstrated a buffer allocation decomposition result for closed production lines, and also provided evidence that optimal buffer allocations in closed lines are less sensitive to bottleneck severity than in open production lines. The placement of buffers in a production line is an old and well-studied problem in industrial engineering research. Vergara and Kim proposed a new buffer placement method for serial production lines [4]. The method is very efficient and uses information generated in a production line simulation whose conceptual representation of job flow and workstation interaction can be described with a network which aims to place buffers in order to maximize throughput. They compared the results of the new method against a method for buffer placement based on a genetic algorithm. Yamashita and Altiok [5] proposed an algorithm for minimizing the total buffer allocation for a desired throughput in production lines with phase-type processing times. They implemented a dynamic programming algorithm that uses a decomposition method to approximate the system throughput at every stage. Gurkan used a simulation-based optimization method to find optimal buffer allocations in tandem production lines where machines are subject to random breakdowns and repairs, and the product is fluid-type [6]. He explored some of the functional properties of the throughput of such systems and derived recursive expressions to compute one-sided directional derivatives of throughput, from a single simulation run. Shi and Gershwin presented an effective algorithm for maximizing profits through buffer size optimization for production lines [7]. They considered both buffer space cost and average inventory cost with distinct cost coefficients for different buffers. To solve the problem, a corresponding unconstrained problem was introduced and a nonlinear programming approach was adopted. Qudeiri et al. used a genetic algorithm for studying the design of serial - parallel production line [8]. They tried to find the nearest optimal design of a serial parallel production line that maximized production efficiency by optimizing buffer size between each pair of work stations, machine numbers in each of the work stations and machine types. Nahas et al. formulated a new optimal design problem of a parallel production line, where parallel machines and in-process buffers are included to achieve a greater production rate [9]. The main objective was to maximize the production rate subject to a total cost constraint. Nourelfath et al. formulated a new problem of the optimal design of a series production line system, and developed an efficient heuristic approach to solve it [10]. The problem was solved by developing and demonstrating a problem-specific system algorithm. Fernandes and Carmo-Silva [11] presented a simulation study of the role of sequence-dependent set-up times in decision making at the order release level of a workload controlled make-to-order flow-shop. They indicated that the local strategy, which has been traditionally adopted in practice and in most of the studies dealing with sequence-dependent set-up times, does not always give the best results. Matta [12] presented mathematical programming representations for the simulation-based optimization of buffer allocation in flow lines.

1.2. Problem specification

In this paper, a simulation method is used to analyse the impact of intermediate buffer capacities and lot sizes on the throughput of a production line. The buffer allocation problem is an NP-hard combinatorial optimization problem which deals with finding optimal buffer sizes to be allocated into buffer areas in a production line [13,14]. In general, the buffer allocation problem is classified into three categories according to its objective function [13]:

1. Maximize the throughput rate of the production line with fixed amount of buffer sizes.
2. Minimize the total buffer size to achieve the desired throughput rate of the production line.
3. Minimize the average amount of work-in-process in the production line.

The presented paper takes into account the combination of the first two problems. The main problem considered in the paper can be formulated as follows: Given a production line with a determined number of manufacturing resources, operation times and set-up times. How does an intermediate buffer capacity and sequence of operation times affect the throughput of the production line? Using a simulation method, the best relation between allocated buffer capacity and throughput for several variants of operation times is examined. On the basis of the proposed simulation experiments, the principles by which solutions are reached can be analyzed and evaluated. In the next chapter, the simulation model of the production system is described and assumptions for simulation experiments are formulated. The third chapter contains the results of the simulation research and an analysis of the behaviour of the manufacturing system. In the last chapter, conclusions and directions for further study are presented.

2. Simulation model of the manufacturing system

The model of the automated production line was prepared on the basis of a real example of a manufacturing system dedicated to metal tooling in an automotive company. The model and simulation experiments are implemented using Tecnomatix PLM simulation software. The studied manufacturing system includes four technological operations: cutting, turning, milling and grinding. The manufacturing process is divided into four stages by technology and between each two stages an intermediate buffer is allocated. Every stage of the manufacturing system encompasses a determined number of manufacturing resources (three CNC machines in every stage). The simulation model of the manufacturing system was prepared with Tecnomatic Plant Simulation Software version 11.0.0 and is presented in Figure 1.

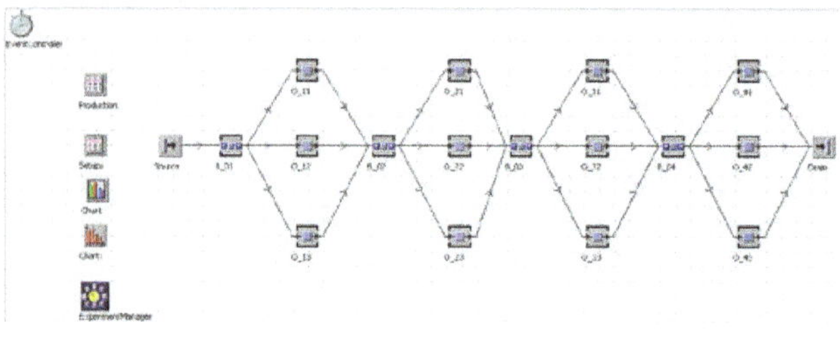

Figure 1. Simulation model of the manufacturing system

The production is divided into four batch sizes of products (A,B,C,D) and realized cyclically. It was assumed that the efficiency of manufacturing resources is approximately 95%. The model of the production system was prepared on the basis of a real manufacturing company. The operation times are based on a lognormal distribution. A lognormal distribution is a continuous distribution in which a random number has a natural logarithm that corresponds to a normal distribution. The realizations are non-negative real numbers. The density of the lognormal distribution *Lognormal(σ,μ)* is calculated as follows [15]:

$$f(x)=\frac{1}{\sigma_0 x\sqrt{2\pi}}\cdot\exp\left[\frac{-\ln(x-\mu_0)^2}{2\sigma_0^2}\right] \tag{1}$$

where σ and μ are respectively mean and standard deviations and are defined as follows:

$$\mu=\exp\left[\mu_0+\frac{\sigma_0}{2}\right] \tag{2}$$

$$\sigma^2=\exp(2\mu_0+\sigma_0^2)\cdot(\exp(\sigma_0^2)-1) \tag{3}$$

The maximum of the density function is defined as:

$$\exp(\mu_0-\sigma_0^2) \tag{4}$$

The example of the density function of lognormal distribution *Lognormal(3,2)* is presented in the Figure 2 [15].

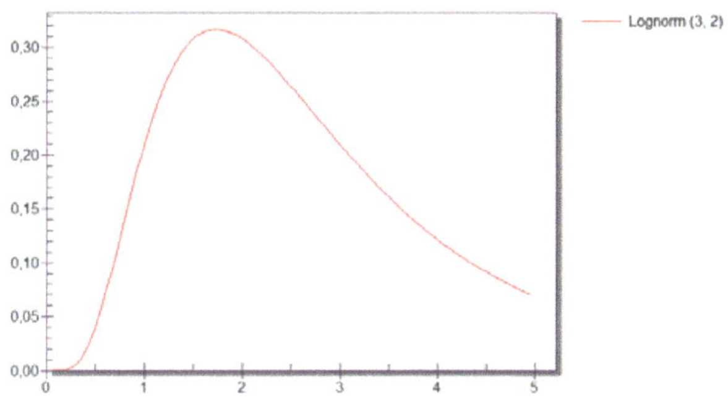

Figure 2. Density function of lognormal distribution – lognormal(3,2)

The variants of operation times are presented in the Table 1.

Table 1. The matrix of variants of operation times

Technological operations	Variant 1	Variant 2	Variant 3	Variant 4
O_11; O_12; O_13;	Lognormal (5;0.5)	Lognormal (5;0.5)	Lognormal (2;0.5)	Lognormal (5;0.5)
O_21; O_22; O_23;	Lognormal (5;0.5)	Lognormal (4;0.5)	Lognormal (3;0.5)	Lognormal (3;0.5)
O_31; O_32; O_33;	Lognormal (5;0.5)	Lognormal (3;0.5)	Lognormal (4;0.5)	Lognormal (3;0.5)
O_41; O_42; O_43;	Lognormal (5;0.5)	Lognormal (2;0.5)	Lognormal (5;0.5)	Lognormal (5;0.5)

The set-up times are defined in a set-up matrix (see Table 2). The numbers presented in the set-up matrix refer to the set-up time of changing the production batch (for example a batch change from product A to product B takes 15 minutes of set-up time).

Table 2. The matrix of setup times

	Product A	Product B	Product C	Product D
	10:00.0000	10:00.0000	10:00.0000	10:00.0000
Product A	10:00.0000	15:00.0000	10:00.0000	10:00.0000
Product B	15:00.0000	10:00.0000	20:00.0000	10:00.0000
Product C	20:00.0000	15:00.0000	10:00.0000	15:00.0000
Product D	10:00.0000	20:00.0000	20:00.0000	10:00.0000

The sequence of lot size is presented in the Table 3.

Table 3. The lot size matrix

	Lot size
Product A	20
Product B	25
Product C	15
Product D	10

The main variable in the simulation experiments was the intermediate buffer capacity. The proposed values of different combinations of intermediate buffers capacities that define the simulation experiments are presented in Table 4. The combination of the buffer capacities was chosen arbitrarily on the basis of author's experiences.

Table 4. The matrix of intermediate buffer capacities

Experiment	B_01	B_02	B_03	B_04
Exp 01	1	1	1	1
Exp 02	3	3	3	3
Exp 03	5	5	5	5
Exp 04	1	5	10	15
Exp 05	15	10	5	1
Exp 06	1	10	10	1
Exp 07	10	1	1	10
Exp 08	10	1	10	10
Exp 09	10	10	1	10
Exp 10	1	2	3	4
Exp 11	4	3	2	1
Exp 12	10	10	10	10
Exp 13	20	20	20	20
Exp 14	30	30	30	30
Exp 15	50	50	50	50

In addition to the system throughput per hour, the average life span of products in the manufacturing system was analyzed. The average life span shows us how long the products stay in the system and enables us to identify the level of work-in-process. In the next chapter, the results of the simulation experiments are presented.

3. The outcomes of computer simulation experiments

Table 5. Font Sizes for Papers

Font Size	Appearance (in Time New Roman)		
	Regular	Bold	Italic
8	cell in a table reference item	table caption, figure caption,	
9	author affiliation		author email address,
10	paragraph abstract body	author name level-3 heading,	Received Line
12		level-2 heading,	
14		level-1 heading,	
20		title	

4. Ease of Use

On the basis of the described simulation model, a set of simulation experiments for four variants of operation times, as presented in Table 1, were performed. The results of the experiments performed for Variant 1 of operation times are presented in Figure 3 a and Figure 3 b.

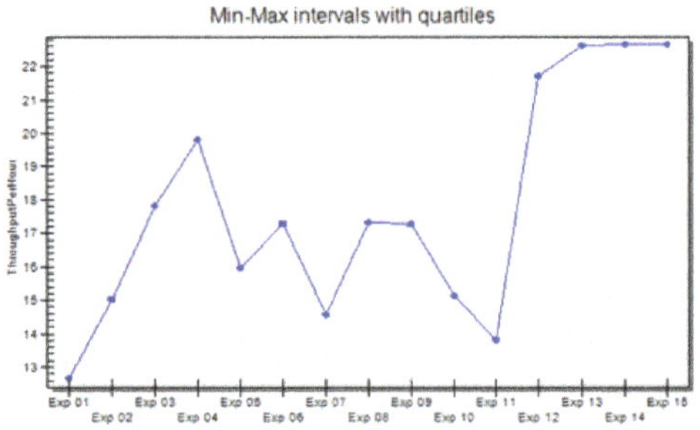

a

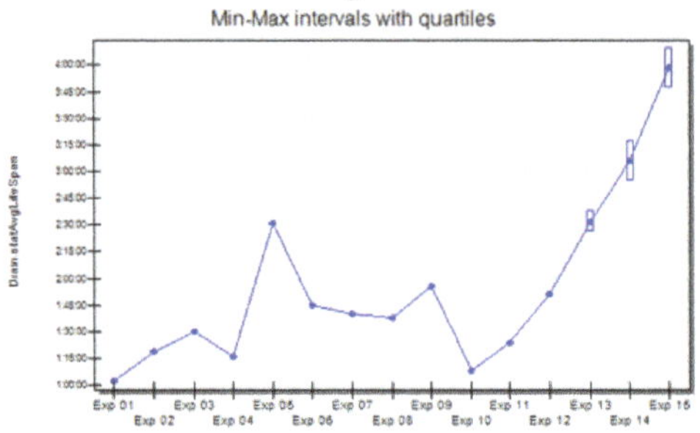

b

Figure 3. a) The throughput per hour for Variant 1, b) The average life span for Variant 1

The minimum values of throughput per hour in the manufacturing system were obtained in experiments 1 and 11 (lower then 14 pieces per hour). Interestingly, the minimum average life span in the system were obtained in experiments 1 and 10. In experiment 1, the total sum of the intermediate buffer capacity is smallest and equal to 4 (single capacity of the buffers). In experiments 10 and 11, the buffer capacities increase or decrease from 1 to 4, respectively. Very good results of throughput were obtained in experiment 4, and the last four experiments in which the buffer capacities were significantly increased.

The results of the experiments performed for Variant 2 of operation times are presented in Figure 4 a and Figure 4 b.

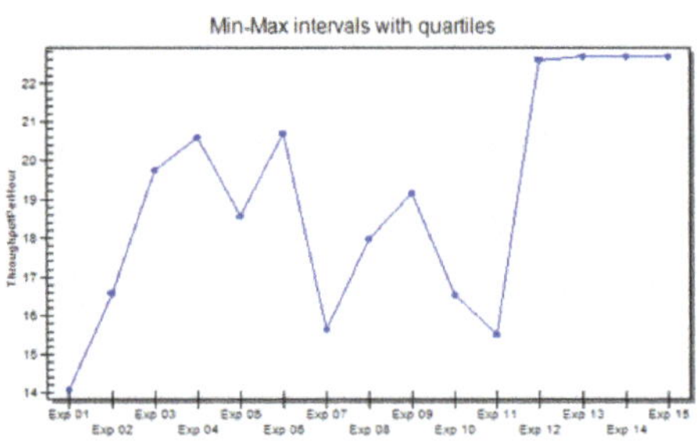

a

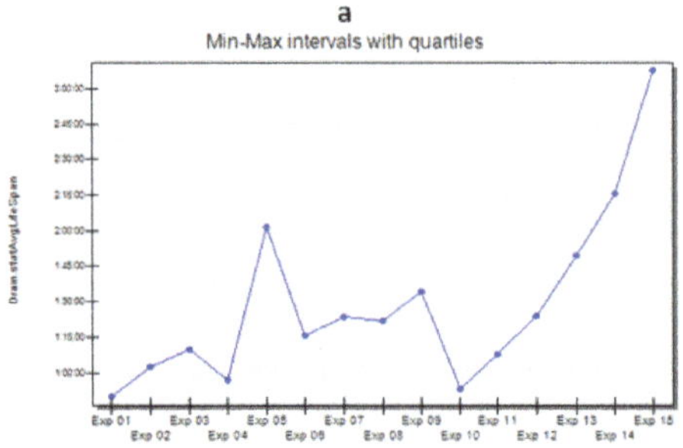

b

Figure 4. a) The throughput per hour for Variant 2, b) The average life span for Variant 2

The minimum value of throughput per hour in the manufacturing system was obtained in experiments 1, 7 and 11 (respectively 14.0; 15.6 and 15.5 products per hour). The minimum average life span in the system was obtained in experiments 1 and 11. Very good results of system throughput by relatively small capacities of intermediate buffers were obtained in experiments 4, 6 and 9.

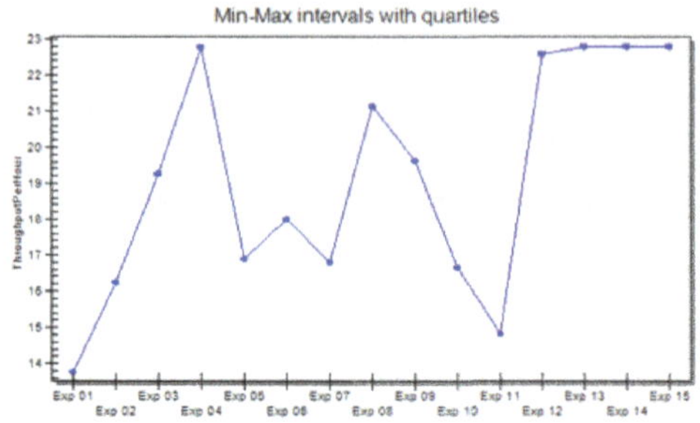

a

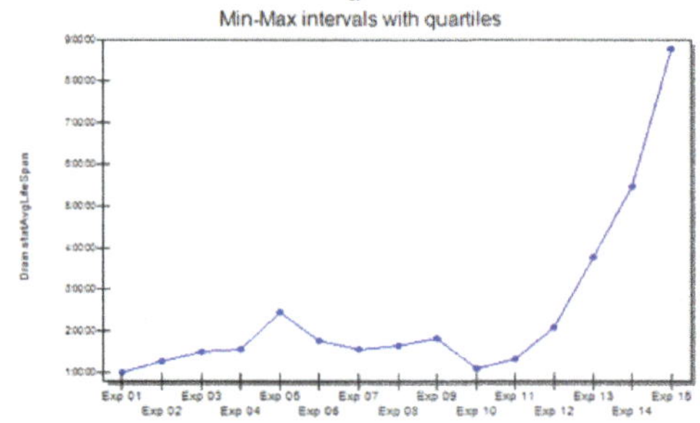

b

Figure 5. a) The throughput per hour for Variant 3, b) The average life span for Variant 3

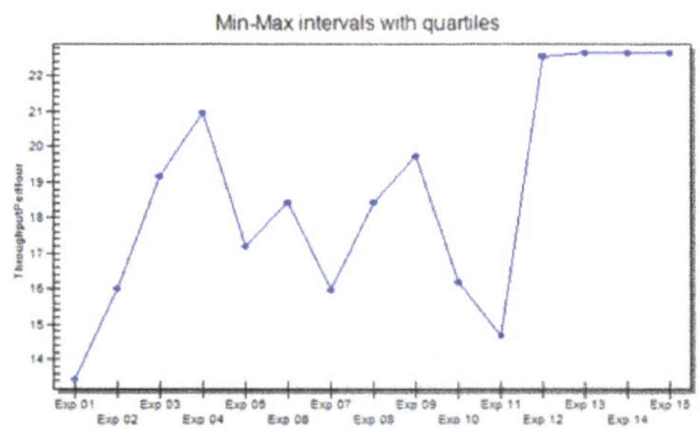

a

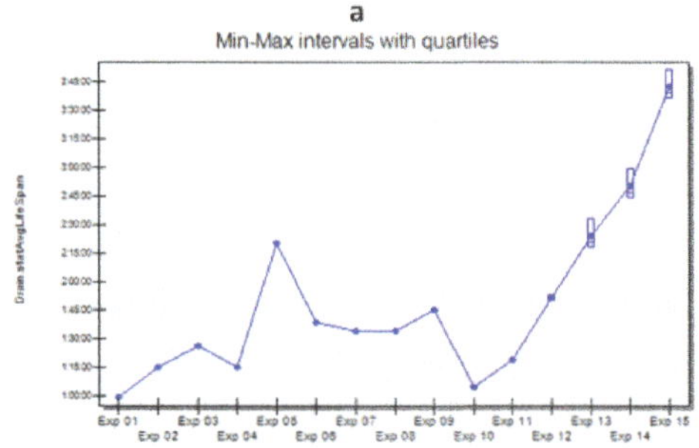

b

Figure 6. a) The throughput per hour for Variant 4, b) The average life span for Variant 4

The results of the experiments performed for Variant 3 of operation times are presented in Figure 5 a and b. The minimum values of throughput per hour were obtained in experiments 1 and 11; and the lowest level of average life span was obtained in experiments 1, 2, 10 and 11. Very high values of throughput for relatively small buffer capacities were obtained in experiments 4 and 8. Similar to the last case, the average life span increased very significantly for the last three experiments when the total buffer capacity was increased.

The results of the experiments performed for Variant 4 of operation times are presented in Figure 6 a and b. The local minimum of system throughput were obtained in experiments 1, 5, 7 and 11. The maximum values of the throughput were obtained in experiments 4, 9 and analogically as in the former cases for the last three experiments. The average life span is minimum in experiments 1, 2, 4 and 10.

The results of the simulation research show that, generally, the throughput of the system increases with increasing buffer capacities, but together with an associated increase in the average life span of products (work-in-process). To find the best compromise between throughput and average life span; flow index θ is proposed. To calculate the index, the value of the throughput of the system is divided by the average life span.

$$\theta = \frac{T}{\Lambda} \tag{5}$$

where T and Λ are throughput and average life span, respectively.

The results for the experiments are presented in Table 5. In the table the best two results for each variants are shown. The greater the value of the index, the better the compromise between the throughput and average life span that can be found. The best values of the flow indexes were obtained in Variant 2 of the operation times. For experiments 4 and 10, the best values of the index are obtained.

	Variant 1	Variant 2	Variant 3	Variant 4
Exp 01	292.72	406.41	331.13	325.94
Exp 02	274.34	381.21	306.11	307.76
Exp 03	285.29	408.47	309.32	319.48
Exp 04	**376.26**	**523.72**	**349.84**	**403.30**
Exp 05	152.47	219.94	166.36	176.27
Exp 06	237.57	394.16	244.18	270.17
Exp 07	210.33	270.68	259.81	244.23
Exp 08	254.46	317.09	310.49	283.30
Exp 09	215.25	293.25	258.38	271.26
Exp 10	**321.87**	**450.00**	**364.99**	**360.03**
Exp 11	239.01	331.22	266.83	267.57
Exp 12	281.30	388.26	259.62	290.60
Exp 13	214.48	299.08	145.01	226.58
Exp 14	175.32	241.21	99.99	191.72
Exp 15	136.87	174.05	62.30	146.69

The chart presented in Figure 7 show the values of the flow indexes for all variants of operation times.

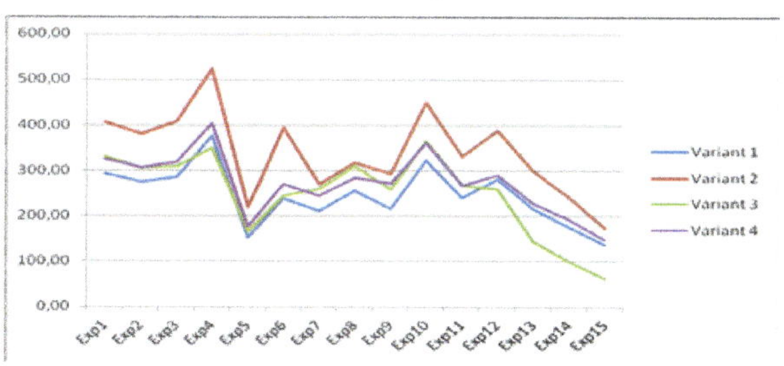

Figure 7. The values of flow indexes for the different variants of operation times

The increase in the intermediate buffer capacities results in an increased average life span and work-in-process. The buffer capacity costs money (place, work-in-process, etc.). To comply with the capacity of the intermediate buffers in the analysis of the efficiency of the investigated manufacturing system, the total buffer productivity index ω is proposed.

$$\omega = \frac{T}{B} \tag{6}$$

where T and B are throughput and total buffer capacity, respectively.

In Figure 8, the total buffer productivity indexes for all variants are presented. There are no significant differences among the values of buffer productivity indexes measured for the various variants. The greatest value in the index was obtained in experiment 1, and it resulted from the small value of the total buffer capacity (single capacity of intermediate buffers). The next best relation between the throughput and buffer capacity for all variants was obtained in experiments 10 and 11.

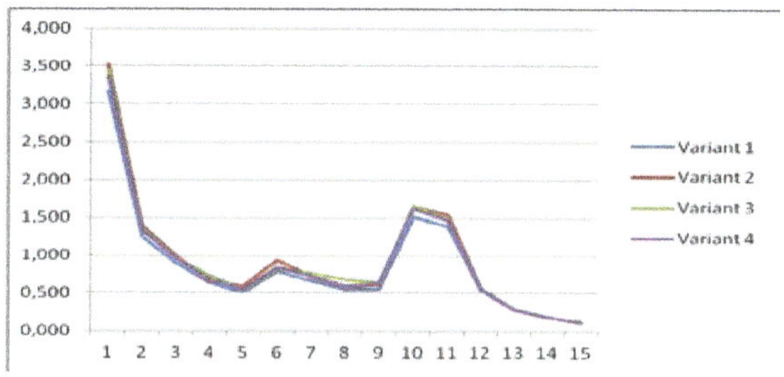

Figure 8. The values of total buffer productivity indexes for the different variants of operation times

5. Conclusions

An analysis of the data collected during the simulation experiments shows us that for each variant of the operation times, in the last four experiments (12, 13, 14 and 15) in which all buffer capacities are equal or greater than 10, the throughput of the manufacturing systems achieved the best results (more than 22 products per hour). For the smallest total capacity of intermediate buffers, the greatest value of throughput was reached in experiments 4, 6, 8 and 9. For flow index, experiments 4, 6 and 10 provided very good results, and for buffer productivity index: experiments 1, 10 and 11. For the manufacturing system, which was investigated in this study, the following general conclusions can be formulated:

1. The intermediate buffer capacity has a significant impact on the throughput and average life span of the system.
2. The proper allocation of buffer capacities provides better values of throughput or average life span of the system.
3. It is possible to find a satisfactory compromise between the throughput and average life span of the system, maximizing the proposed flow index.
4. It is possible to find a satisfactory compromise between the throughput and costs of buffer capacities, maximizing the proposed total buffer productivity index.

Further research will encompass other structures of the manufacturing system and the impact of employees on the throughput of the manufacturing system.

Acknowledgement

This article was created by implementation of the grant projects KEGA 004TUKE-4/2013 Intensification of modeling in teaching II. and III. degree in the field of study 5.2.52 Industrial Engineering and VEGA no. 1/0102/11 Experimental methods and modeling techniques in-house manufacturing and non-manufacturing processes

References

[1] A. Seleima, A. Azaba, T. AlGeddawy, Simulation Methods for Changeable Manufacturing, 45th CIRP Conference on Manufacturing Systems 2012, Procedia CIRP 3, 2012, pp. 179-184.

[2] D. R. Staley, D. S. Kim, Experimental results for the allocation of buffers in closed serial production lines, International Journal of Production Economics, Vol. 137 (2012), pp. 284-291

[3] H. A. Vergara, D. S. Kim, A new method for the placement of buffers in serial production lines, International Journal of Production Research, 2009, 47(16), pp. 4437-4456.

[4] H. Yamashita, T. Altiok, Buffer capacity allocation for a desired throughput in production lines. IIE Transactions 1998; Vol. 30, pp. 883-91.

[5] G. Gurkan, Simulation optimization of buffer allocations in production lines with unreliable machines. Annals of Operations Research 2000, Vol. 93, pp. 177-216.

[6] C. Shi, S.B. Gershwin, An efficient buffer design algorithm for production line profit maximization. International Journal of Production Economics 2009 Vol. 122, pp. 725-40.

[7] J.A. Qudeiri, H. Yamamoto, R. Ramli, A. Jamalim, Genetic algorithm for buffer size and work station capacity in serial-parallel production lines, Artificial Life and Robotics 2008, Vol.12, pp. 102-6.

[8] N. Nahas, D. Ait-Kadi, M. Nourelfath, Selecting machines and buffers in unreliable series-parallel production lines. International Journal of Produc- tion Research 2009;Vol. 47(14), pp. 3741-74. [32].

[9] M. Nourelfath, N. Nahas, D. Ait-Kadi, Optimal design of series production lines with unreliable machines and finite buffers. Journal of Quality in Mainte- nance Engineering 2005, Vol. 11(2), pp. 121-38.

[10] N. O. Fernandes, S. Carmo-Silva, Order release in a workload controlled flow-shop with sequence-dependent set-up times, International Journal of Production Research 2011, Vol. 49(8), pp. 2443-54.

[11] A. Matta, Simulation optimization with mathematical programming repre-sentation of discrete event systems. In: Proceedings of the 2008 winter simulation conference, 2008, pp. 1393-400.

[12] M.G. Huang, P.L. Chang, Y.C. Chou, Buffer allocation in flow-shop-type production systems with general arrival and service patterns. Computers & Operations Research 2002, Vol.29 pp. 103-21.

[13] MacGregor Smith J, Cruz FRB. The buffer allocation problem for general finite buffer queuing networks. IIE Transactions, 2005, Vol. 37(4), pp. 343-65.

[14] L. Demir, S.Tunalı, D. T. Eliiyi, A. Løkketangen, Two approaches for solving the buffer allocation problem in unreliable production lines, Computers & Operations Research, Vol. 40, 2013, pp. 2556-2563.

[15] Tecnomatix Plant Simulation version 11.0.0 on-line documentation.

[16] Trebuňa, P., Straka, M., Rosová, A., Malindžáková, M.: Petri nets as a tool for production streamlining in plastics processing, Przemysl chemiczny, 2015, 94, No. 9, 1000-1003.

[17] Malindžáková, M., Straka, M., Rosová, A., Kanuchová, M., Trebuňa, P.: Modeling the process for incineration of municipal waste, Przemysl chemiczny, 2015, 94, No. 8, 1000-1004.

[18] Straka, M., Trebuňa, P., Straková, D., Kliment, M.: Computer Simulation as means of urban traffic elements design, Theoretical and Empirical Researches in Urban Management, Vol. 10, Issue 4, 2015, p. 40-53.

[19] Kollár, P., Nikitin, Y., Straka, M.: The determination of the shelf mass in the universal shelving stacker by measuring the frequency converter torque generating current of the main drive, Manufacturing technology, Vol. 15, No. 3, p. 363-366.

[20] Malindžák, D., Straka, M., Helo, P., Takala, J.: The methodology for the logistics system simulation model design, 2010, Metalurgija, Vol. 49, No. 4 (2010), p. 348-352.

[21] Straka, M.: System of distribution logistics of enterprise Alfa, a.s., Acta Montanistica Slovaca, Vol. 15, Special issue 1, 2010, p. 34-43.

A New Approach in Design of Model-free Fuzzy Sliding Mode Controller for Multivariable Chemical Processes

Mohammad Khodadadi[1,*], Mahdi Shahbazian[1], Masoud Aghajani[2]

[1]Department of Instrumentation and Automation Engineering, Ahwaz Faculty of Petroleum, PUT, Ahwaz, Iran
[2]Department of Chemical Engineering, Ahwaz Faculty of Petroleum, PUT, Ahwaz, Iran
*Corresponding author: kh.automation@yahoo.com

Received December 31, 2014; Revised January 19, 2015; Accepted January 22, 2015

Abstract Sliding Mode Control (SMC) as a mode of the Variable Structure Control (VSC) is known as an excellent robust control strategy in the field of uncertain and nonlinear control with high performance. However, chattering and process model dependency as main problems of the original SMC constrain its application. Also, in multivariable nonlinear process control, the interaction among the manipulated and controlled variables can degrade the performance of the controller. The objective of this paper is to design a new model-free Fuzzy Sliding Mode Control (FSMC) to solve these problems in MIMO process control. In this strategy, the equivalent –like term of the SMC's control law is estimated using a self-tuned MIMO fuzzy logic inference system in order to overcome the dependency of the SMC to the process model and to reject undesirable effect of interaction toward improvement of robustness. Moreover, in design of linear sliding variables for switching term of the SMC, all states error are associated to lessen the interaction effect. Finally, the proposed strategy has been evaluated through implementation on a warm water tank system equipped with two manipulated servo valve to control temperature and water level. This system is nonlinear and affected by unknown disturbances. The results reveal that the purposed scheme is superior in servo tracking and unknown disturbance rejection performances in comparison with the classical PI control.

Keywords: *model-free fuzzy sliding mode control, robust control, multivariable process control, interaction, chattering, model dependency, unknown disturbance*

1. Introduction

Sliding Mode Control (SMC) is a high performance robust control strategy with simple procedure for control of systems whose dynamics are highly nonlinear and uncertain [1,2,3]. Hence, this methodology has been developed during last three decades in the nonlinear control area. The SMC control law has two components; continuous and switching (discontinuous) elements [2]. The switching element has different gain signs across the predefined sliding surface. This surface divides the phase plane into two regions where the SMC's control action has different structure such that its structure is allowed to change alternatively in order to drive the system's state trajectory onto the manifold-like sliding surface toward the plane's origin. [3,4]. During there stricted sliding motion of the system's state trajectory, its motion will robust against modeling uncertainties and disturbances [2]. However, in original SMC, the problems of *chattering* and model *dependency* can restrict its applications [5]. "*Chattering phenomenon*" is the high frequency oscillation of the state trajectory around the sliding surface because of non-ideal (delayed) switching or non-instantaneous actuating [2,6] which also can cause extreme heat loss and wearing of the moving parts in instrumentation devices [6]. the boundary layer (BL) approach is as common smoothing technique to filter the control signal to lessen or eliminate the chattering but at cost of robustness specially, in the presence of the large range of uncertainties. [7] in which the B.L. results in considerable steady state error [8,9]. To derive he continuous element of the SMC, the *model* of the plant (relation between control and manipulated variables)is required but it is difficult to use the high order complex model or it is hard to modeling in the complex plants or in the lack of some process parametric knowledge. [5,10,11,12]. Fuzzy Logic Control strategy (FLC) based on fuzzy set theory [13] is used to design the control law in a form of inferential rules using some qualitative knowledge about the process under control. This model free method is implemented generally to solve the model dependency problem and improvement of control system's performance [14]. Sliding Mode Fuzzy Control ***(SMFC)*** and Fuzzy Sliding Mode Control (***FSMC)*** [15] are as two integration strategy with features of both original SMC and FLC to remedy the problems of the original SMC which have been widely investigated in recent years such as [5]. The aim of this paper is to design a novel model free-FSMC which can be specially implemented inmulti-variables process control to improve the robustness. For this purpose, a fuzzy continuous element with auto adjustable parameters is proposed to remedy the model dependency problem and removing of steady state error. Furthermore, the sliding variables are developed for each control action using error of all states by consideration the effect of each control action on each state errors to avoid the interaction (likes dynamics decoupling). This paper is organized as follows: section 2 presents the SMC briefly; section 3 illustrates the new model-free FSMC methodology for MIMO processes; in section 4, the proposed FSMC is implemented and its capability is verified; eventually, section 5 concludes the paper.

2. A Brief Review of original SMC

Let an nth order single input nonlinear system defined as [2]

$$x^{(n)} = f(X,t) + b(X,t)u(t) + d(t) \quad (1)$$

Where, $X(t) = \left[x, \dot{x}, \ddot{x}, \dddot{x} \ldots, x^{n-1}\right]^T$ is the state vector and scalar x is the control variable. u(t) is the control action. $f(X,t), b(X,t)$ are generally uncertain nonlinear functions are estimated as $\hat{f}, \hat{b}$ and bounded by continuous functions as follows:

$$\left|\hat{f}(X,t) - f(X,t)\right| \le F(X,t) \quad (2)$$

$$0 \le b_{\min}(X,t) \le b(X,t) \le b_{\max}(X,t) \quad (3)$$

$$\hat{b}(X,t) = \sqrt{b_{\min}(X,t) b_{\max}(X,t)} \quad (4)$$

$$\beta(X,t) = \frac{b_{\max}(X,t)}{b_{\min}(X,t)} \quad (5)$$

Also, the disturbance d(t) is upper bounded as follows:

$$0 \le |d(t)| \le D \quad (6)$$

sliding variable [7,16] is predefined in the state space $R^{(n)}$ as:

$$s(X,t) = \left(\frac{d}{dt} + \lambda\right)^{(n-1)} e(t) \quad (7)$$

$$E(t) = X(t) - X_d(t) = \left[e, \dot{e}, \ddot{e} \ldots, e^{n-1}\right]^T \quad (8)$$

Here, $E(t)$ is the tracking error vector. $X_d(t)$ is the desired state vector and λ is strictly positive tuning parameter. In second order systems (n=2), Eq.(7) yields the sliding variable as:

$$s = \dot{e} + \lambda e \quad (9)$$

In order to move the error vector on sliding surface (null value sliding variable) and to retain on it, the SMC's control law is given as follows which is derived based on reaching condition [2]:

$$u = \frac{-\hat{f} + \ddot{x}_d - \lambda \dot{e}}{\hat{b}} - k_s \operatorname{sgn}(s) \quad (10)$$

Here, first component is the continuous (equivalent) control u_{eq} and second is discontinuous (switching) control k_s respectively [16]. k_s is the positive switching gain [17] bounded as follows:

$$k_s(X,t) \ge \beta(X,t)(F(X,t) + \eta) + (\beta(X,t) - 1)|u_{eq}| \quad (11)$$

Because of existence of non-ideal switching(restricted frequency and non-instantaneous control switching) in practice due to discretized implementation or neglected high frequency un-modeled dynamics (time delays, actuator's dynamics,...) [1,2,6,7,18], the classical SMC controller exhibit oscillatory motion state trajectory with high frequency about the sliding surface so called chattering as depicted in Figure 1. The Boundary Layer technique [2] is one of unti-chatter methods by mechanism of alleviation of the control signal in the thin layer with thickness of ∅ around the sliding surface as follows but at cost of tracking error [2,19]:

$$sat\left(\frac{s}{\varnothing}\right) = \begin{cases} \operatorname{sgn}(s) \text{ if } |s| \ge \varnothing \\ \frac{s}{\varnothing} \text{ if } |s| < \varnothing \end{cases} \quad (11)$$

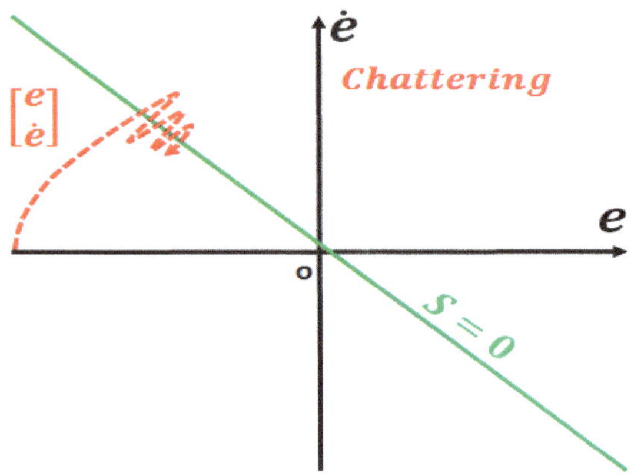

Figure 1. Chattering phenomenon as a result of non-ideal switching [19]

In the systems with large range of uncertainties and disturbances, the switching gain must be chosen high enough but to cancel the chattering induced by this high gain and the thickness ∅ must be chosen large enough which can cause considerable off-set [8].

3. The Proposed FSMC

In this section, we attempt to design a novel model-free FSMC strategy for MIMO systems of control. Firstly, the MISO equivalent-like control of each MISO SMC controllers are designed. They are estimated by auto adjustable MISO fuzzy inferential systems to reject uncertainties, whereby the tracking error which appeared due to boundary layer technique will be eliminated. Secondly, new approach is introduced to design sliding variable using error of all states for each sliding variables for canceling of interaction's effect. The proposed FSMC controllers are constructed as follows (see Figure 2):

$$u^i{}_{(k)} = u^i{}_{eq\,fuzzy(k)} + u^i{}_{s(k)} \quad i = 1:N \quad (12)$$

Where, $u^{i}_{s(k)}$, $u^{i}_{eqfuzzy(k)}$ and $u^{i}_{(k)}$ are the switching control, fuzzy equivalent-like control and the FSMC output signal as the process input respectively in relevant to $y^{1}_{set\,fuzzy}$ manipulated control and N is the number of manipulated variables.

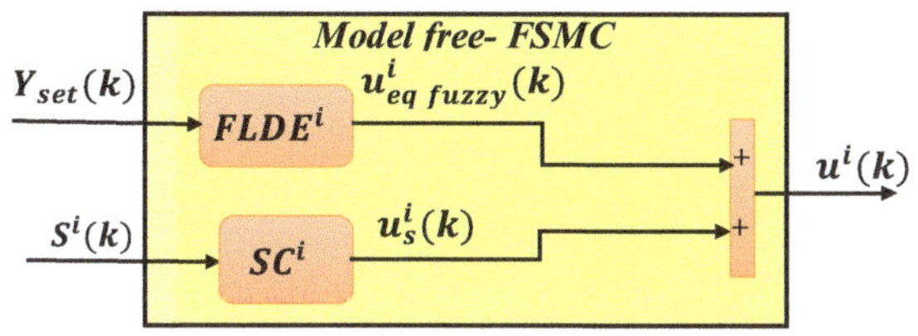

Figure 2. Model free -FSMC controller's elements: Fuzzy Logic Dynamics Estimator (FLDE) and Switching Control (SC) for each manipulated variable

3.1. Equivalent Control Design

Formulation of the equivalent part of SMC controller will be hard or impossible if the model of the chemical process is complex or ill-defined (with unknown states) or if some recognized states are too expensive to be measured (such as concentration). To solve this model dependency problem, the chemical process can be modeled *approximately* such as First Order Plus Dead Time (FOPDT) [20] and it is expected the SMC to be robust in the face of the un-modeled uncertainties which is in this approximated dynamics [17]. It is note that, in the presence of the large range of uncertainties and disturbances, the performance of SMC will be degraded. In this sub-section it is attempted to estimate the equivalent part of the each SMCs in MIMO system by fuzzy logic inference system to solve the model dependency not at cost of robustness.

3.1.1. The Proposed Methodology

To estimate the equivalent control u_{eq} for each manipulated variable, a *Multi inputs-Single outputs Fuzzy Logic Dynamic Estimator (FLDE)* is established using knowledge about system's behavior on the steady state condition. The *steady-state I/O* data of process under control are assigned to centers of *O/I* membership functions as depicted in Figure 3. This method is development of the method which has been verified for SISO process control in previous work [21].

3.1.2. FLDE Design Procedure

To design the FLDE, the steady-state operating points of process (*outputs/inputs* data) are used which are obtained from expert knowledge. ForanN input N output process, each FLDE appropriated to a manipulated variable has N dimensional input and one dimensional output. To design its input membership functions, each dimension appropriated to one of the control variables. Centers of membership functions in each dimension (defined with 4 labels of zero, small, medium and big) are four steady state values of corresponded control variable. Also, the steady state values of each manipulated variables are used as centers of output membership functions of related FLDE. Triangular Membership Functions (MF) are chosen for the fuzzy variables.

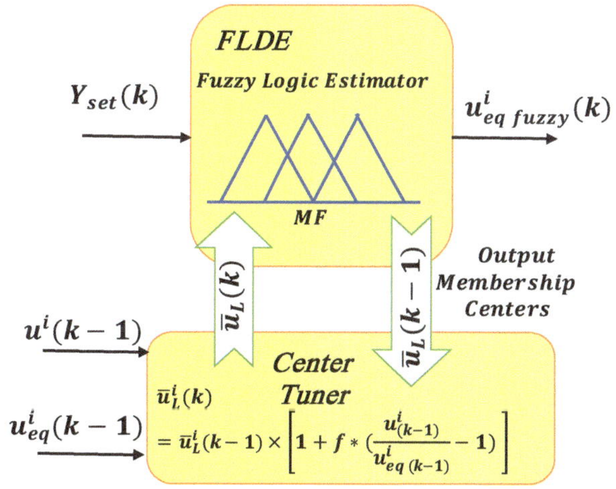

Figure 3. Auto adjustable MISO FLDE for MIMO processes and Center tuner

In the case of two input-two output process, we have two dimensional rule base for each FLDE are given in Table 1, for instance, two typical rules are as follows:

$$\begin{aligned}&IF\ y^{1}_{set\,fuzzy}\ is\ zero\ and\ y^{2}_{set\,fuzzy}\ is\ zero\\&THEN\ u^{i}_{eqfuzzy(k)}\ is\ u_1.\\&IF\ y^{1}_{set\,fuzzy}\ is\ small\ and\ y^{2}_{set\,fuzzy}\ is\ zero\\&THEN\ u^{i}_{eqfuzzy(k)}\ is\ u_2.\end{aligned} \qquad (13)$$

Mamdani product inference engine and Center Average Defuzzifier are used for FLDE estimator. Input and output fuzzy sets are as follows:

Input fuzzy setsfor ith controlled variable:

A^i_1=Zero; A^i_2= Small; A^i_3=Medium; A^i_4=Big; i=1:N

for U^i_L controlled variable.

Output fuzzy sets in relevant to ith controlled variable:

$$U^i_L,\ \mathrm{L}=1:\mathrm{R}$$

Where, R is the number of rules.

Step II) FLDE Self- Tuning

The FLDE estimator is adjusted online based on evanesce of the *difference between output of the SMC and output of its equivalent part (FLDE)* due to existence of uncertainties such that it is eliminated at steady state condition. The centers of output membership functions of the each FLDEs (as free parameters) can be adjusted online and is done once in each sampling interval as follows (see Figure 3):

$$\bar{u}^{i}_{L(k+1)} = \bar{u}^{i}_{L(k)} \times \left[1 + f * reg^{i}_{(k)}\right] L = 1:R \qquad (14)$$

$$reg^{i}_{(k)} = \frac{u^{i}_{(k)}}{u^{i}_{eq(k)}} - 1 i = 1:N \qquad (15)$$

Where, $\bar{u}^i_L$ is the center in relevant to Lth rule and ith manipulated variable. Also, k is the time-interval at time kTs and Ts is the sampling time. f is fixed step size. In the reaching mode, the center tuners will try to update each FLDEs and reject their uncertainties to achieve zero steady state errors.

Table 1. Two dimensional Fuzzy rule base for FLDE[1] and FLDE[2]. Linguistic variables: B=big, M=medium, S=small, Z=zero

		$y_{1setfuzzy}$			
		Z	S	M	B
$y_{2setfuzzy}$	Z	u_1	u_2	u_3	u_4
	S	u_5	u_6	u_7	u_8
	M	u_9	u_{10}	u_{11}	u_{12}
	B	u_{13}	u_{14}	u_{15}	u_{16}

3.2. Design of Switching Control

To design the Switching Control (SC) of the proposed FSMC, the new approach is considered in construction of the sliding variable inspired from dynamic decupling method [20] for cancellation of the interaction effects in control system as follows:

$$S_m = C_{mm}e_m - \sum_{i=1,\neq m}^{N} C_{mi}\,\mathrm{sgn}(K_{mi})e_i \qquad (16)$$

Here, S_m is the sliding variable in relevant to the mth manipulated variable whose second term (term) is devised for reduction or elimination of the influence of the other manipulated variables on the mth control variable (interaction). e_m is the state error correspond to the mth control variable which was paired to mth manipulated variable. K_{mi} is the open-loop steady state gain of the transfer function relating the mth controlled variable to the ith manipulated variable. C_{mi} is positive coefficient. Smoothed switching control of ith manipulated variable is designed using sigma filtering function [17] as follows:

$$u_{si} = -k_{si}\left(\frac{s_i}{|s_i| + \varnothing_i}\right) \qquad (17)$$

Here, k_{si} is the switching gain and $|S_i|$ is absolute value of the sliding function. $\varnothing_i$ is constant positive tuning parameter responsible for the chattering elimination.

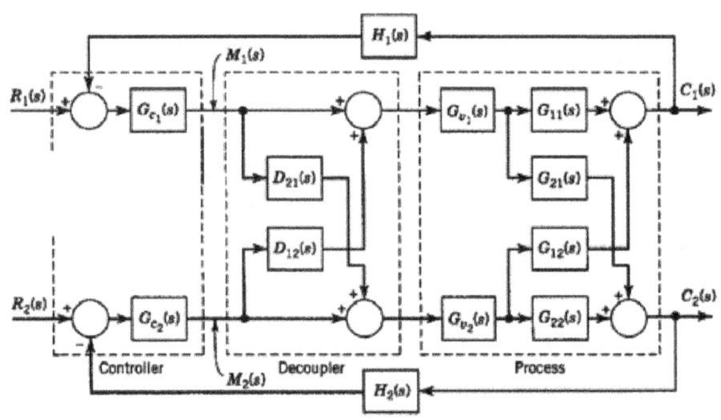

Figure 4. Block diagram of 2x2 control system with decouplers [20]

4. Simulation

4.1. Case Study

In this subsection, performance of the proposed method has been verified through implementation on the warm water production process. As the Figure 5 depicts, in this process, the warm water tank receives two hot water and cold water streams whose contents are stirred by mixer and mixed well to produce uniform warm water for utilization in chemical processes such as washing process. The control of water temperature and water level of this tank is the objective of the controller. The temperature fluctuation of the two inlet streams are considered as unknown disturbances. The discharge stream is function of the water level in the tank. By supposition of constant density and specific heat capacity of the water in this process, the equations of process dynamics (mass balance and energy balance) can be wrote as follows:

$$\frac{AdH}{dt} = F_c + F_h - F_{out} \qquad (18)$$

$$F_{out} = C\sqrt{H} \qquad (19)$$

$$A\frac{d(HT_w)}{dt} = F_cT_c + F_hT_{hot} - F_{out}T_w \qquad (20)$$

Here, volume flow rates F_c, F_h are as manipulated variables in relevant to the inlet cold water and hot water streams respectively and F_{out} is outlet warm water flow rate. T_c, T_h are temperatures of the cold and hot water streams respectively. Warm water temperature T_w and water level H are as controlled variables. A is the tank section area and C is discharge coefficient. It is supposed that, fluctuation in temperatures of the inlet streams are unknown which can disturb the system of control. Each manipulated variable is limited to $0.50m^3\min^{-1}$. The simulation of this process is done using MATLAB software and the above dynamics is solve numerically with sampling time $T_s = 1\sec$. This incremental implementation is supposed as only source of the chattering and for simulation of this process, the actuator's dynamics and temperature and level sensors dynamics are considered ideal and with no delay respectively. The operating parameters of the warm water tank and process variables are given in Table 2.

Table 2. Operating parameters of warm water tank

Parameter	Description	Value
T_h	Hot water stream temperature	90°C
T_c	Cold water stream temperature	10°C
$T_{w\ ss}$	warm water stream temperature	50°C
$F_{c\ ss}$	Cold water stream flow rate	$0.125m^3/\min$
$F_{h\ ss}$	Hot water stream flow rate	$0.125m^3/\min$
H_{ss}	Steady state level of tank	1 m
$H_{\max}$	Maximum level of tank	4 m
C	Discharge coefficient	$0.25m^{2.5}/\min$
A	Cross section area of the tank	2 m^2

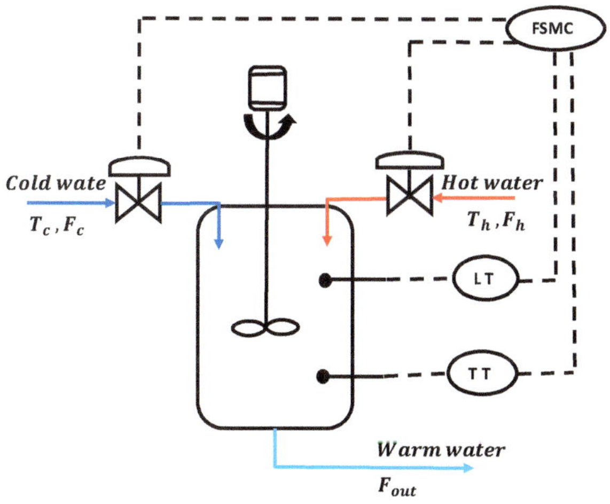

Figure 5. Schematic- warm water tank

4.2. Numerical Parameters

The sliding variable and the state error is defined as follows:

$$s_H = \frac{e_H}{4} - \frac{2e_T}{80}, s_T = \frac{e_H}{4} + \frac{e_T}{80} \tag{21}$$

$$e_H = H - H_{set}, e_T = T_w - T_{set} \tag{22}$$

Here, H_{set}, T_{set} are the set-point of level and temperature.

4.2.1. FLDE's Fuzzy Memory Table

The steady state operating points of the of warm water tank (level and temperature) as the centers of membership functions of the FLDE are given in Table 3.1- 3.2. These date were generated by steady state simulation of the process dynamics by MATLB software.

Table 3.1. Two dimensional Fuzzy rule base of FLDE for cold stream

		$H_{setfuzzy}$			
		0	1	2	4
$T_{setfuzzy}$	10	0	0.1250	0.3536	0.5000
	30	0	0.1875	0.2652	0.3750
	50	0	0.1250	0.1768	0.2500
	90	0	0	0	0

Table 3.2 Two dimensional Fuzzy rule base of FLDE for hot stream

		$H_{setfuzzy}$			
		0	1	2	4
$T_{setfuzzy}$	10	0	0	0	0
	30	0	0.0625	0.0884	0.1250
	50	0	0.1250	0.1768	0.2500
	90	0	0.125	0.3536	0.5000

Also, step size f=0.058

4.2.2. Controller Parameters

I)Proposed FSMC:

The proposed controller parameters were yielded from trial and error manually to achieve the best performance:

$$K_{s1}, K_{s2} = 0.5 \; and \; \varnothing_1, \varnothing_2 = 0.05$$

II)Classical PI Controllers:

Level controller:

$$F_c(k+1) = F_{c\,ss} + \mu_{11}(K_{c1}e_H(k) + K_{I1}\int_0^{kT_s} e_H dt)$$

Temperature controller:

$$F_h(k+1) = F_{h\,ss} + \mu_{22}(K_{c2}e_T(k) + K_{I2}\int_0^{kT_s} e_T dt)$$

The tuning parameters of the PI controllers are computed using the tuning formulas for Dahlin synthesis [11,20] to produce smoother responses than Ziegler-Nichols tuning technique. These parameters at operating point for level controller are $K_{c1} = 0.5359$, $K_{I1} = 0.0518$ and for temperature are $K_{c2} = 0.0140$, $K_{I2} = 0.0042$. The relative gains [20] are computed as $\mu_{11} = 0.430$, $\mu_{22} = 0.375$.

4.3. Performance Evaluation

The servo-tracking and regulatory performances of the proposed FSMC are evaluated in comparison with classical PI controller. Tracking Error (TE), Settling-Time (time taken for the output to reach 95% or 102% of the change in set-point), overshoot percentage and interaction avoiding used for verification.

4.3.1. Servo-Tracking Performance

To verify the tracking performance of the proposed FSMC, the set-point values of temperature and level are changed step-like according to Figure 6(b),(c). From this Figures it is clearly observed that the proposed method improved transient performance effectively. As it is depicted, the settling time is very short and percentage of the over shoot were negligible in comparison with the classical PI controller.

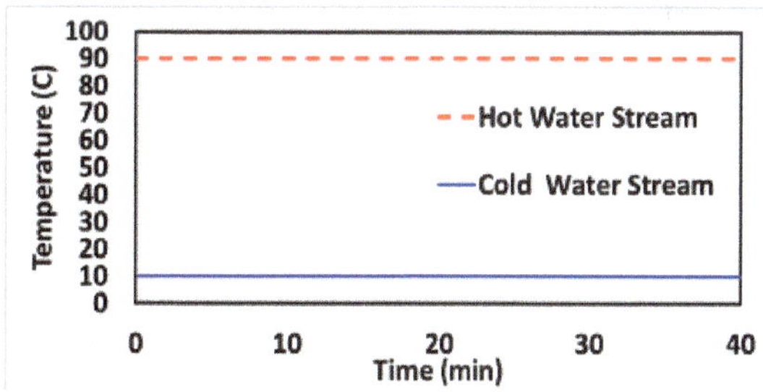

Figure 6 (a). Streams temperatures

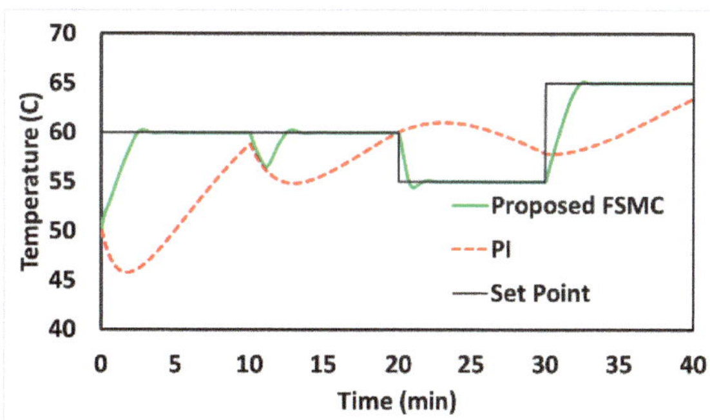

Figure 6(b). Controlled variable: warm water temperature

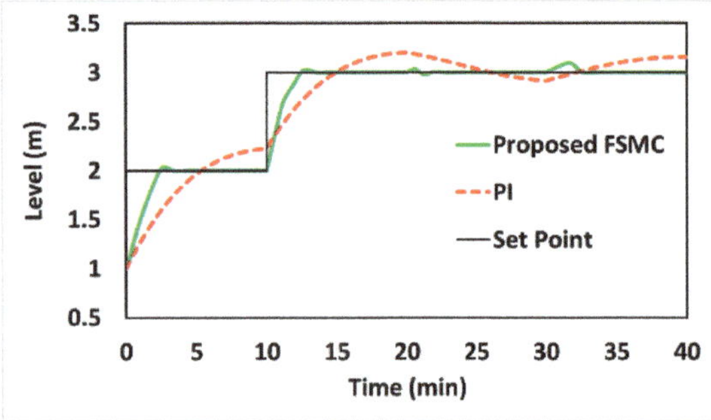

Figure 6 (c). Controlled variable: level

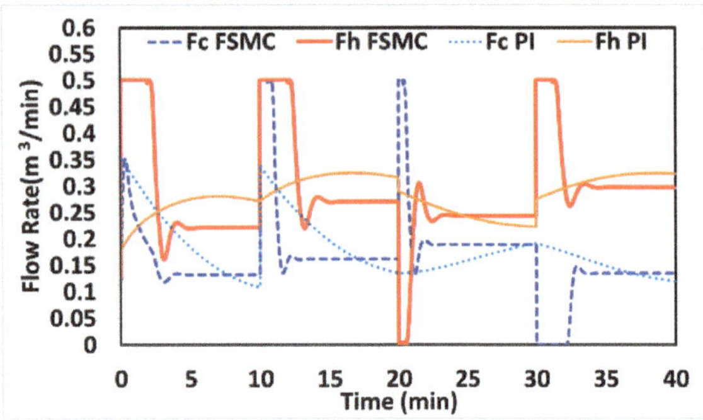

Figure 6 (d). Manipulated variables

Figure 6. Set-point-tracking performance

4.3.2. The Regulatory Performance

To verify the robustness of the proposed FSMC, the temperature of cold and hot water streams is changed step-like while the set-point value is fixed according to Figure 7(a). As well shown in Figure 7(b),(c), the proposed controller succeeded to rejected the disturbance completely by fast action in comparison with the classical PI controller. The proposed controller can canceled the interaction effect powerfully.

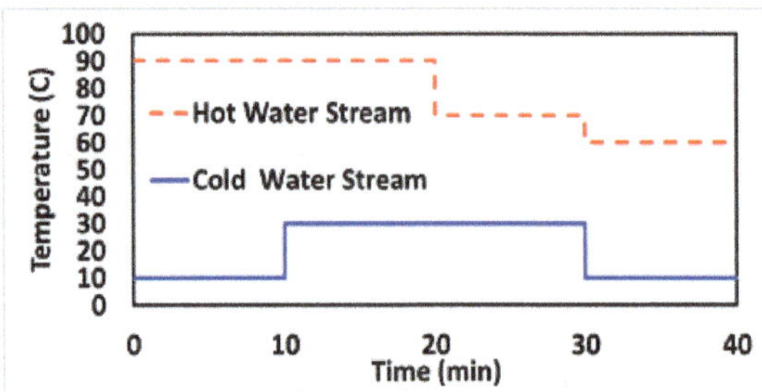

Figure 7 (a). Streams temperatures (Disturbance)

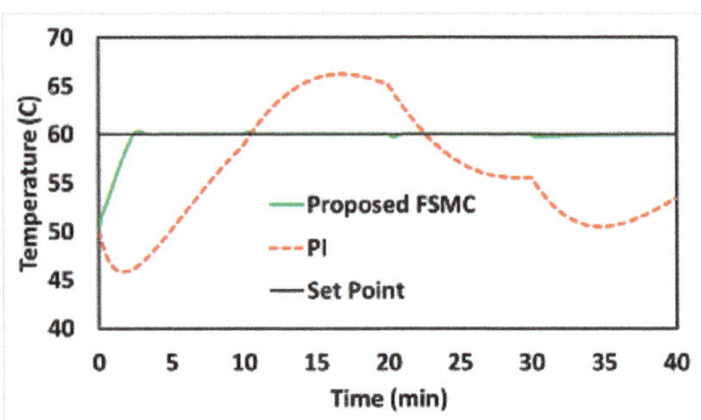

Figure 7(b). Controlled variable: warm water temperature

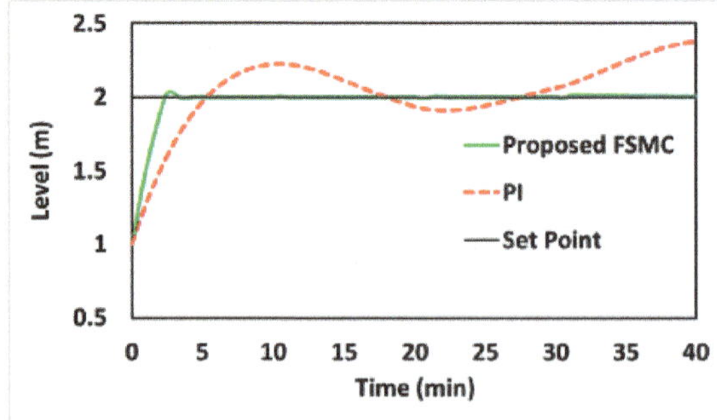

Figure 7 (c). Controlled variable: level

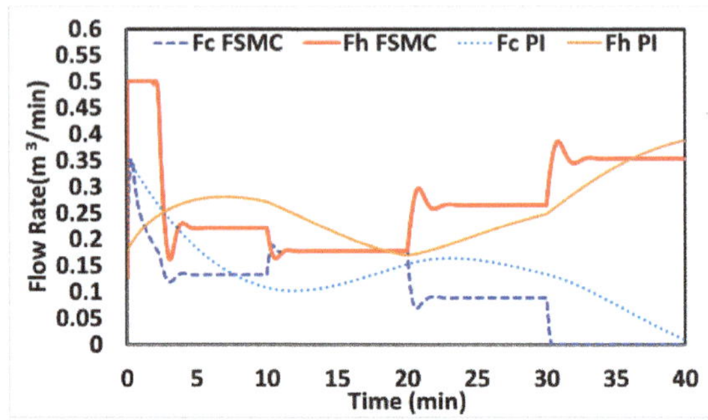

Figure 7 (d). Manipulated variables

Figure 7. Disturbance rejection and system regulation performance

5. Conclusions

New strategy was introduced to design model-free FSMC to control of multi inputs- multi outputs chemical processes. For each of the manipulated variables, Multi inputs-single output fuzzy logic dynamics estimator (FLDE) was designed as equivalent-like component of FSMC to solve model dependency problem based on steady state operating points in wide range. Also to cancel the effect of interaction, the sliding variable of each FLDE was designed using linear combination of error of all control variables inspired from dynamics decoupling. Finally, to verify the proposed method the control of level and temperature of the warm water process was choose as benchmark. The results revealed the merit of proposed controller in comparison with classical PI controller in robustness and servo tracking. The proposed strategy improved effectively the controller performance in terms of successful interaction canceling, disturbance rejection and transient performance with short settling time without overshoot.

References

[1] K. D. D. Young, V. I. V. I. Utkin, and U. Ozguner, "A Control engineer's guide to sliding mode control," *IEEE Trans. Control Syst. Technol.*, vol. 7, no. 3, pp. 328-342, May 1999.

[2] J.-J. E. Slotine and W. Li, *Applied Nonlinear Control*, vol. 199, no. 1. Prentice-Hall Englewood Cliffs, NJ, 1991.

[3] V. I. Utkin, "Variable structure systems with sliding modes: A Survey," *IEEE Trans. Automat. Contr.*, vol. AC-22, no. 2, pp. 212-222, 1977.

[4] O. M. E. El-Ghezawi, S. A. Billings, and A. S. I. Zinober, "Variable structure systems with and system zeros," in *IEE Proceedings D (Control Theory and Applications)*, vol. 130, no. 1, pp. 1-5,1983.

[5] F. Piltan, N. Sulaiman, P. Ferdosali, and I. A. Talooki, "Design model free fuzzy sliding mode control: Applied to internal combustion engine," *Int. J. Eng.*, vol. 5, no. 4, pp. 302-312, 2011.

[6] V. Utkin, J. Guldner, and M. Shijun, *Sliding mode control in electro-mechanical systems*, vol. 34. CRC press, 1999.

[7] B. Bandyopadhyay, F. Deepak, and K.-S. Kim, *Sliding mode control using novel sliding surfaces*, vol. 392. Springer, 2009.

[8] C.-C. Fuh, "Variable-thickness boundary layers for sliding mode control," *J. Mar. Sci. Technol.*, vol. 16, no. 4, pp. 286-294, 2008.

[9] H.-M. H. Chen, J.-C. J. Renn, and J.-P. J. Su, "Sliding mode control with varying boundary layers for an electro-hydraulic position servo system," *Int. J. Adv. Manuf. Technol.*, vol. 26, no. 1-2, pp. 117-123, 2005.

[10] A. Shahraz and R. Bozorgmehry Boozarjomehry, "A fuzzy sliding mode control approach for nonlinear chemical processes," *Control Eng. Pract.*, vol. 17, no. 5, pp. 541-550, 2009.

[11] O. Camacho and C. A. Smith, "Sliding mode control: an approach to regulate nonlinear chemical processes," *ISA Trans.*, vol. 39, no. 2, pp. 205-218, 2000.

[12] E. Cornieles, M. Saad, F. Arteaga, and L. Obediente, "Sliding mode control for multivariable processes," *Rev. Ing. UC*, vol. 11, no. 1, pp. 62-68, 2004.

[13] L. A. Zadeh, "Fuzzy sets," *Inf. Control*, vol. 8, no. 3, pp. 338-353, 1965.

[14] G. Chen and Y. H. Joo, "Introduction to fuzzy control systems," *Handb. Appl. Comput. Intell.*, pp. 1-41, 2001.

[15] W. Guo and D. Liu, "Sliding mode fuzzy dynamic weighting control for a class of nonlinear underactuated systems," *J. Inf. Comput. Sci.*, vol. 13, no.12, pp. 2661-2668, 2011.

[16] P. Kachroo and M. Tomizuka, "Chattering reduction and error convergence in the sliding-mode control of a class of nonlinear systems," *IEEE Trans. Automat. Contr.*, vol. 41, no. 7, pp. 1063-1068, Jul. 1996.

[17] O. Camacho, R. Rojas, and W. García-Gabín, "Some long time delay sliding mode control approaches," *ISA Trans.*, vol. 46, no. 1, pp. 95-101, Feb. 2007.

[18] S. N. Dragan Antic, Marko Milojkovic, "Fuzzy sliding mode control with additional fuzzy control component," *Autom. Control Robot.*, vol. 8, no. 1, pp. 25-34, 2009.

[19] L.-X. Wang, *A Course in Fuzzy Systems and Control*. Prentice-Hall press, USA, pp. 238-248, 1999.

[20] C. A. Smith and A. B. Corripio, *Principles and practice of automatic process control*, vol. 2. Wiley New York, 1985.

[21] M. Khodadadi, M. Shahbazian, and M. Aghajani, "A Novel Approach in Design of Model-free Fuzzy Sliding Mode Controller to SISO Chemical Processes," *J. Autom. Control*, vol. 3, no. 1, pp. 1-9, Jan. 2015.

Control of Stepper Motor by Microcontroller

Ivan Virgala*, Michal Kelemen, Alexander Gmiterko, Tomáš Lipták

Department of Mechatronics, Technical University of Košice, Faculty of Mechanical Engineering, Košice, Slovakia
*Corresponding author: aivan.virgala@tuke.sk

Abstract The paper deals with stepper motor control using microcontroller ATmega8-16PU. In the paper are mentioned two ways of control, namely full-step mode and half-step mode control. At first, bipolar stepper motor is introduced. For stepper motor control it has to be used H-bridge as well as amplifier in one module. Next, there is created measuring stand consisting of bipolar stepper motor SY28STH32-0674A with one step angle 1.8° and revolute encoder ISC3004 with 360 CPR. During the experiment were measured angular velocities by I/O measuring card MF624 cooperating with Matlab/Simulink. The results show differences between used two stepping modes of stepper motor.

Keywords: *encoder, Matlab, microcontroller, stepper motor*

1. Introduction

The main utilization of stepper motor is positioning of stepper motor rotor with required precision. Stepper motor is suitable primarily for tasks where the precision is very important factor. Application areas are computer art (hard disks, printers), tool machines, automobile industry, actuators of industry robots and manipulators etc. For stepper motor control it is necessary control unit, which generates steps of motor. [1]

The stepper motor is special type of synchronous motor. Difference between synchronous and stepper motor is that stepper motor can reaches certain amount of defined positions. The step of motor is reached by suitable activity of control unit. The basic principle of motion is very simple. Electric current flowing through the coil generates magnetic field. This magnetic field pulls opposite pole of rotor magnet, see Fig.1. By suitable generating of pulses to coils, the rotor starts to rotate. [2]

The angular velocity of motor is limited because of transient performance magnetic field. By overshooting of maximum motor speed the motor starts to lose its step. [3]

One of the first advantage of stepper motor is effective positioning. For precise positioning can be determined specific step and motor rotates to required position without encoder. The basic relations of stepper motor are following:

$$\alpha = \frac{2\pi}{spr} \tag{1}$$

$$\theta = n\alpha \tag{2}$$

$$\omega = \frac{\alpha}{dt} \tag{3}$$

where α motor step angle, spr is number of steps per round, θ is position, n is number of steps, ω is angular velocity. The difference of stepper motor in comparison with DC motor is that it doesn´t need encoder, because it operates with open loop, while DC motor operates with closed loop. On the other side, stepper motor positioning is limited to step size of rotor. Disadvantages of stepper motor is for example lower torque in comparison with DC motor [4]. Principle of stepper motor can be seen in the Figure 1.

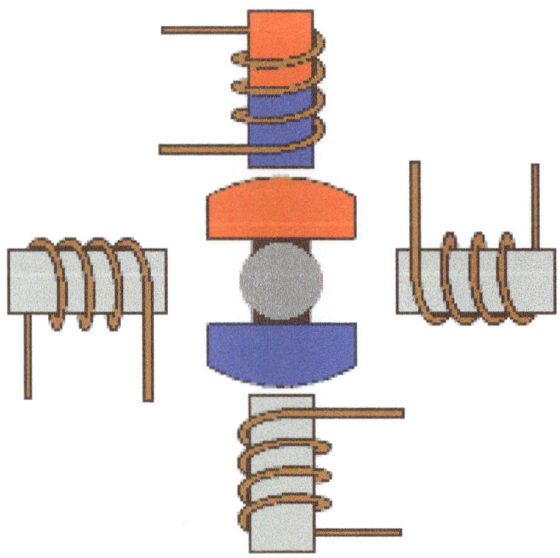

Figure 1. Stepper Motor

2. Bipolar Stepper Motor

Bipolar stepper motors are composed of two windings and have four wires. Unlike unipolar motors, bipolar motors have no center taps. The advantage to not having center taps is that current runs through an entire winding at a time instead of just half of the winding [5].

The torque of bipolar stepper motor is proportional to the magnetic field intensity of the stator windings, which is proportional to the number of turns and the current in the winding [6,7,8]. The motor torque can be increased by increasing of number of turns or by increasing electric current. [9]

For bipolar control it is necessary to ensure changing of voltage polarity so, that coil current can flow to both directions.

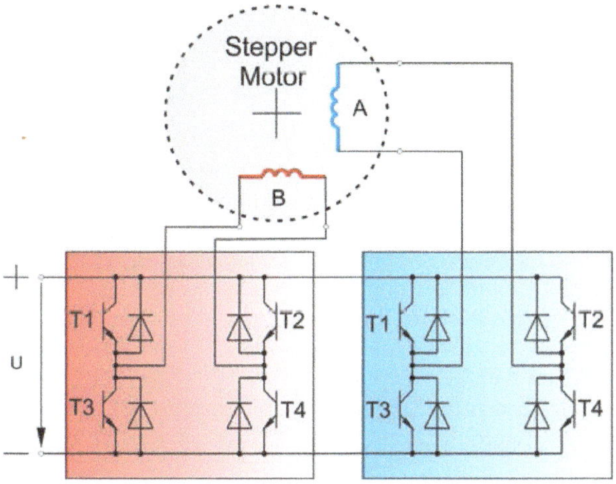

Figure 2. Control of Stepper Motor by H-Bridge

Changing voltage polarity can be reached by H-bridge, see Figure 2. By suitable switching of transistors bases T1, T2, T3 and T4 can stepper motor performs rotational motion. As can be seen on the Figure 2, there are necessary two H-bridges. For our purposes will be used circuit L298N which contains two independent H-bridges as well as amplifier. Amplifier is necessary because of microcontroller´s pins can produce electric current roughly 20 - 40 mA. L298 receives control signals from the system's controller, usually a microcomputer chip, and provides all the necessary drive signals for the power stage. Additionally, it includes two PWM chopper circuits to regulate the current in the motor windings.

3. Control of Stepper Motor

The usual ways of stepper motor control are wave mode, full-step mode, half-step mode and micro-step mode. Ideally stepper motor is controlled by sinusoidal current. Using full-step mode the motor can exhibit vibration, which are in some applications very ineligible. On the other hand, using full-step mode motor can reach higher torque. It depends on particular application what is more important point.

Very frequently used method for stepper motor control are just full-step and half-step mode. This way of control is very simply and easy to realizable by any control unit. However, the problems arise in area of resonant zone and motor can lost step. One way how can be this problem solved is to distribute motor step into micro-steps. This method causes better precise in positioning as well as it limits angular velocity pulsation of rotor in the zone of low step frequencies.

Full-step mode and half-step mode are in the Figure 3 shown.

Thick lines in the Figure 3 denote difference in voltage polarity. By considering Figure 2 and Figure 3, the stepper motor control is following: the signal above the thick line (Figure 3) represents switching of transistor bases T1 and T4 while T2 and T3 are grounded (Figure 2). When the signal falls below thick line (Figure 3) the transistor bases T2 and T3 are switched on, while T1 and T3 are grounded.

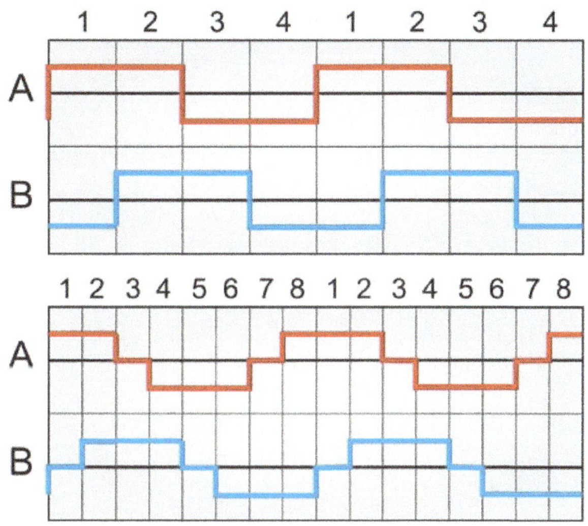

Figure 3. Full step mode and half-step mode

In the Figure 4 and Figure 5 the full step and half-step mode are shown with clockwise rotation.

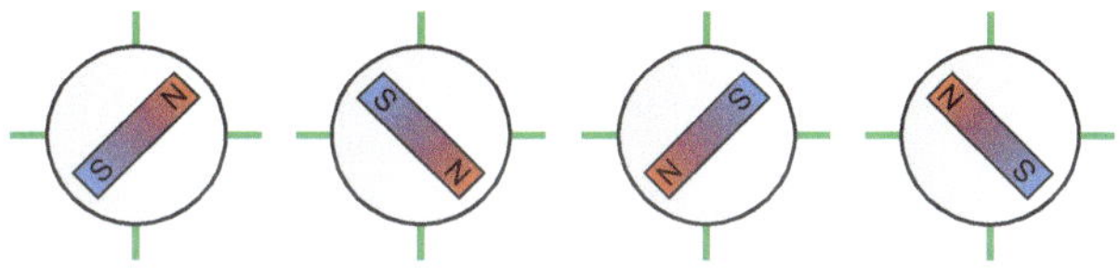

Figure 4. Full step mode and half-step mode

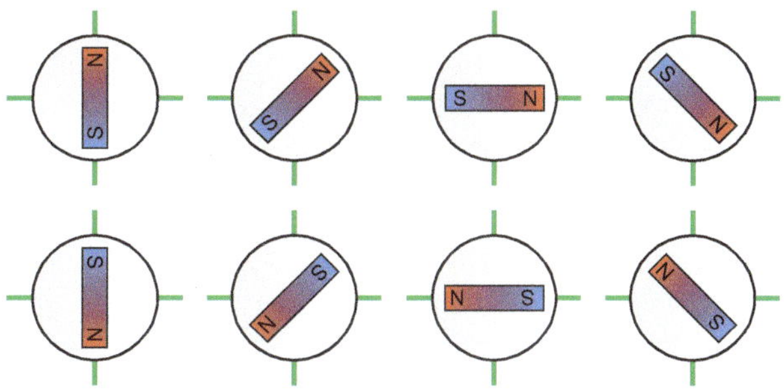

Figure 5. Half-step mode and half-step mode

4. Experiment and Results

The aim of the experiment is to compare full-step mode and half-step mode of stepper motor control. For experiment is used bipolar stepper motor SY28STH32-0674A with 200 steps per revolution, what is 1.8° to one step. For measuring of angular velocity as well as position of motor is used encoder ISC3004 with 360 count per revolution. The measuring stand for experiment in the Figure 6 is shown.

Figure 6. Measuring stand

The motion of stepper motor is controlled by 8-bit microcontroller ATmega8 with frequency of CPU 16 MHz. ATmega8-16PU controls the motor using dual full bridge driver L298. This circuit was working on DC voltage of 7 V. The measuring of angular velocity as well as position of stepper motor is measured by I/O measuring card MF624 which cooperates with Matlab / Simulink. Frequency of angular velocity measuring is 1 kHz. The measuring chain in the following figure is shown.

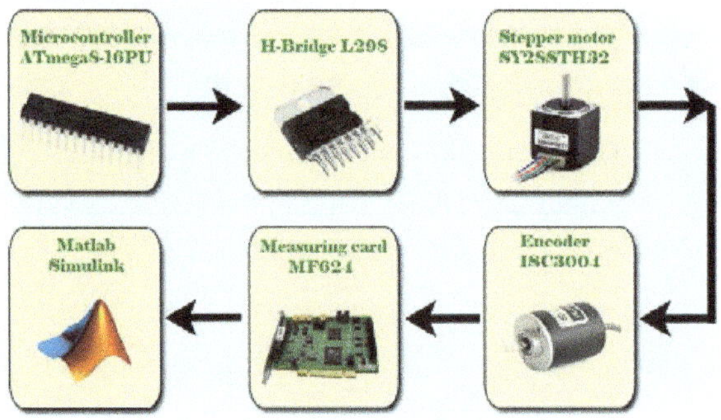

Figure 7. Measuring chain

Example of measured angular velocity is shown in the following figure.

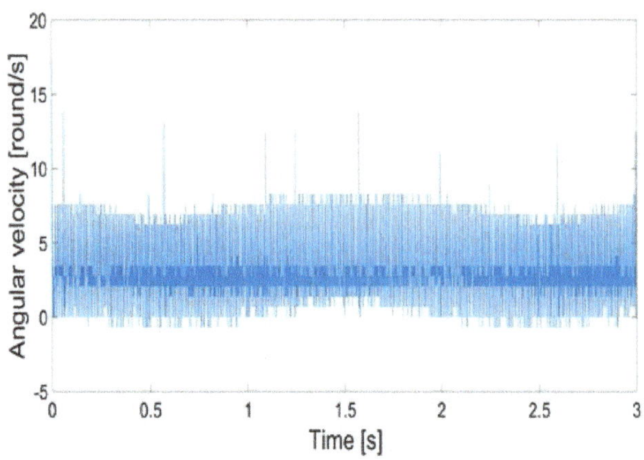

Figure 8. Example of measured angular velocity using time *of* phase 1500 μs

From all measured data was removed extreme peak values. Also for all velocities was determined average values. Then analysis were done from these average values.

In the following figures there are shown results of measurements.

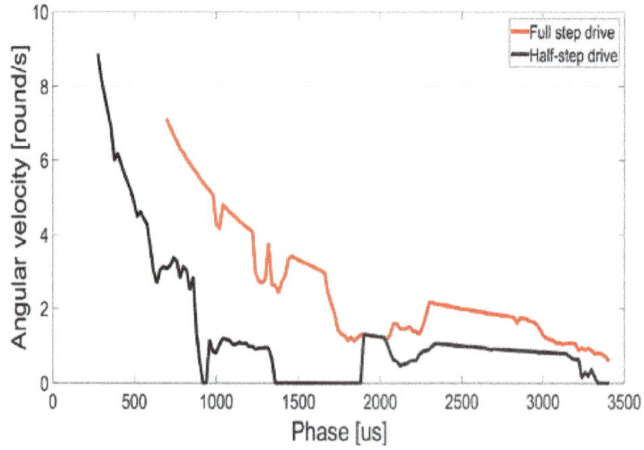

Figure 9. Angular velocity of stepper motor depending on control phase

As can be seen from the Figure 9 using half-step mode motor could start with lower value of phase time. Some problems occurred using phase time 1400 – 1900 μs and motor couldn´t work. On the other hand, using full-step mode motor worked using all phase times between 780 – 3400 μs. Step of phase time was 20 μs. Also the angular velocity of motor was higher using full-step mode in comparison with half-step mode.

During measuring of angular velocities there was also measured electric current. The electric power for both modes is shown in the Figure 10.

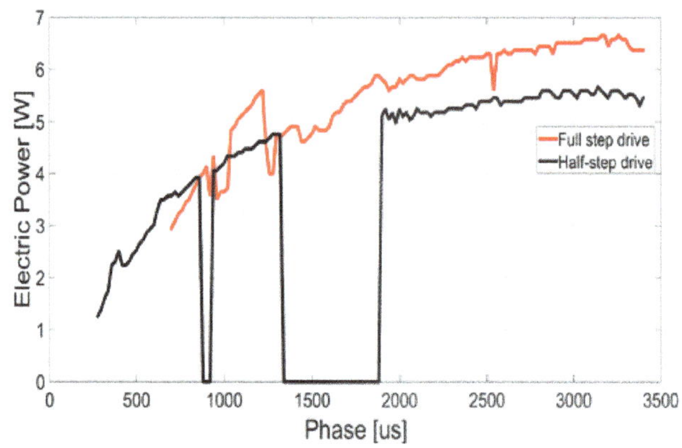

Figure 10. Electric power of stepper motor depending on control phase

Using full-step mode there was used higher amount of electric power. So the motor torque was higher using full-step mode. At the end there is also shown positioning of stepper motor rotor to one round.

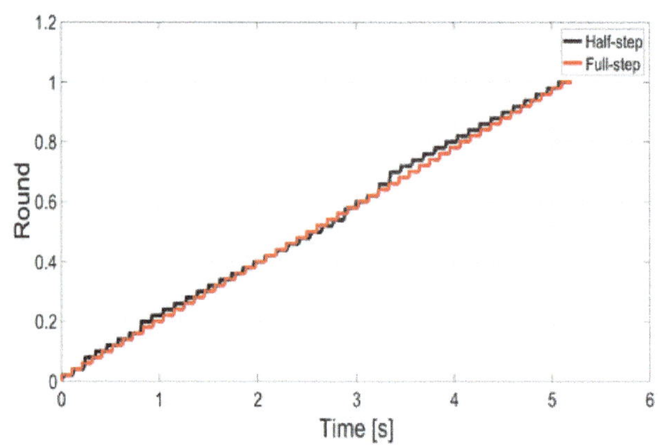

Figure 11. Positioning of stepper motor to one round

5. Conclusion

In the paper was experimentally analyzed bipolar stepper motor SY28STH32-0674A connected to encoder ISC3004 with 360 CPR. Motion was controlled by microcontroller ATmega8-16PU and dual full bridge driver L298 working on 7V. As control method were used full-step mode and half-step mode. Using full-step mode were reached higher angular velocities. Also, rotor was able to rotate using full range of phase time from value 780 up to 3400 μs. Using half-step mode there were occurred some problems using certain phase times from the value 1400 up to 1900 μs and motor was unusable. By measuring of electric current was found out that using full-step mode there is higher electric energy consumption and torque is higher like torque of half-step mode.

Acknowledgement

The authors would like to thank to Slovak Grant Agency KEGA project no. 048TUKE-4/2014 "Increasing of knowledge base of students in area of application of embedded systems in mechatronic systems".

References

[1] M. Kelemen, I. Virgala, Ľ. Miková, P. Frankovský, Experimental Identification of Linear Actuator Properties, Acta Mechanica Slovaca (2015), Vol. 19, No. 1, pp. 42-47.

[2] M. Dababneh, W. Emar, I. TTrad, Chopper Control of a Bipolar Stepper Motor, International Journal of Engineering, Vol. 7, No. 2, pp. 61-73, 2013.

[3] F. Duchoň, M. Dekan, A. Babinec, L. Chovanec, A. Vitko, Detection of Welds in Automated Welding, Applied Mechanics and Materials, Vol. 611 (2014), pp. 519-528.

[4] A. Morar, The modelling and simulation of bipolar hybrid stepping motor by Matlab/Simulink, Procedia Technology, 8th International Conference Interdisciplinarity in Engineering (2015), pp. 576-583.

[5] R. Condit, Stepping Motors Fundamentals, Microchip Technology Inc., (2004).

[6] T. Hopkins, Stepper motor driving, Application note (2012).

[7] Stepanov P., Lagutkin S., Nikitin Y., Development of the Algorithm for Definition of Residual Service Life by a Comprehensive Diagnosis of the Electromechanical Drive, Acta Mechanica Slovaca. Volume 19, Issue 2, Pages 52-56.

[8] P. Stepanov, S. Lagutkin, P. Božek, Y. Nikitin: Comprehensive Approach to Technical Conditions of Electromechanical Units in Mechatronic Systems, American Journal of Mechanical Engineering (2014).

[9] Kollár, P., Nikitin, Y., Straka, M.: The determination of the shelf mass in the universal shelving stacker by measuring the frequency converter torque generating current of the main drive, Manufacturing technology, Vol. 15, No. 3, p. 363-366.

Testing Properties of E-health System Based on Arduino

Róbert Rákay*, Martin Víšňovský, Alena Galajdová, Dušan Šimšík

Department of Automation, Control and Human Machine Interactions, Technical University of Košice, Košice, Slovak Republic
*Corresponding author: robert.rakay@tuke.sk

Abstract This paper deals with health functions-monitoring system for observing patient´s physical characteristics. These physical characteristics of patients are followed by sensors and the output information is broadcasted via Zigbee to PC for further processing. The tested e-health system is based on Arduino e-health kitwith Zigbee for transmitting information and a PC with Arduino and Zigbee for receiving data. Bluetooth is a well-known conventional wireless data transfer protocol, but it´s not much encouraged for long duration applications because of power consumption limitations, so Zigbee offers a better solution for such cases. Basic tasks of the system are to measure main health functions as heart rate, body temperature and body position. The tested system enables quick access to parameters of monitored biosignals, therefore identification of health risks and intervention can be provided in real time [1].

Keywords: *E-health system, Arduino, Zigbee*

1. Introduction

The current monitoring systems used for ambient intelligence applications in home or social facilities have to be solved as mobile wireless devices in opposite to wired hospital solutions [1].

Even though the correctly connected and applied sensors the basic monitoring system requires paramedical assistant or a qualified person to continuously monitor and note down all the information manually. This kind of solution may lead to mistakes in diagnostics and analyzing or even to disaster because of human error. Our experimental system allows creating a wireless sensor network, and provides a good base for further experiments and development. Creating this kind of system allows monitoring vital functions using personal networks at patient´s homes. Applying Zigbee protocol to sensor system creates wireless sensor network suitable for transmission of data and also it allows further data treatment, visualization, and warnings. The power consumption of Zigbee is very low, and the sizes of the sensor nodes are small, therefore it is one of the best options for wireless connection. To develop and improve data collecting and data analyzing processes of above mentioned problems an e-health monitoring system based on Arduino is designed. This paper describes a laboratory sensor system that collects and transfers information about vital characteristics of patients.

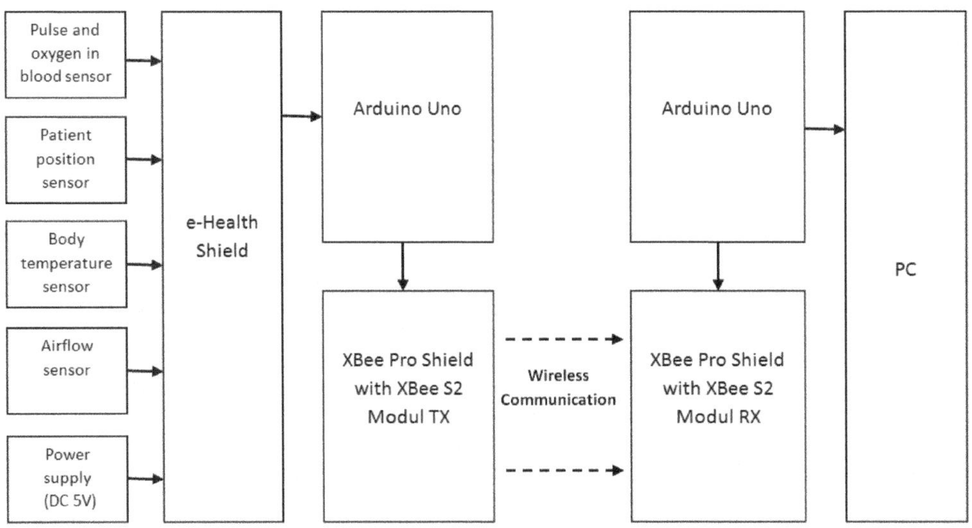

Figure 1. System architecture

2. Hardware and Software Description

Figure 1 displays the block diagram of realised hard ware solution. The system is created with an open-source platform, Arduino. The logic operations in Arduino Uno are executed by computing unit, which is an Atmel Atmega processor. The biggest benefit of this platform is that there is no fixed architecture and defined communication, the individual parts can be connected and divided freely. The devices can work either independently or with superior server. The main idea is to form a system to measure and check biosignals of human body such as heart rate, temperature, respiration and body position. The collected information is processed and sent via Zigbee protocol to a receiver computer. The receiving computer is connected to a second Zigbee module via USB wire. [2,3]

The data can be processed or transferred to a responsible person or web database for long term tracking. If the measured values hit extreme rates an alarm signal may be generated.

The Arduino Software for programming and X-CTU for configuring communication are used. More software details are in the following chapters. [2]

2.1. Pulse and Oxygen in Blood Sensor (SPO2)

Pulse oximetry is used for monitoring of oxygen saturation in blood.It is a non-invasive method based on measuring wavelengths of light waves sent through a thinner body part for example a fingertip or earlobe (Figure 2). The main components of this sensor are an electronic processor and a pair of light emitting diodes (LEDs) and a photodiode, in our case expanded by a display. A photodetector receives light which was not absorbed by the body part. Absorption of light is different forlow and high concentrationsof blood with oxygen (Figure 3) [4].

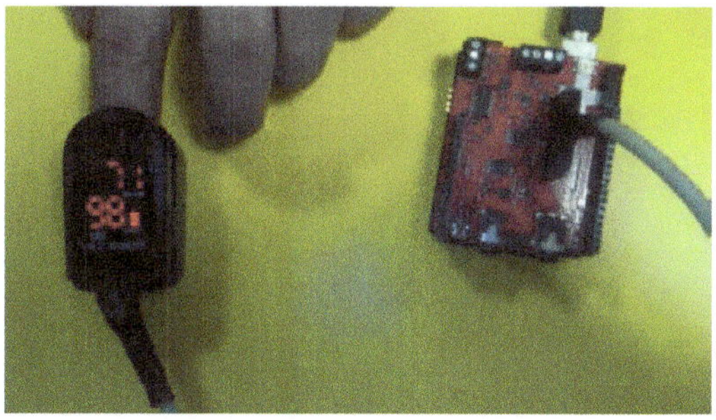

Figure 2. Pulse and oxygen in blood sensor with e-health shield

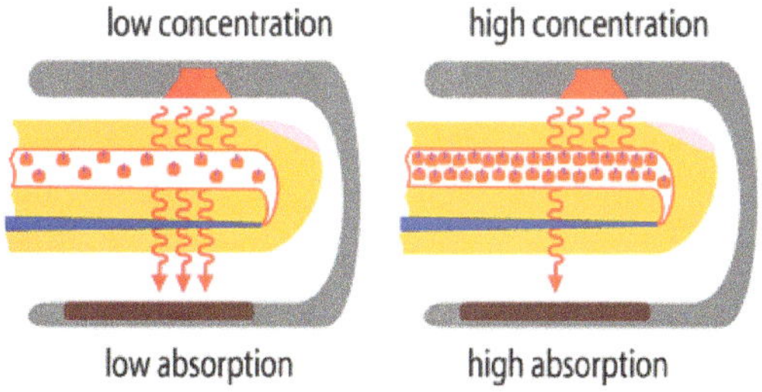

Figure 3. Light absorption

2.2. Temperature Sensor

Sensing of the body temperature is very important in medicine. The reason is that quite many diseases are followed by temperature change.

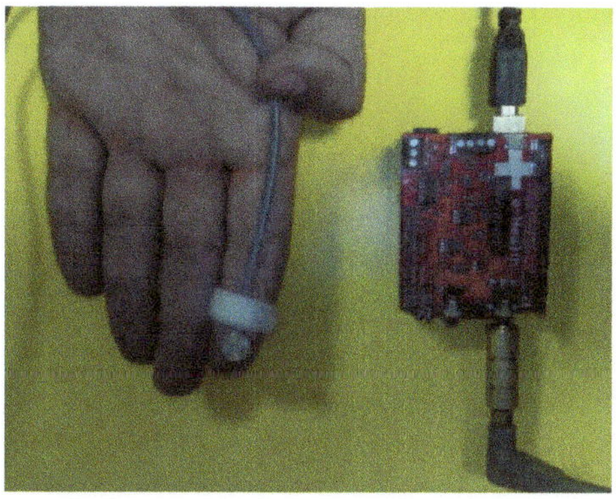

Figure 4. Temperature sensorwith e-health shield

In the same way, monitoring of illness and effects of treatment can be measured using temperature changes (Figure 5). [5]

The measured temperatures can be evaluated and warnings for medical staff can be generated in form of message or light indication. Physiological and pathological limits of temperature are shown in Table 1

Table 1. Illnesses caused by body temperature change

Hypothermy	< 35.0 °C
Normal temperature	36.5 – 37.5 °C
Hyperthermy	> 37.5 – 38.3 °C
Hyperpyrexy	> 40.0 – 41.5 °C

2.3. Body Position Sensor

This sensor uses a triaxial accelerometer which is integrated into a band. The Figure 5 shows the body position sensor and the connection to e-health shield of Arduino. This band has to be fasten on the chest of a patient.

The monitored positions are: standing/sitting. lying on chest, lying on back, lying on left side and lying on right side. In many cases it´s important to monitor body movement and position, because they can indicate health problems, for example persons with sleeping problems and during recoveries after accidents. [6]

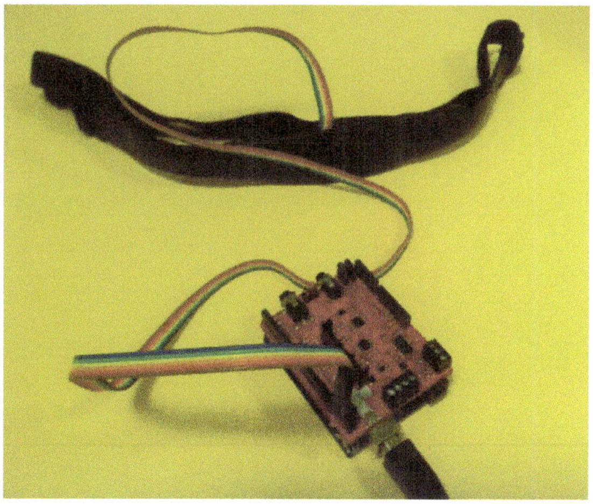

Figure 5. Body position sensor and e-health shield

2.4. Airflow Sensor

The nasal airflow sensor is used to monitor breathing of patients. The measured data can be processed or used as alarm signals in emergency cases. The airflow sensor is shown on Figure 6. [7]

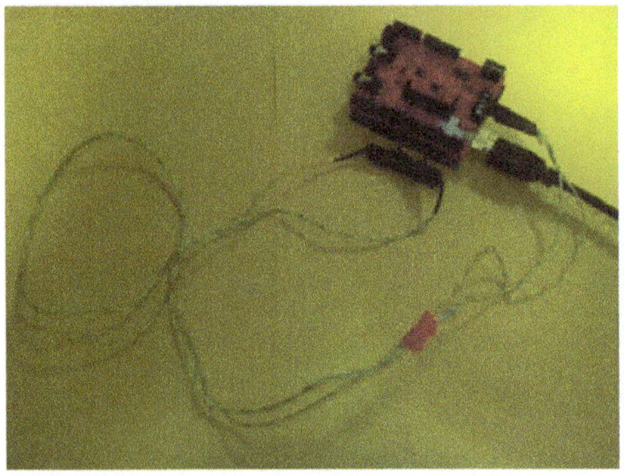

Figure 6. Nasal airflow sensor and e-health shield

2.5. Software

For programming of e-health sensor system the official Arduino Software was used. Part of the program and source code is shown below on the Figure 7.

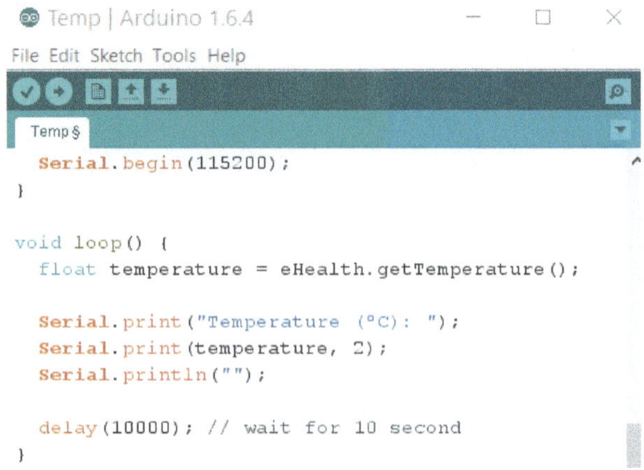

Figure 7. Arduino software

This programming language is based on programming language Wiring, where is no need for using the complicated Assembler. This programming platform was created in Java. The flowchart of the process is shown on Figure 8.

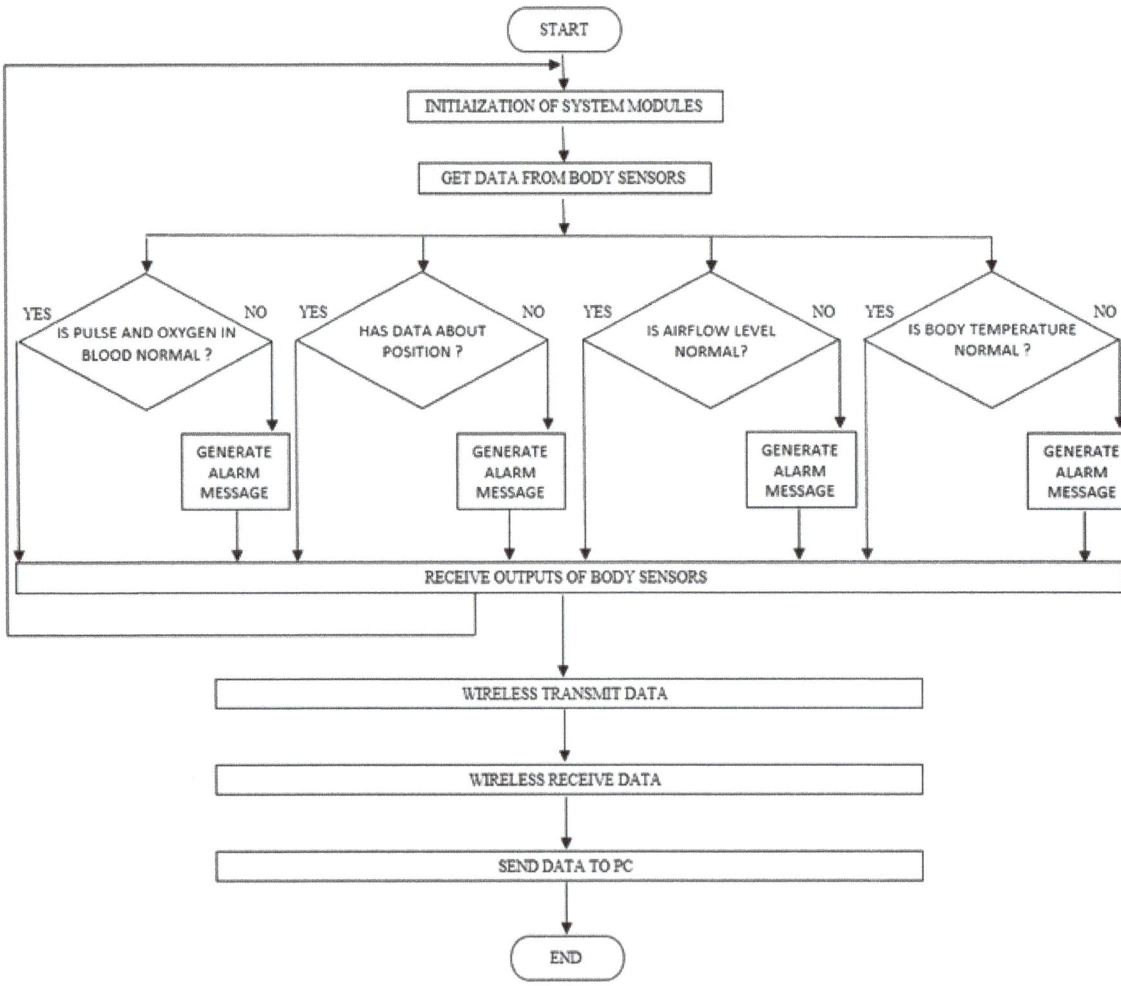

Figure 8. Flowchart of the process

After connecting the main Arduino to a power unit, the program is running continuously in a loop. It begins with initialization of connected modules. This is followed by collecting information from sensors. If the sensors are connected and work correctly, the data are sent by Zigbee to a receiver Zigbee node of another Arduino, which is connected to a PC. After data collecting, they can be processed or stored for longer monitoring. The software can be configured for extreme values or states of measured data and this can be used for generating further alarms and warnings.

3. Communication

The individual parts of e-health system are connected simply trough attributable pins. The communication between the sensors/transfer part and the receiver part is solved by Zigbee protocol. In this case we used Xbee nodes connected to Xbee Pro Shield which can be configured to use this protocol, see Figure 9. The X-CTU software was used for configuring the nodes (Figure 10). The selected characteristics are shown in Table 2. [3]

Table 2. Configured parameters

Baud rate	9600
Flow control	HARDWARE
Data Bits	8
Parity	NONE
Stop Bits	1

Figure 9. Xbee node and Xbee Pro Shield

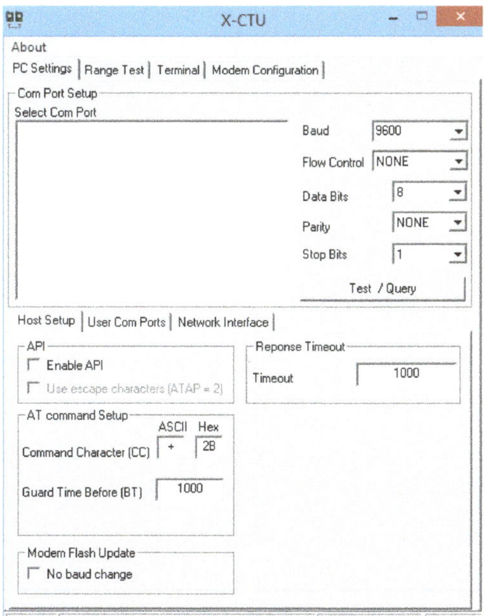

Figure 10. X-CTU software

4. Testing of Properties

We tested system functionalities (Figure 11) by the comparation of measured temperature from the body warn sensor and from the standard manual digital thermometer. Actual body position was measured with body sensor, and the displayed data was compared with real position checked visually by the operator.

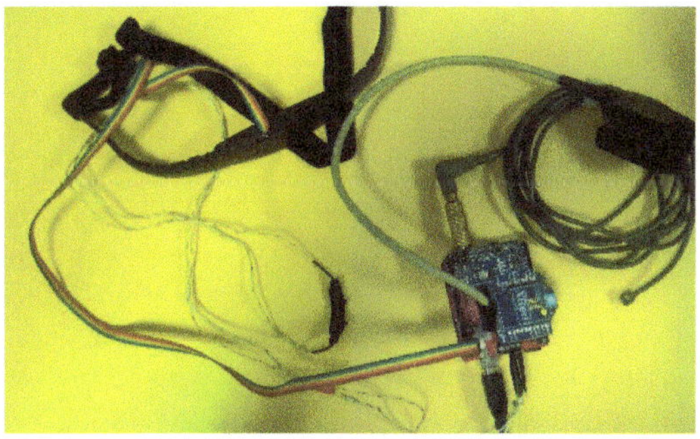

Figure 11. E-health system with sensors and Xbee module

For analyzing and evaluating the received data from sensors we provided 50 measurements of temperature and body position. The testing conditions during the measuring were constant:

- Temperature T= 21 °C
- Atmospheric pressure P_a= 101,325k Pa.

The Table 3 shows results from 10 measurements from sensors and their comparation with reference devices.

Table 3. Measured values of temperature and body positions

No.	Temp.Sensor[°C]	Therm. [° C]	Real Body Position	Body Pos. Sensor
1.	35,76	36,0	Standing	Fowler´s
2.	35,82	36,0	Standing	Fowler´s
3.	35,88	36,0	Standing	Fowler´s
4.	35,78	36,0	Supine	Supine
5.	35,8	36,0	Supine	Supine
6.	35,92	36,0	Sitting	Fowler´s
7.	36,12	36,0	Sitting	Fowler´s
8.	35,95	36,2	LLR *	LLR
9.	35,85	36,2	LLR	LLR
10.	36	36,2	LLR	LLR

*LLR- Left Lateral Recumbent

The graph on Figure 12 shows difference between received data from e-health system and from digital thermometer.

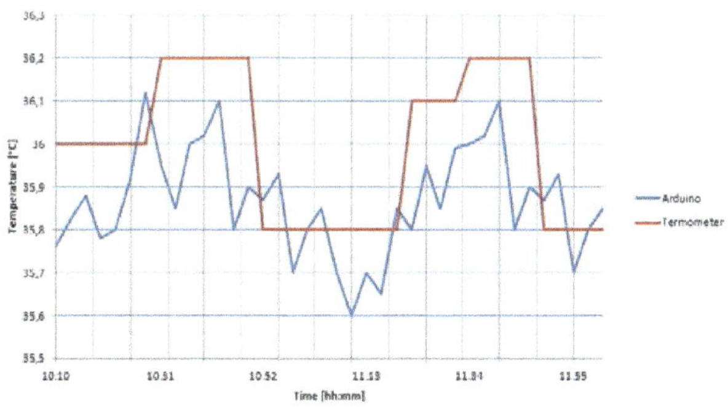

Figure 12. Measured temperatures

Average measured temperatures, the average temperature difference and faulty body positions are shown in Table 4.

Table 4. Average results of measurement

Avg. temp. sensor [°C]	35,87
Avg. thermometer [°C]	35,99
Avg. difference	0,13
Total Body position measurements/False Body position detection	50/5

5. Results

In this paper we described the laboratory system built for monitoring of physiological parameters such as pulse and oxygen in blood, body position, air flow, and body temperature. Two parameters were also tested and compared with referential devices. The system uses the Arduino control and processing unit expanded by the sensor and communication shields, e-health sensors, and Xbee node.

We will continue with further development of the system and experimental testing of data processing using web-based platforms, and reliability of wireless sensor networks.

Experiments proved that the temperature sensor works accurately for standard monitoring of body temperature. However, the body position sensor showed faulty results, which could be caused by non exact positions of the subject during experiments. Some disadvantage of body position sensor was observed, it couldn´t recognize difference between standing and sitting position. We plan to solve that problem by adding another sensor to the system, which can help to recognize the real body position.

Acknowledgement

This work has been supported by the Slovak Grant Agency VEGA contract N. 1/0911/14 Implementation of wireless technologies into the design of new products and services to protect human health.

References

[1] Navya, K., Murthy, M. B. R. A Zigbee Based Patient Health Monitoring System, IJERA,3 (5), Sept-Oct 2013.

[2] Sipos R., Multi-agent systems in sensor networks, Technical University in Kosice, Faculty of Mechanical Engineering, 2015

[3] Zigbee Introduction / ICP DAAS USA Inc. http://www.icpdas-usa.com/zigbeeintro.php.

[4] Pulse and Oxygen in Blood Sensor (SPOP2) for e-Health Platform [Biometric / Medical Applications] < http://www.cooking-hacks.com/pulse-and-oxygen-in-blood-sensor-spo2-ehealth-medical>.

[5] Body Temperature Sensor for e-Health Platform [Biometric / Medical Applications] <http://www.cooking-hacks.com/body-temperature-sensor-ehealth-medical>.

[6] Patient Position Sensor (Accelerometer) for e-Health Platform [Biometric / /Medical Applications] <http://www.cooking-hacks.com/patient-position-sensor-accelerometer-ehealth-medical>.

[7] Airflow Sensor (Breathing) for e-Health Platform [Biometric / Medical Applications] <http://www.cooking-hacks.com/airflow-sensor-breathing-ehealth-medical>.

Regression Analysis and Seasonal Adjustment of Time Series

Eva Ostertagová[1,*], Oskar Ostertag[2]

[1]Department of Mathematics and Theoretical Informatics, Faculty of Electrical Engineering and Informatics, Technical University of Košice, Nemcovej 32, 042 00 Košice, Slovak Republic
[2]Department of Applied Mechanics and Mechatronics, Faculty of Mechanical Engineering, Technical University of Košice, Letná 9, 042 00 Košice, Slovak Republic
*Corresponding author: eva.ostertagova@tuke.sk

Abstract The aim of this article is to demonstrate the dummy variables for estimation seasonal effects in a time series, to use them as inputs in a regression model for obtaining quality predictions. Model parameters were estimated using the least square method. After fitting, special tests to determine, if the model is satisfactory, were employed. The application data were analyzed using the MATLAB computer program that performs these calculations.

Keywords: *seasonal time series, dummy variables, trigonometric regression functions, method of least squares, residual analysis*

1. Introduction

If we analyze the evolution of time series, we are interested not only in the main development trend of the indicators, but also in the course and intensity of any periodic fluctuations, which these time series present. When working with time series, the data must be adjusted seasonally. The aim of seasonal adjustment is to uncover the underlying dynamics in the development of the investigated phenomena and allow a direct comparison of their development in different seasons within the year. There are many methods of seasonal adjustment and their classification is not easy, because in practice the techniques used are a combination of several methods. Often they apply different types of moving averages, which eliminate from the time series the components the frequency of which does not exceed the number of observations forming the moving average length. To eliminate seasonal component regression methods based on the theory of linear regression model are also used. In case, where the nature of the seasonal component may change, e.g. the Winters exponential smoothing is applied.

2. Regression Approaches to the Seasonal Component of Time Series

In the construction of the forecasts of seasonal time series, a regression model with artificial (dummy) variables with simultaneously estimated trend and seasonality parameters can be used. Artificial variable is used to quantify the effect of the respective period on the estimated value of the investigated variables. The trend component is modeled via suitable regression function, for example line, parabola, and so on. The seasonal component is expressed using artificial (zero unit) variables that assign a value to the time series unit in case it is found in the considered season and zero otherwise.

Let us assume an additive time series model in which the value of the indicator y_t in the t-period is given by the sum $y_t = T_t + S_t + \varepsilon_t$, where T_t is the trend component, S_t is the seasonal component and ε_t is a random component. In the presence of free parameter (constant) in the model trend, in order to avoid multicollinearity, seasonality is modeled as a qualitative variable using the $s-1$ of artificial variables, where s is the length of the season included in the time series. Furthermore, we assume that the time series has a linear trend and quarterly seasonality. The relevant regression model can then be formulated for $t = 1, 2, \ldots, n$ in the form of

$$y_t = \beta_0 + \beta_1 t + \alpha_2 d_{t2} + \alpha_3 d_{t3} + \alpha_4 d_{t4} + \varepsilon_t, \qquad (1)$$

where the artificial variables are defined as vectors

$$d_{t2} = (0, 1, 0, 0, 0, 1, 0, 0, \ldots),$$
$$d_{t3} = (0, 0, 1, 0, 0, 0, 1, 0, \ldots),$$
$$d_{t4} = (0, 0, 0, 1, 0, 0, 0, 1, \ldots), \qquad (2)$$

of length of which is equal to n number of the time series of observations.

Since the artificial variable attains the value of one in a particular observation, we declare that in this period, to the value generated from a linear trend we shall add the value of seasonal fluctuations, which is calculated compared to the base period, which is in this case the first quarter of the year.

The artificial variable $d_{t1} = (0, 0, 0, 0, 0, 0, 0, 0, \ldots)$ is a zero vector and the effect of the first quarter is included in the intercept β_0 of the linear trend, which is interpreted in terms of the base level of the studied variables.

Model (1) contains a trend, seasonal and random component. Model parameters can be estimated using the least square method. The estimated model will take the form of:

$$\hat{y}_t = \hat{\beta}_0 + \hat{\beta}_1 t + \hat{\alpha}_2 d_{t2} + \hat{\alpha}_3 d_{t3} + \hat{\alpha}_4 d_{t4}. \quad (3)$$

The verification of the suitability of the regression model (1) is analogous to that in any other regression model. Particularly important is to test the heteroscedasticity and autocorrelation of the random component.

The estimated regression model (3) can be used for the construction of point and interval forecasts. Forecasting requires us to choose the time variables in the horizon of $h > 0$ and for the seasonal variables, substitute the unit values of the respective seasons in the horizon h.

In case of the regression model with artificial variables we shall adjust the estimated trend $\hat{T}_t$ and the seasonal factors I_j, $j = 1, 2, 3, 4$, to the form of [1]:

$$\hat{T}_t = (\hat{\beta}_0 + \overline{a}) + \hat{\beta}_1 t, \quad (4)$$

$$I_1 = -\overline{a}, I_2 = \hat{\alpha}_2 - \overline{a}, I_3 = \hat{\alpha}_3 - \overline{a}, I_4 = \hat{\alpha}_4 - \overline{a}, \quad (5)$$

where

$$\overline{a} = (\hat{\alpha}_2 + \hat{\alpha}_3 + \hat{\alpha}_4)/4 \quad (6)$$

is the "average" of seasonal regression parameters.

In an analogous manner we shall proceed in case of twelve month seasonality.

Another regression method for eliminating seasonal component is based on the fact that this component is estimated by means of a suitably selected mathematical function. The most commonly used are trigonometric functions with the period length equal to the number of periods s in the year, or a fraction of this number.

Provided that the trend of the considered time series is linear, the model may have e.g. this shape for $t = 1, 2, \ldots, n$:

$$y_t = \beta_0 + \beta_1 t + \beta_2 \cos(2\pi t/s) + \beta_3 \sin(2\pi t/s) + \varepsilon_t. \quad (7)$$

Since it is a general linear regression model, estimates of the parameters may be obtained by the least square method.

In case the coefficient of determination R^2 for the stated model is too small, we can continue to add to the model further unit values in the form of the considered trigonometric functions, with a half, fourth, or even smaller period, e.g. $\beta_4 \cos(4\pi t/s)$, $\beta_5 \sin(4\pi t/s)$, etc. From the models listed we select the one for which we achieved, for example, the maximum value of the coefficient of determination and also which best meets the other criteria imposed on the linear regression model.

3. The Application of Regression Models with artificial Variables and trigonometric Functions at Selected Time Series

We have data available on the number of sold pieces of selected articles of a business company engaged in the Internet sales of automotive accessories for individual quarters of the year, during the period of 2008 – 2014.

Figure 1 displays the time series presented in a form of plot via line chart.

The presented graph makes clear, that the stated time series has in the respective period an increasing, approximately linear trend and quarterly seasonality. The proposed regression model with artificial variables will have the form of (1).

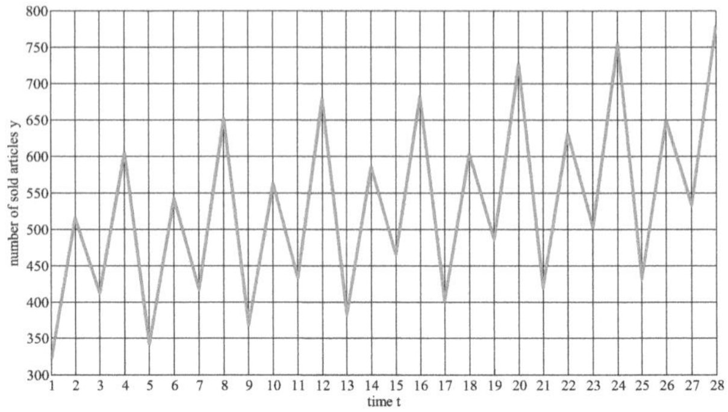

Figure 1. The development of the number of articles sold in the period of 2008-2014

The model estimated by the least squares method is:

$$\hat{y}_t = \hat{T}_t + \hat{S}_t = 308.5201 + 5.5424t$$
$$+197.8862 d_{t2} + 71.6295 d_{t3} + 299.3728 d_{t4}. \quad (8)$$

For two-sided 95% confidence intervals for regression coefficients applies, that:

$$\beta_0 \in \langle 298.7445, 318.2957 \rangle,$$

$$\beta_1 \in \langle 5.0682, 6.0166 \rangle,$$

$$\alpha_2 \in \langle 187.1461, 208.6262 \rangle,$$

$$\alpha_3 \in \langle 60.8580, 82.4009 \rangle,$$

$$\alpha_4 \in \langle 288.5493, 310.1963 \rangle.$$

To test the statistical significance of individual coefficients of the regression model the t-tests were used, where we received the following result values of test statistics and p-values:

[65.2874, 24.1790, 38.1151, 13.7565, 57.2181],

$[1.2863 \cdot 10^{-27}, 7.4491 \cdot 10^{-18}, 2.7269 \cdot 10^{-22}, 1.3820 \cdot 10^{-12}, 2.6265 \cdot 10^{-26}]$.

Since p-values are in all cases below the significance level $\alpha = 0.05$, all regression coefficients are considered statistically significant. The same result has also been provided by the confidence intervals for regression coefficients, since none of them contains zero value.

Based on the resulting value of the coefficient of determination $R^2 = 0.9953$ we can conclude that the model explained the variability of the number of units sold of selected articles to 99.53%.

The least squares method provides unbiased point estimates of parameters of the linear regression model while meeting certain assumptions about the probability distribution of random errors ε_t, for $t = 1, 2, \ldots, n$, within the model.

We assume, that the random errors ε_t [2,3]:

- have normal distribution,
- have zero mean values, i.e. $E(\varepsilon_t) = 0$;
- have constant variance (homoscedasticity), i.e. $V(\varepsilon_t) = \sigma^2$;
- are not correlated to each other (in case of the normality of the distribution are independent), i.e. the covariance $K(\varepsilon_i, \varepsilon_l) = 0$ for each $i \neq l,\ i, l = 1, 2, \dots, n$.

The most important methods of regression model analysis include residual analysis. It is based on the assumption that the residuals e_t represent the point estimate of random errors ε_t. The equation $e_t = y_t - \hat{y}_t$ applies, i.e. (classical) residual is the difference of empirical and theoretical values.

The assumptions, on which the model is based, are generally verified by simple graphs, respectively, using known statistical tests [4,5].

In case of the graph displaying standardized residuals versus the theoretical values (see Figure 2), i.e. of a scatter plot $\left(\hat{y}_t, e_{St}\right)$ applies, that the model is good if approximately 95% of residuals lies in the interval $(-2, 2)$. Also residuals have to be randomly distributed around zero and the plot must not show any indication of a potential trend or pattern of development [2,6].

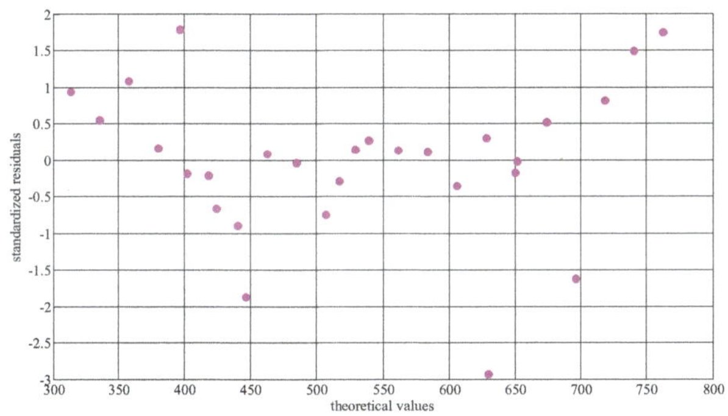

Figure 2. The dependence of standardized residuals on the theoretical values

The normality of the distribution of random errors we verified using the Anderson-Darling test of goodness-of-fit. On the significance level of $\alpha = 0.05$ we have tested the null hypothesis $H_0 : F(x) = F_0(x)$ against the alternative hypothesis $H_1 : F(x) \neq F_0(x)$, where $F(x)$ is the distribution function of random selection (residuals) and $F_0(x)$ is the distribution function of the normal distribution. We attained these results: Anderson-Darling statistics $AD = 0.5801$, p-value $= 0.1217 > 0.05$. Thus with a 95% reliability we can claim that random errors have a normal distribution.

Further, we tested the null hypothesis H_0: *random errors are uncorrelated* compared to the alternative hypothesis H_1: *random errors are correlated.* We have applied the Durbin-Watson test on the significance level of $\alpha = 0.05$ with the following results: statistics $DW = 2.6948$, whilst the p-value $= 0.1068 > 0.05$. We therefore do not reject the hypothesis on the no correlation of the random errors.

Based on these results it can be stated that model has good quality, and therefore it can be used to calculate extrapolations of the quarterly changes in the number of sold pieces of the selected articles of goods in the year 2015.

We shall use the equations (5) and (6). For the "average" of the seasonal regression parameters applies that $\bar{a} = 142.2221$.

For seasonal factors, we get the following results:

$$I_1 = -142.2221, I_2 = 55.6641,$$
$$I_3 = -70.5926, I_4 = 157.1507. \quad (9)$$

The presented results can be interpreted so as the average number of sold pieces of goods annually in the first quarter decreased by about 142 pieces, in the second quarter increased by about 56 pieces, in the third quarter decreased by about 71 pieces and in the fourth quarter increased by about 157 pieces compared to the trend.

Based on the equation (4) we get for the trend estimate the relation:

$$\hat{T}_t - 450.7422 + 5.5424t. \quad (10)$$

Figure 3 shows a plot of the stated time series and the estimated trend.

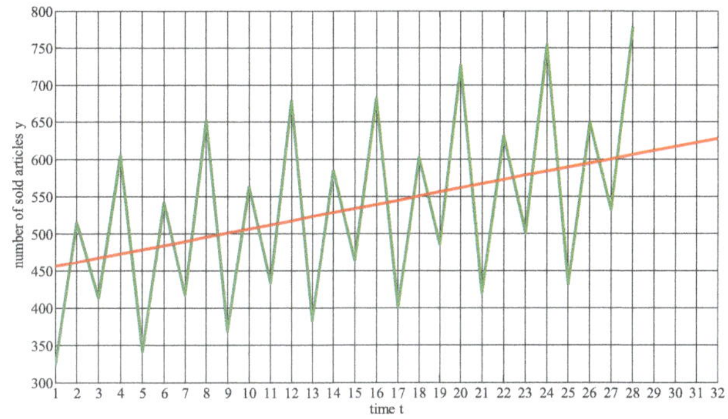

Figure 3. The plot of the stated time series together with the estimated trend

The extrapolated values of the original time series for each quarter of the year 2015 can be obtained based on the basis of the estimated model (8), respectively, on the basis of the respective seasonal factors (9) and the estimated trend (10).

The prediction for individual quarters of the year is as follows:

$$\hat{y}_{29} = 308.5201 + 5.5424 \cdot 29 = 469.2500,$$

or $\hat{y}_{29} = 450.7422 + 5.5424 \cdot 29 - 142.2221 = 469.25$,

$$\hat{y}_{30} = 308.5201 + 5.5424 \cdot 30 + 197.8862 = 672.6786,$$

or $\hat{y}_{30} = \hat{T}_{30} + I_2 = 672.6786$,

$$\hat{y}_{31} = 308.5201 + 5.5424 \cdot 31 + 71.6295 = 551.9643,$$

or $\hat{y}_{31} = \hat{T}_{31} + I_3$,

$$\hat{y}_{32} = 308.5201 + 5.5424 \cdot 32 + 299.3728 = 785.25,$$

or $\hat{y}_{32} = \hat{T}_{32} + I_4$.

Now follow the application of the linear regression model (7). The good model estimated by the least squares method is:

$$\hat{y}_t = \hat{\beta}_0 + \hat{\beta}_1 t + \hat{\beta}_2 \cos\frac{\pi t}{2} + \hat{\beta}_3 \sin\frac{\pi t}{2} + \hat{\beta}_4 \cos \pi t, \quad (11)$$

where $\hat{\beta}_0, \hat{\beta}_1, \ldots, \hat{\beta}_4$ are unbiased estimators of the true regression coefficients $\beta_0, \beta_1, \beta_2, \beta_3, \beta_4$.

Least squares parameter estimates for this model are $\hat{\beta} = (450.7422, 5.5424, 50.7433, -35.8147, 106.4074)^T$.

The predictions for individual quarters of the year 2015 are the same as in the case of application of the regression model with artificial variables: $\hat{y}_{29} = 469.25$, $\hat{y}_{30} = 672.6786$, $\hat{y}_{31} = 551.9643$, $\hat{y}_{32} = 785.25$. The coefficient of determination is in this case $R^2 = 0.9953$ too.

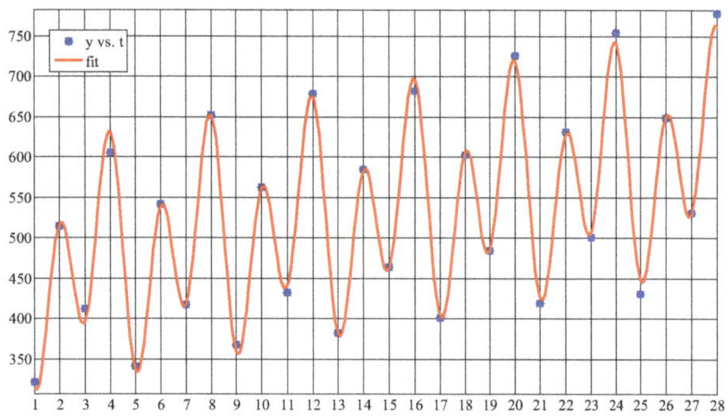

Figure 4. The plot of the measured data with the estimated trend (11)

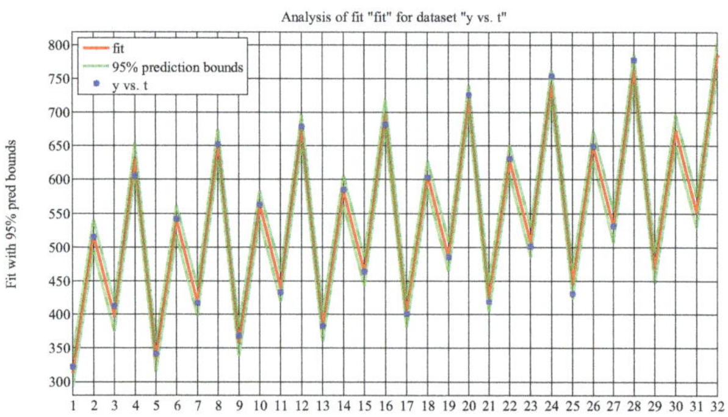

Figure 5. The plot of the stated time series together with the estimated trend (11) and 95 % prediction interval

Figure 4 presents a scatter diagram of the measured data with the least squares fitted trend (11). Figure 5 shows more so the 95 % prediction interval for sold pieces of selected articles of a business company.

4. Conclusion

The current paper presents the analysis of time series with linear growing trend and additive seasonal component. To determine the seasonal component, a method based on the theory of linear regression model with artificial variables, i.e., variables that are discrete or qualitative in nature, so they cannot be directly quantified, was used. For eliminating seasonal component was used regression model with trigonometric functions too.

The analysis of the seasonal component allows us to increase our knowledge about the patterns of behavior of a given effect, respectively phenomenon, and contribute to the construction of better forecasts of the considered time series.

Acknowledgement

This work was supported by the VEGA grant scheme no. 1/1205/12 Numerical Modeling of Mechatronic Systems and the VEGA grant scheme no. 1/0393/14 Analysis of Causes of Mechanical System Failures by the Quantification of Strains and Stress Fields.

References

[1] Arlt, J., Arltová, M., Rublíková, E., *The Analysis of Economic Time Series with Examples* (in Czech), VŠE Prague, 2002.

[2] Montgomery, D.C., Runger, G.C., *Applied Statistics and Probability for Engineers*, John Wiley & Sons, 2003.

[3] Ostertagová, E., *Modelling Using Polynomial Regression*, Procedia Engineering, 48 (2012), p. 500-506.

[4] Ostertagová, E., *Applied Statistics* (in Slovak), Elfa, Košice, 2011, 161 pp.

[5] Ostertagová, E., *Applied Statistics in the Computational Environment of the MATLAB software* (in Slovak), TU Košice, 2015, 175 pp.

[6] Ostertagová, E., Ostertag, O., *Time Series Modelling*, The 4th International Conference on Modelling of Mechanical and Mechatronic Systems, Technical University of Košice, Slovak Republic, Proceedings of conference,2011, p. 380-384.

Methodology for Workplace Design at Food Production Process

Marek Vagaš[1,*], Vladimír Baláž[1], Lucia Koukolová[1], Jozef Putala[2]

[1]Department of Robotics, Faculty of mechanical engineering, Technical University of Košice, Slovakia
[2]Matador Industries, a. s., General engineering & Automatization, Dubnica n. Váhom, Slovakia
*Corresponding author: marek.vagas@tuke.sk

Abstract This paper aims to the methodology for workplace design at food production process. Based on implementation of industrial robot into the workplace and object specification is suitable to create an effective methodology for supporting attractive workplace design with emphasis to individual phases during solution process. Article also offers mathematical output, which can be generally applied to all workplaces with regards to food production process.

Keywords: *palletization, production, methodology, process, food*

1. Introduction

Approach to an optimal methodology for design of workplace structure with industrial robot for palletization must be taken into account as a philosophy of understanding of the problem as its objective function. That can be characterized as performance capability of individual elements to change position and orientation of an object over time. A systematic approach includes phases from the perspective of the object for palletizing to its possible modifications as need for robotic manipulation through setting of various handling operations, choice of robot and peripherals, spatial distribution, decomposition handling process to programmable sections, fittings workplace conditions of operation, and operator safety to learning curve for staff to operate and maintain [1].

Use of industrial robots for palletizing is characterized by two areas. First is palletizing processes which require less capacity of the robots, but higher speed of their movements.

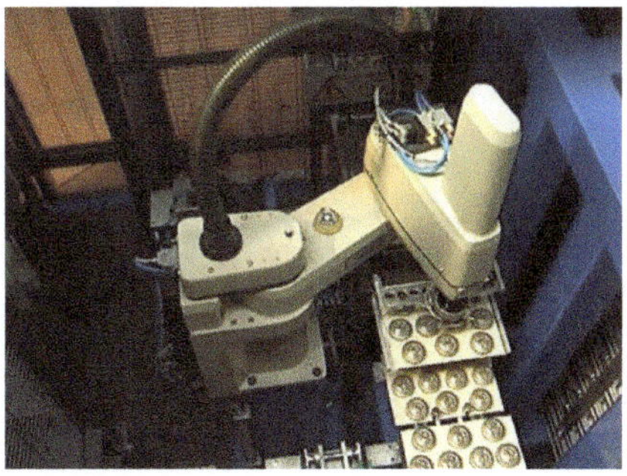

Figure 1. Configuration of industrial SCARA robot

This is particularly aimed to initial handling and palletizing operations i.e. consumer storage facilities to transport of packaging respectively to a pre-molded or like relocations. Here can be applied configuration of industrial robot arm to move in a horizontal plane, SCARA type, see Figure 1 to Figure 5 kg capacity [2].

Configuration of industrial robots with arm in a vertical plane are used for palletizing of heavy objects such as transport pallet, see Figure 2.

Figure 2. Configuration of industrial robot for palletizing of heavy objects to transport pallet

Characteristic feature for proposal solutions of robotic workstations for palletizing food is use such size and kinematics of robot arm movements to contain the palletizing zone. Table 1 characterizes the extent of their performance.

Table 1. Characteristics of robotic workstations for palletizing

Characteristics	Value
Payload [kg]	5 - 140
DoF	4 - 5
Speed for linear movements [mm.s-1]	0.6 - 1
Speed for rotary movements [$^{O}.s^{-1}$]	90 - 140
Positioning accuracy [mm]	± 0,2 - 5
Number of programmed steps	1500
Performance of palletization [objects.h^{-1}]	300 - 1440

2. Design Phase of Workplace

In applying design phase to palletizing workplace with industrial robot, we must realize that such workplace is an essential component of a broader handling-transport system. Problem for creation such workplaces to handling is essentially driven by the general methodology for design of automated and robotic systems for engineering technology with regard to specify handling operations of palletizing facility [3].

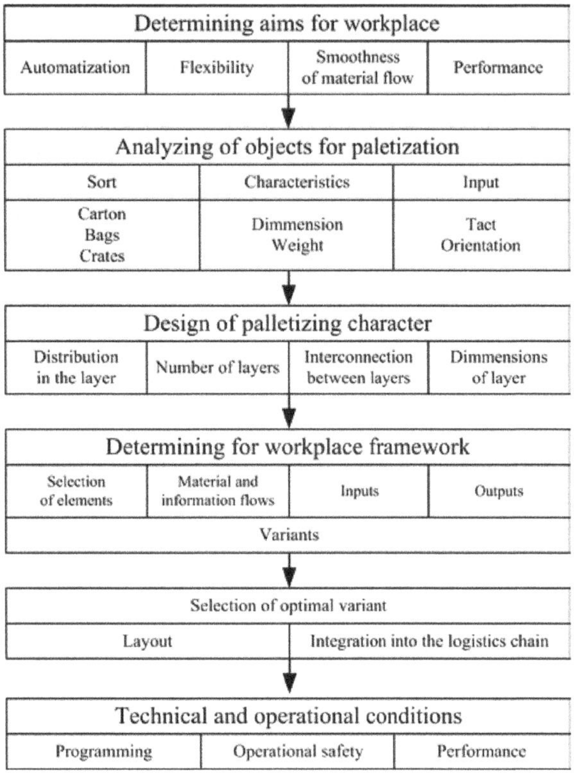

Figure 3. Etapy navrhovania paletizačného pracoviska s priemyselným robotom

Actually are currently being developed automated-based projection systems with compatible program modules based on creation of layouts, generation of control signals and programming of robot control language at robot to simulation and adaptation of data for actual workplace conditions. An effort to applying of design phases at workplace with a robot for palletizing is a one view to achieving optimal function so that on the one hand to avoid having to refill manual action and, second, that having too much untapped respectively under-utilized capacity.

The industrial robot is an essential component for deployment of palletizing workplaces with frequent changes of objects palletizing, their complex positioning and orientation, the irregular imposition and implementation respectively control of various other functions in workplace as an example is evaluation of damaged object and its disposal, orientation sensing, etc. [4].

Palletizing basis is a concentration of material in a single integrated unit called handling unit. Handling unit represents a packaged and unpackaged material stored in bulk or as a bundle on palette with which it can be handled as one coherent piece. It is determined by size (L x W x H) weight and volume of stored material. Pallet load space is limited to 1 m^3, above this threshold we are already talking about containers.

On basis of generalization at previous experience and knowledge for design and operation of robotic workstations for group packaging and palletizing as well as in other workplaces with robotic technology, it is possible to create phases of designing robotic workplace for palletizing with the following points:

- Setting of targets to be achieved by palletizing,
- Strategic decisions on transforming for manipulate-transport process,
- Analysis of object palletization and handling-transporting conditions process,
- Selection of elements workplace taking into account with regard to compatibility to its surroundings,
- Design of structure and spatial distribution of workplace elements,
- Determination of requirements for inputs,
- Providing of links to surroundings.

A determinant step in design phases for designing of robotic workplace for palletizing are unwinding from the proposed bond for palletizing and knowledge of basic types of structures with palletizing workplaces See Figure 3.

3. Methodology for Designing Workplace

In process of the methodology for designing workplace with palletizing robot is a mutually interconnected phases, where it is necessary to solve following aspects:

1. Method of attachment of object by palletizing grippers,
2. Determining the number of objects in the layer sequence, distribution and storage,
3. Determination of max. number of objects in the layer.

Method of attachment of object by palletizing grippers (1):

During applied method of attachment of object by palletizing grippers is the effort towards use of pallet storage areas to maximum extent possible. Advantage is usually where it is possible to make a regular pattern of a palletizing object.

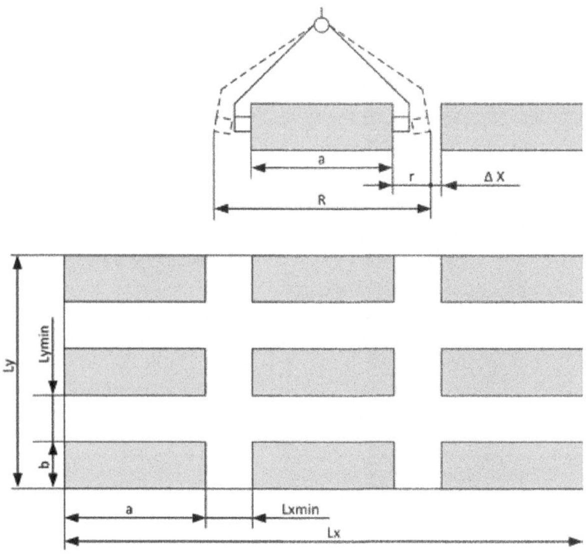

Figure 4. Proposal for distributing of the object and method of attachment by using claw gripper

For irregular pattern would increase demand for management activities, kinematics of gripper of an industrial robot, but on the other hand, in addition to achieving high utilization of pallet surface reaches the so-called interconnection facilities by cross-linking, thereby increasing the stability of transport [5].

Proposal for distributing of the object and method of attachment by using claw gripper is shown in Figure 4.

Determining the number of objects in the layer sequence, distribution and storage (2):

The calculation of distribution for quadratic objects on a pallet in X-axis we can calculate by equations:

$$M_x = \frac{L_x - a}{a + l_{x\min}} + 1 \quad (1)$$

Where

$$L_{x\min} = r + \Delta x = \frac{R}{2} - \frac{a}{2} + \Delta x \quad (2)$$

The calculation of distribution for quadratic objects on a pallet in Y-axis we can calculate by equations:

$$M_y = \frac{L_y - b}{b + l_{y\min}} + 1 \quad (3)$$

Where

$$l_{y\min} = \Delta y \quad (4)$$

Where:

M_x	Number of objects (direction X)
M_y	Number of objects (direction Y)
L_x	Length of pallet (direction X)
L_y	Length of pallet (direction Y)
$L_{x\min}$	Min. distance between objects (direction X)
$L_{y\min}$	Min. distance between objects (direction Y)
R	Max. opened gripper of released object
r	Distance of gripped surface from gripper shape during its opening
Δx, Δy	Accuracy for robot positioning

Determination of max. number of objects in the layer (3):

In this step, total maximum number of layers of objects based on the permissible load pallets and on this basis calculated allowable number of layers. In these calculations are based on the weight of objects palletizing and distribution of selected objects in the layer. The maximum number of objects in the layer depending on load pallets calculated as follows:

$$p_M = \frac{Q_p}{m_o} \quad (5)$$

Where:

p_M	Max. number of objects in layer
Q_p	Payload of pallet
m_o	Weight of object

Determination of number of layers:

$$v = \frac{p_M}{p_V} \quad (6)$$

Where:

V	Number of layers
p_M	Max. number of objects in layer
p_v	Number of objects at one layer

The number of layers can be rounded up, i.e. the last layer is not complete. The number of layers calculated according to the equation:

$$H_z = V \times C \quad (7)$$

Where:

H_z	Height of layer
V	Number of layers
C	Height of object for palletization

The follow-up is determination of max. number of layers as is a layer stability control and testing capacity and limit values set out from surrounding heights. During testing of layers according to above criteria arising from environment is a fact or a set number of objects or the height fits in terms of load transport system. Capacity transport system can be calculated on the basis of:

$$p_V = \frac{Q_V - m_p}{m_o} \quad (8)$$

Where:

p_v	Number of objects at one layer (regards to conveyor system)
Q_v	Payload of conveyor system
m_p	Weight of pallet

The final step to determination max. number of layers according to the above criteria resulting from temperature control is to state of capacity at the outlet conveyor respectively load at rack stackers or other equipment. With regard to this we fix the amount of cells in a warehouse respectively the height of the means of transport:

$$V_B = \frac{H_B}{C} \quad (9)$$

Where:

V_B	Number of layers (regards to height of cell storage system)
H_B	Height of cell storage system

The calculated values and determined requirements for number of objects in the layer and its amount is compared with the value P_m and V respectively Hz. In case some are smaller than in the p_m and after reference point for number of objects and allowing height of the layer is taken to be the smallest value. The stability control is generally based on the principle of the practice verified that the object is relatively small, several layers palletizing and we choose the cross-linking, if possible.

The exact calculation requires knowledge of many situations and parameters such as speed of traffic, acceleration, deceleration, material and coefficient of friction, vibration etc. In fact palletized layer is almost always fixed in order to protect what is sufficiently ensured the stability.

With manual disassembly of layers need to be designed layers in terms of compliance with ergonomic criteria such as easy accessibility to the building, easy to grip, i.e.

security that the layer does not fall apart during withdrawal of the object and so on.

4. Conclusion

Article offers a methodology for proposed structure of palletizing workplace with industrial robot according to the calculations of input characteristics of object for palletizing, palletizing design and arrangement of layers in the layer or number of layers and their linkages to the palletization objects.

The typical structures of palletizing workplaces with industrial robots are either with one or two palletizing places. The decisive criteria for the choice of the variants are particularly time aspect.

The main core of workplace is industrial robot whose function is to carry out handling operations due to a change of position respectively position and orientation i.e. take or object respectively objects more of homogeneous species or species-diverse from a place and move them to another location by a predetermined spatial mosaic with a specific sequence.

In determining the structure of the palletizing workplace with industrial robot should be given more scope and issues associated with determining type of structure palletizing of workplace, selection of elements of workplace with special attention to the selection of an industrial robot and shape of the workplace in terms of its spatial arrangement of elements and determination of material and information bonding connections.

Acknowledgement

This contribution is the result of the project implementation: 059TUKE-4-2014 Rozvoj kvality života, tvorivosti a motoriky hendikepovaných a starších osôb s podporou robotických zariadení.

References

[1] Tolnay, M., Bachratý, M.: Prevádzka a údržba výrobných systémov. 1. vyd., Bratislava: FX s.r.o., 2008.

[2] Semjon, M., Hajduk, M., Vagaš, M.: Koncipovanie výrobných systémov na báze zlučiteľných modulov, in: Robotics and Intelligent Manufacturing Systems: zborník príspevkov z medzinárodnej konferencie: 27. jún 2013, Košice., TU, 2013 pp. 19-24.

[3] Binka, J., Hricko, J.: Compliant Mechanisms Exploited as Micro-Grippers. In: State-of-the-Art in Mechatronics, Volume II. Alpen aan den Rijn, 2010. Pp.119-144.

[4] Taranenko G, Taranenko W, Świć A, Szabelski J. Modelling of dynamic systems of low-rigidity shaft machining. Eksploatacja i Niezawodnosc – Maintenance and Reliability 2010; 4 (48): 4-15.

[5] Beniak, J., Križan, P.,Matúš, M., Kováčová, M.: The operating load of a disintegration machine. In: Acta Polytechnica. Vol. 54, No. 1 (2014), s. 1-5.

Remaining Life Assessment of Power Transformer

Sonali Mayuresh Akolkar*, **Bansidhar Eknath Kushare**

Electrical Engineering Department, K.K.W.I.E.E. & R, Nashik, India
*Corresponding author: akolkar_sonali@yahoo.co.in

Abstract Power Transformers are the most vital components in a sub-station / Receiving station. Failure of a Transformer leads to loss of revenue besides affecting reliability of power supply to consumers. It can lead to non-availability of the transformer for long durations. In this paper, an approach to evaluate transformer's aging condition is introduced based on multi-parameters. Firstly, the different types of insulations in transformers like solid insulation and liquid insulation have been discussed. Later, the ageing process of these insulations used in transformer has been illustrated in detail. Different condition monitoring techniques used for transformer are included to assess the life of insulating material. A program is developed in Visual Basic Software which is very user friendly software. Through the developed program, we can predict the remaining life of transformer. Moreover, a software system based on transformer's electrical and thermal parameters is developed correspondingly, by using a multi- parameters analytic approach. This system is expected to help in planned maintenance of transformer on fields. This can help the utilities in making optimum use of the transformers and also taking timely decisions regarding refurbishment / replacement of transformers. Various transformer insulation properties like electrical properties, oil quality and temperature are considered in assessment of remaining life of transformer.

Keywords: *transformer insulation, insulation aging, condition monitoring, dissolve gas analysis*

1. Introduction

Being one of the most expensive and important equipment in electrical power transmission and distribution (T&D) system, power transformer plays a vital role in maintaining a reliable and efficient electricity supply. Since transformer is continuously subjected to thermal, electrical, mechanical and chemical stresses etc. during its operation, it is necessary to include comprehensive parameters when establishing aging assessment system [1].

In order to ensure that Power Transformers provide long and trouble-free service, several diagnostic tests are carried out and remedial actions initiated throughout their operational lifetime. For the oil-filled Transformers, more particularly which are in service for more than fifteen years, it is advisable that we should also estimate the residual life of the transformers. It can be done by assessing the extent of degradation of the insulation used in transformer. This calculation considers not only typical test results such as dissolved gas analysis (DGA), oil quality, and dissipation factor, but also other parameters such as top oil temperature and hot spot temperature estimation.

The insulation system of a power transformer is understood as the complete internal assembly of dielectric insulating materials. This includes parts and supporting structures that cover the winding wires, insulate the turns from each other in each winding, separate different winding bodies from each other and from the core and tank.

The transformer insulation system can be categorized into major insulation system & and minor insulation system. The major insulation consists of insulation between windings, between windings and limb/yoke, and between high voltage leads and ground. The minor insulation consists of basically internal insulation within the windings, viz. inter-turn and inter-disk insulation.

The gap between low voltage (LV) and high voltage (HV) windings is subdivided into many oil ducts by means of solid insulating barriers. The insulation system of oil cooled power transformers consists of combination of oil and solid insulations (paper and pre-compressed board).

1.1. Transformer Oil

Transformer oil comes under liquid insulation. Mineral oil is the major type of liquid insulation used in electrical equipment. Oil used for insulation in transformers is mineral oil and it is obtained by refining crude petroleum. Mineral oils were in use as liquid dielectrics in electrical equipments for over hundred years now. Despite the availability of variety of synthetic oil, with far more superior properties, mineral oil held its way due to their abundant availability and economy. Particular oil will contain a mixture of many different molecular species and types of carbon atoms. The differences in the chemical composition will result in differences in physical properties and in the chemical behavior of the oils. The chemical composition has profound effects on the physical characteristics of the oil.

2. Composition of Oil

Mineral oil can vary greatly in its composition. All mineral oils are mixtures of hydrocarbon compounds with about 25 carbon atoms per molecule. Crude oils from some sources are higher in *paraffinic* compounds, whereas others are higher in *naphthenic* compounds. Crude oils also contain significant amounts of aromatic and poly aromatic compounds. The refining of crude oil for the production of dielectric fluids reduces the aromatic and poly aromatic content to enhance the dielectric properties and stability of the oil.

3 Insulating Paper

Oil impregnated paper, Kraft paper, pressboards, etc come under solid insulation of transformer. Power transformer windings are most commonly insulated with multiple layers of Kraft paper immersed in mineral oil. The paper insulation is required to withstand both electrical and mechanical stress. The oil provides both insulation and cooling. Paper and pressboard insulations are made from organic materials called cellulose fibers obtained from vegetable sources including cotton, hemp, manila, straw, wood and coniferous/ deciduous trees because the cell of such plants consist mainly of cellulose. These sources therefore felted to form a fabric, a mat, a web or a sheet of paper according to their application. The Kraft process reduces unwanted constituents of wood pulp such as lignin and hemicelluloses. The main constituent of Kraft paper is cellulose and the remaining components of the paper consist of lignin, hemicelluloses. The insulation is dried after winding. The dried paper is impregnated with insulating oil, which increases its dielectric strength and also serves to cool the windings.

4. Cellulose

The cellulose fibers consist of a bundle of cellulose molecules of different lengths which lay side by side. They are held together by hydrogen bonds involving the hydroxyl group (-OH group) on the adjacent molecules as in Figure 1. Cellulose itself is a linear polymer construct by a long straight chain of anhydrous glucose units which are linked together through the glycoside bond. The letter 'n' from Figure 1 represents the length of cellulose molecule and is measured in terms of the degree of polymerization (DP).

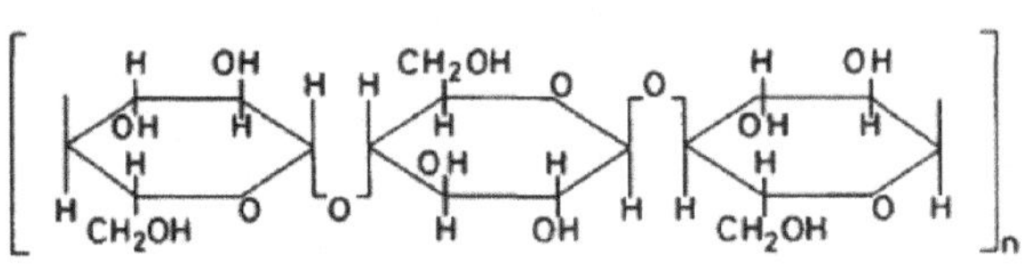

Figure 1. Structure of Cellulose

5. Ageing of Transformer Insulation

Most transformer failures can eventually be related to the deterioration of insulation with time. For liquid-filled transformers, the traditional insulation system is oil impregnated (cellulose) paper. The dielectric strength of the oil-impregnated paper is greatly affected by water contamination. The cellulose insulation must also maintain its tensile strength to be able to withstand the stresses that occur with surges, and must be able to withstand detrimental chemical transformations. Oxygen and water are the primary agents that degrade cellulose insulation. Heat acts as catalysts and accelerates the reactions in producing various oil-degradation products.

Paper aging in transformers is not uniform and follows thermal and moisture gradients. In addition, external layers are more exposed to higher concentrations of oxygen and the byproducts of aging in the oil. There are three main factors which promote paper degradations, which are thermal, oxidative and hydrolytic i.e.:

1. Hydrolysis
2. Oxidation
3. Pyrolysis

Historically, the loss of life of insulating paper has been determined by the reduction of its tensile strength. A 50% reduction of tensile strength has been used as the criteria for a long period of time. However, this has produced very conservative results. An alternative method using the degree of polymerization (DP) of the insulating paper has also been used to determine the electrical insulation life. Most recently, the relative dielectric constant and loss tangent methods are also being investigated by monitoring changes in capacitance and conductance as a function of frequency to determine the thermal age of a power as stated in the IEEE Standard C57.91-1995[41. Transformer insulation deteriorates as a function of time and temperature. Since the temperature distribution in most transformers is not uniform, the most common practice is to consider the aging effects produced by the hottest-spot temperature (θ_{hs}). Various transformer temperatures such as top oil, radiator surfaces, and cooling medium can be measured directly to estimate the transformer hottest spot temperature. Normal life expectancy is said to occur when transformers with a rated average winding rise of 65°C are operated continuously with hottest spot temperatures of 110°C. This value, (θ_{hs}) can be expressed in two different ways. It is based on the sum of the ambient temperature (θ_a) and the average winding temperature rise ($\Delta\theta_w$) and the hottest spot temp. rise margin above the average winding temp. rise ($\Delta\theta_1$). Hence,

$$\theta_{hs} = \theta_a + \Delta\theta_w + \Delta\theta_1 \tag{1}$$

$\Delta\theta_1$ = 15°C for 65° average winding temperature rise.

To estimate accurately the hot-spot, winding and oil temperatures, one needs to consult the loading guides. There are two popular guides: the IEEE guide and the IEC guide. The IEC guide is applicable mainly for the non-thermally upgraded paper and the hot spot temperature is limited to 98°C at 20°C ambient temperature (78°C hot spot rise). The guide briefly mentions, without any further details, the use of thermally upgraded paper for 110°C hot spot temperature. Both standards utilize the top-oil temperature and the hotspot rise over top-oil for the calculation of the hot-spot temperature. However, the IEC uses the bottom-oil instead of the top-oil for direct-forced-oil flow transformers. The IEC has a limited discussion on loss-of-insulation-life and there is only relative-aging rate that is used to compare aging due to overload [6].

6. Condition Monitoring Methods for Power Transformer

The condition monitoring of transformer cab be carried out using following methods:

6.1. Measurement of dissipation factor (tan δ) and capacitance

The dissipation factor of power factor is an important source of data to monitor transformer and bushing conditions. This test is performed to determine the condition of capacitive insulation between different winding and compartments.

6.2. Dissolved Gas Analysis (DGA)

It is widely used to detect incipient faults in transformers. The main gases formed by decomposition of oil and paper are hydrogen, methane, ethane, ethylene, acetylene, carbon monoxide, carbon dioxide, oxygen and nitrogen. These gases dissolve in oil or accumulate above it and are analyzed by DGA. Some laboratories also report the contents of C3 and C4 hydrocarbon gases formed. DGA is the most widely used technique for detecting and monitoring faults in electrical equipment.

6.3. Insulation Resistance

Insulation resistance measurement may be the simplest field testing but also useful one that could determine whether tested object is integrally moistened or inter-connectively deteriorated. Generally, R_{60}, which is equal to resistance value of insulation after applied potential is subjected for 60 seconds, is an integrant; additionally absorptance (Ka) and polarization index (PI) are also needed to have an overall judge. It is worth noting that temperature translation is necessary before ranking to get normalized values.

6.4. Leakage Current

Leakage current measurement has the same testing circuit as insulation resistance measurement, but with much higher applied voltage and higher sensitivity which are easier to detect insulating defects and deterioration.

7. System Flow Chart

Figure 2 provides a summary of the scoring system and a flowchart of the main condition parameters that are used in this study for aging condition assessment. Figure 3 depicts result of developed software system for assessment of remaining life of new power transformer whose expected remaining life is 100% ideally. In the software developed in Visual Basic, standard acceptable values of different parameters related to testing of transformer insulation are considered. The testing considers electrical testing, oil testing, oil quality, temperature. An algorithm is able to calculate remaining life of transformer in hours or in percentage considering different parameters related to its insulation. It is observed that the remaining life of a new transformer is 99.99% (ideally 100%) considering all the insulation testing parameters as per standard requirement.

The same software system can be used to judge remaining life of any transformer on site. As well, same system is able to predict remaining life of same transformer even if some of testing parameters are not available considering the weighting factors for other available parameters in increased proportion.

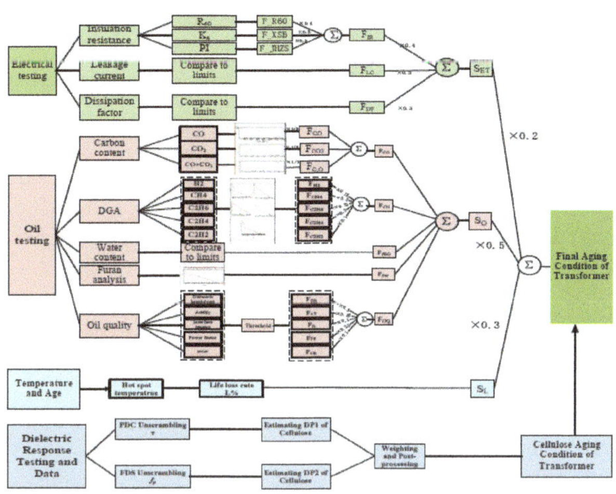

Figure 2. Flowchart for assessment of transformer aging with multi parameters [1]

Table 1. Standard Value Of Various Parameters Related To Power Transformer

Properties	U≤66kV	66<U≤22kV	U>220kV	Score
Dielectric strength(kV)	≥40	≥47	≥50	100
	35-40	42-47	45-50	80
	30-35	35-40	40-45	60
	<30	<35	<40	40
Acidity (mg(KOH/g))	≤.05	≤0.04	≤0.03	100
	0.05-0.1	0.1-0.15	0.03-0.07	80
	0.1-0.2	0.1-0.15	0.07-0.1	60
	≥0.2	≥0.15	≥0.1	40
IFT(25®C)	≥25	≥30	≥35	100
	20-25	23-30	25-35	80
	15-20	19-23	20-25	60
	<15	<19	<20	40
Tan δ	<1.5	<1	<0.5	100
	1.5-3	1-1.5	0.5-1	80
	3-4	1.5-3	1-2	60
	≥4	≥3	≥2	40
Colour	Light yellow			100
	Yellow			80
	Dark yellow			60
	Chocolate brown			40
Leakage current	100µA			
Power factor	0.8			

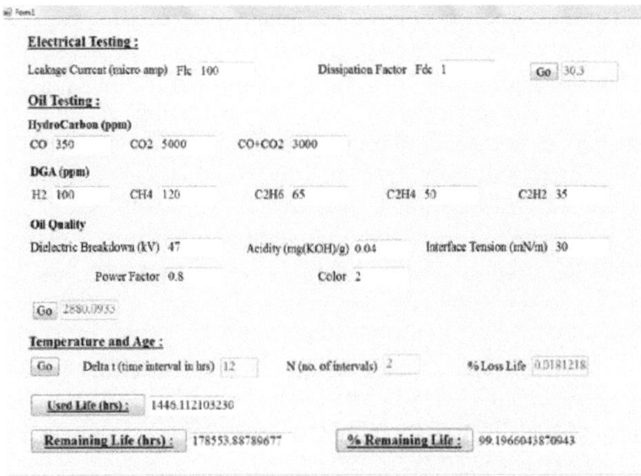

Figure 3. Result for remaining life of new transformer

8. Conclusion

In this paper, an approach to evaluate aging condition of power transformer is introduced based on multi-parameters. 'Aging Index' as a useful tool combines not only electrical testing and oil quality testing but also hot spot temperature to estimate transformer's aging rate. A corresponding software system is developed to help make condition based maintenance plan and replacement of transformers. Various transformer insulation properties like electrical properties, oil quality and temperature are considered in assessment of transformer's life. Weighting factors is a useful tool which combines not only electrical testing, oil quality testing but also hot spot temperature to estimate transformer's aging rate. The final total score represents aging condition of transformer. This assessment of remaining life of transformer helps us to predict the time duration for which a transformer has left in service.

Acknowledgment

The author is thankful to Prof. Dr. B.E. Kushare, Head of Electrical Engineering Department - K.K.W.I.E.E.& R., Nasik, for his contributions, time to time encouragement and guidance in the project work and also to the Principal and Management of college for providing facilities to present this work.

References

[1] Yuan Li, Ming-jie Tang, Jun-bo Deng, Guan-jun Zhang, Shu-hong Wang, "An Approach to Aging Assessment of Power Transformer Based on Multi-parameters", 2012 IEEE International Conference on Condition Monitoring and Diagnosis, 23-27 September 2012, Bali, Indonesia.

[2] Qiang fu, Zhi li, Vongping lin, "Recent advances on analysis and explanation of Aged insulation in operating power transformers".

[3] Hannah Ahmad Rosli, Mohd Aizam Talib,"Condition and life assessment of transformers with specific application to power station transformers", ICEE - 2006.

[4] Sen, P.K. Dept. of Eng., Colorado Sch. of Mines, Golden, CO, USA, "Overloading and loss-of-life assessment guidelines of oil-cooled transformers", IEEE conference 2001.

[5] D.M. Allan "Practical life-assessment technique for aged transformer insulation", IEEE Proceedings-A, Vol. 140, No. 5, September I993.

[6] IEEE Guide for Loading Mineral-Oil-Immersed Transformer up to and Including 100 MVA with 55°C and 65°C Winding Rise, IEEE Standard C57.91-1995.

Analysis and Design of a Three Inputs Fuzzy System for Maintaining the Cane Level during Sugar Manufacturing

Yogesh Misra[1,*], **H R Kamath**[2]

[1]Mewar University, Chittorgarh, Rajasthan, India
[2]Malwa Institute of Technology, Indore, Madhya Pradesh, India
*Corresponding author: yogeshmisra@yahoo.com

Received September 20, 2014; Revised November 01, 2014; Accepted November 04, 2014

Abstract The amount of cane fiber carried by cane carrier in a sugar mill varies due to non-uniformity of cane supply. The continuous variation of cane in Donnelly chute during the cane juice extraction inversely affects the cane juice extraction efficiency of mill. This paper deals with the development of algorithm for a three input fuzzy controller with an aim to maintain the cane level in Donnelly chute during cane juice extraction. The developed controller generates the required rake carrier motor speed in rpm depending upon the value of cane level in Donnelly chute, quantity of cane on rake carrier and the roll speed. The three inputs fuzzy controller is developed and simulated for six cases by using fuzzy logic toolbox of 'MATLAB® version 7.11.0.584 (R2020b).

Keywords: *sugar mill, cane carrier, Donnelly chute, juice extraction, fuzzy algorithm, fuzzy controller*

1. Introduction

Sugar is one of the oldest commodities originated from India in 4th century AD. Indians first discovered how to manufacture sugar. Now a days the sugar industry processes sugar cane and sugar beet to manufacture edible sugar. More than sixty percent of the world's sugar production is from sugar cane; the balance is from sugar beet [1]. Sugarcane plays a very important role in the economy of Indian farmers as it is the only input used by sugar mills to produce sugar in India. There is similar process for sugar making adopted in the entire sugar mill in India.

The sugar making process flow is shown in Figure 1. During sugar manufacturing the cane billets are converted into cane fiber and this cane fiber is crushed for the extraction of cane juice. The residual of fiber after juice extraction is used as fuel in boiler house for the generation of steam. The cane juice extracted contains impurities and the clarification of the cane juice is the next step. The juice clarification process converts the impurities into a thick paste of mud which settle down in the clarifier vessel while clarified juice overflows the clarifier vessel. The separated mud is called as filter cake and used as fertilizers. The thin clarified juice is concentrate to heavy syrup by boiling it from a series of evaporators at high vacuum. The heavy syrup contains 65% of sugar. The heavy syrup, which is also called as mother liquor, is seeded with small sugar crystal and further concentrated by boiling in vacuum pan. This process of crystal growth by adding seed crystal in syrup is called crystallization. The mixture of mother liquor and grown crystals in pan is called as massecuite. Crystals of required size are grown typically in four hours. After the growth of sugar crystals the massecuite is discharged in receiving tank. The mother liquor separated from sugar crystals during centrifugal process is called as molasses. The raw sugar from the centrifugal is dried by tumbling through a stream of air in a rotating drum. The dried sugar is then stored and ready to use [2,3].

Cane juice extraction is a very important process during sugar making. The supply of cane for processing is very uneven and this uneven supply of cane during juice extraction adversely affects the sugar mill efficiency and causes mill breakdown, stoppage and jamming [2].

In this paper we have developed an intelligent three input fuzzy controller which maintains the cane level at desired height in Donnelly chute during cane crushing by nullifying the variation in cane supply and roll speed. The fuzzy controller for maintaining the cane level in Donnelly chute generates the signal for controlling the speed of motor driving the rake carrier depending upon the cane amount on rake, level of cane in Donnelly chute and roll speed.

In Section II we have discussed the cane juice extraction process and some important terminology used in sugar mills. In Section-III we discussed the three inputs fuzzy controller design philosophy and algorithm. The three inputs fuzzy controller is simulated for six different cases by using fuzzy toolbox of 'MATLAB® version

7.11.0.584 (R2020b). In Section-IV the simulation results of the fuzzy controller is discussed and finally Section-V contains the concluding remark.

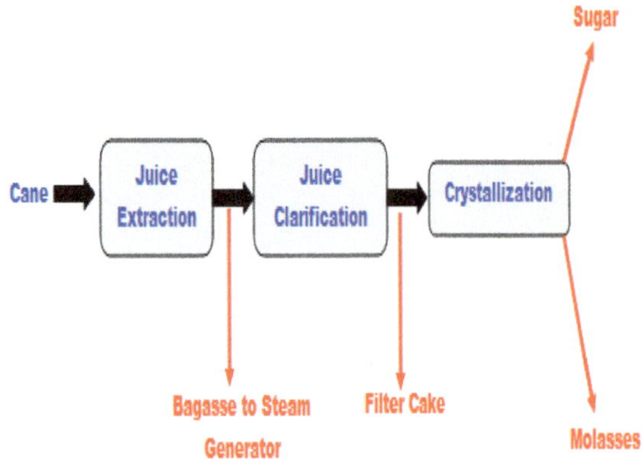

Figure 1. Sugar Making Process

2. Cane Juice Extraction and Important Mill Terminology

A schematic of cane juice extraction process is shown in Figure 2. Cane juice extraction is very important process of a sugar mill which affects the sugar mill's efficiency. The cane billets received by sugar mill from cane growers are weighted and dumped in cane yard. From the cane yard the cane billets are dumped in the cane carrier with the help of a crane. The cane carrier moves continuously and is responsible for bringing the cane to the factory for sugar production. The cane is first passed through two sets of rotating knives. The first one is called cane knives and it cut the cane into pieces. The second set is called as shredder knives and it prepare the fiber of cane billets. The small fibers are around 1-2 cm. The cane fibers are feed to Donnelly chute via a rake carrier and cane juice is extracted by crushing fiber in two or three rolls of the mill. This process is repeated through sets of five/six mills until last mill is reached. Before the crushing of cane fiber in last mill it is immersed in a diluted mixture of final mill cane crushing juice and hot water. The residual cane fiber after the last mill is called bagasse is conveyed to bagasse bin and finally to the boiler station where it is used as fuel [2,3,4].

In a sugar mill for optimum juice extraction and better sugar production efficiency the cane level in Donnelly chute is to be maintained constant. The height of cane level above the mill depends upon various parameters like roll diameter, work opening, surface speed of roll, feed depth, contact angle between the roll, escribed volume and chute inclination with respect to horizontal.

Cane fiber is carried by rake carrier and finally dumped in Donnelly chute from where it is crushed by the rollers. The amount of cane fiber carried by rake carrier varies due to non-uniformity of cane supply. If the level of cane fiber falls below the desired level then the speed of rake carrier motor is to be increased and if the level of cane fiber rises above the desired level the speed of rake carrier motor is to be decreased for maintaining the desired level. The variation of desired level of cane in chute is also affected by the variation of roll surface speed. The steam turbines are used to rotate the rolls in sugar mill. The final product left out after the extraction of juice from the milling train is called baggasse. This baggase is used as fuel to produce steam and this steam is used to run turbines. In a sugar mill the supply of steam to run turbine is not uniform therefore the rotational speed of rolls vary.

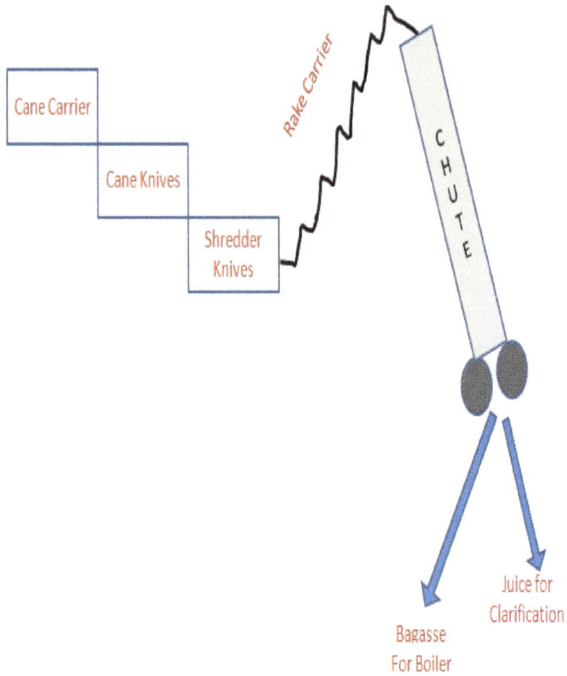

Figure 2. Schematic of Juice Extraction Process

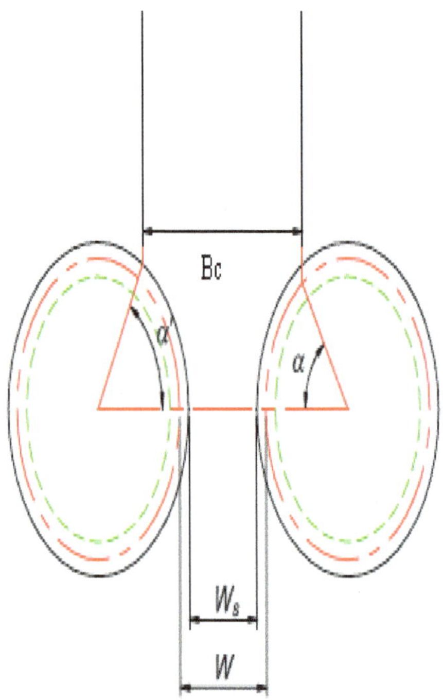

Figure 3. Two Rolls and Chute Arrangement of a Mill

Two rolls and the chute arrangement used for cane crushing are shown in Figure 3. It has been investigated that the physical structure of mill effect the feed depth at which maximum crushing rate can be achieved [5]. The diameter of roll when measured from the tip of groves is Do and Dg is the length of groves and D is the average diameter of roll. The mean diameter of roll is given as:

$$D = D_o - D_g \quad (1)$$

Where D_o is the outside diameter of roll and D_g is Groove Depth. The opening measured between the two rolls outside diameter is called as nib opening or set opening. The opening measured between the mean diameters of two rolls is called work opening and is given as:

$$W = W_s + D_g \quad (2)$$

Where W_s is nib opening. The surface speed of roll is given as:

$$S = (\pi \times D \times N)/60 \quad (3)$$

Where S is surface speed of roll in cm/s and N is roll shaft speed in rpm. The thickness of cane blanket at the feed opening of the mill effects the juice extraction from the mill. The optimum feed depth is investigated and found as follows:

$$B_c = (W + D)/2 \quad (5)$$

Where B_c is the optimum feed depth in cm [11]. The contact angle is the angle between the line joining the center of the two rolls and the line joining the center of roll to the point where chute touches the roll [12]. The contact angle is given as:

$$\text{Cos } \alpha = (D + W - B_c)/D \quad (6)$$

The escribed volume is the volume of prepared cane passing through the work opening of the mill [12] and is given as:

$$V_e = L_r \times D \times S[1 + (W/D) - \text{Cos } \alpha) \text{ Cos } \alpha \quad (7)$$

Where V_e is escribed volume in (m^3/s) and L_r is roll length in cm. The average speed of cane blanket at the point where chute touches the rolls is given as:

$$S_f = S \text{ Cos } \alpha \quad (8)$$

Where S_f is the average speed of cane blanket in cm/s. At the entry of chute the volume of cane passing the entry plane is given as [13]:

$$V_e = L_r \times B_c \times S \text{ Cos } \alpha \quad (9)$$

The fibre rate is the amount of fibre crushed by mill in one second and it is given as:

$$Q_f = (Q_c \times f)/\ 100 \quad (10)$$

Where Q_f is the fibre rate in Kg/s, Q_c is cane crushing rate in Kg/s and f is the percentage of fibre present in cane [6,7].

The prepared cane is carried by cane carrier and dumped in chute. The chute is inclined to horizontal and the angle of inclination and the dimension of chute vary from one mill to other. The length of chute (L_c) is 180cm, width (W_c) is 43.5cm and depth (D_c) is 183cm. The Roll length (Lr) is 183cm, roll diameter (D) is 75.5cm and work opening (W) is 11.45cm. The optimum angle (α) calculated from (6) is 61°. The optimum feed depth (B_c) calculated from (5) is 43.5cm.

It is required to select parameters for a mill which can crush 2000 tons cane per day (tcd). Allowing about 10% excess crushing, the maximum mill capacity should be 2200 tcd. For achieving 2200 tcd crushing of cane the mill must be able to crush 26.6Kg/s cane [8]. The amount of cane crushed by mill in one second is termed as flow rate and denoted as Q_c in this paper. We can relate flow rate with cane density and escribed volume as follows:

$$Q_c = \rho_c V_e \quad (11)$$

Where V_e is escribed volume and ρ_c is density of cane (350Kg/m^3). The escribed volume calculated from (11) is 0.076m^3 /s. The surface speed of roll is calculated from (9) is 16.6cm/s. The average speed of cane blanket when it touches the roll surface is calculated from (8) is 9.5cm/s. If the cane crushing rate is 26Kg/s and the fibre percentage in cane is 15% then the fibre rate calculated from (10) is 4Kg/s. The Pressure required to feed mill is given as follows [5]:

$$p_2 = \left[\int_{\beta}^{\alpha} pv \cos\theta \tan(\theta - \beta) d\theta \right] / (Bc/D) \quad (12)$$

Where p_2 is the pressure required at chute exit, p_v is the pressure applied to the cane, θ is the angle with which chute is inclined to horizontal, β is the angle with which mill will be feed without the application of external force [11]. β is given as:

$$\text{Where } \beta = \tan^{-1}\mu \quad (13)$$

Where μ is the coefficient of friction and its value selected in this application is 0.3. Solving (12) after putting α = 61°, θ=61° and β = 20° gives as:

$$p_2 = p_v(2.042) \quad (14)$$

The p_v is given as:

$$p_v = 36.9\ (100\ C_f - 3.3)^2 \quad (15)$$

Where C_f is filling ratio and is given as:

$$C_f = \gamma / 1260 Kg/m^3 \quad (16)$$

Where γ is called compaction [9] and its value in this application is 52Kg/m^3. The value of C_f calculated from (16) is 0.041. The value of the pressure applied to the cane (p_v) is calculated as 23.6lb/ft^2 (0.01152Kg-force/cm^2) and the pressure required at chute exit (p_2) is calculated as 0.02352Kg-force/cm^2 (= 2.3KPa). In an open chute, the pressure due to fiber is given as:

$$p_2 = \rho c\ (\text{Sin } \theta - \mu \text{Cos } \theta)\ L \quad (17)$$

Where L is the height of cane in chute and the value of θ in this application is 61°. The height of cane in open chute is calculated from (17) as 92cm. In order to minimize the failure rate of fiber the cane must be maintained at 90cm in chute. The cane fiber is assumed to fail in a similar was to soils. The 'failure ratio' [10] in fiber is the ratio of maximum shear stress to shear strength of fiber. A volume of prepared cane contains fiber, air and juice. When fiber is compressed in a pair of roll then air is expressed until fiber contains only fiber and juice. Any further compression of fiber expresses juice. It has been investigated that failure rate of fiber decreases with the increment of pressure applied on fiber at mill opening but beyond certain value the failure rate starts increasing with the increment of pressure. The failure rate is minimum (0.04) when the feed pressure is 2kPa [11].

3. Three Inputs Fuzzy Controller Design Philosophy and Algorithm

The proposed three input fuzzy controller based system to control the cane level is shown in Figure 4. The three variables in juice extraction process are weight of prepared cane on rake carrier, level of cane in Donnelly chute and the rotational speed of rolls. The prepared cane is dumped in Donnelly chute of height 180cm. The Rake Carrier which carries the prepared cane up to Donnelly chute is of length 800cm, width 150cm and its weight is 500Kg. The rake carrier is run by a motor whose speed can be varied from 17rpm to 116rpm. The amount of prepared cane on rake carrier varies from 500Kg to 1000Kg. This variation of prepared cane can be measured with a load cell. Due to uneven supply of cane billets the level of prepared cane varies in Donnelly chute. This variation of cane level is measured with the help of a light sensor. The two rolls rotate in anti-clockwise direction with the surface speed in the range from 12cm/s to 16.6cm/s. The steam turbines are used to rotate the rolls in sugar mill. In a sugar mill the supply of steam to run turbine is not uniform therefore the rotational speed of rolls vary. A tachometer can be used to calculate the speed of rolls. The details of interfacing circuits with the fuzzy controller are not covered in this paper. Rather we are concentrating on the development of three inputs fuzzy control algorithm with an aim of changing the speed of rake carrier depending upon the values of the three variables so that the cane level in Donnelly chute will remain constant. It has been calculated that for maintaining the 2200tcd crushing the flow rate should be 26.6Kg/cm. It has also been calculated that for achieving 26.6Kg/s flow rate the roll surface speed should be 16.6cm/s. If we assume that the roll speed will remain same then depending on the height of cane in chute and amount of cane available in rake carrier the speed of rake carrier is controlled to vary the feed rate which ultimately maintains the cane level at 90cm. If due to the variation in steam pressure the roll speed vary then the flow rate will also change and controller must be able to nullify these variations so that the cane level should remain constant at 90cm.

In order to design a Fuzzy Inference System (FIS) to control the level of cane fibers in Donnelly chute, we must describe the operation of system linguistically.

The four main components of a FIS [13,14] are as follows:

i. Fuzzifier – It performs the function of fuzzification. During fuzzification the crisp inputs (input values from real world) is converted into a set of membership values in the interval [0,1] in the corresponding fuzzy sets.

ii. Rule Base – In the rule base the fuzzy inputs are compared based on the membership functions of each input and the rule base developed.

iii. Inference Engine – It generates the output of the fuzzy controller into linguistic variable.

iv. Defuzzifier – It performs the function of defuzzification. During defuzzification the fuzzy outputs of the inference engine is again converted into crisp form so that it can be used in real world [12,13].

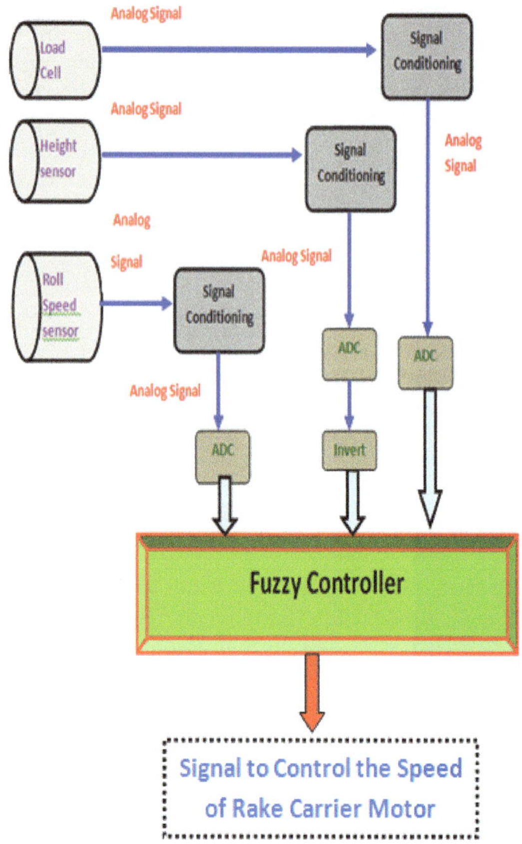

Figure 4. Three Inputs Fuzzy Controller to Maintain Cane Level

The input (crisp or numerical) data are fed into fuzzy logic rule based system where physical quantities are represented into linguistic variables with appropriate membership functions. These linguistic variables are then used in the (IF-Part) of a set of fuzzy. "IF THEN" rules within an inference engine to result in a new set of fuzzy linguistic variables or consequent (THEN-Part). Fuzzifier maps crisp numbers into fuzzy sets. Fuzzifier is used to activate the particular rules having the linguistic variable associated with specific fuzzy set. Inference Engine work just as we humans think before coming to decisions.

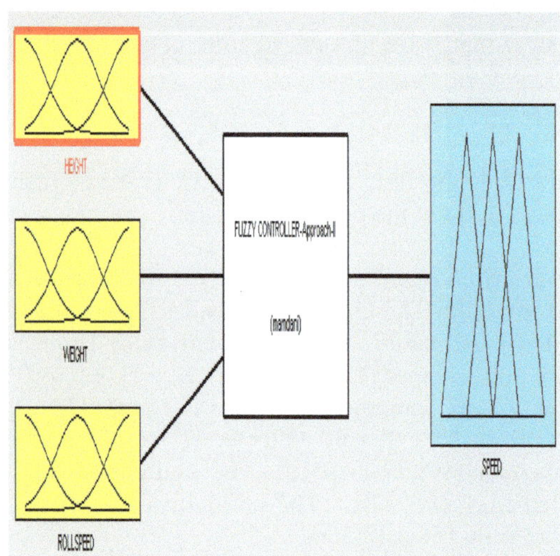

Figure 5. Text Block of Three Inputs FIS

A text block of FIS is shown in Figure 5. Three input parameters and one output parameters selected for the modeling of fuzzy controller are **'HEIGHT'**, **'WEIGHT'**, **'ROLLSPEED'** and **'SPEED'** respectively. All the membership functions in input and output parameters in this Algorithm are triangular.

Input parameter **'HEIGHT'** represents the level of prepared cane in Donnelly Chute in centimeter, **'WEIGHT'** represents the weight of cane in cane carrier in Kilogram and **'ROLLSPEED'** represents the speed of roll of the mill in cm/s. The output parameter **'SPEED'** represents the speed of the rake carrier motor in rpm.

A. Fuzzification of 'HEIGHT'

The universe of discourse of input parameter **'HEIGHT'** is in the range 0 to 180cm and it is fuzzified into seven triangular linguistic variables as shown in Figure 6 and represented in Table 1. It has been calculated that the level of cane in the chute should be maintained at 90cm for proper feeding of prepared cane in the mill. Therefore during the selection of membership function it is decided that the 90cm height of cane in chute must possess membership value 1 for linguistic variable JR (Just Right).

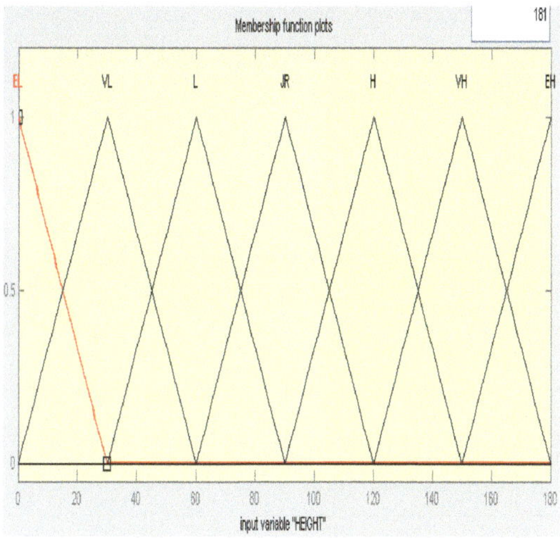

Figure 6. Membership function of 'HEIGHT'

Table 1. Linguistic Variables of Input Parameter 'HEIGHT

S.No.	Linguistic Variable	Range (cm)	Interpretation
1	EL	[0 30]	Extreme Low
2	VL	[0 30 60]	Very Low
3	L	[30 60 90]	Low
4	JR	[60 90 120]	Just Right
5	H	[90 120 150]	High
6	VH	[120 150 180]	Very High
7	EH	[150 180]	Extreme High

B. Fuzzification of 'WEIGHT'

The universe of discourse of input parameter **'WEIGHT'** is in the range 500Kg to 1000Kg and it is fuzzified into eleven triangular linguistic variables as shown in Figure 7 and represented in Table 2.

C. Fuzzification of 'ROLLSPEED'

The universe of discourse of input parameter **'ROLLSPEED'** is in the range 12cm/s to 16.6cm/s and it is fuzzified into three triangular linguistic variables as shown in Figure 8 and represented in Table 3. The triangular membership functions of each linguistic variable of all input parameters are selected such that each one overlaps 50% with its neighbor.

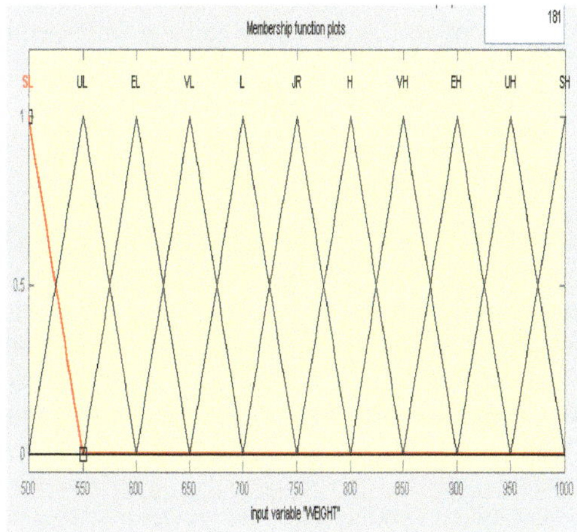

Figure 7. Membership function of 'WEIGHT'

Table 2. Linguistic Variables of Input Parameter 'WEIGHT

S.No.	Linguistic Variable	Range (Kg)	Interpretation
1	SL	[500 550]	Super Low
2	UL	[500 550 600]	Ultra Low
3	EL	[550 600 650]	Extreme Low
4	VL	[600 650 700]	Very Low
5	L	[650 700 750]	Low
6	JR	[700 750 800]	Just Right
7	H	[750 800 850]	High
8	VH	[800 850 900]	Very High
9	EH	[850 900 950]	Extreme High
10	UH	[900 950 1000]	Ultra High
11	SH	[950 1000]	Super High

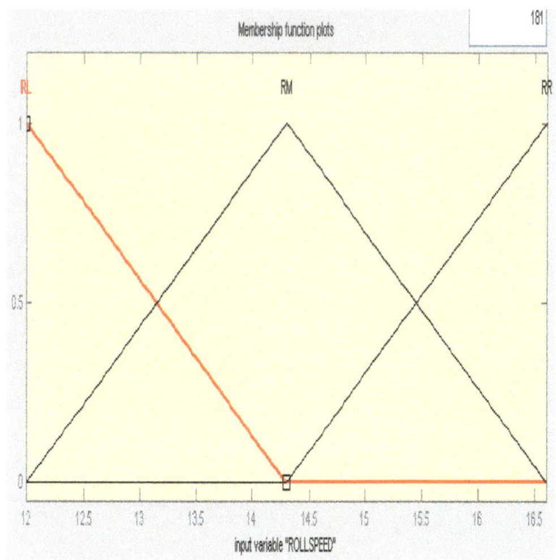

Figure 8. Membership function of 'ROLLSPEED'

Table 3. Linguistic Variables of Input Parameter 'ROLLSPEED

S.No.	Linguistic Variable	Range (cm/s)	Interpretation
1	RL	[12 14.3]	Roll Low
2	RM	[12 14.3 16.6]	Roll Medium
3	RR	[14.3 16.6]	Roll High

D. Fuzzy Rule Base Development Algorithm

The algorithm used to develop the rule base of proposed fuzzy controller is shown in Figure 9. Two hundred thirty one rules are defined for the proposed fuzzy controller which generates the value at which the rake carrier motor should run for maintaining the cane level at 90cm height. The fuzzy rules are developed for three conditions of roll speed i.e. when the roll speed is 12cm/s (RL), 14.3cm/s (RM) and 16.6cm/s (RR).

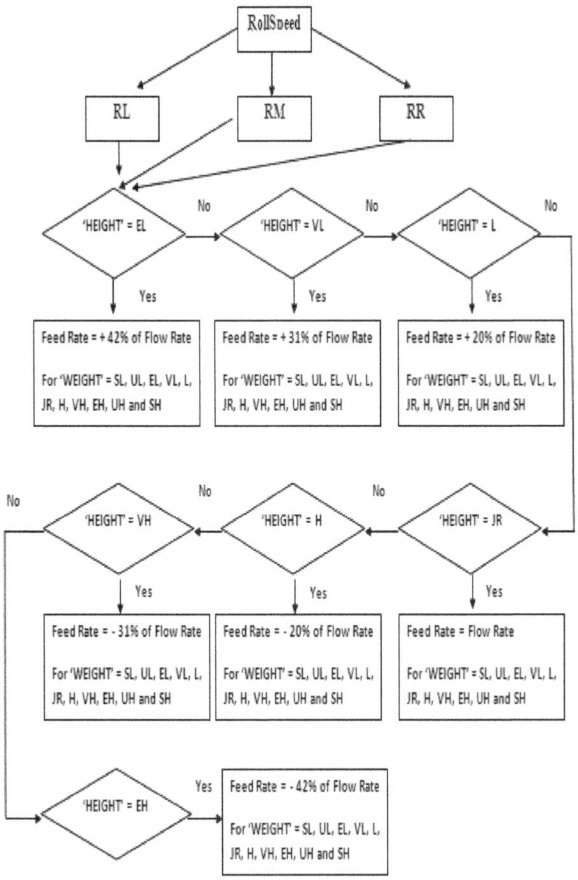

Figure 9. Three Inputs Fuzzy Controller Development Algorithm

When the cane level in chute is at 0cm, 30cm and 60cm then the feed rate of cane is increased by 42%, 31% and 20% of flow rate respectively. If the cane level in chute is at 90cm then the feed rate of cane should be same as flow rate of cane. If the cane level in chute is at 120cm, 150cm and 180cm then the feed rate is decreased by 20%, 31% and 42% of flow rate respectively. The rule base development matrix when roll speed is 12cm/s (RL), 14.3cm/s and 16.6cm/s are shown in Table 4, Table 5 and Table 6 respectively.

Table 4. Three Inputs Fuzzy Controller Design Algorithm When the Roll Speed is 12.0cm/s (RL)

S.No.	Cane weight (Kg)	Variation of Feed Rate (%)	Feed rate (Kg/s)	Carrier Speed (cm/s)	Motor Speed (rpm)
Cane Level in Chute = 0cm (EL)					
1	500 (SL)	+42	27.4	43.8	83.7
2	550 (UL)	+42	27.4	39.8	76.0
3	600 (EL)	+42	27.4	36.5	69.7
4	650 (VL)	+42	27.4	33.7	64.4
5	700 (L)	+42	27.4	31.3	59.8
6	750 (JR)	+42	27.4	29.2	55.8
7	800 (H)	+42	27.4	27.4	52.3
8	850 (VH)	+42	27.4	25.8	49.3
9	900 (EH)	+42	27.4	24.4	46.6
10	950 (UH)	+42	27.4	23.1	44.1
11	1000 (SH)	+42	27.4	21.9	41.8
Contd....					
Cane Level in Chute = 30cm (VL)					
12	500 (SL)	+31	25.3	40.5	77.4
13	550 (UL)	+31	25.3	36.8	70.3
14	600 (EL)	+31	25.3	33.7	64.4
15	650 (VL)	+31	25.3	31.1	59.4
16	700 (L)	+31	25.3	28.9	55.2
17	750 (JR)	+31	25.3	27	51.6
18	800 (H)	+31	25.3	25.3	48.3
19	850 (VH)	+31	25.3	23.8	45.5
20	900 (EH)	+31	25.3	22.5	43
21	950 (UH)	+31	25.3	21.3	40.7
22	1000 (SH)	+31	25.3	20.2	38.6
Cane Level in Chute = 60cm (L)					
23	500 (SL)	+20	23.2	37.1	70.9
24	550 (UL)	+20	23.2	33.7	64.4
25	600 (EL)	+20	23.2	30.9	59
26	650 (VL)	+20	23.2	28.5	54.4
27	700 (L)	+20	23.2	26.5	50.6
28	750 (JR)	+20	23.2	24.7	47.2
29	800 (H)	+20	23.2	23.2	44.3
30	850 (VH)	+20	23.2	21.8	41.6
31	900 (EH)	+20	23.2	20.6	39.3
32	950 (UH)	+20	23.2	19.5	37.2
33	1000 (SH)	+20	23.2	18.6	35.5
Cane Level in Chute = 90cm (JR)					
34	500 (SL)	0	19.3	30.9	59
35	550 (UL)	0	19.3	28.1	53.7
36	600 (EL)	0	19.3	25.7	49.1
37	650 (VL)	0	19.3	23.7	45.3
38	700 (L)	0	19.3	22.1	42.2
39	750 (JR)	0	19.3	20.6	39.3
40	800 (H)	0	19.3	19.3	36.9
41	850 (VH)	0	19.3	18.2	34.8
42	900 (EH)	0	19.3	17.2	32.9
43	950 (UH)	0	19.3	16.2	30.9
44	1000 (SH)	0	19.3	15.4	29.4
Cane Level in Chute = 120cm (H)					
45	500 (SL)	-20	15.4	24.6	47
46	550 (UL)	-20	15.4	22.4	42.8
47	600 (EL)	-20	15.4	20.5	39.2
48	650 (VL)	-20	15.4	18.9	36.1
49	700 (L)	-20	15.4	17.6	33.6
50	750 (JR)	-20	15.4	16.4	31.3
51	800 (H)	-20	15.4	15.4	29.4
52	850 (VH)	-20	15.4	14.5	27.7
53	900 (EH)	-20	15.4	13.7	26.2
54	950 (UH)	-20	15.4	13	24.8
55	1000 (SH)	-20	15.4	12.3	23.5
Cane Level in Chute = 150cm (VH)					
56	500 (SL)	-31	13.3	21.3	40.7
57	550 (UL)	-31	13.3	19.3	36.9
58	600 (EL)	-31	13.3	17.7	33.8
59	650 (VL)	-31	13.3	16.4	31.3
60	700 (L)	-31	13.3	15.2	29
61	750 (JR)	-31	13.3	14.2	27.1
62	800 (H)	-31	13.3	13.3	25.4
63	850 (VH)	-31	13.3	12.5	23.9
64	900 (EH)	-31	13.3	11.8	22.5
65	950 (UH)	-31	13.3	11.2	21.4
66	1000 (SH)	-31	13.3	10.6	20.2
Cane Level in Chute = 180cm (EH)					
67	500 (SL)	-42	11.2	17.9	34.2
68	550 (UL)	-42	11.2	16.3	31.1
69	600 (EL)	-42	11.2	14.9	28.5
70	650 (VL)	-42	11.2	13.8	26.4
71	700 (L)	-42	11.2	12.8	24.4
72	750 (JR)	-42	11.2	11.9	22.7
73	800 (H)	-42	11.2	11.2	21.4
74	850 (VH)	-42	11.2	10.5	20.1
75	900 (EH)	-42	11.2	10	19.1
76	950 (UH)	-42	11.2	9.4	18
77	1000 (SH)	-42	11.2	9	17.2

Table 5. Three Inputs Fuzzy Controller Design Algorithm When the Roll Speed is 14.3cm/s (RM)

S.No.	Cane weight (Kg)	Variation of Feed Rate (%)	Feed rate (Kg/s)	Carrier Speed (cm/s)	Motor Speed (rpm)
Cane Level in Chute = 0cm (EL)					
1	500 (SL)	+42	32.4	51.8	98.9
2	550 (UL)	+42	32.4	47.1	90
3	600 (EL)	+42	32.4	43.2	82.5
4	650 (VL)	+42	32.4	39.9	76.2
5	700 (L)	+42	32.4	37	70.7
6	750 (JR)	+42	32.4	34.5	65.9
7	800 (H)	+42	32.4	32.4	61.9
8	850 (VH)	+42	32.4	30.5	58.3
9	900 (EH)	+42	32.4	28.8	55
10	950 (UH)	+42	32.4	27.3	52.1
11	1000 (SH)	+42	32.4	25.9	49.5
Cane Level in Chute = 30cm (VL)					
12	500 (SL)	+31	29.9	47.8	91.3
13	550 (UL)	+31	29.9	43.5	83.1
14	600 (EL)	+31	29.9	39.9	76.2
15	650 (VL)	+31	29.9	36.8	70.3
16	700 (L)	+31	29.9	34.2	65.3
17	750 (JR)	+31	29.9	31.9	60.9
18	800 (H)	+31	29.9	29.9	57.1
19	850 (VH)	+31	29.9	28.1	53.7
20	900 (EH)	+31	29.9	26.6	50.8
21	950 (UH)	+31	29.9	25.2	48.1
22	1000 (SH)	+31	29.9	23.9	45.6
Cane Level in Chute = 60cm (L)					
23	500 (SL)	+20	27.4	43.8	83.7
24	550 (UL)	+20	27.4	39.8	76
25	600 (EL)	+20	27.4	36.5	69.7
26	650 (VL)	+20	27.4	33.7	64.4
27	700 (L)	+20	27.4	31.3	59.8
28	750 (JR)	+20	27.4	29.2	55.8
29	800 (H)	+20	27.4	27.4	52.3
30	850 (VH)	+20	27.4	25.8	49.3
31	900 (EH)	+20	27.4	24.4	46.6
32	950 (UH)	+20	27.4	23.1	44.1
33	1000 (SH)	+20	27.4	21.9	41.8
Cane Level in Chute = 90cm (JR)					
34	500 (SL)	0	22.8	36.5	69.7
35	550 (UL)	0	22.8	33.1	63.2
36	600 (EL)	0	22.8	30.4	58.1
37	650 (VL)	0	22.8	28	53.5
38	700 (L)	0	22.8	26.1	49.9
39	750 (JR)	0	22.8	24.3	46.4
40	800 (H)	0	22.8	22.8	43.5
41	850 (VH)	0	22.8	21.4	40.9
42	900 (EH)	0	22.8	20.3	38.8
43	950 (UH)	0	22.8	19.2	36.7
44	1000 (SH)	0	22.8	18.2	34.8
Cane Level in Chute = 120cm (H)					
45	500 (SL)	-20	18.2	29.1	55.6
46	550 (UL)	-20	18.2	26.5	50.6
47	600 (EL)	-20	18.2	24.3	46.4
48	650 (VL)	-20	18.2	22.4	42.8
49	700 (L)	-20	18.2	20.8	39.7
50	750 (JR)	-20	18.2	19.4	37.1
51	800 (H)	-20	18.2	18.2	34.8
52	850 (VH)	-20	18.2	17.1	32.7
53	900 (EH)	-20	18.2	16.2	30.9
54	950 (UH)	-20	18.2	15.3	29.2
55	1000 (SH)	-20	18.2	14.6	27.9
Cane Level in Chute = 150cm (VH)					
56	500 (SL)	-31	15.7	25.1	47.9
57	550 (UL)	-31	15.7	22.8	43.5
58	600 (EL)	-31	15.7	20.9	39.9
59	650 (VL)	-31	15.7	19.3	36.9
60	700 (L)	-31	15.7	17.9	34.2
61	750 (JR)	-31	15.7	16.7	31.9
62	800 (H)	-31	15.7	15.7	30
63	850 (VH)	-31	15.7	14.8	28.3
64	900 (EH)	-31	15.7	14	26.7
65	950 (UH)	-31	15.7	13.2	25.2
66	1000 (SH)	-31	15.7	12.6	24.1
Contd....					
S.No.	**Cane weight (Kg)**	**Variation of Feed Rate (%)**	**Feed rate (Kg/s)**	**Carrier Speed (cm/s)**	**Motor Speed (rpm)**
Cane Level in Chute = 180cm (EH)					
67	500 (SL)	-42	13.2	21.1	40.3
68	550 (UL)	-42	13.2	19.2	36.7
69	600 (EL)	-42	13.2	17.6	33.6
70	650 (VL)	-42	13.2	16.2	30.9
71	700 (L)	-42	13.2	15.1	28.8
72	750 (JR)	-42	13.2	14.1	26.9
73	800 (H)	-42	13.2	13.2	25.2
74	850 (VH)	-42	13.2	12.4	23.7
75	900 (EH)	-42	13.2	11.7	22.3
76	950 (UH)	-42	13.2	11.1	21.2
77	1000 (SH)	-42	13.2	10.6	20.2

Table 6. Three Inputs Fuzzy Controller Design Algorithm When the Roll Speed is 16.6cm/s

S.No.	Cane weight (Kg)	Variation of Feed Rate (%)	Feed rate (Kg/s)	Carrier Speed (cm/s)	Motor Speed (rpm)
Cane Level in Chute = 0cm (EL)					
1	500 (SL)	+42	37.8	60.5	115.6
2	550 (UL)	+42	37.8	54.9	104.9
3	600 (EL)	+42	37.8	50.4	96.3
4	650 (VL)	+42	37.8	46.5	88.8
5	700 (L)	+42	37.8	43.2	82.5
6	750 (JR)	+42	37.8	40.3	77
7	800 (H)	+42	37.8	37.8	72.2
8	850 (VH)	+42	37.8	35.6	68
9	900 (EH)	+42	37.8	33.6	64.2
10	950 (UH)	+42	37.8	31.8	60.7
11	1000 (SH)	+42	37.8	30.2	57.7
Cane Level in Chute = 30cm (VL)					
12	500 (SL)	+31	34.8	55.7	106.4
13	550 (UL)	+31	34.8	50.6	96.6
14	600 (EL)	+31	34.8	46.4	88.6
15	650 (VL)	+31	34.8	42.8	81.7
16	700 (L)	+31	34.8	39.8	76
17	750 (JR)	+31	34.8	37.1	70.9
18	800 (H)	+31	34.8	34.8	66.5
19	850 (VH)	+31	34.8	32.7	62.5
20	900 (EH)	+31	34.8	30.9	59
21	950 (UH)	+31	34.8	29.3	56
22	1000 (SH)	+31	34.8	27.8	53.1
Cane Level in Chute = 60cm (L)					
23	500 (SL)	+20	31.9	51	97.4
24	550 (UL)	+20	31.9	46.4	88.6
25	600 (EL)	+20	31.9	42.5	81.2
26	650 (VL)	+20	31.9	39.2	74.9
27	700 (L)	+20	31.9	36.5	69.7
28	750 (JR)	+20	31.9	34	64.9
29	800 (H)	+20	31.9	31.9	60.9
30	850 (VH)	+20	31.9	30	57.3
31	900 (EH)	+20	31.9	28.4	54.2
32	950 (UH)	+20	31.9	26.9	51.4
33	1000 (SH)	+20	31.9	25.5	48.7
Cane Level in Chute = 90cm (JR)					
34	500 (SL)	0	26.6	42.6	81.4
35	550 (UL)	0	26.6	38.7	73.9
36	600 (EL)	0	26.6	35.5	67.8
37	650 (VL)	0	26.6	32.7	62.5
38	700 (L)	0	26.6	30.4	58.1
39	750 (JR)	0	26.6	28.4	54.2
40	800 (H)	0	26.6	26.6	50.8
41	850 (VH)	0	26.6	25	47.8
42	900 (EH)	0	26.6	23.6	45.1
43	950 (UH)	0	26.6	22.4	42.8
44	1000 (SH)	0	26.6	21.3	40.7

Contd....					
S.No.	Cane weight (Kg)	Variation of Feed Rate (%)	Feed rate (Kg/s)	Carrier Speed (cm/s)	Motor Speed (rpm)
Cane Level in Chute = 120cm (H)					
45	500 (SL)	-20	21.3	34.1	65.1
46	550 (UL)	-20	21.3	31	59.2
47	600 (EL)	-20	21.3	28.4	54.2
48	650 (VL)	-20	21.3	26.2	50
49	700 (L)	-20	21.3	24.3	46.4
50	750 (JR)	-20	21.3	22.7	43.4
51	800 (H)	-20	21.3	21.3	40.7
52	850 (VH)	-20	21.3	20	38.2
53	900 (EH)	-20	21.3	18.9	36.1
54	950 (UH)	-20	21.3	17.9	34.2
55	1000 (SH)	-20	21.3	17	32.5
Cane Level in Chute = 150cm (VH)					
56	500 (SL)	-31	18.4	29.4	56.2
57	550 (UL)	-31	18.4	26.7	51
58	600 (EL)	-31	18.4	24.5	46.8
59	650 (VL)	-31	18.4	22.6	43.2
60	700 (L)	-31	18.4	21	40.1
61	750 (JR)	-31	18.4	19.6	37.4
62	800 (H)	-31	18.4	18.4	35.1
63	850 (VH)	-31	18.4	17.3	33
64	900 (EH)	-31	18.4	16.4	31.3
65	950 (UH)	-31	18.4	15.5	29.6
66	1000 (SH)	-31	18.4	14.7	28.1
Cane Level in Chute = 180cm (EH)					
67	500 (SL)	-42	15.4	24.6	47
68	550 (UL)	-42	15.4	22.4	42.8
69	600 (EL)	-42	15.4	20.5	39.2
70	650 (VL)	-42	15.4	18.9	36.1
71	700 (L)	-42	15.4	17.6	33.6
72	750 (JR)	-42	15.4	16.4	31.3
73	800 (H)	-42	15.4	15.4	29.4
74	850 (VH)	-42	15.4	14.5	27.7
75	900 (EH)	-42	15.4	13.7	26.2
76	950 (UH)	-42	15.4	13	24.8
77	1000 (SH)	-42	15.4	12.3	23.5

E. Flow Rate Calculations for 12.0cm/s Roll Speed

When the surface speed of roll is 12.0cm/s then average speed of cane blanket when it touches the roll surface (S_f) is 6.9cm/s from (8). The escribed volume (V_e) at the entry of chute is 0.055m^3/s from (9). The mass flow rate of cane through the rolls (Q_c) is 19.3Kg/s from (11) and the fibre rate of cane through the rolls (Q_f) from (10) when cane has 15% fibre percentage is 2.9Kg/s.

F. Flow Rate Calculations for 14.3cm/s Roll Speed

When the surface speed of roll is 14.3cm/s then average speed of cane blanket when it touches the roll surface (S_f) is 8.2cm/s from (8). The escribed volume (V_e) at the entry of chute is 0.065m^3/s from (9). The mass flow rate of cane through the rolls (Q_c) is 22.8Kg/s from (11) and the fibre rate of cane through the rolls from (10) is 3.5Kg/s.

G. Flow Rate Calculations for 16.6cm/s Roll Speed

When the surface speed of roll is 16.6cm/s then average speed of cane blanket when it touches the roll surface (S_f) is 9.5cm/s from (8). The escribed volume (V_e) at the entry of chute is 0.076m^3/s from (9). The mass flow rate of cane through the rolls (Q_c) is 26.6Kg/s from (11) and the fibre rate of cane through the rolls from (10) is 4.0Kg/s.

H. Fuzzification of 'SPEED'

The universe of discourse of output parameter **'SPEED'** is in the range 17.2rpm to 115.5rpm and it is fuzzified into nine triangular linguistic variables as shown in Figure 10 and represented in Table 7. It has been calculated that for maintaining 2,200tcd the roll speed should be 16.6cm/s. It is shown in S. No. 39 of Table 6 that for 750Kg of cane in rake the rake carrier motor should run at 54.2rpm to maintain cane level at 90cm. Therefore during the selection of membership function it is decided that the 54.2rpm must possess membership value 1 for linguistic variable JR (Just Right).

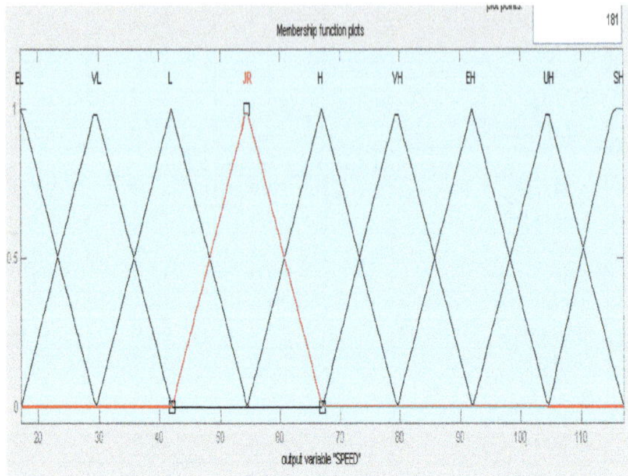

Figure 10. Membership function of 'SPEED'

Table 7. Linguistic Variables of Output Parameter 'SPEED"

S.No.	Linguistic Variable	Range (rpm)	Interpretation
1	EL	[17.2 29.5]	Extreme Low
2	VL	[17 29.5 42]	Very Low
3	L	[29.5 42 54.5]	Low
4	JR	[42 54.5 67]	Just Right
5	H	[54.5 67 79.5]	High
6	VH	[67 79.5 92]	Very High
7	EH	[79.5 92 104.5]	Extreme High
8	UH	[92 104.5 117]	Ultra High
9	SH	[104.5 115.5]	Super High

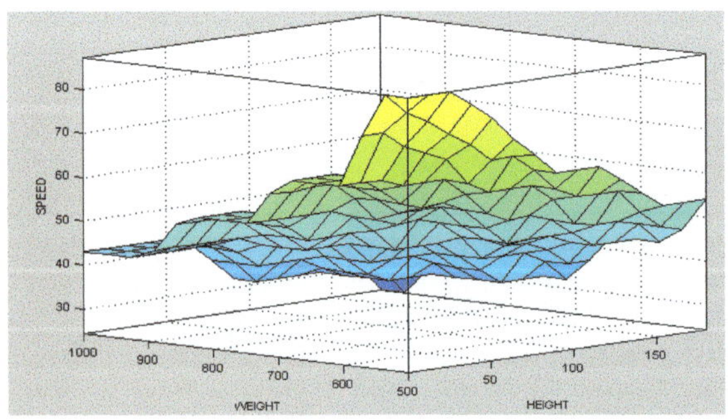

Figure 11. Surface Viewer between 'HEIGHT', 'WEIGHT' and 'SPEED'

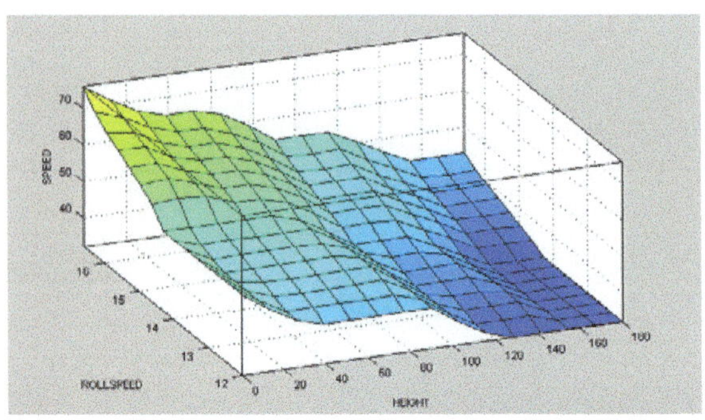

Figure 12. Surface Viewer between 'ROLLSPEED', 'HEIGHT' and 'SPEED'

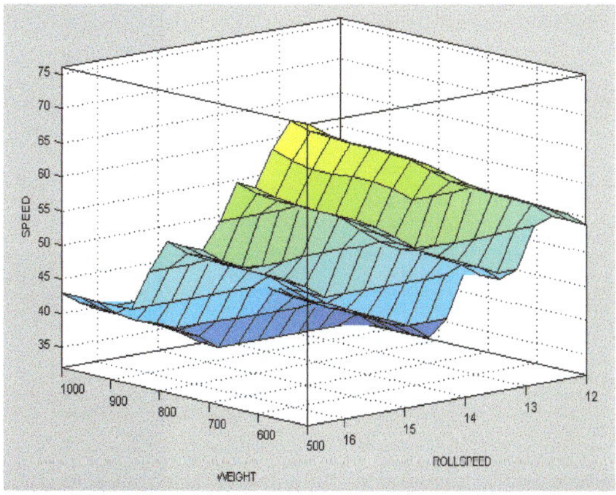

Figure 13. Surface Viewer between 'WEIGHT', 'ROLLSPEED', 'HEIGHT' and 'SPEED'

The surface viewer of FIS representing the relationship between input parameters 'HEIGHT' and 'WEIGHT' and output parameter 'SPEED' is shown in Figure 11. The surface viewer of FIS representing the relationship between input parameters 'ROLLSPEED' and 'HEIGHT' and output parameter 'SPEED' is shown in Figure 12. The surface viewer of FIS representing the relationship between input parameters 'WEIGHT' and 'ROLLSPEED' and output parameter 'SPEED' is shown in Figure 13.

I. Explanation of S. No. 1 of Table 4

The cane level of 0cm in Donnelly chute, prepared cane of 500Kg in rake carrier and roll speed of 12cm/s are represented by EL, SL and RL linguistic variables of input parameters 'HEIGHT', 'WEIGHT' and 'ROLLSPEED' respectively. It has been calculated that if the roll speed is 12cm/s then the flow rate of cane through the rolls will be 19.3Kg/s. According to the algorithm under these conditions the feed rate should be increased by 42% of flow rate and it comes to be 27.4Kg/s. The weight of cane in 1cm of rake carrier when it has 500Kg cane is 0.625Kg. The carrier speed is calculated from (9) which comes to be 43.8cm/s and motor speed in rpm to run carrier at 43.8cm/s is calculated from (11) which comes to be 83.7rpm.

J. Explanation of S. No. 39 of Table 5

The cane level of 90cm in Donnelly chute, prepared cane of 750Kg in rake carrier and roll speed of 14.3cm/s are represented by JR, JR and RM linguistic variables of input parameters 'HEIGHT', 'WEIGHT' and 'ROLLSPEED' respectively. It has been calculated that if the roll speed is 14.3cm/s then the flow rate of cane through the rolls will be 22.8Kg/s. According to the algorithm under these conditions the feed rate should be equal to the flow rate. The weight of cane in 1cm of rake carrier when it has 750Kg cane is 0.938Kg. The carrier speed is calculated from (9) which comes to be 24.3cm/s and motor speed in rpm to run carrier at 24.3cm/s is calculated from (11) which comes to be 46.4rpm.

K. Explanation of S. No. 57 of Table 6

The cane level of 150cm in Donnelly chute, prepared cane of 550Kg in rake carrier and roll speed of 16.6cm/s are represented by VH, UL and RR linguistic variables of input parameters 'HEIGHT', 'WEIGHT' and 'ROLLSPEED' respectively. It has been calculated that if the roll speed is 16.6cm/s then the flow rate of cane through the rolls will be 26.6Kg/s. According to the algorithm under these conditions the feed rate should be decreased by 31% of flow rate and it comes to be 18.4Kg/s. The weight of cane in 1cm of rake carrier when it has 550Kg cane is 0.688Kg. The carrier speed is calculated from (9) which comes to be 26.7cm/s and motor speed in rpm to run carrier at 26.7cm/s is calculated from (11) which comes to be 51.0rpm.

L. Examples of Rule Base Development

Example-1: Rule 1 Antecedent is "If (HEIGHT is EL) and (WEIGHT is SL) and (ROLLSPEED is RL)"

For the antecedent of Rule 1 the motor speed should be 83.7rpm (S. No. 1 of Table 4). The motor speed 83.7rpm lies in VH and EH linguistic variable of 'SPEED' but it's membership value is higher in VH as compared to EH (Figure 10.) therefore the consequent of Rule 1 is "then (SPEED is VH)". So, the antecedent and consequent of Rule 1 is given as:

Rule 1: If (HEIGHT is EL) and (WEIGHT is SL) and (ROLLSPEED is RL) then (SPEED is VH)

Example II: Rule 91 Antecedent is "If (HEIGHT is VL) and (WEIGHT is EL) and (ROLLSPEED is RM)"

For the antecedent of Rule 91 the motor speed should be 76.2rpm (S. No. 14 of Table 5). The motor speed 76.2rpm lies in H and VH linguistic variable of 'SPEED' but it's membership value is higher in VH as compared to H (Figure 10.) therefore the consequent of Rule 91 is "then (SPEED is VH)". So, the antecedent and consequent of Rule 91 is given as:

Rule 91: If (HEIGHT is VL) and (WEIGHT is EL) and (ROLLSPEED is RM) then (SPEED is VH)

Example III: Rule 201 Antecedent is "If (HEIGHT is H) and (WEIGHT is EL) and (ROLLSPEED is RR)"

For the antecedent of Rule 201 the motor speed should be 54.2rpm (S. No. 47 of Table 6). The motor speed 54.2rpm lies in L and JR linguistic variable of 'SPEED' but it's membership value is higher in JR as compared to H (Figure 10) therefore the consequent of Rule 201 is "then (SPEED is JR)". So, the antecedent and consequent of Rule 91 is given as:

Rule 201: If (HEIGHT is H) and (WEIGHT is EL) and (ROLLSPEED is RR) then (SPEED is JR)

Table 8. Rule Matrix "If 'ROLLSPEED' is RL"

		HEIGHT						
		EL	VL	L	JR	H	VH	EH
	SL	VH	VH	H	JR	L	L	VL
	UL	VH	H	H	JR	L	L	VL
	EL	H	H	JR	JR	L	VL	VL
	VL	H	JR	JR	L	L	VL	VL
W	L	JR	JR	JR	L	VL	VL	VL
E	JR	JR	JR	L	L	VL	VL	EL
I	H	JR	JR	L	L	VL	VL	EL
G	VH	JR	L	L	VL	VL	VL	EL
H	EH	L	L	L	VL	VL	EL	EL
T	UH	L	L	L	VL	VL	EL	EL
	SH	L	L	VL	VL	VL	EL	EL

Table 9. Rule Matrix "If 'ROLLSPEED' is RM"

		HEIGHT						
		EL	VL	L	JR	H	VH	EH
	SL	UH	EH	VH	H	JR	L	L
	UL	EH	VH	VH	H	JR	L	L
W	EL	VH	VH	H	JR	L	L	VL
E	VL	VH	H	H	JR	L	L	VL
I	L	H	H	JR	JR	L	VL	VL
G	JR	H	H	JR	L	L	VL	VL
H	H	H	JR	JR	L	VL	VL	VL
T	VH	JR	JR	JR	L	VL	VL	VL
	EH	JR	JR	L	L	VL	VL	EL
	UH	JR	L	L	L	VL	VL	EL
	SH	JR	L	L	VL	VL	VL	EL

Table 10. Rule Matrix "If 'ROLLSPEED' is RR"

		HEIGHT						
WEIGHT		EL	VL	L	JR	H	VH	EH
	SL	SH	UH	EH	VH	H	JR	L
	UL	UH	EH	EH	VH	JR	JR	L
	EL	EH	EH	VH	H	JR	L	L
	VL	EH	VH	VH	H	JR	L	L
	L	VH	VH	H	JR	L	L	VL
	JR	VH	H	H	JR	L	L	VL
	H	H	H	H	JR	L	VL	VL
	VH	H	H	JR	L	L	VL	VL
	EH	H	JR	JR	L	L	VL	VL
	UH	JR	JR	JR	L	VL	VL	VL
	SH	JR	JR	JR	L	VL	VL	VL

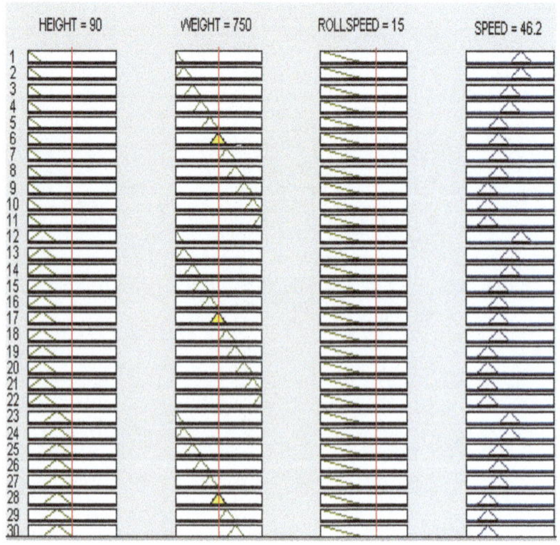

Figure 14. Case-I "Cane Level=90cm, Cane Weight=750Kg and Roll Speed=15.0cm/s Then Rake Carrier Motor rpm=46.2

4. Result and Discussion

Three inputs fuzzy controller developed is simulated for six different cases by using fuzzy toolbox of 'MATLAB® version 7.11.0.584 (R2020b). The controller generates the required rake carrier motor speed in rpm depending upon the value of cane level in chute, quantity of cane on rake carrier and the roll speed. The sampling period is 10 seconds and total simulation duration is 210 seconds.

In Case-I and Case-II the cane level and cane weight on carrier at the start of simulation is 90cm and 750Kg respectively. The roll speed at the start of simulation is 15cm/s for Case-I and 15.4cm/s for Case-II. The roll speed variation is less in Case-I and in Case-II it varies at the time of each sampling. The Matlab fuzzy tool box simulation result for specified condition for Case-I and Case-II is shown in Figure 14 and Figure 15 respectively. The simulation result is given in Table 11 and Table 12, the graphical representation of simulation result is given in Figure 20 and Figure 21 for Case-I and Case-II respectively.

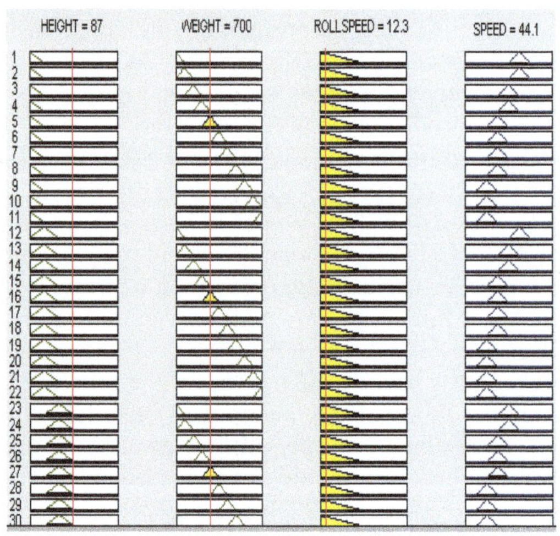

Figure 15. Case-II "Cane Level=87cm, Cane Weight=700Kg and Roll Speed=12.3cm/s Then Rake Carrier Motor rpm=44.1"

Table 11. Case-I Cane Level is at 90cm and Less Variation in Roll Speed

Parameters		Cane Level (cm)	Cane Weight (Kg)	Motor Speed (rpm)	Carrier Speed (cm/s)	Cane in Carrier (Kg/cm)	Feed Rate (Kg/s)	Data for next sampling		Level for next sample (cm)
Time (s)	Roll speed (cm/s)							Kg	cm	
0	15	90	750	46.2	24.2	0.938	22.7	-13	-4.6	85.4
10		85.4	729	50.2	26.3	0.911	24.0	0	0	85.4
20		85.4	792	49.2	25.8	0.990	25.5	+15	+5.4	90.8
30		90.8	908	41.5	21.7	1.135	24.6	+6	+2.1	92.9
40	12.6	92.9	965	33.3	17.3	1.206	20.9	+7	+2.5	95.4
50		95.4	720	43.0	22.5	0.900	20.3	+1	+0.4	95.8
60		95.8	760	39.0	20.4	0.950	19.4	-8	-2.9	92.9
70		92.9	790	40.3	21.1	0.988	20.8	+6	+2.1	95.0
80		95.0	820	36.8	19.3	1.025	19.8	-4	-1.4	93.6
90	16.2	93.6	555	71.4	37.4	0.694	26.0	0	0	93.6
100		93.6	609	61.6	32.3	0.761	24.6	-14	-5.0	88.6
110		88.6	578	70.7	37.0	0.723	26.8	+8	+2.9	91.5
120		91.5	598	63.8	33.4	0.748	25.0	-10	-3.6	87.9
130		87.9	700	55.7	29.2	0.875	25.6	-4	-1.4	86.5
140		86.5	679	62.1	32.5	0.849	27.6	+16	+5.7	92.2
150	15.4	92.2	800	46.7	24.5	1.000	24.5	-2	-0.7	91.5
160		91.5	845	43.0	22.5	1.056	23.8	-9	-3.2	88.3
170		88.3	835	48	25.1	1.044	26.2	+15	+5.4	93.7
180		93.7	874	39.5	20.7	1.093	22.6	-21	-7.5	86.2
190		86.2	900	44.5	23.3	1.125	26.2	+15	+5.4	91.6
200		91.6	924	40.8	21.4	1.155	24.7	0	0	91.6

Table 12. Case-II Cane Level is at 90cm and Roll Speed Vary during Each Sample

Parameters		Cane Level (cm)	Cane Weight (Kg)	Motor Speed (rpm)	Carrier Speed (cm/s)	Cane In Carrier (Kg/cm)	Feed Rate (Kg/s)	Data for next sampling		Level for next sample (cm)
Time (s)	Roll speed (cm/s)							Kg	cm	
0	15.4	90	750	48	25.1	0.938	23.5	-12	-4.3	85.7
10	15.8	85.7	729	52.1	27.3	0.911	24.9	-4	-1.4	84.3
20	15.0	84.3	792	49.7	26.0	0.990	25.7	+17	+6.1	90.4
30	16.2	90.4	908	41.8	21.9	1.135	24.9	-11	-3.9	86.5
40	16.6	86.5	965	44.0	23.0	1.206	27.7	+11	+3.9	90.5
50	13.4	90.5	720	48.9	25.6	0.900	23.0	+15	+5.4	95.9
60	13.8	95.9	760	39.0	20.4	0.950	19.4	-27	-9.6	86.3
70	13.4	86.3	790	44.3	23.2	0.988	22.9	+14	+5	91.3
80	15.4	91.3	820	47.2	24.7	1.025	25.3	+6	+2.1	93.4
90	16.2	93.4	555	71.5	37.4	0.694	26.0	0	0	93.4
100	13.0	93.4	609	51.3	26.9	0.761	20.5	-3	-1.1	92.3
110	14.3	92.3	578	58.8	30.8	0.723	22.3	-6	-2.1	90.2
120	14.6	90.2	598	57.5	30.1	0.748	22.5	-9	-3.2	87.0
130	12.3	87.0	700	44.1	23.1	0.875	20.2	+5	+1.8	88.8
140	12.6	88.8	679	47.1	24.7	0.849	21.0	+8	+2.9	91.7
150	15.4	91.7	800	47.0	24.6	1.000	24.6	-1	-0.4	91.3
160	12.0	91.3	845	31.2	16.3	1.056	17.2	-20	-7.1	84.2
170	14.3	84.2	835	45.1	23.6	1.044	24.6	+17	+6.1	90.3
180	14.6	90.3	874	41.8	21.9	1.093	23.9	+5	+1.8	92.1
190	15.0	92.1	900	40.7	21.3	1.125	24.0	0	0	92.1
200	15.4	92.1	924	40.5	21.2	1.155	24.5	-2	-0.7	91.4

In Case-III and Case-IV the cane level at the start of simulation is 0cm whereas cane weight on carrier and roll speed at the start of simulation for Case-III is same as Case-I and for Case-IV is same as Case-II. The roll speed variation is less in Case-III and in Case-IV it varies at the time of each sampling. The Matlab fuzzy tool box simulation result for specified condition for Case-III and Case-IV is shown in Figure 16 and Figure 17 respectively. The simulation result is given in Table 13 and Table 14, the graphical representation of simulation result is given in Figure 22 and Figure 23 for Case-I and Case-II respectively.

Table 13. Case-III Cane Level is at 0cm and Less Variation in Roll Speed

Parameters		Cane Level (cm)	Cane Weight (Kg)	Motor Speed (rpm)	Carrier Speed (cm/s)	Cane in Carrier (Kg/cm)	Feed Rate (Kg/s)	Data for next sampling		Level for next sample (cm)
Time (s)	Roll speed (cm/s)							Kg	cm	
0		0	750	71.2	37.3	0.938	35.0	+110	+39.3	39.3
10	15	39.3	729	66.9	35.0	0.911	31.9	+79	+28.2	67.5
20		67.2	792	55.2	28.9	0.990	28.6	+46	+16.4	83.9
30		83.9	908	45.2	23.7	1.135	26.9	+29	+10.4	94.3
40		94.3	965	33.3	17.4	1.206	21.0	+8	+2.9	97.2
50		97.2	720	42.2	22.1	0.900	19.9	-3	-1.1	96.1
60	12.6	96.1	760	38.9	20.4	0.950	19.4	-8	-2.9	93.2
70		93.2	790	40.2	21.0	0.988	20.7	+5	+1.8	95.0
80		95.0	820	36.8	19.3	1.025	19.8	-4	-1.4	93.6
90		93.6	555	71.4	37.4	0.694	26.0	0	0	93.6
100		93.6	609	61.6	32.3	0.761	24.6	-14	-5.0	88.6
110	16.2	88.6	578	70.7	37.0	0.723	26.8	+8	+2.9	91.5
120		91.5	598	63.8	33.4	0.748	25.0	-10	-3.6	87.9
130		87.9	700	55.7	29.2	0.875	25.6	-4	-1.4	86.5
140		86.5	679	62.1	32.5	0.849	27.6	+10	+3.6	90.1
150		90.1	800	48	25.1	1.000	25.1	+4	+1.4	91.5
160		91.5	845	43.0	22.5	1.056	23.7	-10	-3.6	95.1
170	15.4	95.1	835	43.6	22.8	1.044	23.8	-9	-3.2	91.9
180		91.9	874	40.6	21.3	1.093	23.3	-14	-5.0	86.9
190		86.9	900	44.1	23.1	1.125	26.0	+13	+4.6	91.5
200		91.5	924	40.9	21.4	1.155	24.7	0	0	91.5

Table 14. Case-IV Cane Level is at 0cm and Roll Speed Vary during Each Sample

Parameters		Cane Level (cm)	Cane Weight (Kg)	Motor Speed (rpm)	Carrier Speed (cm/s)	Cane In Carrier (Kg/cm)	Feed Rate (Kg/s)	Data for next sampling		Level for next sample (cm)
Time (s)	Roll speed (cm/s)							Kg	cm	
0	15.4	0	750	73	38.2	0.938	35.8	+111	+39.6	39.6
10	15.8	39.6	729	67.9	35.5	0.911	32.3	+70	+25	64.6
20	15.0	64.6	792	56.3	29.5	0.990	29.2	+32	+11.4	76.0
30	16.2	76.0	908	47.9	25.1	1.135	28.5	+25	+8.9	84.9
40	16.6	84.9	965	44.8	23.5	1.206	28.3	+17	+6.1	91.0
50	13.4	91.0	720	48.6	25.4	0.900	22.9	+14	+5.0	96.0
60	13.8	96.0	760	39	20.4	0.950	19.4	-27	-9.6	86.4
70	13.4	86.4	790	44.2	23.1	0.988	22.8	+13	+4.6	91.0
80	15.4	91.0	820	47.4	24.8	1.025	25.4	+7	+2.5	93.5
90	16.2	93.5	555	71.4	37.4	0.694	26.0	0	0	93.5
100	13.0	93.5	609	51.3	26.9	0.761	20.5	-3	-1.1	92.4
110	14.3	92.4	578	58.7	30.7	0.723	22.2	-7	-2.5	89.9
120	14.6	89.9	598	57.7	30.2	0.748	22.6	-8	-2.9	87.0
130	12.3	87.0	700	44.1	23.1	0.875	20.2	+5	+1.8	88.8
140	12.6	88.8	679	47.2	24.7	0.849	21.0	+8	+2.9	91.7
150	15.4	91.7	800	47	24.6	1.000	24.6	-1	-0.4	91.3
160	12.0	91.3	845	31.2	16.3	1.056	17.2	-20	-7.1	84.2
170	14.3	84.2	835	45.1	23.6	1.044	24.6	+17	+6.1	90.3
180	14.6	90.3	874	41.8	21.9	1.093	23.9	+5	+1.8	92.1
190	15.0	92.1	900	40.7	21.3	1.125	24.0	0	0	92.1
200	15.4	92.1	924	40.5	21.2	1.155	24.5	-2	-0.7	91.4

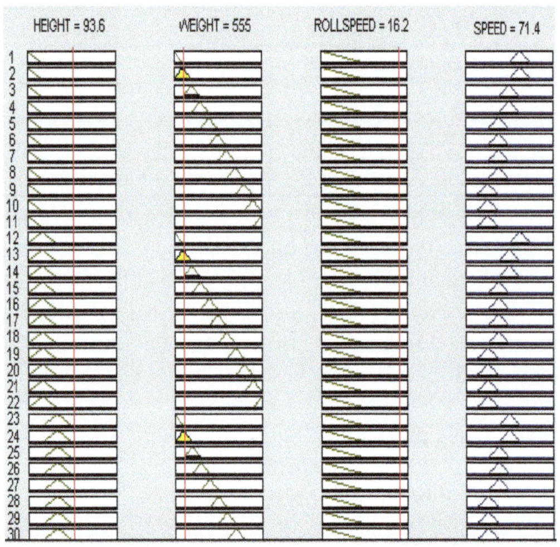

Figure 16. Case-III "Cane Level=93.6cm, Cane Weight=555Kg and Roll Speed=16.2cm/s Then Rake Carrier Motor rpm=71.4"

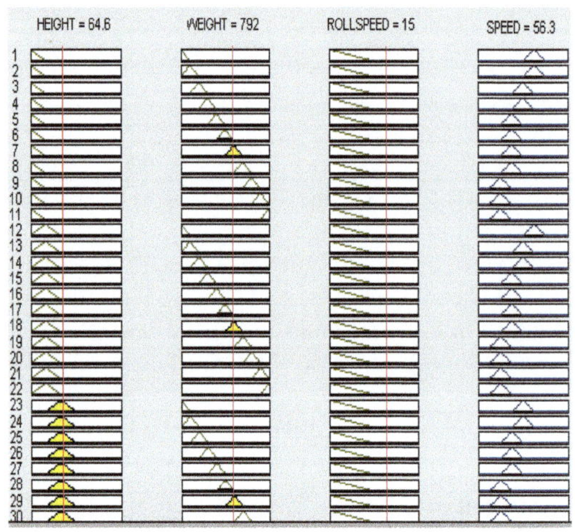

Figure 17. Case-IV "Cane Level=64.6cm, Cane Weight=792Kg and Roll Speed=15cm/s Then Rake Carrier Motor rpm=56.3"

Table 15. Case-V Cane Level is at 180cm and Less Variation in Roll Speed

Parameters		Cane Level (cm)	Cane Weight (Kg)	Motor Speed (rpm)	Carrier Speed (cm/s)	Cane in Carrier (Kg/cm)	Feed Rate (Kg/s)	Data for next sampling		Level for next sample (cm)
Time (s)	Roll speed (cm/s)							Kg	cm	
0		180	750	35.7	18.7	0.938	17.5	-65	-23.2	156.8
10	15	156.8	729	36.8	19.3	0.911	17.6	-64	-22.9	133.9
20		133.9	792	40.2	21.0	0.990	20.8	-32	-11.4	122.5
30		122.5	908	35.7	18.7	1.135	21.2	-28	-10	112.5
40		112.5	965	32.0	16.8	1.206	20.3	+1	+0.4	112.9
50		112.9	720	41.6	21.8	0.900	19.6	-6	-2.1	110.8
60	12.6	110.8	760	35.7	18.7	0.950	17.8	-24	-8.6	102.2
70		102.2	790	35.6	18.6	0.988	18.4	-18	-6.4	95.8
80		95.8	820	35.6	18.6	1.025	18.6	-16	-5.7	90.1
90		90.1	555	74.5	39.0	0.694	27.1	+11	+3.9	94.0
100		94.0	609	61.4	32.1	0.761	24.4	-16	-5.7	88.3
110	16.2	88.3	578	70.9	37.1	0.723	26.8	+8	+2.9	91.2
120		91.2	598	64.0	33.5	0.748	25.1	-9	-3.2	88.0
130		88.0	700	55.7	29.2	0.875	25.6	-4	-1.4	86.6
140		86.6	679	62.0	32.5	0.849	27.6	+16	+5.7	92.3
150		92.3	800	46.6	24.4	1.000	24.4	-3	-1.1	91.2
160		91.2	845	43.2	22.6	1.056	23.9	-8	-2.9	88.3
170	15.4	88.3	835	48	25.1	1.044	26.2	+15	+5.4	93.7
180		93.7	874	39.5	20.7	1.093	22.6	-21	-7.5	86.2
190		86.2	900	44.5	23.3	1.125	26.2	+15	+5.4	91.6
200		91.6	924	40.8	21.4	1.155	24.7	0	0	91.6

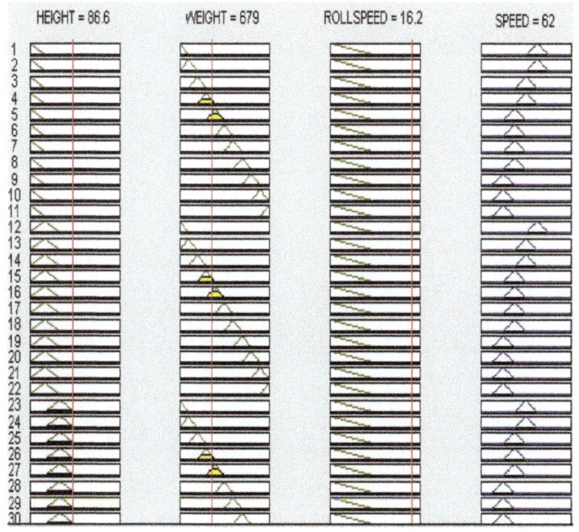

Figure 18. Case-V "Cane Level=86.6cm, Cane Weight=679Kg and Roll Speed=16.2cm/s Then Rake Carrier Motor rpm=62"

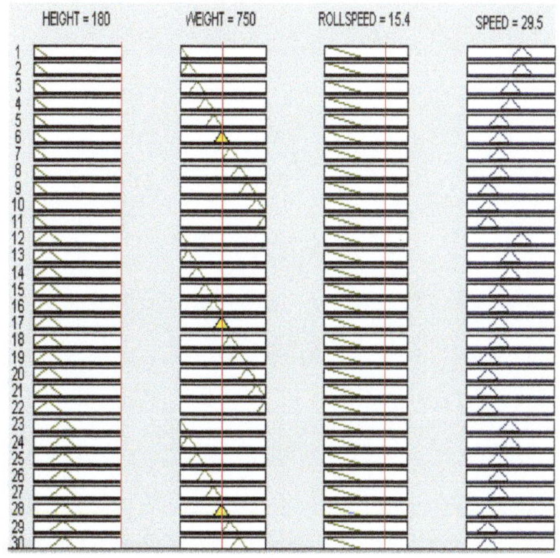

Figure 19. Case-VI "Cane Level=180cm, Cane Weight=750Kg and Roll Speed=15.4cm/s Then Rake Carrier Motor rpm=29.5"

Table 16. Case-VI Cane Level is at 180cm and Roll Speed Vary during Each Sample

Parameters		Cane Level (cm)	Cane Weight (Kg)	Motor Speed (rpm)	Carrier Speed (cm/s)	Cane In Carrier (Kg/cm)	Feed Rate (Kg/s)	Data for next sampling		Level for next sample (cm)
Time (s)	Roll speed (cm/s)							Kg	cm	
0	15.4	180	750	29.5	15.4	0.938	14.4	-103	-36.8	143.2
10	15.8	143.2	729	37.1	19.4	0.911	17.7	-76	-27.1	116.1
20	15.0	116.1	792	36.3	19.0	0.990	18.8	-52	-18.6	97.5
30	16.2	97.5	908	39.2	20.5	1.135	23.3	-27	-9.6	87.9
40	16.6	87.9	965	43.3	22.7	1.206	27.4	+8	+2.9	90.8
50	13.4	90.8	720	48.7	25.5	0.900	23.0	+15	+5.4	96.2
60	13.8	96.2	760	38.9	20.4	0.950	19.4	-27	-9.6	86.6
70	13.4	86.6	790	44.1	23.1	0.988	22.8	+13	+4.6	91.2
80	15.4	91.2	820	47.3	24.8	1.025	25.4	+7	+2.5	93.7
90	16.2	93.7	555	71.3	37.3	0.694	25.9	-1	-0.4	93.3
100	13.0	93.3	609	51.6	26.9	0.761	20.5	-3	-1.1	92.2
110	14.3	92.2	578	58.8	30.8	0.723	22.3	-6	-2.1	90.1
120	14.6	90.1	598	57.5	30.1	0.748	22.5	-9	-3.2	86.9
130	12.3	86.9	700	44.1	23.1	0.875	20.2	+5	+1.8	88.7
140	12.6	88.7	679	47.2	24.7	0.849	21.0	+8	+2.9	85.8
150	15.4	85.8	800	50.5	26.4	1.000	26.4	+17	+6.1	91.9
160	12.0	91.9	845	31.2	16.3	1.056	17.2	-20	-7.1	84.8
170	14.3	84.8	835	44.8	23.5	1.044	24.5	+16	+5.7	90.5
180	14.6	90.5	874	41.6	21.8	1.093	23.8	+4	+1.4	91.9
190	15.0	91.9	900	40.8	21.4	1.125	24.1	+1	+0.4	92.3
200	15.4	92.3	924	40.4	21.2	1.155	24.5	-2	-0.7	91.6

In Case-V and Case-VI the cane level at the start of simulation is 180cm whereas cane weight on carrier and roll speed at the start of simulation for Case-V is same as Case-I (or Case-III) and for Case-VI is same as Case-II (or Case-IV). The roll speed variation is less in Case-V and in Case-VI it varies at the time of each sampling. The Matlab fuzzy tool box simulation result for specified condition for Case-V and Case-VI is shown in Figure 18 and Figure 19 respectively. The simulation result is given in Table 15 and Table 16, the graphical representation of simulation result is given in Figure 24 and Figure 25 for Case-V and Case-VI respectively.

The comparison between Case-I to Case-VI is given in Table 17 and the results of Case-I to Case-VI are shown in Figure 26 to Figure 29.

Table 17. Comparison between Case-I to Case-VI of Conventional Controller

	Cane Level in Between 85cm-95cm (% Time)	Time Required to Reach Cane Level at 90cm (sec)	Lowest Level of Cane in chute (cm)	Highest Level of Cane in Chute (cm)
Case-I	93.1	NA	85.4	95.8
Case-II	94.7	NA	84.2	95.9
Case-III	74.7	35.9	NA	97.2
Case-IV	78.9	48.4	NA	96.0
Case-V	61.2	107	86.2	NA
Case-VI	82.6	37.8	84.8	NA

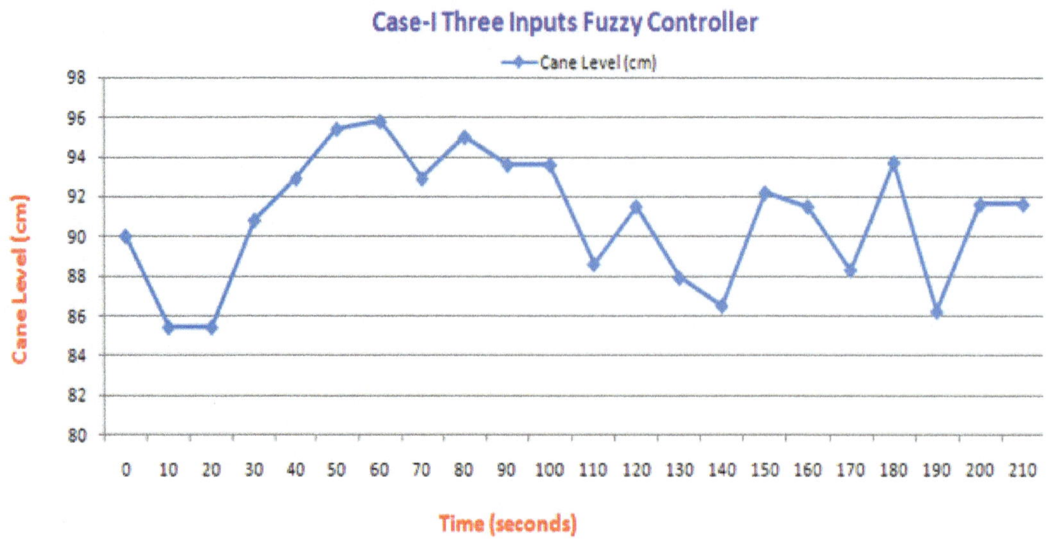

Figure 20. Case-I

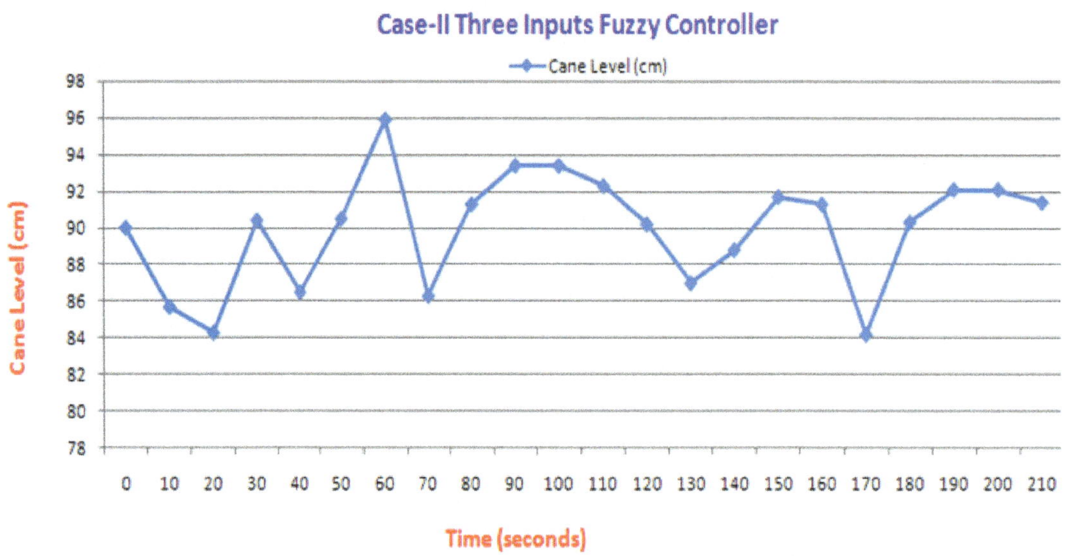

Figure 21. Case-II

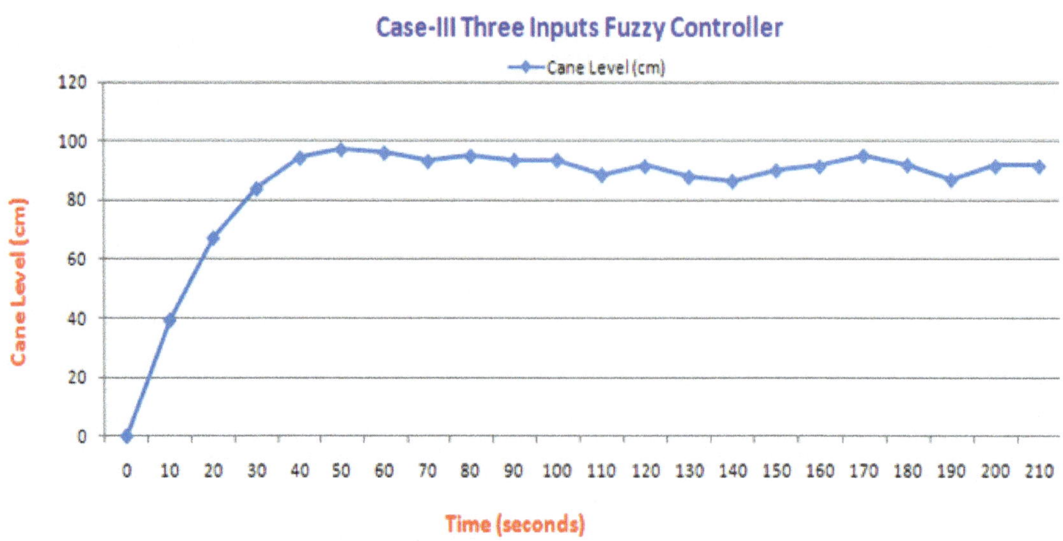

Figure 22. Case-III

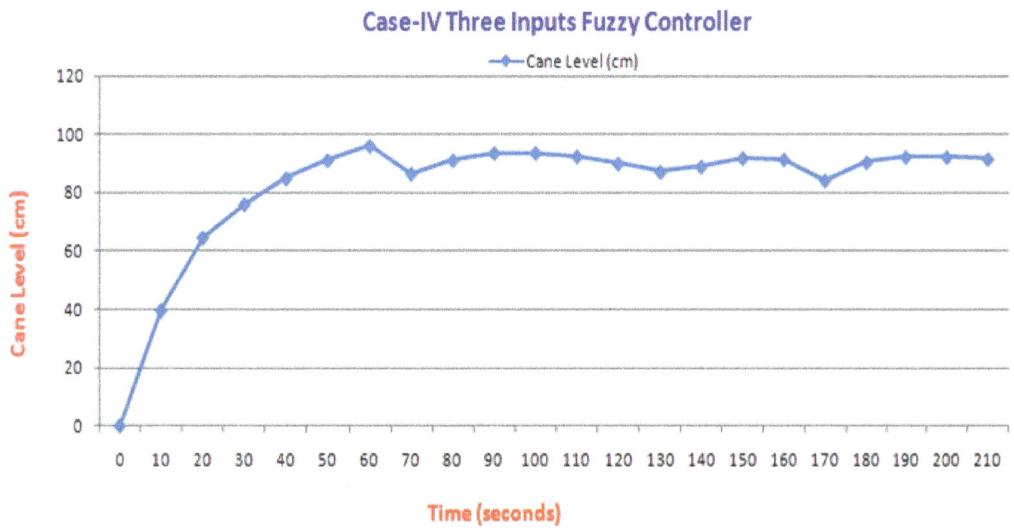

Figure 23. Case-IV

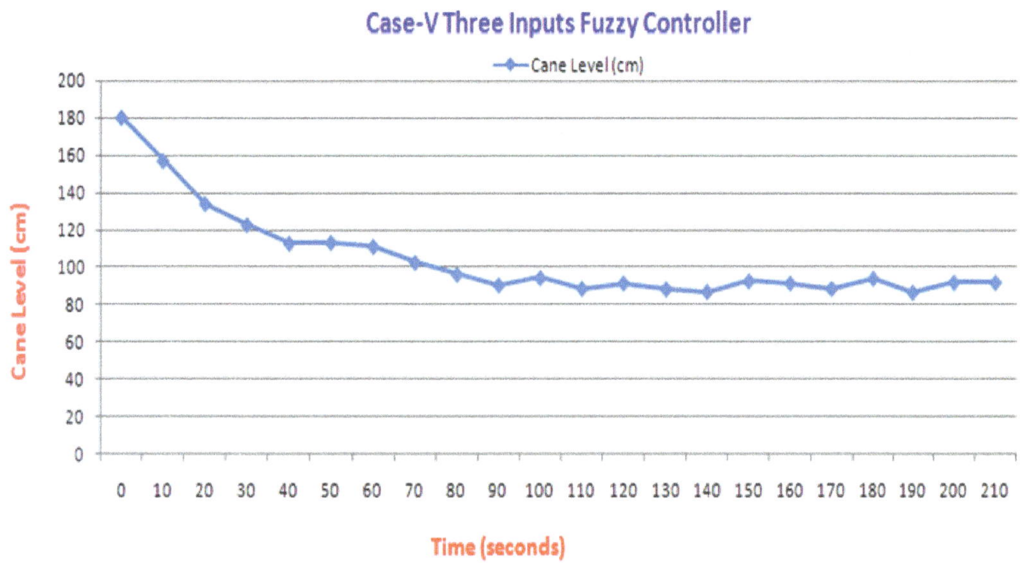

Figure 24. Case-V

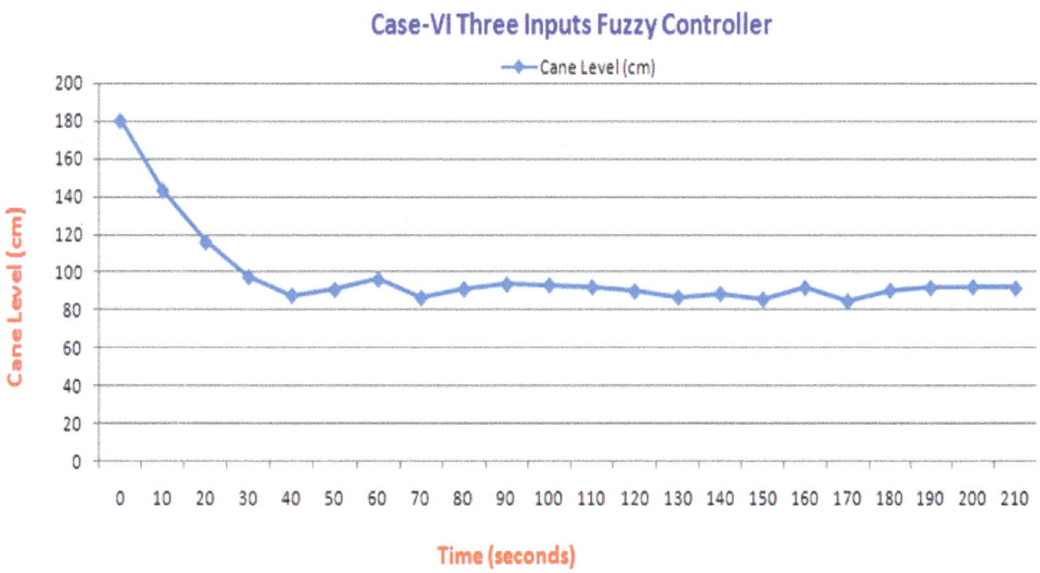

Figure 25. Case-VI

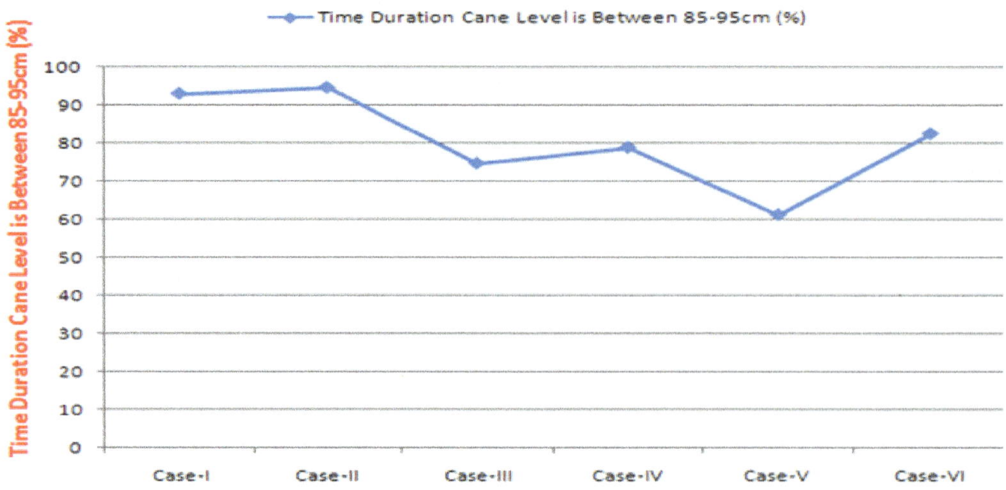

Figure 26. Percentage of Time Cane Level is Between 85-95cm

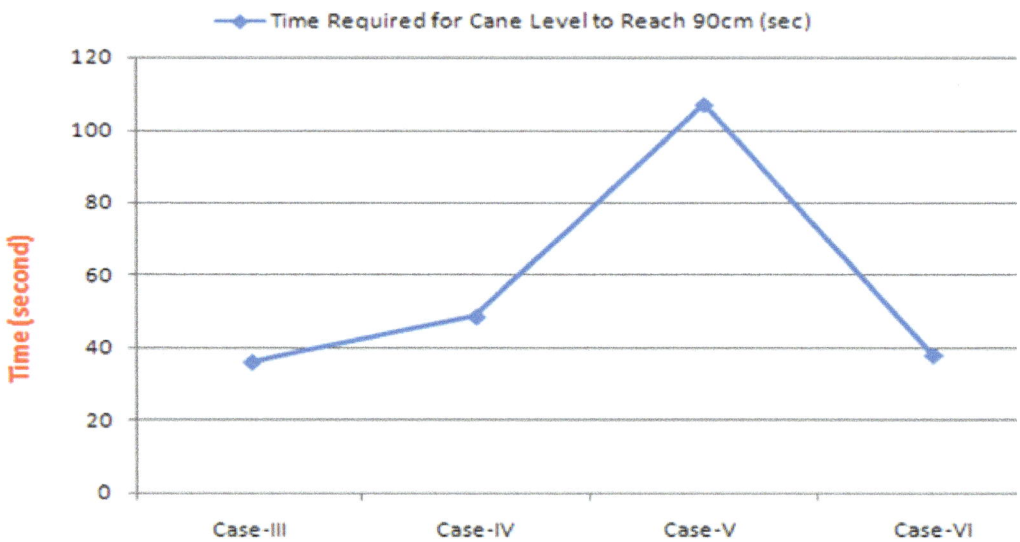

Figure 27. Time Required to Reach Cane Level at 90cm

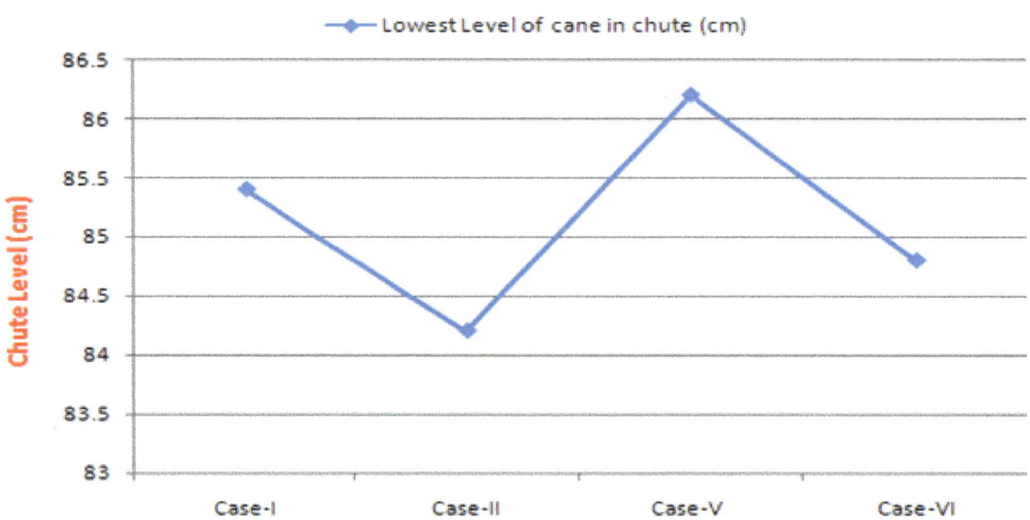

Figure 28. Lowest Level of Cane

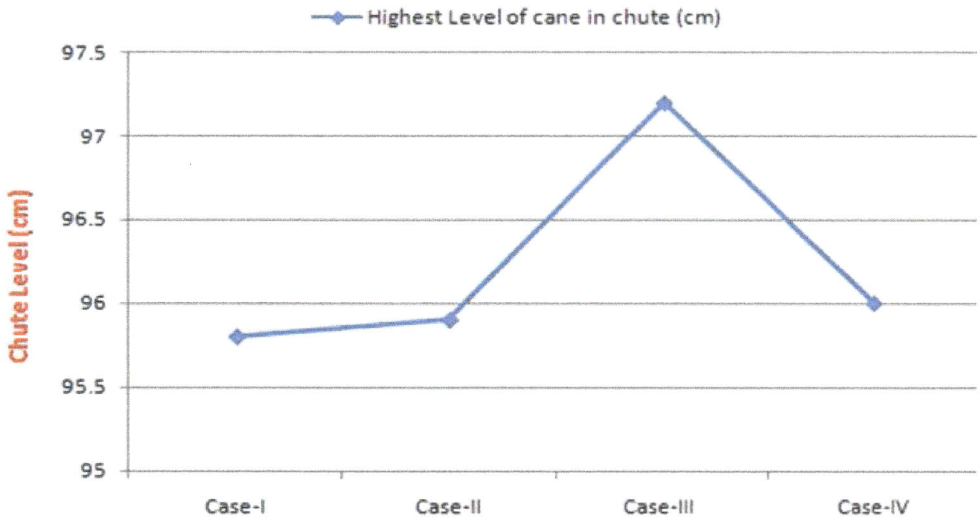

Figure 29. Highest Level of Cane

5. Conclusion

The comparison between Case-I to Case-VI gives rise following conclusions:

(i). It is evident that in Case-I and Case-II when the cane level is at 90cm at the start of simulation then fuzzy controller maintains the cane level between 85-95cm 1.6% time more if the roll speed has more variations i.e. in Case-II.

(ii). In Case-III and Case-IV when the cane level is at 0cm at the start of simulation then fuzzy controller maintains the cane level between 85-95cm 4.2% time more if the roll speed has more variations i.e. in Case-IV.

(iii). In Case-V and Case-VI when the cane level is at 180cm at the start of simulation then fuzzy controller maintains the cane level between 85-95cm 21.4% time more if the roll speed has more variations i.e. in Case-VI.

(iv). In Case-I the lowest level of cane is more near to 90cm desired level and highest level of cane is more or less same in both cases.

(v). In Case-III fuzzy controller brings the cane level at 90cm 12.5 seconds before as compared to Case-IV. The highest level of cane is more near to 90cm desired level in Case-IV.

(vi). In Case-VI fuzzy controller brings the cane level at 90cm 69.2 seconds before as compared to Case-V. The lowest level of cane is more near to 90cm desired level in Case-V.

From (i), (ii) and (iii) it can be concluded that a three input fuzzy controller maintains the cane level between 85-95cm more time if the roll speed vary during each sampling. If there are less frequent variations in roll speed then the performance of three inputs fuzzy controller degraded.

From (iv), (v) and (vi) it can be concluded that fuzzy controller performs better if there is more variation in roll speed.

Finally it can be concluded that if the roll speed is more uneven than fuzzy controller performance is better.

References

[1] D.P Kulkarni, Cane Sugar Manufacture, The sugar technologists association of India, New Delhi, 2002-03.

[2] A Jha, "India's sugar policy and world sugar economy" in *proc. FAO International sugar conference*, Fiji, August 2012.

[3] Indian Sugar Mills Association (ISMA), 2010, "Important Events" (online) available at: http://www.indiansugar.com/EventDetails

[4] S. Ahmed, *Indian sugar industry,* Centre for management studies, Jamia Millia Islamia, New Delhi, India, Unpublished.

[5] Sugar Knowledge International Ltd., "How sugar is made", (online) available at: http://www.sucrose.com/learn.html.

[6] D.J Tayfield, P.W Rein and S.R Proome, "The application of a microprocessor based system for automatic pan boiling control", *Proc. The South African Sugar Technologists Association Conf;* pp 56-62, June 1980.

[7] C.R Murry, "The Pressure Required to Feed Cane Mills". *International Sugar Journal*, 62, p. 346-349, 1960-61.

[8] T. Ozkocak and G.C Goodwin, "A Nonlinear Modeling Approach to The Sugar Cane Crushing Process". *IEEE conference on Decision and Control*, Florida, USAI.S.J., pp. 3144-3149, 1998.

[9] T. Ozkocak, M. Fu and G.C Goodwin, "Maceration control of a sugar cane crushing mill", *Proc. American Control Conf;* pp 2255-259, June 2000.

[10] Misra, Y., *Digital System Design using VHDL*, Dhanpatrai & sons (P) Ltd., India, 2006.

[11] Bhasker, J., *A VHDL Primer*, Revised Edition Prentice-Hall, 1992.

[12] Brown, S., Vranesic, Z., Fundamentals of Digital Logic with VHDL Design, McGraw-Hill, 2000.

[13] T Mitsuishi, N Endou and Y Sidhama, "The concept of fuzzy set and membership function and basic properties of fuzzy set operation" *Journal of formalized mathematics*, pp 1-6, Volume 12, 2003.

[14] Mathworks, "*Fuzzy Logic Toolbox User's Guide* ", Mathworks, Inc., 2010.

[15] R.M. Hilloowala and A.M. Sharaf, "A rule-based fuzzy logic controller for a PWM inverter in a standalone wind energy conversion scheme", in *Industry Applications, IEEE Transactions*, pp. 57-65, January/February 1996.

[16] L.Reznik, *Fuzzy Controllers,* Newnes-Butterworth-Heinemann, Oxford-Boston, 1997.

[17] B Mahdad, T Bouktir and K Srairi, "Dynamic compensation of reactive energy usinf fuzzy controller" Leonardo Electronic *Journal of practice and technologies*, pp 1-16, Issue 7, July – December, 2005.

Model of Stabilization of Helicopter in Hover Mode over a Given Point Object under Destabilizing Action of Weather Conditions

Vasiliy Kazak, Dmitry Shevchuk, Nataliia Tymoshenko*

National Aviation University, Kiev, Ukraine
*Corresponding author: t_nataly@ukr.net

Abstract The design problem of the high precision stabilizer for hovering of the helicopter about the point object as a part of the automatic control system (ACS) is considered. Here we substantiate method and models for hover mode of helicopter as well as measurements of parameters of this flight mode that requires smaller volume of calculation and preserves high speed and accuracy of stabilization. Also we investigate reduction of number of state vectors and volume of calculation via usage of extended Kalman filter (EKF) and standard sensors.

Keywords: *control, control system, helicopter, filter, perturbation, stabilization*

1. Introduction

Due to the increased usage of helicopters to perform works over the point objects: discharge of cargos into the holes of pipelines (for example, in nuclear power stations), the mounting of sections of the television and radio towers and their antennas, rescue operations, landing on a limited area etc, there is a need to stabilize the helicopter with precision accuracy in the hover mode over a point object under the negative impact of environmental factors (Figure 1) [1,2,3].

It's difficult for the crew to solve this problem. The 1st reason is fast change of conditions in the system "helicopter-point object- weather". Second reason is slow response of the system "helicopter-point object- weather" comparing with speed of the variations of perturbations. That's why it is necessary to create stabilizer as part of ACS with high speed of reaction and precision accuracy. Classic theory of control used for mentioned issue can't provide necessary accuracy and speed-of-response. LQG theory is the most popular now. Information, which is necessary for the LQG theory application for solution of the stabilization problems, could be achieved in two ways:

- usage of the additional sensors and essential improvement of their accuracy. This way is not reasonable because of increasing of the weight, seizes and cost of equipment;
- usage of standard navigation equipment and improved program product. Second way is chosen for this work.

The design of the stabilizer with precision accuracy for the hover mode of the helicopter as part of ACS requires development of the model of stabilization of the system "helicopter-point object- weather" and also the model of measurement of parameters of helicopter hovering above point objects.

Figure 1. Example of precision stabilization of the helicopter

2. Methodology

2.1. Model of System with Zero Acceleration Relatively to a Point Object

The aggressive action of weather conditions on the helicopter in flight, and especially in hovering mode, causes a violation of stabilization conditions relatively to the trajectory and this is the most important reason of organization of the precision stabilization relatively to the center of the given point object. (Figure 2, point O). Random character of inputs: internal and external perturbations, that affect helicopter in the hover mode, requires the implementation of the management strategy based on probabilistic characteristics of the measured parameters [4]:

$$z(t_k) = z\left[x(t_k), u(t_k), g(t_k), \phi(t_k), k\right],$$

where $z(t_k)$ - vector of measured reactions; $x(t_k)$ - variables of the state; $u(t_k)$ - variable input of managing actions; $g(t_k), \phi(t_k)$ - measurable and not measurable internal and external perturbations respectively. So to determine the appropriate control action $u_r(t_k)$, that would ensure the satisfaction of the conditions of steady following the actual parameter values of the helicopter hovering $X_r(t_k)$ in the changing of predetermined parameters $X_p(t_k)$ [5]:

$$\lim_{t\to\infty} X_r(t_k) = \lim_{t\to\infty} X_p(t_k). \quad (1)$$

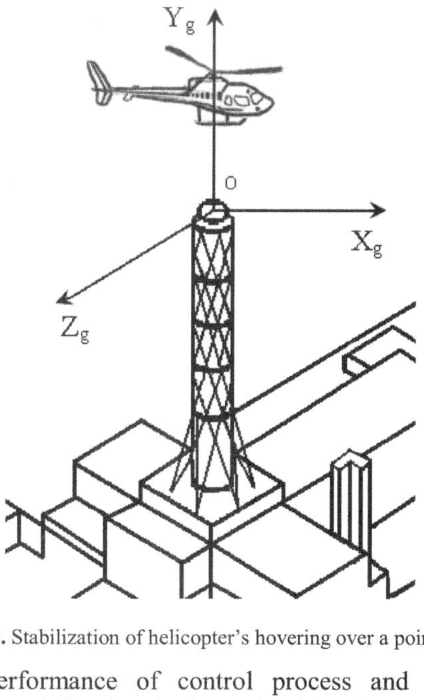

Figure 2. Stabilization of helicopter's hovering over a point object

The performance of control process and in flight helicopter stabilization is determined by the error that is caused by the mismatch $r(t_k)$ between given $x_p(t_k)$ and current $x_r(t_k)$ values of the controlled/stabilized value:

$$r(t_{k+1}) = x_p(t_k) - x_r(t_{k+1}). \quad (2)$$

Under the influence of perturbations stabilization parameters would deteriorate and mismatches (2) would not have zero value and the helicopter would be shifted from the point O (Figure 2)

$$r(t_{k+1}) = x_p(t_k) - x_r(t_{k+1}) \neq 0. \quad (3)$$

The process of observation of the requirements violation (1) can be formalized as following:

$$\overline{x}(t_{k+1}) = A\hat{x}(t_k) + Bu(t_k); \quad (4)$$

$$\hat{x}(t_{k+1}) = \overline{x}(t_{k+1}) - Kr(t_{k+1}); \quad (5)$$

$$r(t_{k+1}) = z(t_{k+1}) - C\overline{x}(t_{k+1}), \quad (6)$$

where $\overline{x}(t_{k+1})$ - estimated evaluation $x(t_{k+1})$ before measuring at time t_{k+1} ; $\hat{x}(t_{k+1})$ - evaluation of "helicopter-ACS-point object" system state parameters, that includes measurement at time t_{k+1} ; $r(t_{k+1})$ - mismatch; K - matrix of Kalman coefficients filter.

The error of the "Helicopter-ACS-Point object" system response on perturbations has also random nature and it is defined as the difference between the system state at time t_k and its change at time t_{k+1}

$$\overline{\xi}(t_k) = x(t_k) - \overline{x}(t_k); \quad (7)$$

$$\hat{\xi}(t_k) = x(t_k) - \hat{x}(t_k). \quad (8)$$

After subtracting from the system state equation (4) another equation that includes external factors and internal processes destabilizing action

$$x(t_{k+1}) = Ax(t_k) + Bu(t_k) + B\Delta u(t_k), \quad (9)$$

we obtain:

$$\overline{\xi}(t_{k+1}) = A\xi(t_k) + B\Delta u(t_k). \quad (10)$$

After combining measurements equations:

$$z(t_k) = Cx(t_k) \quad (11)$$

with equations (5), (6), (7) we obtain:

$$\hat{x}(t_{k+1}) = \overline{x}(t_{k+1}) + KC\overline{\xi}(t_{k+1}). \quad (12)$$

Let us represent the equation (12) in the form of an error

$$\hat{\xi}(t_{k+1}) = \xi(t_{k+1}) - KC\overline{\xi}(t_{k+1}). \quad (13)$$

Taking into account that "helicopter-ACS-point object" measurement of reaction of system error (13) is random and has normal distribution law, it can be characterized by mean square error. The optimality condition of such system can be defined as follows [6]:

$$\eta\left[\hat{\xi}(t_{k+1})\right] = M\left[r^2(t_{k+1})\right] = \min, \quad (14)$$

and for hovering mode

$$\eta\left[\hat{\xi}(t_{k+1})\right] = M\left[r^2(t_{k+1})\right] = 0. \quad (15)$$

The value of $\eta\left[\hat{\xi}(t_{k+1})\right]$, as the initial covariance of "helicopter-ACS-point object" system error of $\hat{\xi}(t_{k+1})$ (Figure 2) can be defined through the expected error and error variance. Generalized criterion (14) is the criterion of some function of mathematical expectation and error variance (13):

$$f\left[M(\hat{\xi}(t_{k+1})), D(\hat{\xi}(t_{k+1}))\right] = \text{extremum}. \quad (16)$$

If helicopter deviation from point O (Figure 2) in hovering mode is considered as an error, then stabilization system task is to ensure the minimum of the function (16), i.e.

$$f\left[M(\hat{\xi}(t_{k+1})), D(\hat{\xi}(t_{k+1}))\right] = 0. \quad (17)$$

Requirements (17) represent the offset of the adjustable parameter under mathematical expectation

$$\varepsilon_{\mathrm{M}}(t_k) = M_p(t_{k+1}) - M_r(t_{k+1}) \to 0 \quad (18)$$

and variance

$$\varepsilon_{\mathrm{D}}(t_k) = D_p(t_{k+1}) - D_r(t_{k+1}) \to 0, \quad (19)$$

that are written with consideration (13) and should tend to the minimal allowable values, for our task it is zero.

If there is a possibility to evaluate the performance of stabilization on the value of deviation of mathematical expectation $\varepsilon_{\mathrm{M}}(t_k)$ and variance $\varepsilon_{\mathrm{D}}(t_k)$, that provide reaching given performance parameters of the stabilization process, we will minimize the corresponding functional [7]:

$$J_\varepsilon = \frac{1}{N}\sum_{i=1}^{N}\left(\varepsilon_M^2 + \varepsilon_D\right) \to 0. \quad (20)$$

Performance criterion (20) gives an opportunity to carry out structural synthesis of stabilization and control system, with which you can implement automatic stabilization of the helicopter, including hovering mode, in the random weather conditions. Let's define the task of the structural synthesis of automatic control system (ACS) as follows: for the helicopter, given by an appropriate mathematical model, it is necessary to determine control

$$u(t_k) = f\left(x(t_k), g(t_k), \phi(t_k), V, t\right), \quad (21)$$

as a function of specified arguments that provides satisfying of condition (20). In function (21) V – is the measurement error.

Control actions (21) are formed by two parallel channels (Figure 3), one of which is a quick-action channel based on the accelerations vector $\overline{J}$, and the second – inertial channel based on the speed vector $\overline{V}$. Moreover, the signal of the inertial channel is subtracted from the quick-action channel signal.

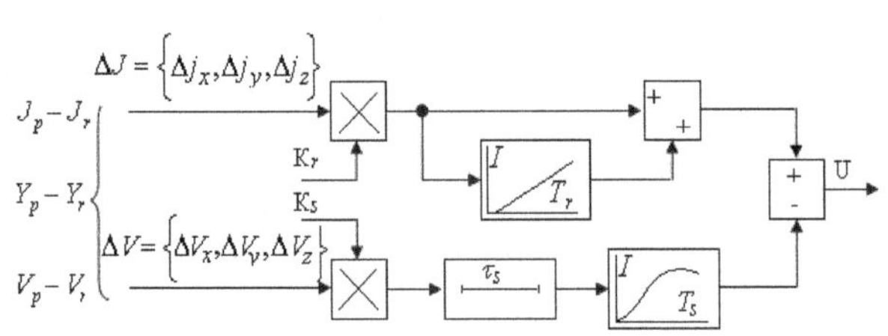

Figure 3. Block diagram of control actions formation.

The developed stabilizer (Figure 3), which is a part of ACS, solves the problem of simultaneous performance and storage stability of the automatic control system that is impossible to implement in standard schemes, where there is a contradiction, namely the growth of amplification coefficient, causes a decrease the stability margin of the automatic stabilization and control.

Acceleration sensors j_x, j_y, j_z, that characterize short-period oscillations of the helicopter center of masses, are first to react on the violations of conditions (20). The second sensors are the sensors of the displacement rate of the helicopter center of masses O (Figure 2) $\dot{H}, \dot{X}, \dot{Z}$, which characterize long-period of the helicopter center of mass oscillations. Therefore, to solve the task of helicopter on given point object O (Figure 2) point hovering stabilization, you need to make a model of the system with zero accelerations relative to the centre of the point object. It should be noted, that in case of using of such model the system (9) state vector, and therefore a vector of measurements (11), taking into account that accelerations vector is the first derivative of the speed and the second from the distance, are significantly reduced (up to 6 parameters). In view of this the differential equation of "helicopter-point object" system takes the form [8]:

$$\begin{aligned}&\Delta\dot{S}(t_k) = \Delta V(t_k),\ \Delta S = \{\Delta X, \Delta Y, \Delta Z\},\\ &\Delta V = \{\Delta V_x, \Delta V_y, \Delta V_z\};\\ &\Delta\dot{V}(t_k) = -J_h(t_k) + \dot{W}_h,\ J_h = \{j_x, j_y, j_z\},\\ &\dot{W}_h = \{\dot{W}_x, \dot{W}_y, \dot{W}_z\},\end{aligned} \quad (22)$$

where $J_h = \{j_x, j_y, j_z\}$ - the accelerations vector, $\dot{W}_h = \{\dot{W}_x, \dot{W}_y, \dot{W}_z\}$ - the perturbation accelerations vector.

1st Results

Therefore, "helicopter-point object" system model condition evaluation using extended Kalman EKF filter acquires following form:

$$\begin{bmatrix}\Delta\dot{S}(t_k)\\ \hline \Delta\dot{V}(t_k)\end{bmatrix}_{6\times1} = \begin{bmatrix}0 & | & I\\ \hline 0 & | & 0\end{bmatrix}_{6\times6}\begin{bmatrix}\Delta S(t_k)\\ \hline \Delta V(t_k)\end{bmatrix}_{6\times1} + \begin{bmatrix}0 & | & 0\\ \hline 0 & | & I\end{bmatrix}_{6\times6}\begin{bmatrix}0\\ \hline -J_h(t_k)\end{bmatrix}_{6\times1} + \begin{bmatrix}0\\ \hline W_h(t_k)\end{bmatrix}_{6\times1}. \quad (23)$$

If "helicopter-point object" system model is given by the relation of type (22), then you can make the state transition matrix for it as follows [5]:

$$\Phi(t_k, t_{k+1}) = \begin{bmatrix}I & | & \Delta t I\\ \hline 0 & | & I\end{bmatrix} \quad (24)$$

Thus, using the information from acceleration sensors, that are the part of inertial navigation system of the helicopter, you can significantly reduce the state vector without loss of accuracy helicopter's hovering stabilization.

2.2. Measuring Model of Helicopter Hovering Stabilization Parameters with Respect of the Point Objects

To organize the functioning of the linear-quadratic Gauss (LQG) stabilizer (Figure 3) based on the extended Kalman filter EKF (23), it can be used the measurement model consisting of two measurements: speed $V(t_k)$ and acceleration $J(t_k)$ in the inertial frame. Taking into account that helicopter speed $\dot{S}(t_k)$ has nature of long-period oscillations, and the rate of this speed changes (acceleration), which incidentally is a derivative of speed $dV(t_k)/dt_k$, has nature of the short-period oscillations, they can be divided and considered as a case of nested models of measurements $J_h(t_k)$ and $\dot{S}(t_k)$, i.e. $J_h(t_k)$ is a nesting in $\dot{S}(t_k)$, or that is the same $J_h(t_k)$ is a subset of a set $\dot{S}(t_k)$. It is proposed to form an identification of helicopter acceleration changes $J_h(t_k)$ via usage of the errors prediction of the parameters measurement $V(t_k)$, which is obtained from the inertial navigation system (INS), GPS receiver or Doppler radar for measuring velocity and drift angle (DMVD), as the data for evaluation procedure. The procedure proposed has two stages: measured data with error prediction method is displayed by the model in $\dot{S}(t_k)$ space, in turn $\dot{S}(t_k)$ is displayed in space $J_h(t_k)$ [9].

Therefore, the way of evaluation based on the use of parametric vector $J_h(t_k)$ and parametric vector $V(t_k)$ error prediction method $\overline{\xi}_J$ can be set as follows:

$$\hat{\xi}_J = \arg\min_{\xi_J} \sum_{t=1}^{N} \varepsilon_1^2(t_k, \overline{\xi}_J); \qquad \hat{\xi}_v = \arg\min_{\xi_v} \sum_{t=1}^{N} \varepsilon_2^2(t_k, \overline{\xi}_v), \tag{25}$$

where N is the number of measurements (data recording points); $\varepsilon_1(t_k, \overline{\xi}_J)$ – error of the prediction.

$$\varepsilon_1(t_k, \overline{\xi}_J) = j(t_k) - \hat{j}(t_k / t_{k-1}, \overline{\xi}_J)$$

in the model of prediction $J_h(t_k)$.

In a further we enter two assumptions regarding the structures of the models $J_h(t_k)$, $V(t_k)$ and data [8]:

1. Structures $J_h(t_k)$ and $V(t_k)$ are attached

$$J_h(t_k) \subset V(t_k), \tag{26}$$

so that there is a smooth displaying

$$\xi_v(\overline{\xi}_J): S_1 \to S_2; S_1 \subset R^{\dim \xi_J},\ S_2 \subset R^{\dim \xi_v}, \tag{27}$$

so that

$$\varepsilon_2(t_k, \xi_v(\overline{\xi}_J)) = \varepsilon_1(t_k, \overline{\xi}_J), \tag{28}$$

where S_1, S_2 - nonzero sets; $\varepsilon_1(t_k, \xi_J)$ is the stationary process for any $\xi_J \in S_1$; $\frac{\partial \xi_v(\overline{\xi}_J)}{\partial \xi_J} = \max$ on S_1.

2. Structures $J_h(t_k)$ and $V(t_k)$ allow to identify the parameters that confirms the existence of single "real" vectors of parameters ξ_v^* and ξ_J^* such that:

$$\varepsilon_1(t_k, \xi_J^*) = \varepsilon_2(t_k, \xi_v^*)$$

-white noise with mean zero and σ^2-dispersion.

Taking into account the assumptions one and conditions (28) we can record a conclusion [9]:

$$\xi_v^* = \xi_v(\xi_J^*). \tag{29}$$

The first assumption is a general condition for attached models J_h and $\dot{S}$. In turn the model $\dot{S}(t_k)$, in addition to the above, includes also the characteristics of the data. In view of the fact that $J_h(t_k)$, represents an attached into $\dot{S}(t_k)$ subset, i. e. the model of more simplified structure, it should provide a clear identification model than $\dot{S}(t_k)$. Indeed, in view of the fact that $\ddot{S}(t_k) = \{\ddot{S}_x(t_k), \ddot{S}_y(t_k), \ddot{S}_z(t_k)\}$, measured with sensors of linear accelerations SLA_x, SLA_y, SLA_z, their signals create the measurement model of more simplified structure with clearer information. In addition, the implementation of such model is more affordable to construct, because these sensors are the part of the regular inertial navigation system.

In the practical implementation there are cases where $\hat{J}_h$ evaluation is more difficult to calculate than $\hat{\xi}_v = \arg\min_{\xi_v} \sum_{t=1}^{N} \varepsilon_2^2(t_k, \overline{\xi}_v)$ evaluation, so it is advisable to determine $\hat{\xi}_v$ and on these evaluations to determine the evaluations $\hat{J}$. If S_2 set in not confluent (not zero) in space $R^{\dim \xi_v}$, then for $N \to \infty$, one could argue that under $\xi_v^* \in S_2$, $\hat{\xi}_v$ it also belongs to the S_2 and enables to calculate $\overline{J}_h$ under the ratio:

$$\overline{J}_h = J_h(\hat{\xi}_v), \tag{30}$$

where $J_h(\hat{\xi}_v)$ is the inverse function

$$J_h(\xi_v): S_2 \to S_1.$$

For our case, we can accept that $\overline{J}_h = \hat{J}_h$. Under the assumption that $\hat{\xi}_v \in S_2$ can be shown as [9]:

$$\sum_{t=1}^{N} \varepsilon_1^2(t_k, J_h) = \sum_{t=1}^{N} \varepsilon_2^2(t_k, \overline{\xi}_v(J_h)) \geq \sum_{t=1}^{N} \varepsilon_2^2(t_k, \hat{\xi}_v) = \sum_{t=1}^{N} \varepsilon_1^2(t_k, J_h),\ J_h \in S_1. \tag{31}$$

Condition (31) requires the implementation of equality $\overline{J}_h = \hat{J}_h$ and that the dimension J_h was lower than dimension ξ_v, which is not always done. If these conditions are not met, i.e. $\hat{\xi}_v \notin S_2$, condition (30) is not met too. So, in the future during evaluation of predicting errors that condition cannot be used. In such cases, the vector $\overline{J}_h$ can be defined as a solution of an optimization task [9].

$$\overline{J}_h = \arg\min_{J_h} G(J_h),$$

$$G(J_h) = 0,5\left[\hat{\xi}_v - \xi_v(J_h)\right]^T Q_{\xi_v}^{-1}\left[\hat{\xi}_v - \xi_v(J_h)\right], \quad (32)$$

where $\hat{Q}_{\xi_v}$ - reasoned evaluation

$$Q_{\xi_v} = \left\{ E\left(\left.\frac{\partial \varepsilon_2(t_k,\xi_v)}{\partial \xi_v}\right|_{\bar{\xi}_v = \xi_v^*} \right)^T \left.\frac{\partial \varepsilon_2(t_k,\xi_v)}{\partial \xi_v}\right|_{\bar{\xi}_v = \xi_v^*} \right\}^{-1}. \quad (33)$$

Natural evaluation Q_{ξ_v} may be determined by the ratio:

$$\hat{Q}_{\xi_v} = \left\{ \frac{1}{N}\sum_{t=1}^{N} \left(\left.\frac{\partial \varepsilon_2(t_k,\xi_v)}{\partial \xi_v}\right|_{\bar{\xi}_v = \hat{\xi}_v} \right)^T \times \right.$$
$$\left. \times \left.\frac{\partial \varepsilon_2(t_k,\xi_v)}{\partial \xi_v}\right|_{\bar{\xi}_v = \hat{\xi}_v} \right\}^{-1} \quad (34)$$

Optimization task (32) can be formed by the following rule:

$$\begin{cases} \tilde{\xi}_v = \arg\min\limits_{\alpha \in S_2} (\hat{\xi}_v - \xi_v)^T Q_2^{-1} (\hat{\xi}_v - \xi_v); \\ \bar{J}_h = J_h(\tilde{\xi}_v). \end{cases} \quad (35)$$

Regarding the challenge of stabilization of helicopter in hover mode over a point object with a limited area, a geometric representation can be provided. While hovering, the helicopter is influenced by destabilizing the random external weather conditions and internal processes, so the condition of the system "helicopter-point object" (Figure 2) is not determined. Under these conditions, to provide precision accuracy of helicopter stabilization, it is necessary to know the error of the prediction of evaluation of parameter vector $\bar{J}_h(t_k)$ of the enclosed model and parametric vector in model $V(t_k)$.

2nd Results

The error of the prediction of evaluation of helicopter accelerations in hovering mode j_x, j_y, j_z, that appear under false weather conditions, will greatly depend on the amount and quality of information about its speed $V(t_k)$, which comes from the inertial navigation system, satellite navigation system and Doppler velocity meter and drift angle, so on the errors arising from the violation of the conditions of precision accuracy of hovering

$$\Delta V_x = \frac{dS}{dt} \neq 0,\ \Delta V_y = \frac{dH}{dt} \neq 0,\ \Delta V_z = \frac{dz}{dt} \neq 0, \quad (36)$$

$$\Delta j_x = \frac{d^2S}{dt^2} \neq 0,\ \Delta j_y = \frac{d^2H}{dt^2} \neq 0,\ \Delta j_z = \frac{d^2z}{dt^2} \neq 0. \quad (37)$$

3. Conclusions

Using the information from the accelerations sensors, which are part of the inertial navigation system of the helicopter, you can essentially reduce the state vector $x(t_k)$ without perturbation of the requirements for accuracy of stabilization of helicopter in hover mode over a point object. Besides, changes under the external influences of helicopter accelerations $J(t_k)$ are characterized by short-period oscillations (requirement 32) of the helicopter centre of mass and significantly ahead in time changes in its velocity vector $V(t_k)$, which are characterized by long-period oscillations of the helicopter in space relatively to the given point O (Figure 2) of the limited area object (requirement 30). Therefore completing the prediction for the given method of inaccuracies, arising from the perturbation of the requirements (31), we can identify the state of the «helicopter-point object» system, in another words the degree of fulfillment of the requirements (25). The degree of the requirements (36) and (37) perturbations can serve as a signal for the functioning of the stabilization system for a helicopter in the hovering mode.

References

[1] Beltsov V. Holdin the Sky In the Strong Hands. All-Russian aerospace journal "Herald of Aviation and Cosmonautics", No.1, 2000, pp. 8-10.

[2] Raymond W. Prouty and H. C. Curtiss. *"Helicopter Control Systems: A History"*, Journal of Guidance, Control, and Dynamics, Vol. 26, No. 1, 2003, pp. 12-18.

[3] Prouty R.W. *Helicopter Performance, Stability, and Control.* first ed. Krieger Publishing Company, London, United Kingdom, 2002, 746 pp.

[4] Kazak V.N., Tunick A.A., Salimon V.I. System of Automatic and Semi-automatic Flight Control. Monograph. / Ed. by V.N Kazak K.: ed. NAU, 2002, 201 pp.

[5] Livneh L., Slater G. *Adaptive controller for wing flutter with unmeasurable excitations and unmodelled dynamics.* "Int.J.Cont", 1991, 53, No. 2, pp. 293-309.

[6] Kazak V.M. Fundamentals of control and technical diagnostics. K.: NAU, 2013, 300 pp.

[7] Kazak V.M. *System recovery methods survivability of aircraft in special situations in flight*: monograph / V.M. Kazak. – K.: NAU, 2010, 284 pp.

[8] Paul Verges, Sammons, J.M., Balakrishman S., Speyer J.L., and Hull, D.G. *"Development and Comparison of Optimal Filters"* AFATL – TR-79-87, Oct. 1979.

[9] Soderstrom T., Stoica P., Friedlander B. *An indirect prediction error method for system identification.* "Automatica", 1991, 27, No.1, pp. 183-188.

[10] Wentzel, E.S. *Probability Theory*: Proc. for universities. - 6th ed. sr. - M.: Higher. sch., 1999. 576 pp.

35

Experimental Testing of Wireless Sensors Network Functionality

Alena Galajdová*, Dušan Šimšík

Department of Automation, Control and Human Machine Interactions, Technical University of Košice, Slovakia
*Corresponding author: alena.galajdova@tuke.sk

Abstract Paper deals with wireless communication networks. Goal is to describe the most commonly used wireless protocols and test their reliability, range and coexistence with others wireless protocol, especially WiFi. The biggest problem in sensor networks using wireless networks PAN (e.g, IEEE 802.15.4 , ZigBee , ZWave ...) is their range, especially in an urban area. In most cases it is very difficult to achieve distance stated by the manufacturer in the technical specification. XBee platform has been tested on common building where monitored variables are number of sent packets, number of received and lost packets and lost ratio for different distance from server and several time differences between sending messages from server.

Keywords: *wireless sensor network, reliability, XBee, indoor testing*

1. Introduction

The advantages of a wireless sensors network take effect in their implementation especially in older buildings. During the design of majority older buildings the realisation of system for physical parameters measuring such as temperature, humidity, intensity of lighting was not assumed. So the introduction of such a wireless system in older building will not require construction work, thereby substantially reduce the cost of its implementation.

Being reliable and real-time are the two most challenging requirements of wireless sensor network (WSN) technology. As IEEE 802.15.4 has relatively short range and low bandwidth, but is widespread in home automation it poses a significant limitation in application of WSN in bigger old buildings. Performance of WSN should be evaluated for every implementation of WSN in older building. Depending on the operating environment, significant signal loss can occur, particularly when the radios require line-of-sight for optimal performance, with 2.4 GHz more susceptible than 900 MHz. Among the different performance parameters, the packet lost ratio should be evaluated for any application.

Our goal was to describe the most commonly used wireless protocols and test their reliability, range and coexistence with others protocols.

Verification of a range for different communication protocols based on XBee platform and their coexistence with other wireless protocols (especially WiFi) implemented in three stored house was realised [1].

2. Methods

IEEE 802.15.4 is included in group of standards LR WPAN (Low rate - Wireless Personal Area Network). These networks are designed to transmit data over short distances. Unlike conventional WLANs (e.g. WiFi), WPAN networks have simple infrastructure, hence are suitable for cheap and effective solution with low power consumption. IEEE 802.15.4 defines two layers - physical and MAC sublayer (Medium Access Control) for low power and short distances (typically up to 10 m). Standard supports multiple frequencies for its operation.

Communication based on this standard consists of two types of devices FFD (Full-function device) and RFD (reduced-function device). FFD is capable communicate as coordinator of PAN (Personal Area Network). PAN coordinator is the main elements of the network that initiates and sets up networks. RFD is a device that is designed as a terminal and communicates only with one superior FFD device. RFD devices send only small amounts of data to have the requirements for energy as small as possible, and they can be long term supplied by the battery [2].

Depending on the needs of different applications, the standard offers two types of topologies: star and peer-to-peer topology with mesh support. In the standard there are three types of transmissions. The mechanism for each type of transmission depends on whether the network supports a beacon frames and super frames, i.e. networks, where synchronization is required or networks where fast response is required. The first type of transmission is to the coordinator from device, where the device is sending data and coordinator receiving data. The second type of transmission is from the coordinator to the device. The third type of transmission is peer-to- peer network where each device communicates with each device in an accessible range [2]. Technologies based on IEEE

802.15.4 operate in a non licensed ISM band (Industrial Scientific and Medical).

XBee from Digi International offers several models to support the ZigBee protocol and its XBee protocol on open ISM band 868 MHz, 2.4 GHz and 900 MHz. Communication chips are divided into two main categories: Standard and PRO. Chips marked as PRO have higher transmition power and are also larger in size.

XBee 802.15.4 These modules use a standard implementation of the IEEE 802.15.4 protocol. They are produced in two versions XBee 802.15.4 and 802.15.4 XBee PRO. The speed of modules is 250 kbps and operate in an open 2.4 GHz ISM band. They support point-to-multipoint or peer-to- peer communication. Chip has no support for mesh networking. XBee 802.15.4 IEEE are mainly used as terminal devices for different areas, especially for home and building automation and industry. Their main advantage is low cost and power of 1 mW with range of 100 meters. PRO version range is 1.6 km with direct visibility, but also much more power (60 mW). [3], [4].

XBee ZB Radio communication chip, which has an implementation of an open Zigbee protocol. The baud rate is 250 kbps. Support type of communication is mesh, point-to-point, point-to-multipoint and repeater. Open transmission operates in the 2.4 GHz band, 2 mW power and reach of 120 meters. In the PRO version has power of 50 mW and range 1.6 km. This communication chip advantage is that the ZigBee protocol is included in, and is compatible with equipment from other manufacturers operated with Zigbee protocol.

XBee Pro 868 The radio module for wireless communication with a long range. XBee PRO 868 OEM RF module operates in the 868 MHz ISM band, with better RF signal propagation, for example, in urban areas. Larger transmission distances can be achieved thanks to higher radiated power compared to modules, operating in the 2.4 GHz band. Radiated power can be adjusted. The current version supports peer-to-peer, point-to-point and point-to-multipoint topology. Typical areas of application are telemetry, monitoring and industrial control systems and data acquisition systems. Transmission speed is 24 kbps. The outdoor range with line sight is up to 40 km and at built-up area 550 m. Adjustable transmitting power is up to 500 mW. [4]

XBee WiFi On-chip is implemented communication network protocol WiFi – standard IEEE 802.11. The modules can easily communicate with existing 802.11 infrastructures, with security WPAPSK or WPA2PSK. Supported standards are 802.11a /b/g /n. Modules provide 10 digital input/output lines, 4 analogue inputs and during standby mode their consumption is only 2uA. The main applications are sensor networks, automation of processes, buildings and household energy management.

2.1. Testing

We evaluated the working range of sensor network in a common older building. Testing was realised by queries randomly selected for a particular pin of each tested device from a static server location. Server was placed on 1st floor in an edge room with balcony, marked with black dot at Figure 2. We created a single application for testing (Figure 1).

During testing were monitored variables as follows:

- number of sent packets - in each test for each device it was sent 1000 reports,
- number of received packets – answers,
- the number of lost packets - the difference between the sent and received messages,
- lost ratio - the ratio of lost and outgoing messages.

Parameters for test performing were:

- distance from the server,
- time difference between sending messages from the server: 0.3s, 0.15s, 0.1s, 0.05s, 0.03s, 0.025s.

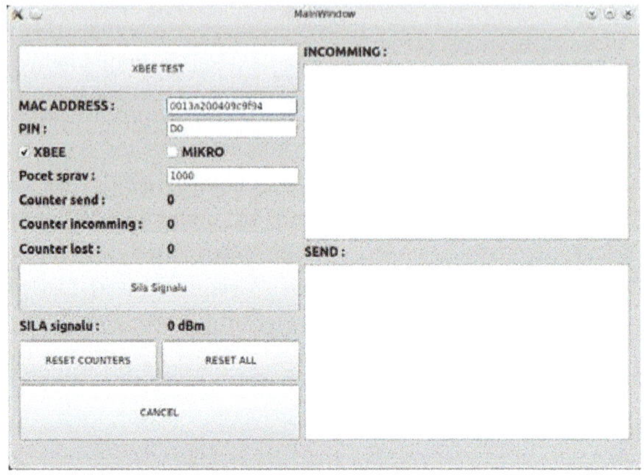

Figure 1. Testing application

Testing was done with chips:

- XBee IEEE 802.15.4
- Digi XBeePRO 868
- XBee DigiMesh 2.4
- XBee DigiMesh 2.4 PRO
- XBee ZB

Figure 2, Figure 3 and Figure 4 describe floor plans of individual rooms around the house. In the upper room is placed main server – black dot, from which were all tests of communication triggered. Red ring determines the location of the test units.

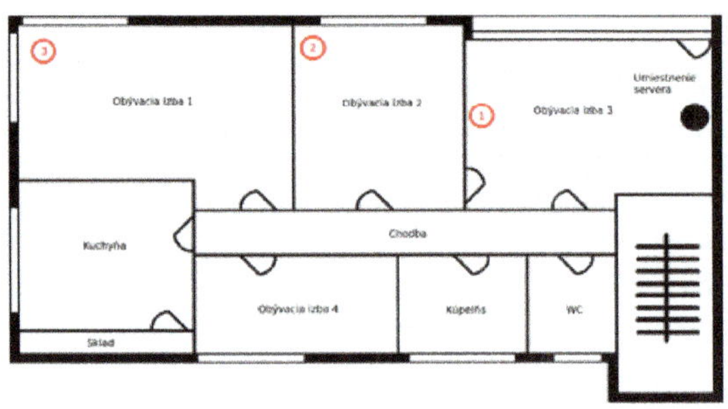

Figure 2. 1st floor plan

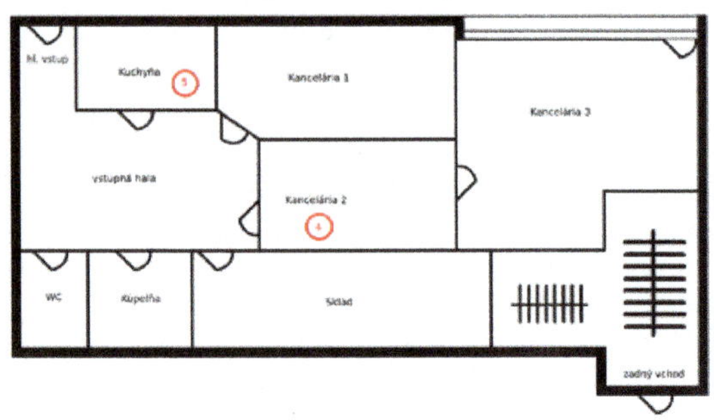

Figure 3. Ground floor plan

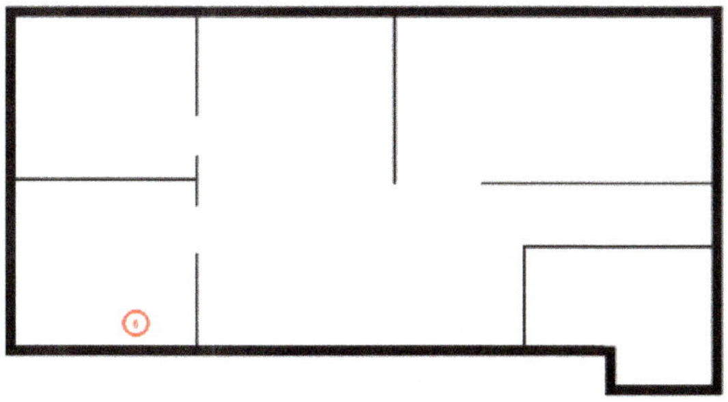

Figure 4. Plan of basement

2.1.1. Test XBee 802.15.4

Testing took place only with node number 1 because, it was not possible to reach with this device to another nodes placed further.

Table 1. Measurement Protocol for XBee IEEE 802.15.4 and compatibility with WiFi

Device XBee IEEE 802.15.4		
Measurement Protocol		
MAC Coordinator 0013a2004069822c	MAC- End device 0013a20040698219	
Delay	Received	% Packed lost ratio
0,3	994	0,006
0,15	995	0,005
0,1	990	0,01
0,05	899	0,101
0,03	733	0,267
0,025	620	0,38

Compatibility between XBee 802.15.4 and WiFi			
Distance	Send packets	Received packets	% Packed lost ratio
0	1000	98	90,2
30	1000	150	85
70	1000	547	45,3
150	1000	993	0,7

This test shows that the IEEE 802.15.4 chips are suitable for use in the room. Communication to another room through the wall was impossible.

2.1.2. Test XBeePRO 868

Table 2. Measurement Protocol for XBeePRO 868

Device XBeePRO868				
Measurement Protocol				
	MAC Coordinator 0013a20040670c72		MAC End device 0013a20040670be4	
	End device placement -1		End device placement -2	
Delay	Received	% Packed lost ratio	Received	% Packed lost ratio
0,3	994	0,006	992	0,008
0,15	992	0,008	997	0,003
0,1	995	0,005	991	0,009
0,05	991	0,009	995	0,005
0,03	688	0,312	655	0,345
0,025	598	0,402	588	0,412
	End device placement -3		End device placement -4	
Delay	Received	% Packed lost ratio	Received	% Packed lost ratio
0,3	997	0,003	992	0,008
0,15	992	0,008	996	0,004
0,1	995	0,005	991	0,009
0,05	993	0,007	990	0,010
0,03	647	0,353	628	0,372
0,025	530	0,470	520	0,480
	End device placement -5		End device placement -6	
Delay	Received	% Packed lost ratio	Received	% Packed lost ratio
0,3	991	0,009	993	0,007
0,15	987	0,013	992	0,008
0,1	986	0,014	982	0,018
0,05	988	0,012	971	0,029
0,03	671	0,329	670	0,330
0,025	510	0,490	520	0,480

Compatibility between XBee PRO 868 and WiFi			
Distance	Send packets	Received packets	% Packed lost ratio
0	1000	759	24,1
30	1000	991	0,9
70	1000	994	0,6

XBeePRO had no problem to communicate with the most delayed node in building without the difference in packet lost from nearest to most farthest node.

Its disadvantage is high energy consumption, but its high transmission power is suitable for automation of bigger buildings or in industry, where more robust solution is necessary for bigger distances.

2.1.3. Test XBee DigiMesh 2.4

DigiMesh protocol is similar to the ZigBee protocol and is designed for the same applications - home and building automation and sensor networks. It is a proprietary (description of the implementation is not public)

Table 3. Measurement Protocol for XBee DigiMesh

Device XBee DigiMesh 2.4				
Measurement Protocol				
	MAC Coordinator 0013a20040b65e78		MAC End device 0013a20040b65e30	
	End device placement -1		End device placement -2	
Delay	Received	% Packed lost ratio	Received	% Packed lost ratio
0,3	992	0,008	987	0,013
0,15	991	0,009	984	0,016
0,1	992	0,008	962	0,038
0,05	991	0,009	846	0,154
0,03	740	0,260	715	0,285
0,025	612	0,388	611	0,389

Compatibility between XBee DigiMesh and WiFi			
Distance	Send packets	Received packets	% Packed lost ratio
0	1000	236	76,4
30	1000	992	0,8
70	1000	995	0,5

2.1.4. Test XBee DigiMesh 2.4 PRO

Table 4. Measurement Protocol for XBee PRO DigiMesh

Device XBee PRO DigiMesh				
Measurement Protocol				
	MAC Coordinator 0013a20040a6bdd9		MAC End device 0013a20040a6bddb	
	End device placement -1		End device placement -2	
Delay	Received	% Packed lost ratio	Received	% Packed lost ratio
0,3	993	0,007	991	0,009
0,15	994	0,006	994	0,006
0,1	994	0,006	993	0,007
0,05	993	0,007	994	0,006
0,03	680	0,320	688	0,312
0,025	560	0,440	547	0,426
	End device placement -3		End device placement -4	
Delay	Received	% Packed lost ratio	Received	% Packed lost ratio
0,3	996	0,004	990	0,010
0,15	991	0,009	994	0,006
0,1	994	0,006	993	0,007
0,05	995	0,005	996	0,004
0,03	688	0,312	667	0,333
0,025	549	0,451	570	0,430
	End device placement -5		End device placement -6	
Delay	Received	% Packed lost ratio	Received	% Packed lost ratio
0,3	986	0,014	993	0,007
0,15	983	0,017	902	0,098
0,1	970	0,030	899	0,101
0,05	995	0,005	911	0,089
0,03	636	0,364	588	0,422
0,025	536	0,464	492	0,508

Compatibility between XBee DigiMesh 2.4 PRO and WiFi			
Distance	Send packets	Received packets	% Packed lost ratio
0	1000	236	76,4
30	1000	992	0,8
70	1000	995	0,5

2.1.5. Test XBee ZB

Test shows that the XBee ZB is suitable for home automation and one chip embraces without a problem two rooms.

Table 5. Measurement Protocol for XBee DigiMesh

Device XBee DigiMesh				
Measurement Protocol				
	MAC Coordinator 0013a20040		MAC End device 0013a20040	
	End device placement -1		End device placement -2	
Delay	Received	% Packed lost ratio	Received	% Packed lost ratio
0,3	997	0,003	992	0,008
0,15	996	0,004	995	0,005
0,1	998	0,002	994	0,006
0,05	999	0,001	874	0,126
0,03	789	0,211	769	0,231
0,025	656	0,344	645	0,355

Compatibility between XBee ZB and WiFi			
Distance	Send packets	Received packets	% Packed lost ratio
0	1000	178	82,2
30	1000	875	12,5
70	1000	996	0,4

3. Results

The reliability of chips regarding to distance and coexistence with other wireless protocols - primarily with the protocol WiFi was tested.

Devices XBee PRO DigiMesh and XBee PRO 868 had very similar performance regarding the distance from server and time difference between sending messages from the server. They were able communicate reliable with all nodes even with the farthest node number 6 placed in basement.

Packets lost rate increased with smaller time difference between sending messages from server and it was not influenced by distance from server (Figure 5, Figure 6).

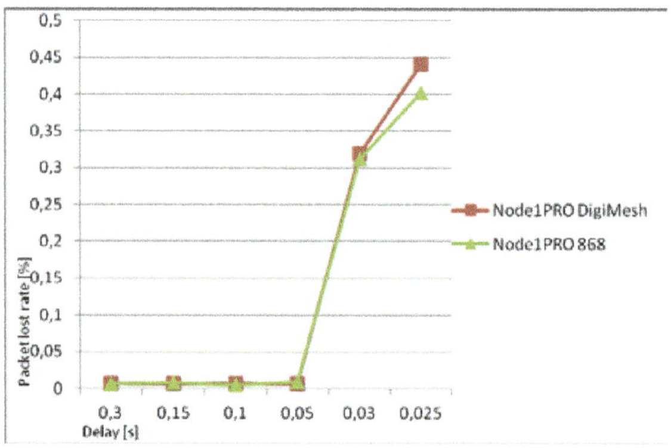

Figure 5. Packet lost rate with delay between sending messages to next node

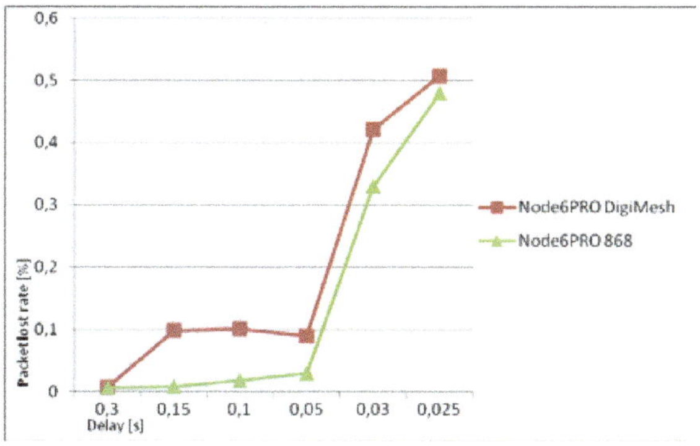

Figure 6. Packet lost rate with delay between sending messages to farthest node

The test results differed significantly from statements of manufacturer. According results of these tests with XBEE platform, several recommendations for installation and use in older buildings environment were created.

In the frame of availability, reliability and coexistence of XBEE network testing, the interaction between the individual chips was also tested. During testing it was found that the chips do not influence each other.

4. Conclusions

As a result of testing we worked out followed recommendations for the use of wireless protocols in older buildings:

- Verify the coexistence of selected protocol with other wireless networks already used in buildings - especially WiFi.
- The maintaining of sufficient distance from other wireless devices (for example for XBEE chips)
- The maintaining distance from the interference of high-voltage power components, which can affect any wireless networks, such as electric motors, power cables above 1 kV.
- When installing in buildings or houses, it is necessary to know the current electrical installation and to place chips out of main wiring routes.
- Ensure a sufficient distance (maximum and minimum) between communication chips for reliable communication (distance needs to be experimentally verified).
- Chips placing away from large metal surfaces - metal surfaces cause interference, attenuation and signal reflections.
- Ensure of network partial redundancy - in the case of communication chip loss, the chips must be linked to other alternative route.
- Appropriate management of asynchronous communication - as in the case of an insufficient response to the request, this request has to be repeated, or possible transmission errors have to be detected.

Acknowledgement

This work has been supported by the Slovak Grant Agency VEGA contract N. 1/0911/14 Implementation of wireless technologies into the design of new products and services to protect human health

References

[1] Biroš, O., Karchňák, J., Šimšík, D., "Multiagent approach in home care based on open architecture" *American Journal of Mechanical Engineering,* Vol. 1, no. 7 (2013), p. 217-220.

[2] *IEEE Standard for Local and metropolitan area networks— Part 15.4: LowRate Wireless Personal Area Networks (LRWPANs)* Available: www.ieee.org/.[Accessed Sept. 2, 2015]

[3] *Typy modelov XBEE, Available : http://www.macro.sk/index.php?p=DigiRFmoduly/*.[Accessed Sept. 2, 2015]

[4] *Opis produktu XBEE 802.15.4, Avaliable: http://www.digi.com/products/wirelesswiredembeddedsolutions/zigbeerfmodules/pointmultipointrfmodules/XBeeseries1module#overview/*.[Accessed Sept. 2, 2015]

Design and Tuning of Parallel Distributed Compensation-based Fuzzy Logic Controller for Temperature

Snejana Yordanova*, Yavor Sivchev

Faculty of Automation, Technical University of Sofia, Sofia, Bulgaria
*Corresponding author: sty@tu-sofia.bg

Abstract The parallel distributed compensation (PDC) gains popularity in designing of simple fuzzy logic controllers (FLCs) for nonlinear plants taking advantage of the well-developed linear control theory. The established design approaches suffer difficulties in derivation of standard TSK plant models for processes characterized by time delays, model uncertainties, nonlinearities, inertia, etc., needed for the PDC design. The standard PDC structure is also unfit for parameter optimization and embedding in programmable logic controllers which ensures a broad industrial application. The aim of the investigation is to develop an approach for the design and tuning of a modified process PI PDC-FLC and an on-line fuzzy logic supervisor via off-line GA optimization techniques in order to ensure closed loop stability, robustness, desired performance and energy efficiency. The approach is developed on the example of air temperature control in a laboratory dryer using MATLAB™. Simulations and real time control shows a decreased settling time and increased system robustness and energy efficiency compared to linear PI control.

Keywords: *fuzzy logic supervisor, genetic algorithms, process parallel distributed compensation, temperature real time control, tsk models*

1. Introduction

The parallel distributed compensation (PDC) gains popularity in designing of simple fuzzy logic controllers (FLCs) for nonlinear plants taking advantage of the well-developed linear control theory [1,2]. The nonlinear plant is represented by a TSK plant model with local linear plant models in the conclusions of the fuzzy rules, each described by a state vector x_{nx1}, an output vector y_{mx1}, an input control vector u_{dx1} and the corresponding matrices A_{nxn}, B_{nxd}, C_{mxn} and D_{mxn}:

If z_1 is Lz_1 **And** … **And** z_n is Lz_n

Then $\dot{x}(t) = Ax(t) + Bu(t)$ **And** $y(t)=Cx(t)+Du(t)$.

The premise variables z with linguistic values Lz recognize the sub-domain for the corresponding local linear plant model. The overall nonlinear plant output is obtained as a fuzzy blending of individual rules (local plants) outputs via the fuzzy inference and defuzzification mechanisms.

According to the principle of PDC a linear controller (state feedback, dynamic compensator, PID, internal model controller, Smith predictor, etc. [2,3,4]) is designed for each local linear plant to ensure stability, robustness and desired performance of the local linear closed loop system using methods from the linear control theory thus compensating a corresponding conclusion in the rules of TSK plant model. The PDC-FLC rules have the same premise as in the TSK plant model and conclusions with the state space descriptions of the linear controllers, for local state feedback the rules take the form:

If z_1 is Lz_1 **And** … **And** z_n is Lz_n **Then** $u=-Fx$.

The final nonlinear control is a fuzzy blending of the individual rules control actions. The global nonlinear system stability is studied mainly by the help of Lyapunov methods and linear matrix inequalities (LMIs) numerical technique [1,2,3,4,5]. The stability conditions and the corresponding LMIs have been already derived for a great number of TSK-PDC systems in which various system peculiarities such as plant uncertainties, time-delays, signal constraints, constraints on system performance, etc. and conservatism reduction techniques are considered [2,6-10].

Though attractive the PDC-FLC systems design and performance depends on the local linear plant models and the plant dynamics partitioning which are approximate, use average estimations and change with the operation conditions and time. This makes necessary the optimization of tuning of the controllers' parameters. Two basic intelligent techniques in improving system performance are widely spread in industrial applications because of their simplicity and effectiveness - fuzzy logic supervisors (FLSs) for on-line auto-tuning and off-line optimization by genetic algorithms (GAs) [11-29]. The FLS tunes in a nonlinear manner usually the FLC gains introducing adaptive resolution [11] and keeping desired accepted performance indices. Most companies develop extensions for fuzzy assisted auto-tuning - FLSs that minimize the overshoot and undershoot in a shortest time [12,13,14,15] applied mainly to linear controllers. The GAs as a random derivative-free search technique for multi-objective global optimization [16,17] of a non-analytically defined multimodal function of many variables (parameters), are used to tune both linear controllers and FLCs and also their auto-tuning FLS in order to improve different system performance and energy efficiency indices [18]. The GA optimization are not fit for on-line application as it interferes the plant operation, is slow - a great number of experiments are required, inaccurate because of the many disturbances from the industrial environment, restricted by the system stability and parameter constraints. The off-line GA optimization is based on an accurate plant model, an accepted fitness function and a sample of experiments/simulations used in its evaluation.

The two intelligent techniques complement to improve linear controllers [19,20,21] and ordinary FLCs of Mamdani and Sugeno type [11,18,22,23,24,25] which dominate in industrial applications. Lately few applications on PDC-FLCs [23,26,27,28,29] appeared in literature:

- particle swarm optimization - PSO (another similar to GA optimization technique) of tuning of PDC-FLC for load frequency control in interconnected power systems based on fuzzy model of the plant uncertainty [26];
- PSO of tuning of the membership functions (MFs) of the PDC-FLC for a constantly stirred tank reactor – a plant with a time delay in the state variables [27];
- performance-oriented PDC-FLS [28,29] with state feedback gains for each of the linearized local plants computed as fuzzy blending of several gains, which are FLS tuned from accepted performance criteria, with application to a flexible joint robot [28] and an inverted pendulum [29] - the improvements are tested in simulation and conclude in faster system response despite the restrictions on the control;
- simple two-input FLS design in real time auto-tuning of the output scaling factor of a PI PDC-FLC – the FLS inputs are performances easily estimated from on-line measurements $-y/y_r$ (relative with respect to reference y_r plant output y) and magnitude of system error $|e|$, with application to temperature control [23].

The main deficiencies of the few existing approaches for PDC-FLC design are: 1) complex and obscure derivation of the required TSK plant model from identification data for processes with inertia, time delay and model uncertainties – most approaches are developed for mechatronic systems (robots, inverted pendulum, etc.) or plants with no time delay (tank level, frequency in power system, etc.) which facilitates the plant dynamics partitioning based on existing nonlinear plant model; 2) intuitive and application oriented design; 3) lack of tuning or difficult and improper tuning that results in control oscillations responsible for problems with the valve and the energy efficiency; 4) lack of GA and FLS based tuning design methodology for selection of performance indices and of corresponding tuning parameters that best influence them; 5) control algorithms demanding high computational effort, unfit for embedding in programmable logic controllers (PLC) and real time implementations.

The aim of the present investigation is to develop an approach for the design and tuning of a modified process PDC-FLC and a fuzzy logic supervisor via GA optimization techniques in order to ensure closed loop stability and robustness, desired performance and energy efficiency. The approach will be developed on the example of air temperature control in a laboratory dryer. The design and tuning procedure should be easy to apply using MATLAB™ facilities [30,31] and the obtained PDC-FLC-FLS has to be simple in structure for industrial completion in PLCs and real time operation [32,33].

The research is organized as follows. Experimental study of the plant and derivation of its TSK plant model based on a suggested Sugeno-dynamic structure with GA optimized parameters is presented in Section 2. In Section 3 a PID-Sugeno FLC structure is developed for a PDC-FLC and a GA optimization is applied to tune the PDC-FLC parameters. The design and the GA tuning of the FLS is described in Section 4. Simulation and real time experiments are explained in Section 5. The analysis of the main results, the future work and the conclusions is included in Section 6.

2. Experimental Investigations and TSK Modeling of the Plant

The aim of this section is to obtain and validate a TSK model of the plant on the basis of experimental data. It is needed for determination of the structure of the PDC-FLC and in the off-line GA based optimization of the PDC-FLC parameters and later of the designed FLS parameters via closed loop system simulations.

The plant is the air temperature y in a laboratory dryer [2,18,23] which is controlled by changing the voltage u to a Pulse-Width Modulator (PWM) thus varying the duty ratio of switching of an electrical heater and a fan. The dryer is equipped with industrial sensors, transmitters, solid state relay, fan, electrical heater and interfacing board to a computer with analog-to-digital converter and digital outputs. The controller, the PWM, the graph recorders and the noise filters are accomplished as software in the Simulink model.

Six experimental step responses of the temperature y in various operation points to equal in magnitude step changes of u are depicted in Figure 1. One of the input steps is reverse in order to estimate the plant dynamic asymmetry to heating and cooling. Each i of the step responses (i=1÷6) is approximated by Ziegler-Nichols (ZN) model $P_i(s)=K_i.e^{-\tau_i s}.(T_i s+1)^{-1}$ and the model parameters – plant gain K_i, time constant T_i and time delay τ_i, are graphically estimated. The ZN model is simple with clear physical meaning and describes most of the industrial processes, also is well studied and widely accepted in engineering practice. Adjacent step responses with similar ZN model parameters can be grouped in j zones. This is a prerequisite in favour of the selection of a PDC-FLC. There can be distinguished three overlapping linearization zones (j=1÷3) as shown in Figure 1 – Zone 1 for y=20÷50°C, Zone 2 for y=40÷57°C and Zone 3 for y=50÷80°C. In each zone a mean ZN model is computed with mean values for the parameters of the ZN models. The TSK plant model consists of three rules – one for each zone. The conclusion in each rule is the state space representation of the mean ZN model for the corresponding zone – time lag with time delay with mean values for the parameters. The premise variable is the temperature $z=y$, which distinguishes the zones and the plant behavior respectively.

Instead of a standard TSK plant model an equivalent Sugeno plant model structure of clear physical nature is developed, shown in Figure 2. It consists of a Sugeno soft switch (Smodel) with input y in the range [20,80] °C partitioned by 3 MFs [S M B] for Small, Medium and Big allocated in the range according to the definition of the three zones in Figure 1. The outputs of the Smodel are singletons that yield specific values of the three MFs of belonging μ_j of the current plant output y to each of the fuzzy zones, defined by S, M and B. The dynamic part is divided into three parallel branches – one dominating for each zone. Each branch consists of two time lags to represent a high order local linear plant. The first time lag in a branch has the physical meaning of the time lag in the ZN model, while the second time lag – the linear term of the Taylor's series expansion of the time delay element in the ZN model $e^{-\tau_i s} \approx (\tau_i s+1)^{-1}$. Input to the plant is u, which is passed to the inputs of the three branches whose outputs are weighted by the corresponding μ_j and then summed to accomplish defuzzification and yield the plant model output. The inertia of the plant is increased by the common time lag - (K=1, T=1) at the plant model output, where the initial condition (initial or ambient air temperature $y(0)=y_{amb}$) in the dryer is added.

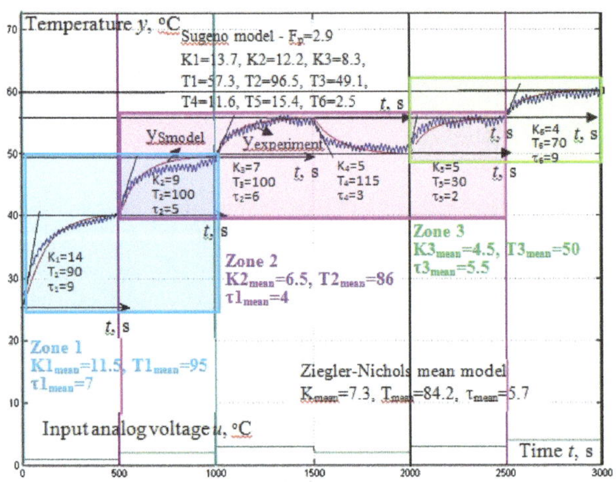

Figure 1. Experimental (blue) and GA optimized Sugeno model (red) plant step responses

A GA algorithm is used to objectively optimize the parameters of the time lag elements in the suggested Smodel structure. The optimal parameters computed are $\mathbf{q_p}$=[K_1=13.71, K_2=12.2, K_3=8.3, T_1=57.3, T_2=96.5, T_3=49.1, T_4=30, T_5=11.6, T_6=15.4, T_7=2.5]. The GA optimization minimizes the integral squared relative error accepted as a fitness function

$$\mathbf{F_p} = \int [E(t)/y_e(t)]^2 dt, \qquad (1)$$

where $E(t)=y_{Smodel}(t)-y_e(t)$ is the difference between the model output $y_{Smodel}(t)$ from simulations and the real plant output from experiments $y_e(t)$ – both as responses to one and the same pattern of input step changes $u(t)$. The GA optimization reaches minimum of $\mathbf{F_p}$=2.9. The pattern of input signal $u(t)$ is selected according to the requirements to be rich in magnitudes and frequencies and to cover the whole range of the input signal, so that the Smodel can learn the real plant nonlinearity. The ready Sugeno plant model is validated comparing its response to the real plant experimental response for input signals, different from the used in modeling. The red line in Figure 1 shows the Smodel responses which are close to the experimental ones in blue from the real time operation of the plant. The parameters of the GA are: population size=20; number of generations=20; elite=2; crossover rate = 0.8 and method – single point; mutation operator –adapt feasible; fitness scaling – rank based; selection – roulette wheel; binary coding.

3. Design and GA Optimization of PID PDC-FLC

When the plants dynamics can be partitioned in a finite number of linear models, a TSK plant model with a small number of rules can be derived to allow a PDC-FLC design. The standard PDC-FLC representation is a single fuzzy block with linear controllers' state space descriptions in the conclusions of the same number of rules with the same premise as in the TSK plant model.

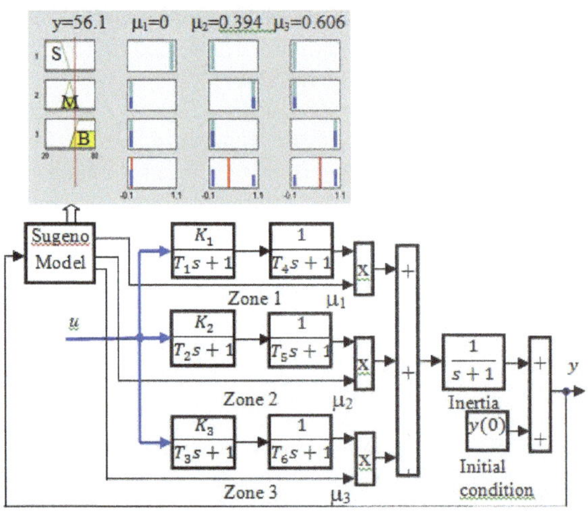

Figure 2. Modified TSK plant model

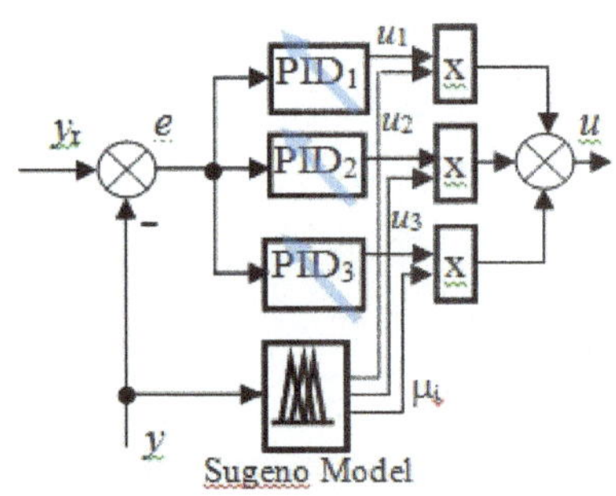

Figure 3. Modified PDC-FLC

Here, first this PDC-FLC is equivalently substituted by a Sugeno-dynamic structure similar to the new plant Sugeno modification of TSK model, shown in Figure 2. The new structure is depicted in Figure 3. It contains the same Sugeno model block used in producing the degrees μ_i of belonging of the current temperature y to each of the three linearization zones, and j PID linear controllers with input the system error e. The controller output u is computed as weighted by μ_j sum of the individual PID controllers' outputs u_j thus performing fuzzy blending and defuzzification.

The modified PDC-FLC shows a simple way for turning linear controllers into nonlinear fuzzy as well as easy embedding of the PDC-FLC into a PLC algorithm for industrial real time operation. Several linear PID controllers are smoothly mixing outputs when y is moving along different linearization zones, characterized by different linear plant models. The fuzzy blending is ensured by a simple Sugeno fuzzy model of j rules with singletons as outputs –available as a fuzzy block in most PLCs [12,13,14,15,32,33]. This modified PDC-FLC facilitates the design – each of the PID controllers is tuned considering the mean linear plant model in the corresponding linearization zone without turning plant and PID controller description into a state space representation. Different tuning criteria and techniques can be selected from the linear control theory – robust stability, robust performance, engineering approaches, pole placement, desired performance indices, etc. In this research it is accepted that the PID block operates as PI algorithm.

The complexity of tuning of a number of linear controllers considering local linear plant model uncertainties can be decreased using a GA based optimization. The parameters to be searched are the parameters of the three PI controllers $\mathbf{q}_{PI}=[K_{p1}, T_{i1}; K_{p2}, T_{i2}; K_{p3}, T_{i3}]$.

The fitness function combines four objectives – minimization of: 1) mean absolute error (MAE) ; 2) mean control effort (MCE); 3) maximal relative output with respect to reference y_r and the range of the step changes in reference D_r - an estimate of the maximal overshoot $\sigma = y/y_r - 1$; 4) relative with respect to D_r maximal deviation y_m from reference y_m:

$$\mathbf{F} = \frac{1}{N}\sum_{k=1}^{N}|e_k| + \frac{1}{N}\sum_{k=1}^{N}u_k + \mathbf{max}\left(\frac{y}{y_r D_r}\right) + \frac{\mathbf{max}(y_m)}{D_r} \rightarrow \min_{\mathbf{q}_{PID}}, \quad (2)$$

where N is the total number of samples, collected from the experiment/simulation.

The fitness function is computed using (2) and the recorded values for the corresponding variables from simulation of the closed loop system with an initial set of parameters for the three PI controllers. The initial parameters are taken from the user defined bounds for the parameters, which are determined using some engineering tuning approach. The closed system model used in the simulation with the GA optimizer for tuning of the modified PDC-FLC parameters is shown in Figure 4, where a Limiter bounds the control signal to correspond to the laboratory set-up for operation of the dryer. The GA performs selection, crossover and mutation to build a new generation of offspring thus provides a new set of controller's parameters. After a new simulation the fitness (2) is computed. This cycle continues till the end condition is met – the final number of generations of 20 is reached. The simulation is fast, more realistic, with full control on the experiment, without unplanned noise or disturbances, safe for the plant which ensures the best solution with respect to system stability and gradual parameter changes without restrictions on parameters and signals. It allows considering of various inputs – reference changes and disturbances with different magnitude and frequency that cover the realistic industrial environment impact range. However, a reliable plant model is required for the whole range of operation conditions.

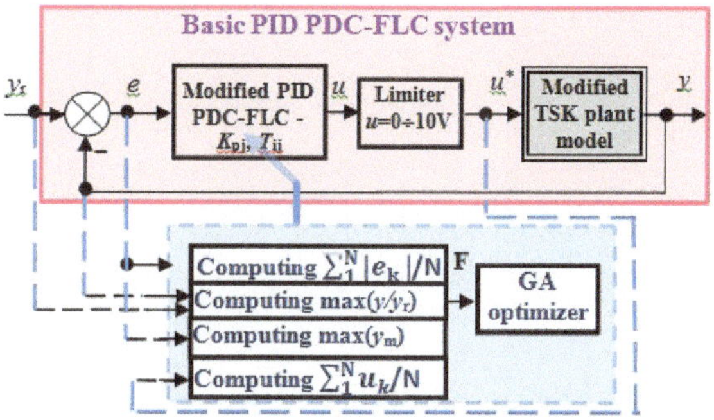

Figure 4. Closed loop system of modified PID PDC-FLC and TSK plant model with GA optimizer

The optimal PI controllers' parameters that minimize (2) in the GA optimization are $\mathbf{q}_{PI}^{opt}=[K_{p1}=0.27, T_{i1}=99s; K_{p2}=0.36, T_{i2}=120s; K_{p3}=0.83, T_{i3}=200s]$. A linear PI controller for comparison is tuned by GA optimization and its optimal parameters are $K_p=0.25$, $T_{i3}=85s$.

The stability of the designed global nonlinear system can be proven using the methodology, developed in [2,5].

4. Design and GA Optimization of Fuzzy Logic Supervisor

A further improvement of the closed system performance is possible by connecting a properly designed FLS. The FLS operates on-line in real time and can compensate for signals, difference in plan of experiments and changes of plant parameters which are not included in the sample used by the GA for optimization.

Here it is suggested that the FLS has an input the bounded control u^*_{k-1} in previous time moment t_{k-1} and output u_c which to provide a multiplicative correction of the PDC-FLC output u according to Figure 5. The correction is assumed to vary in the range [0.3, 1.7] not to influence system stability but also for the purpose of its tuning a scaling gain K_u is introduced. The input and output MFs of the FLS fuzzy block and the rules are selected standard and are shown also in Figure 4. The meaning of this embedding of a FLS is to keep a reasonable control action thus ensuring energy efficiency in control. The bigger the control – the longer the period when the dryer and the fan are connected to the nets supply, so the greater the consumed electrical energy for heating. Oscillations with great peaks of control action are not good for the heater durability and reliability, and also show energy inefficiency – too much heating followed by a necessary cooling.

The scaling gain is optimized by GA minimization of the fitness function (2) using a modified simulation model from Figure 4 with the connection of the FLS from Figure 5. Besides, the PDC-FLC parameters remain fixed to their optimal values and the tuning parameter is $\mathbf{q}_{FLS}=[K_u]$. As a result the optimal FLS parameter computed is $K_u=0.71$ and the minimal value for $\mathbf{F}$ reached is $\mathbf{F}=1$.

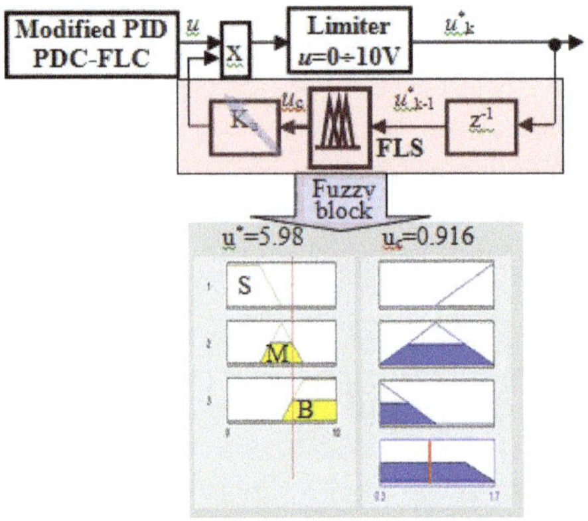

Figure 5. Fuzzy logic supervisor block diagram

5. Simulations and Real Time Control of Temperature in a Laboratory Dryer

The designed PDC-FLC, consisting of a Sugeno model and three local PI linear controllers (PDC-3PI) with GA optimized parameters, and the GA optimized FLS are applied for the control of the air temperature in a laboratory dryer. The plan of the experiments is to prove first in simulations and then in real time control:

- expected improvement of closed loop system performance (settling time and overshoot reduction) and increased robustness to plant model uncertainty - similar step responses in different operation points despite the different plant parameters;
- improvements in the step responses of the system with the PDC-3PI with respect to the system with the linear PI controller, and improvements in the step responses of the system with the PDC-3PI with the FLS with respect to the system without the FLS;
- improvements in the control action $u(t)$ (smooth, without peaks and oscillations) and in the energy efficiency, defined as $\text{EEF} = \sum_{k=1}^{N} u_k$;
- closeness between the step responses in simulations and in the real time control;
- preservation of good performance for step inputs different from the used in the GA optimization.

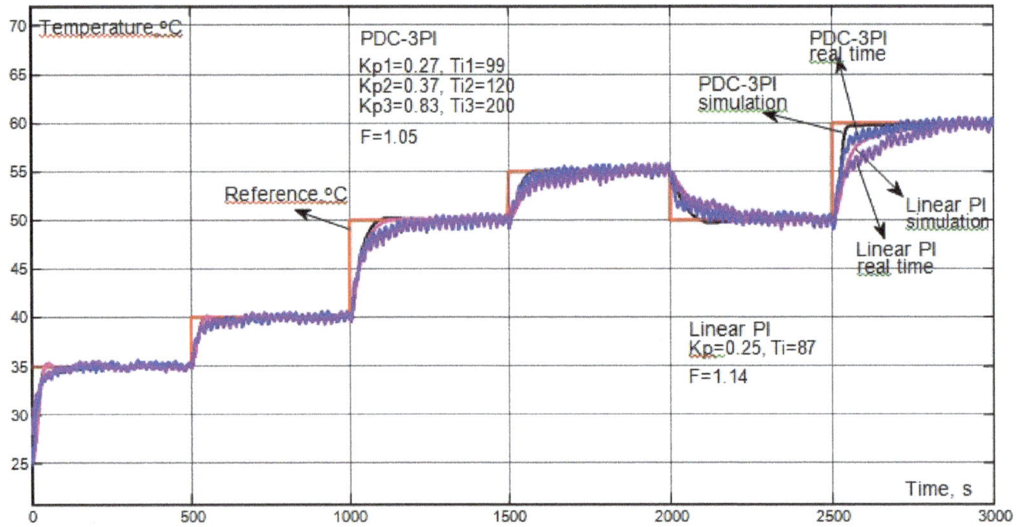

Figure 6. Step responses of systems with PDC-3PI and linear PI controller from simulation and real time control

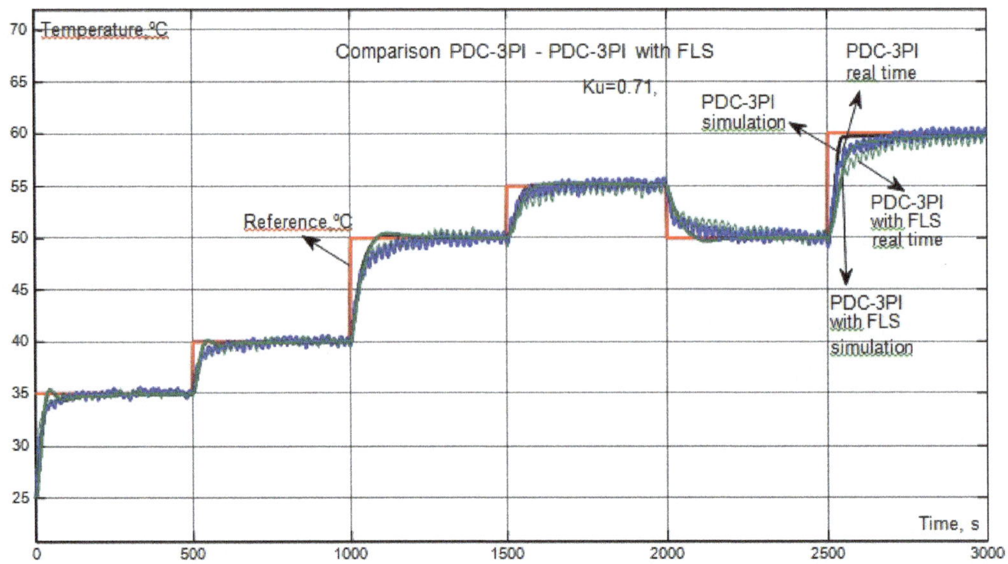

Figure 7. Step responses of systems with PDC-3PI with FLS and without FLS from simulation and real time control

The step responses of temperature in systems with PDC-3PI and linear PI controller from simulation and real time control are depicted in Figure 6. Simulation and real time results are close and lead to the same conclusion:

– the PDC-3PI system shows faster settling for high reference, where the plant has the smallest gain for the whole temperature range, which makes the settling time of all system step responses almost the same - an evidence for high robustness;

- the settling time of the PI control system step responses differ in the different operation points -increase with the increase of reference, so the this system is less robust to the plant parameter changes with the operation point;

- both systems have very good performance for small references, PDC-3PI system improves performance where the linear PI system is slow at high references.

The step responses of temperature of the PDC-3PI system and the PDC-3PI system with FLS from simulation and real time control are shown in Figure 7. The FLS does not lead to significant improvements in temperature responses since it is designed to reduce control effort and GA optimized to minimize a fitness function that unites several criteria – three related to system performance (MAE, σ, y_m), and one - to the control action (MCE). Both in simulations and in real time the two systems have close settling time – only for the last reference the FLS system has a slower response.

The improvements due to the FLS are observed in Figure 8, where the EEF and the control action take smaller values – the control is more economic and smooth (small peaks and oscillations) and the system more energy efficient.

The GA optimization of the controllers' parameters is carried out for the developed plant model and the accepted fitness function and sample of step responses. The fact that simulation results are close to the obtained in real time control in all previous investigations gives grounds to test the three systems by simulation to a different pattern of step reference changes. The results lead to the same conclusions and this confirms their validity for all possible step inputs in the given range.

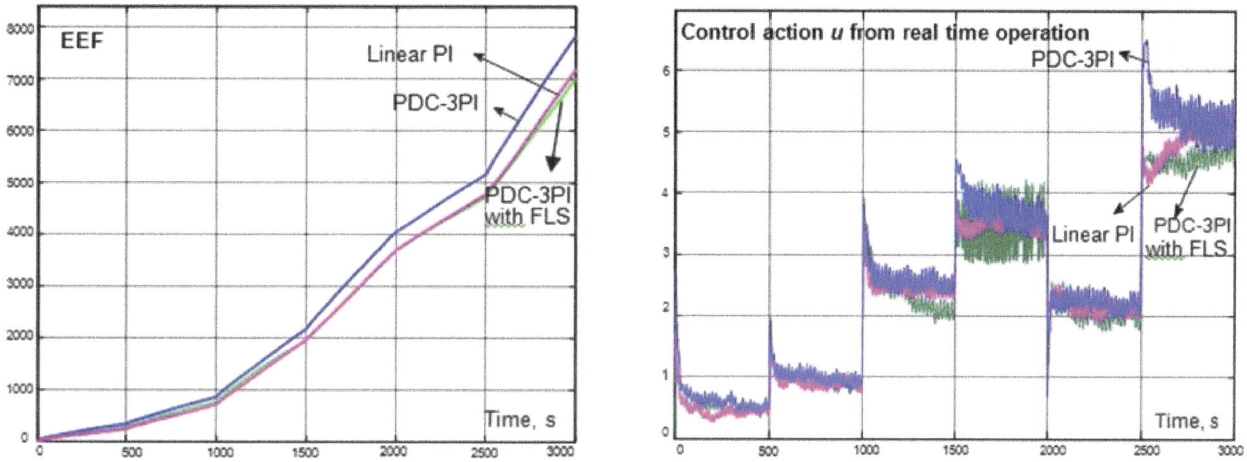

Figure 8. Energy efficiency (up) and control action (down) from real time control with linear PI controller and with PDC-3PI with and without FLS

6. Analysis of Results, Conclusion and Future Work

The main results of the present research can be summarized as follows.

A modification of the PDC design approach is developed and applied for the control of temperature. It concludes in building of a modified TSK plant model from experimental plant step responses in order to obtain a simple model structure of clear physical nature that is fit for easy GA optimization used to objectively and accurately determine the model parameters. The model is validated and further used in off-line GA optimization of different controllers via simulation of the closed loop system. A modified structure of a PDC-FLC is suggested on the basis of the modified TSK plant model for easy design and embedding in industrial PLCs. It consists of the same fuzzy block used in the modified TSK plant model to distinguish the linearization zones and parallel linear standard PI/PID controllers – PDC-PID which parameters are tuned via off-line GA optimization. Then a FLS is designed with GA optimized output scaling factor to reduce the PDC-FLC system control effort and improve system energy efficiency. The approach suggested is applied for the control of the air temperature in a laboratory furnace. The simulations and real time control show improved system performance and robustness with respect to linear PI control system. The implementation for temperature control demonstrates the simple design and tuning technique which together with the simple structure of both the PDC-PID and the FLS aims to facilitate the industrial application of the approach for the control of a great number of other process variables.

The future work will focus on completion of the PDC-FLS on Siemens PLCs and its implementation for the control of other industrial plants.

References

[1] Tanaka, K. and Wang, H.O., Fuzzy Control Systems Design and Analysis: A Linear Matrix Inequality Approach, John Wiley & Sons, Inc., 2001.

[2] Yordanova, S., Methods for Design of Fuzzy Logic Controllers for Robust Process Control, KING, S., 2011. (in Bulgarian).

[3] Yordanova, S. and Tashev, T., "Fuzzy Internal Model Control of Nonlinear Plants with Time Delay based on Parallel Distributed Compensation", WSEAS Transactions on Circuits and Systems, 11 (2), 2012, 56-64.

[4] Yordanova, S., "Fuzzy Smith Predictor for Nonlinear Plants with Time Delay based on Parallel Distributed Compensation", Proceedings of TU-Sofia, 62 (2), 2012, 15-24.

[5] Yordanova, S., "Lyapunov Stability and Robustness of Fuzzy Process Control System with Parallel Distributed Compensation", Information Technologies and Control, Bulgarian Union of A&I, Year VII, No 4, 2009, 38-48.

[6] Fang, C.H., Y.S. Liu, S.W. Kau, Hong, L., C.H. Lee, A new LMI-based approach to relaxed quadratic stabilization of T-S fuzzy control systems, IEEE Trans. Fuzzy Syst. 14 (2006) 386-397.

[7] Gu, Y., Wang, H.O. and Tanaka, K., "Fuzzy Control of Nonlinear Time-Delay Systems: Stability and Design Issues", Proceedings of the 2001 American Control Conf., Arlington, VA, June 25-27, 2001.

[8] Lam, H.K. and Leung, F.H.F., "LMI-Based Stability and Performance Conditions for Continuous-Time Nonlinear Systems in Takagi–Sugeno's Form", IEEE Trans. on Systems, Man, and Cybernetics, Part B, 37 (5), 2007, 1396-1406.

[9] Li, L. and Liu, X., "New Approach on Robust Stability for Uncertain T–S Fuzzy Systems with State and Input Delays", Chaos, Solitons & Fractals, 40 (5), 2007, 2329-2339.

[10] Yoneyama, J., "New Robust Stability Conditions and Design of Robust Stabilizing Controllers for Takagi–Sugeno Fuzzy Time-Delay Systems", IEEE Trans. on Fuzzy Systems, 15 (5), 2007, 828-839.

[11] Li, H.-X. and Tso, S.K., "A Fuzzy PLC with Gain-Scheduling Control Resolution for a Thermal Process – A Case Study", Control Eng. Practice, 7, 1999, 523-529.

[12] Future Design Controls, Inc. – http://www.futuredesigncontrols.com/index.htmn.

[13] NI LabVIEW 2012 PID and fuzzy logic toolkit for Windows-http://sine.ni.com/nips/cds/view/p/lang/en/nid/1394.

[14] Auto-tune PID temperature controller. Fuzzy logic plus PID microprocessor-based controller series, BRAINCHILD Electronic Co., Ltd., 2003, http://www.thetacontrols.com/pdf/PID_Controller/Catalogue/CT9100.pdf.

[15] Manual TEC-220/920 Revision 12/2012 – TEC-220 and TEC-920 Auto-tune fuzzy PID process temperature controller, TEMPCO Electrical Heater Corporation – http://www.tempco.com/Engineering/TEC220Manual.pdf.

[16] Goldberg, D.E., Genetic algorithms in search, optimization and machine learning, Addison-Wesley, 1989.

[17] Haupt, R.L. and Haupt, S.El., Practical genetic algorithms, NY, John Wiley & Sons, 1998.

[18] Yordanova, S., "Intelligent Approaches for Linear Controllers Tuning with Application to Temperature Control", Journal of Intel. and Fuzzy Systems, IOS Press, NL, 2014 (to appear).

[19] Kumar, B. and Dhiman, R., "Optimization of PID Controller for Liquid Level Tank System Using Intelligent Techniques", Canadian Journal on EEE, 2 (11), 2011, 531-535.

[20] Bandyopadhyay, R. and Patranabis, D., "A New Auto-tuning Algorithm for PID Controllers Using Dead-Beat Format", ISA Transactions, 40, 2001, 255-266.

[21] Wu, C-J., Ko, C-N., Fu, Y-Y. and Tseng C-H., "A Genetic-based Design of Auto-tuning Fuzzy PID Controllers", Int. Journal of Fuzzy Systems, 11 (1), 2009, 49-58.

[22] Chung, H-Y., Chen, B-C. and Lin, J-J., "A PI-type Fuzzy Controller with Self-tuning Scaling Factors", Fuzzy Sets and Systems, 93, 1998, 23-28.

[23] Yordanova, S. and Yankov, V., "Fuzzy Supervisor for Nonlinear Self-Tuning of Controllers in Real Time Operation", Proceedings of TU-Sofia, 1 (64), 2014, 131-140.

[24] Mudi, R. K. and Pal, N.R., "A Self-Tuning Fuzzy PI Controller", Fuzzy Sets and Systems, 115 (2), 2000, 327-338.

[25] Tsai, M-T., Tung, P-Ch. and Chen, K-Y., "Experimental Evaluations of Proportional–Integral–Derivative Type Fuzzy Controllers with Parameter Adaptive Methods for an Active Magnetic Bearing System", Expert Systems, 28 (1), 2011, 5-18.

[26] Shayanfar, H.A., Shayeghi, H. and Jalili, A., "Takagi -Sugeno Fuzzy Parallel Distribution Compensation based Three-Area LFC Design", Int. Journal on "Technical and Physical Problems of Engineering", 8 (3), No 3, 2011, 55-64

[27] Alavi, S. M. S., Zarif, M. H., Sadeghi, M. S ha and Barzegar, A. R., "Particle Swarm Optimization of Robust Fuzzy Parallel Distributed Controller", J. Basic. Appl. Sci. Res., 2 (7), 2012, 6385-6391.

[28] Seidi, M. and Markazi, A. H.D., "Performance-oriented Parallel Distributed Compensation", Journal of the Franklin Institute, 7 (348), 2011, 1231-1244.

[29] Vafaee, K. and Geranmehr, B., "Controlling Inverted Pendulum Using Performance-Oriented PDC Method", International Journal of. Automation and Control, 2 (2), No 2, 2014, 39-44.

[30] Fuzzy Logic Toolbox User's Guide for Use with MATLAB, MathWorks Inc., 1998.

[31] MATLAB – Genetic Algorithm and Direct Search Toolbox. User's Guide, MathWorks, Inc., 2004.

[32] SIMATIC S7. Fuzzy Control. User's Manual, SIEMENS AG, 2002.

[33] IEC 61131–Programmable controllers. 1999.

The Application of the Matrix Tactile Sensor (MTS) for Measuring Contact Pressures between the Trunk Orthosis and a Patient Body and a Mechanical Testing of the Measuring System

Peter Sedlačko[1,*], RadovanHudák[1], Tatiana Kelemenova[1], Jozef Živčák[1], Marek Penhaker[2]

[1]Department of Biomedical Engineering and Measurement, Faculty of Mechanical Engineering, Technical University of Košice, Košice 042 00, Slovakia

[2]Department of Cybernetics Biomedical Engineering, Faculty of Electrical Engineering and Computer Science, VŠB- Technical University of Ostrava, Ostrava 70800, Czech Republic

*Corresponding author: apeter.sedlacko@tuke.sk

Abstract Matrix tactile sensors (MTS) as the tactile sensors provide us with information about the measurement object by simply touching the object. Sensitive part of the sensor array consists of the tactile elements arranged in the matrix, which tell us about the distribution of normal forces acting on the tactile elements. Identifying, establishing and monitoring the pressure between the trunk and correcting pad of the trunk orthosis can inform us about undesirable compression actions, progression of the treatment or deformation of the production material of the trunk braces. Identification of the values distribute pressure between the body and pad of the trunk braces serve some information about effects on the human organism.

Keywords: *matrix tactile sensor, MTS, Scoliosis, pad, trunk orthosis, measuring system*

1. Introduction

The implementation of tactile sensing systems into medicine, prosthetics, orthotics and diagnostics was in the past researched by several research teams. The main object of the research was to determine selected parameters by non-invasive methods. [1] In orthotics was a demand for the progressive replacement of the unnecessary irradiation of the patient by CT examination. [2] Identification of the treatment progression was by the comparison of pressures between the contact correction pad of trunk braces and the patient's body.

In clinical practice was the application of the trunk braces associated with the birth defects in adolescent patients up to the age of 18. Due to the age of the patients, the use of CT (x-ray) examine as the main tool of diagnostics was very limited because of the potential risk from radiation to the human body. [3] For the application of the measurement system to determine the progress of the chosen treatment was needed verification of accuracy of the measurements used for comparison of data about compression actions in the various stages of the carrying trunk braces [4,5].

In the evaluation of the therapeutic effect we needed consider some following parameters and measuring methods, for example: The methods of the assessing comparative measurements, measurement methodology of the exercise acts according to the proposal of the previous research activities was mentioned sensor system subjected to a variety of the test measurements in the laboratory to determine temporal characteristics and the verification of the interpretation and presentation of the measurement data [6,7].

2. Materials and Methods

2.1. The Methodology of the Applications

Determine of the optimum methodology of the work with measuring system, its temporal characteristics and accuracy of the data interpretation verified comparison of the compression actions in the ongoing monitoring of the progress and the setting of the chosen treatment of the scoliosis spine [8,9].

2.1.1. The Tactilus Measuring System

The Tactilus measuring system (Sensorprod USA) was customized as a tool to measure contact pressures between correcting pad of the trunk braces and the patient's body. Sensory pad as a direct measuring tool had a range from 0 to 97.16 kPa, which was a range that could be detected between the body and the brace and the patient body with surface of 400 cm^2 (20 x 20 cm).[10]

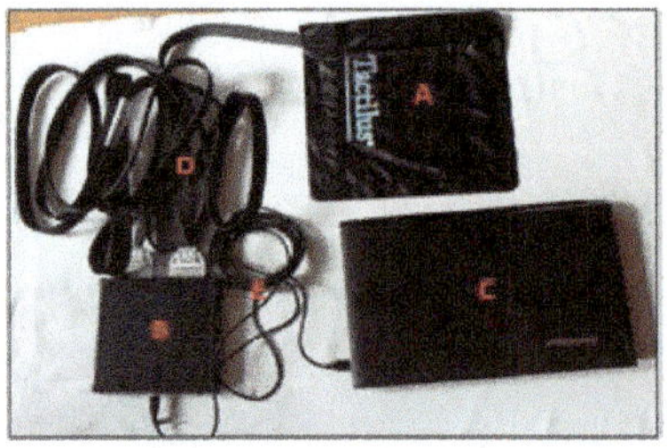

Figure 1. MTS system [11]

The measurement system consisted of the five parts (Figure 1) which were sensory pad (sensor) (A) with dimensions 20 x 20 cm, a density of touch-sensitive points 1 per 1 cm^2 and a 400 scan points with a range of 0-724 psi (0 - 729 mmHg, . 0 - 97.19 KPa). Tactilus converter for converting analog signals to digital (B). Notebook (C) and interface cables with dimension of 10 m for a comfortable measuring and USB cable to connect transmitter to a computer (D, E).

CT scanning of measurement matrix was realized by Metrotom (Carl Zeiss Germany) (Figure 2) .

Figure 2. CT scan of the sensor

Table 1. Technical specifications of sensor

Working principle	Piezoresistive / resistive
Pressure range	0.1 to 200 PSI (0.007 to 14.1 kg/cm²)
Sensory area	20 x 20 cm
Thickness	(0.3 mm)
Thickness of pack	23 x 23 cm
Scanning speed	Up to 1,000 hertz
Spatial resolution	0.04 - 2 mm
Accuracy	± 10%
Repeatability	± 2%
Hysteresis	± 5%
Nonlinearity	± 1.5%
Calibration	Pre-calibrated

2.2. Mechanical Testing

In determining of the testing methodology of the pressure system in the laboratory, we worked with already realized experiments and studies which dealed with the similar testing. [2] The testing methodology was created with regard to the repeatability and reproducibility of the measurement, to eliminate possible measurement errors, using the available hardware (Figure 4).

The measurement of time-load relationship dependence of the continuous load (A). Because of the lack of the results of the measurements and the assumption of changes in the behavior of the signal in time, was the first type of measurement set identification of time course signal time on the continuous load. Measurement of the continual loaded in a pre-selected position of the measuring sensor(B).

During the implementation of the test measurements with the system Tactilus at orthosis were observed variation in the load display with software Tactilus under peripheral parts of sensor pad.

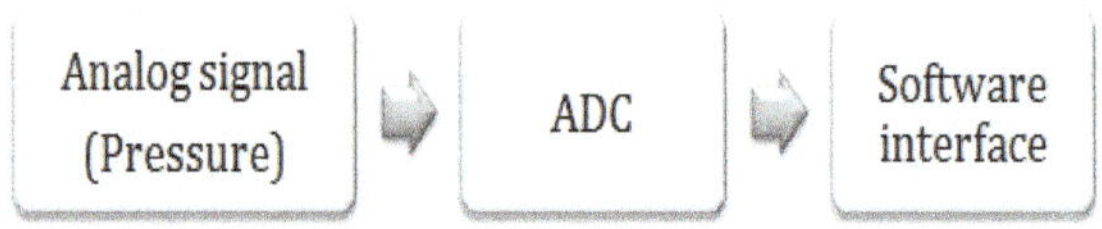

Figure 3. Scanning principle

The proposal of the measurement matrix

Measurement matrix represented the auxiliary measuring device. Its design and converting was designed to ensure accurate measurements with repeated placing the object on the same place and also defined five test positions on the pressure sensor. Production of measurement matrix was realized by 3D printer, Eden 260V (Stratasys Inc. USA) [19] according to designed in format .stl.

Figure 4. Measuring progress

The measurement A

The principle of this kind of the measurement was behavior of the monitoring signal overthe time with continuous load. Weights of 1 and 2 kg were placed on position 5 (center position) in the pre-formed matrix (Figure 4). Scanning programs turns on after the weight was on position. Dynamic action of the weight was not monitored but the progress of the application of the load sensor in time for 20 minutes

The Measurement B

The principle of this kind of measurement was monitoring signal behavior over time with continuous load on 5 position. Weights of 1 and 2 kg were placed in position 1-5 in the pre-formed matrix. Scanning program turned on after the weight was on position. Dynamic action of the weight was not monitored but the progress of the application of the load sensor in time for 1 minute.

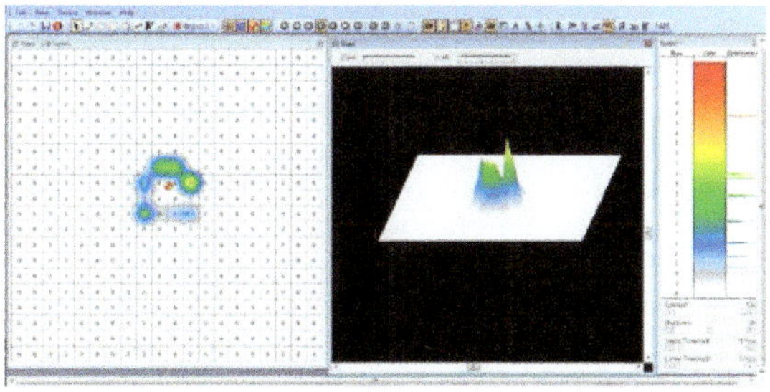

Figure 5. The graphical display of the load

Graphical user interface of the software Tactilus represented on Figure 5, offering information about the location and size of the pressure action, but this information is not sufficient for further statistical processing

2.3. Testing Measurements

Before applications of the measuring system in orthotic practice and various phases of the testing of the braces it would be necessary to create a methodology of the work with measuring system Tactilus and their verification .

The test measurements ministered to establish mentioned the methodologies were implemented in the Department of Biomedical Engineering and Measurement and consisted of the series of the measurements of the contact pressures between the body and the correction pad. The measured subjects did not show indications for corrective treatment with the trunk orthosis, and it was a test measurement which did not serve to determine the exact values of the operating pressures. The measured subjects on the basis of the procedural instruction did individual exercise acts of the chosen methodology applied to the trunk braces.

2.3.1. The Prescribed Method of the Exercise Acts

Unification and determination a uniform method of the measuring compression force appeared to be very important in the progression of the treatment. On the base of the recent research were created the training acts methods that represented the movements which the patient performed daily for self-guided activities. In creating the series of the exercise acts which we related to studies which dealt with biomechanical analysis of the trunk braces measuring of the contact pressure between the body and the brace taking into account movements that a person normally performed during the day because the daily human activities had an effect on the brace, and progress of the treatment.

The subsequent illustration shows the exercise acts performed by the patient during the measurement, which are divided into two categories. Act suitable for taking pressure at all pads - Type 1. Acts for useful when reading pressure on the side and rear pads - Type 2 (Figure 6)

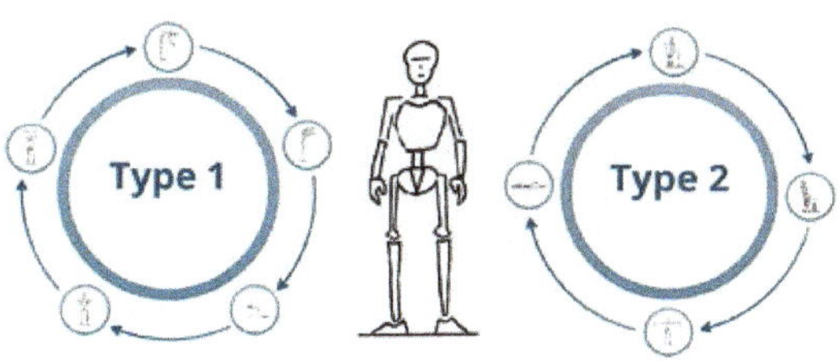

Figure 6. Types of exercise acts

2.3.2. The Determination of the Exercise Controlled Operations

Measured exercises represented movements that everybody uses in dressing, walking or sitting. The purpose of these measurements lay in finding and defining forces with corrective pad acting on the patient's body while the Exercise acts simulated his normal day.(Table 2) This measurement should be done at the stage of testing brace which on detection of pressure is too high for certain activities can prevent the occurrence of the abrasions, or pathologies conditions of the skin, an appropriate adjustment of the brace before its release. Subsequent test runs for a period of wearing the brace will give us information on the course of treatment the scoliosis curve, which is achieved due the compression brace.

Table 2. Exercise acts

A	Standing in the basic position	F	Rotation
B	Escape from the brace	G	Stretched out
C	Bending forward	H	Pull down
D	Kyphosis	I	Sit down
E	Kyphosis on four	J	Lay down

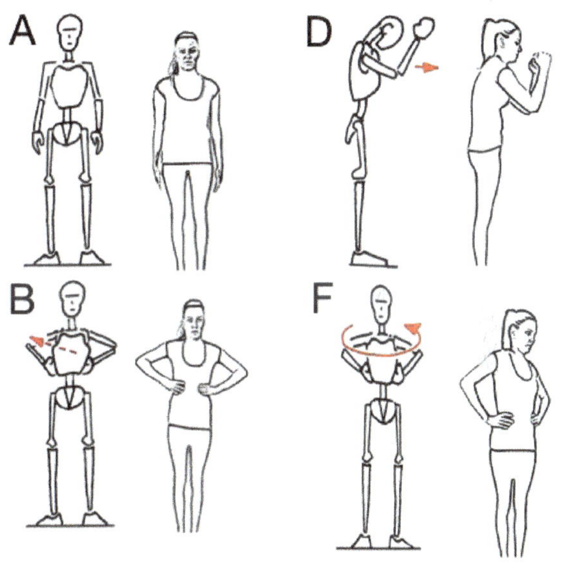

Figure 7. Selected exercise acts

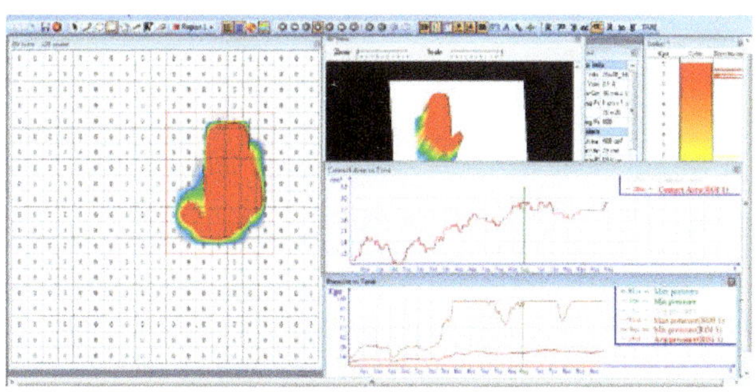

Figure 8. Tactilus software interface

Record of the measurements gave us information about the measurement object and it was possible to apply the editing functions of the software program which recorded adjusted according to the desired output.(Figure 8).

3. Result and Discussion

3.1. Mechanical Testing

In the work with the measuring system was setting a frame rate of 30 Hz. With the image frequency and extent of the measurements it appeared necessary evaluation and selection of the data generated using automated software applications. For this aim in collaboration with the Faculty of Aeronautics, Technical University there were created two software applications.

3.1.1. Time-load Dependence

The waveform chart showed the similarity of the shape of the force in time characterized by a sharp increase in the first minute of the measurement - The leading phase and steady rise curves after 600 seconds measurement (Figure 9).

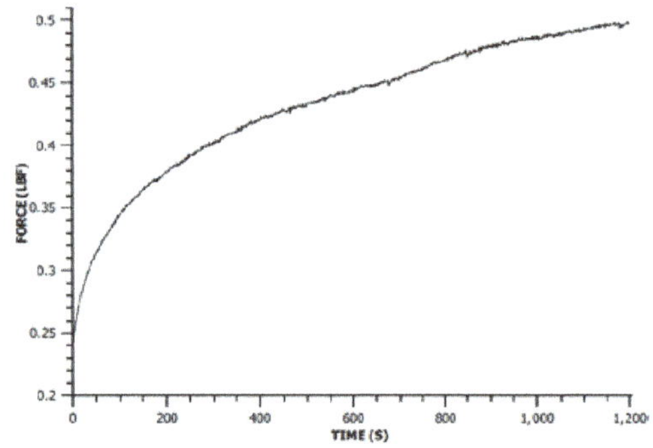

Figure 9. Load-timedependance

The realized measurements were set at 1200 seconds during which the load curve should buoyant. By realized experiment hypothesis was not confirmed about stabilization of the measuring system during the measurement (Figure 10).

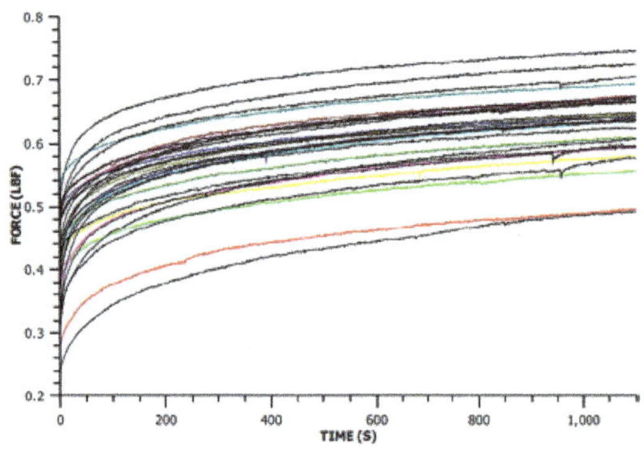

Figure 10. Load in time of 30 measurements

The nature of the curves depicting the history of the force in time after the subjective assessment of the observed shape similarity. For verifying the hypothesis and to define precisely the waveform display load was determined the comparative method.

3.1.2. The Mathematical Definition of the Measurement

Ambiguity graphical outputs of the measurement showed us a need for the using the mathematical regression techniques to compare different measurements .

These methods included the trendcurvesdrawing during the measurement, approximation of the measured values of the polynomial of the least squares (Figure 11).

Second order polynomial regression

$$y = ax^2 + bx + c \tag{1}$$

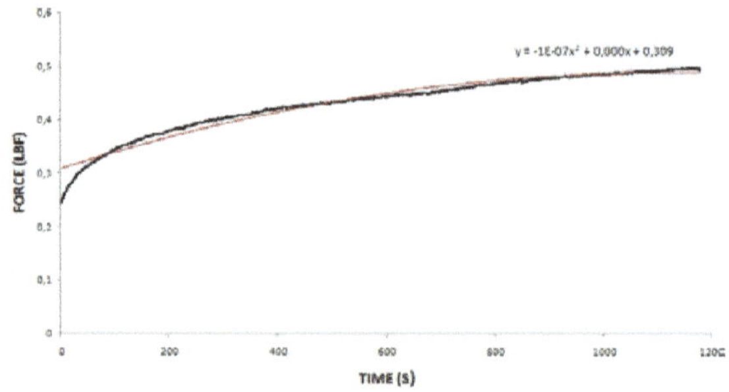

Figure 11. Applied regresionon load in time

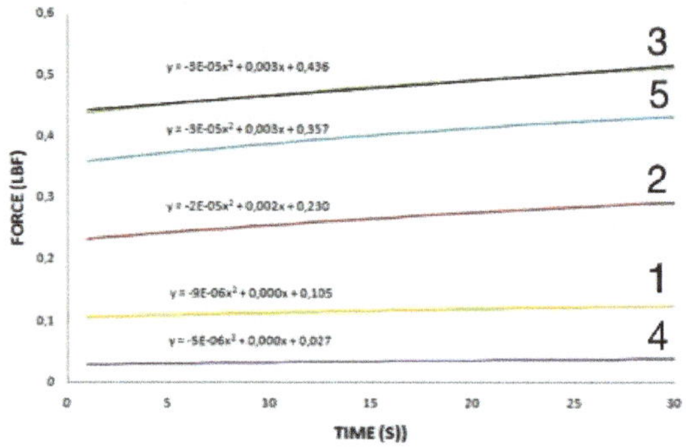

Figure 12. Applied regression on load in 5 places

Statistical processing of the individual parameters of the polynomial equations were developed the reference equations with parameter values equal to the median of 30 measurements.

Measurement B told us presumption of the unequal interpretation of the load in peripheral locations of the sensory pads. Individual curves on Figure 12 presented the time course of load on 5 measuring points pads during the first 30 seconds of the measurement. Mathematical analysis of measured values were confirmed upward trend of signal. Non-linear behavior was observed in parameter C-start parameter determining start positioning of curve.

3.2. The Clinical Application

Measurement was carried out in the workplace of The Prosthetic Center in Kosice in the process of testing the trunk orthosis for scoliosis treatment (Figure 8).

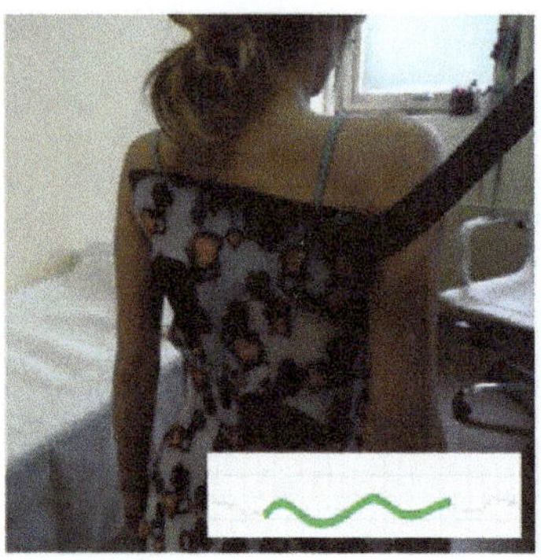

Figure 13. Clinical applications of Tactilussystem

4. Conclusions

Identification of the values distributed pressure between the body and pad of the trunk braces served an information about effects on the human organism. Excessively high pressure values at the interface body - orthosis might indicate in testing time of the brace skin wounds and pressure sores on congested areas, while low pressure tell us about bad function of the brace and the insufficient corrective action. Continuous monitoring of the pressure values during treatment with trunk braces can provide information about its progression. Methodology of measurement consists of the series of the acts necessary for the proper implementation of the repeatable measurement. The object of measurement is a measure of the exercise operations. Designed exercise is the series of the movements that a person performs in a typical day, and the exercise can be supplemented by specific movements according to the individual interests of the patient. The results of these measurements is data about the values of contact pressure at the interface body - orthosis them in a graph of daily activities. In the future, such information which was used in the diagnosis of the scoliosis treatment using on-line monitoring system of the pressure, which should record the pressure during all phases of wearing the trunk braces.

The results of the comparative measurements on the positions were intended to establish the impact of the placing the burden of the representation of the measurement results. The comparison was carried out with the hypothesis that the correct positioning of the load is in the middle of the sensory pads. Results of the analysis confirmed the working hypothesis of the different representation of the results by measurement system according to the location of the loading. Found that the situation can be attributed to the architecture of the sensory pads. Further research dedicated to diagnostic applications of the pressure measurement systems operating in a continuous pressure orthotics may be aimed by creating the wireless online measurement system applicable to continuous measurement of the compression actions.

Acknowledgement

This paper has been elaborated in the framework of the projects: VEGA 1/0182/15:Research strategies, of the coordinate measuring surfaces of free shapes touching the contactless measuring system, KEGA 014STU-4/2015: Improving the knowledge level and students skills in the design, application methods and means for measuring of geometrical quantities, New Diagnostic Methods in Invasive Implantology under Grant MŠSR-3625/2010-11, and in part by the Stimuli for the Research and Development, Ministry of Education, Science, Research and Sport of Slovakia and „Support research and development in the Moravian-Silesian Region 2014 DT 1 - Research Teams“(RRC/07/2014). Financed from the budget of the Moravian-Silesian Region.

References

[1] Herrán,J et al.: Flexible and large area pressure sensors for human-neuroprostheses and human-neurorobotic interface assessment. In: Microsyst Technol. 2012, no. 542, s. 1155-1161.

[2] Park, Y.L et al.: Bio-inspired Active Soft Orthotic Device for Ankle Foot Pathologies In: Intelligent Robots and Systems (IROS). 2011.

[3] Wong, M. S. – Evans, J. H.: Biomechanical evaluation of the Milwaukee brace. In: Prosthetics and Orthotics International.1998, s. 54-67.

[4] Chou, Wen-kai et al.: Using Finite Element Method to Determine Pad Positions in a Boston Brace for Enhancing Corrective Effect on Scoliotic Spine: A Preliminary Analysis. In: Journal of Medical and Biological Engineering. 2010, s. 29-35.

[5] Clin, Julien et al.: Biomechanical study of the Charleston Brace for the Treatment of Scoliosis. In: SPINE. vol. 35, 2010, no. 19, s. E940-E947.

[6] Tiwana, Mohsin et al.: A review of tactile sensing technologies with applications in biomedical engineering. In: Sensors and Actuators. 2012, s. 17-31.

[7] Antonelli, M.G. et al.: AN INNOVATIVE BRACE WITH PNEUMATIC THRUSTS FOR SCOLIOSIS TREATMENT In: Design & Nature and Ecodynamics. vol. 5, 2010, no.4 , s. 354-367.

[8] Fader, Mandy et al.: Effects of absorbent incontinence pads on pressure management mattresses. In: Journal of Advanced Nursing. 2004, s. 569-574.

[9] Hatzivasiliou V, Tzafestas SG. Analysis and design of a new piezoresistive tactile sensor system for robotic applications. J Intell Robot Syst 1994;10:243-56.

[10] Sensor Products INC. Tactile pressure expert, Internet: http://www.sensorprod.com/static/biomechanics.php (2.10.2012).

[11] Hudák, R et al. The use of matrix tactile sensors (MTS) for diagnostics of the efficiency of production, testing and application of a trunk orthosis. In:13th IEEE International Symposium on Computational Intelligence and Informatics. 2012.

Model Order Reduction Using Routh Approximation and Cuckoo Search Algorithm

D. K. Sambariya*, **Omveer Sharma**

Department of Electrical Engineering, Rajasthan Technical University, Kota, India
*Corresponding author: dsambariya_2003@gmail.com

Abstract In this paper, a large order system is reduced by using the Cuckoo Search Algorithm (CSA) to a reduced order approximate model. The denominator coefficients of a desired reduced order system are determined by Routh approximation method while the numerator coefficients are determined using CSA based on integral square error minimization as an objective function pertaining to a unit step as input. The efficacy of the proposed method is tested with three SISO test systems to get a corresponding reduced order system and extended to a MIMO system. The results are satisfactory in terms of minimum error with the proposed method as compared to Routh Pade approximation and weighted sum multi-objective harmony search based reduced models.

Keywords: *ISE minimization approach, large scale systems, Model order reduction, Optimization using Cuckoo search algorithm, Routh approximation*

1. Introduction

To use this template, you will need to (1) apply the embedded styles to each paragraph-level item in your manuscript or (2) use the specifications shown in Table 1 to format your manuscript, with this template as a visual guide. Information about paper submission is available from the Journal website.

A system is a large scale when its dimensions are so high, such that conventional techniques of modeling, analysis control, design and computation fail to give accurate solutions with reasonable computational efforts. Thus, Model reduction of a high order system is an important problem in analysis, as well as in controller synthesis of a practical system [1]. To reduce large system to its low-order many techniques are available in literature using time domain and frequency domain methods [2,3,4,5].

In a frequency domain, one important order reduction method is Pade approximation reported in [4,6,7]. It is reported that the Pade approximation method does not guarantee the stability of the reduced-order model; therefore, many other alternative methods, such as the Mihailov stability criterion [8], Routh approximation [6,9], stability equation method [10,11,12], Routh stability array method [11,13] and Hurwitz polynomial approximation [14] have been reported time to time. Shamash [15] have considered the effect of including Markov parameters along with time moments to ensure the initial time response. A modified version Routh Pade approximation method is reported in [3]. The stability equation method application is presented in [16] and Routh array method as in [17]. Stability preservation using differentiation method is presented in [18] and extended to MIMO systems in [19].

Soloklo and Farasangi in 2013 [1], have introduced the multiobjective weighted sum approach for model reduction by Routh Pade approximation using Harmony search algorithm. The advantage of stability equation method along with genetic algorithm (GA) is used to reduce the order of a system in [20]. A 10th order two input and two output linear time invariant power system model is reduced using PSO in [21]. Other evolutionary methods based reduced order model is reported in [22] , H∞ Model Reduction in [23,24,25] and Hankel singular based in [26,27]. Moreover, the application of model order reduction methods has been applied in the field of power system [9,28,29,30]. The application of bat algorithm for model order reduction is presented in [31].

The organizations of the contents are as problem formulation in section 2, review on Cuckoo Search algorithm and the modifications made in it are included in section 3, the test systems considered for the reduction and comparison to existing reduced models are incorporated in section 4 and the concluding remarks are made in section 5, followed by references.

2. Problem Formulation

Consider a higher order system of order n and is represented by equation (1).

$$G(s)=\frac{Y(s)}{R(s)}=\frac{a_1 s^{n-1}+a_2 s^{n-2}+\ldots+a_n}{s^n+b_1 s^{n-1}+b_2 s^{n-2}+\ldots+b_n} \qquad (1)$$

Where, a_i and b_i are constants for $i=1,2,\ldots,n$.

If r represents a reduced order as of lesser order than n, then, the reduced order model of the system in equation (1) is represented as in equation (2). The important and principal requirement of the reduced order model is to posses all important specifications of the original system.

$$G_r(s) = \frac{Y_r(s)}{R_r(s)} = \frac{c_1 s^{r-1} + c_2 s^{r-2} + \ldots + c_r}{s^r + d_1 s^{r-1} + d_2 s^{r-2} + \ldots + d_r} \quad (2)$$

Where, c_r and d_r are unknown constants and are to be determined by using cuckoo search algorithm subjected to minimization of integral square error defined in equation (3). Error denotes difference of the unit step responses by original system and the reduced model to a unit-step function as shown in Figure 1.

$$J = \int_{t=0}^{t=\infty} |e(t)|^2 dt \quad (3)$$

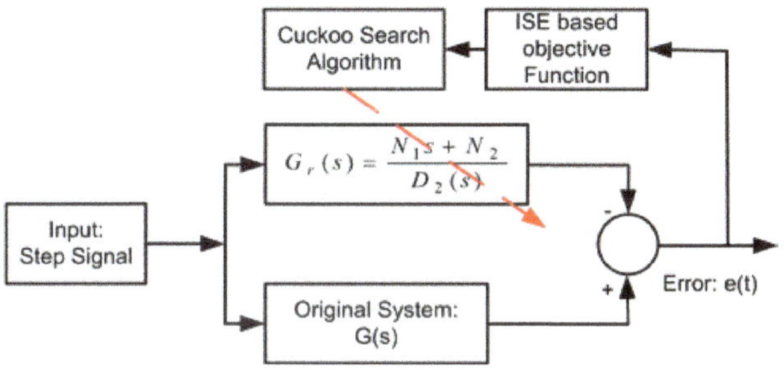

Figure 1. Scheme for optimization of free coefficients of reduced order model

3. Cuckoo Search Algorithm

The literature on cuckoo search is expanding rapidly. There has been a lot of attention and recent studies using cuckoo search with a diverse range of applications. Walton et *al.* improved the algorithm by formulating a modified cuckoo search algorithm [32], while Yang and Deb extended it to multi-objective optimization problems [33].

As the Cuckoos lay their eggs in the nest of other birds and respective host birds take care of the cuckoo's chicks [34]. It is mainly inspired by the obligate brood parasitism of cuckoos by laying their eggs in the nests of other host birds. The infringing cuckoos are in direct contest with the host birds. The host bird discovers the eggs of other birds and may throw these out of nest or may construct another nest elsewhere. The Parasitic cuckoos generally selects a nest in which the host bird just laid its own eggs [34]. The Cuckoo eggs generally hatch somewhat earlier than their host eggs [35]. As soon as, cuckoo chick is hatched starts to evict y blindly propelling the eggs out of the nest to reduce the share of food. Cuckoo chick starts to mimic the voice call of host chicks to gain more opportunity of feeding [34].

An algorithm provides a set of output variables on application of input variables. An optimization algorithm generates/produces a new set of solution x^{t+1} to a given problem from a given solution x^t at time t or iteration.

$$x^{t+1} = A\{x^t, p(t)\} \quad (4)$$

Where, the new solution vector x^{t+1} is nonlinearly mapped through A to given d-dimensional vector x^t . Let the variables of the problem are k and are represented as $p(t) = p_1, p_2, \ldots, p_k$ which may be time dependent and can be tuned by A. Let an optimization problem is S with states as ψ then according to predefine criterion D, the optimal solution x_{os} selects the desired states as φ as in equation (5).

$$S(\psi) \xrightarrow{A(t)} S\{\varphi(x_{os})\} \quad (5)$$

Thus, the final found/converged state φ represents to an optimal solution x_{os} of the problem of interest. Here, the system states are selected in the design space by running the optimization algorithm A. Thus, the performance of the algorithm is depended /controlled by the initial solution $x^{t=0}$, the parameters p, and stopping criterion D.

3.1. Procedural Steps

The cuckoo search algorithm is based on the brood parasitism of some cuckoos such as the ani and Guira and is enhanced by use of *Levy flights* [36], not just by simple isotropic random walks. The Cuckoos are special birds not only because of the beautiful sounds but also because of their aggressive reproduction strategy. Cuckoos engage the obligate brood parasitism by laying their eggs in the nests of other host birds. The ani and Guira as the species of cuckoos used to lay their eggs in other bird's nests and they may remove others' eggs to increase the *hatching probability* of their own eggs. It is necessary to make assumptions as followings:

Assumptions

- At a time each cuckoo lays one egg and dumps it in a randomly selected nest
- The nests with high-quality eggs are selected and being carried over to the next generations
- The available number of nests (of hosts) is kept fixed (as n), and the probability of cuckoo egg detection by the host bird is fixed as $P_a \in [0,1]$. As above, the host bird may get rid of the egg or may even abandon the nest to build a new nest i.e. a fraction Pa of the n host nests that are replaced by new nests [34].

Further, as an implementation, it should be assumed that the solution refers to an egg in a nest, and each cuckoo can lay only one egg. Thus, there is no distinction between cuckoo, egg or nest because as each nest consists one egg which corresponds to one cuckoo. CS algorithm uses a combination of a local random walk (for local search) and the global random walk (for global search) and is controlled by a switching parameter P_a.

3.1.1. Local Random Walk

Let two different solutions selected by random permutation are as x_j^t and x_k^t , Heaviside function as $H(P_a - \epsilon)$, random number drawn from a uniform distribution as ϵ, and the step size as '**s**'. Then, the local random walk can be represented as.

$$x_i^{t+1} = x_i^t + \alpha\, s \otimes H(P_a - \in) \otimes (x_j^t - x_k^t) \quad (6)$$

Here, $\alpha > 0$ is the step size related to the scales of the problem of interests. It is generally selected as $\alpha = 1$.The product "$\oplus$" means entry-wise walk during multiplications.

3.1.2. Global Random Walk

The global random walk is carried out by using Levy flights in which the step-lengths are distributed according to a heavy-tailed probability distribution [34]. On completion of large number of steps the random walk tends to a stable distribution as compared to its origin. The final solution can be represented by equation (7) as following.

$$x_i^{t+1} = x_i^t + \alpha\, L(s,\lambda) \quad (7)$$

Where,

$$L(s,\lambda) = \frac{\lambda \Gamma(\lambda) \sin(\pi\lambda/2)}{\pi} \frac{1}{s^{1+\lambda}} \quad (8)$$
$$(s >> s_0 > 0)$$

The equation (7) is the stochastic representation for a random walk. The random walk is a Markov chain; whose next location directly depends on the current location and the transition probability. An appropriate value of new solutions generated by randomization and their locations should be far enough from the best solution (current) to make sure not be trapped in a local optimum [35,37]. The local search exists about to 1/4 of the search time (with Pa = 0.25), while global search exists for 3/4 of the total search time.

3.1.3. Levy Distribution

Levy flights are characterized by infinite mean and variance therefore, CSA can explore the search space more efficiently as compared to standard Gaussian process. Thus, CSA guaranteed global convergence and highly efficient [37,38,39].

In Lévy flight the step-lengths are distributed according to the probability distribution as in equation (9), which provides a random walk while the random step length is drawn from a Levy distribution [35].

$$Levy(u) = t^{-\lambda}, (1 < \lambda \le 3) \quad (9)$$

3.1.4. Improved Cuckoo Search

As above the α introduced in the CSA is to find locally improved solutions, while P_a and λ to find global solution. P_a and α parameters play a vital role in tuning of solution vectors. In original CSA, P_a and α are kept fixed and cannot be altered during new generations, therefore, the number of iterations kept large to get optimal solutions. With large value of P_a and small value of α, the convergence speed is high but unable to find required solutions. To mitigate the problem of adjusting the value of P_a and α, these are considered as variables in improved CSA. The values of P_a and α must be large enough to make capable the algorithm to increase the diversity of solution vectors during early generations and decreased in final generations to result in a better fine-tuning of solution vectors [34]. Thus, P_a and α are dynamically changed with the number of generation and expressed in equations (10) – (12), where NI and gn are the number of total iterations and the current iteration, respectively [34].

$$P_a^{gn} = P_{a\max} - \frac{(P_{a\max} - P_{a\min})}{NI}.\, gn \quad (10)$$

$$\alpha\,(gn) = \alpha_{\max}.\exp\,(c.gn) \quad (11)$$

$$c = \frac{Ln\left(\alpha_{\min} / \alpha_{\max}\right)}{NI} \quad (12)$$

The performance of the algorithm may deteriorate by an increase in the maximum value of α as in [34], therefore, the suitable values are $0.005 \le P_a \le 1.0$ and $0.05 \le \alpha \le 0.5$. The considered values of P_a and α are 0.25 and 0.25, respectively.

4. Results and Discussions

4.1. Example-1: SISO System

Considering a fourth order single input and single output (SISO) system [20,40,41] and is described by the transfer function as in equation (13). The second order ROM as proposed is represented by equation (14) with free coefficients as N_1 and N_2. The range of these free coefficients are considered as $0.1 \le N_1 \le 1.0$ and $0.1 \le N_2 \le 2.0$, thus the lower and upper bounds are defined in CSA accordingly in vector format. The other necessary parameters of CSA as number of nests i.e. different solutions (n) as 25, discovery rate of alien eggs (P_a) as 0.25 and maximum iterations as 200 are selected to compute the free coefficients of equation (14). The computed coefficients N_1 and N_2 are obtained as 0.8130 and 0.7945, respectively and resulting equation (15) and the variation of these free coefficients for 200 iterations by CSA is shown in Figures 2(b) - 2(c). The performance of CSA during coefficient computation in terms of fitness function value for 200 iterations is shown in Figure 2(a).

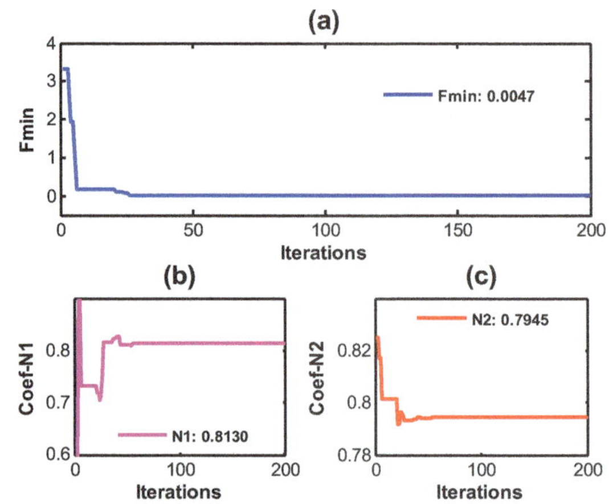

Figure 2. Performance of Cuckoo search algorithm for test system - 1 for (a) Plot of fitness function, (b) Variation of free coefficient N_1, and (c) Variation of free coefficient N_2

The reduced second order model by Desai [41] using Routh approximation and Big Bang Big Crunch is represented by equation (16). In this method the numerator coefficients are determined by Big Bang Big Crunch algorithm and the denominator elements are determined by Routh approximation. The original system is reduced to second order, where in [20], the denominator is reduced using stability equation method and the numerator coefficients are found using Genetic algorithm and represented in equation (17). Another approach proposed by Boby and Philip in [42], where in denominator found by dominant pole method and the numerator by Big Bang Big Crunch algorithm and represented by equation (18).

$$G(s)|_{Original} = \frac{s^3+7s^2+24s+24}{s^4+10s^3+35s^2+50s+24} \quad (13)$$

$$G_r(s) = \frac{N_1 s + N_2}{s^2+1.6560s+0.7947} \quad (14)$$

$$G(s)|_{Proposed} = \frac{0.8130s+0.7945}{s^2+1.6560s+0.7947} \quad (15)$$

$$G(s)|_{Desai} = \frac{0.8058s+0.7944}{s^2+1.65s+0.7944} \quad (16)$$

$$G(s)|_{Parmar} = \frac{0.7442575s+0.6991576}{s^2+1.45771s+0.6997} \quad (17)$$

$$G(s)|_{Boby} = \frac{0.9315s+1.6092}{s^2+2.75612s+1.6092} \quad (18)$$

Step response of original system, proposed reduced, and other published reduced models are compared in Figure 3 and other performance parameters as overshoot, rise time, settling time and performance indices such as integral time multiplied absolute error (ITAE), integral absolute error (IAE) and integral square error (ISE) are enlisted in Table 1 and defined in [43,44,45]. All three performance indices (PIs) for proposed reduced order model (ROM) are lesser in magnitude as compared to other published ROMs [20,40,41], proving superior performance. In Desai and Prasad, 2013[11], the 2nd order reduced model of the considered system in equation (13) is $(0.8s+0.686)/(s^2+1.47s+0.686)$ with ISE value 3.5×10^{-4} which is greater than the proposed 2.15×10^{-4}; proving effectiveness of the proposed methodology.

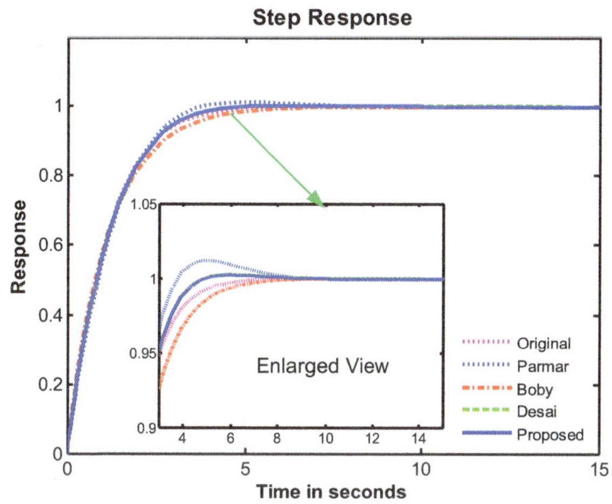

Figure 3. Step response of Original system, 2nd order reduced (Proposed) and other comparing reduced models

Table 1. Performance parameters of original test system -1 and reduced models (proposed ROM and as in [20,40,41])

Para-meters	Original [20,40,41]	Pro-posed	Desai [41]	Parmar [20]	Boby [42]
Over-shoot	0	0	0.2738	1.3007	0.2738
Rise Time	2.2602	2.2767	2.2787	2.189	2.2787
Settling Time	3.9307	3.6120	3.6199	3.222	3.6199
ISE	-	2.15E-4	2.17E-4	1.64E-3	1.7E-3
ITAE	-	0.1758	0.1770	0.5226	0.2575
IAE	-	0.0430	0.0447	0.1147	0.0933

4.2. Example-2: SISO System

Considering a test system of order six as in [1] and presented in equation (19). Considering second order reduced order model as in equation (20), where, N_1,and N_2 are the free parameters to be optimized by CSA while the denominator is obtained using Routh approximation as in [41,46]. The parametric bound as lower and upper value for CSA are considered as $0.1 \leq N_1 \leq 1$and $0.1 \leq N_2 \leq 1.0$ respectively. The performance of CSA in optimization for 200 iterations is shown in Figure 4. The fitness function variation over 200 iterations is shown in Figure 4(a), while the free coefficient variation is shown in Figure 4(b) for N_1 and Figure 4(b) for N_2. The resulting value of fitness function at completion of 200 iterations is 0.0791 and the values of N_1 and N_2 parameters are 0.0910 and 0.0109, respectively. Thus resulting 2nd order reduced model (proposed) is represented in equation (21).

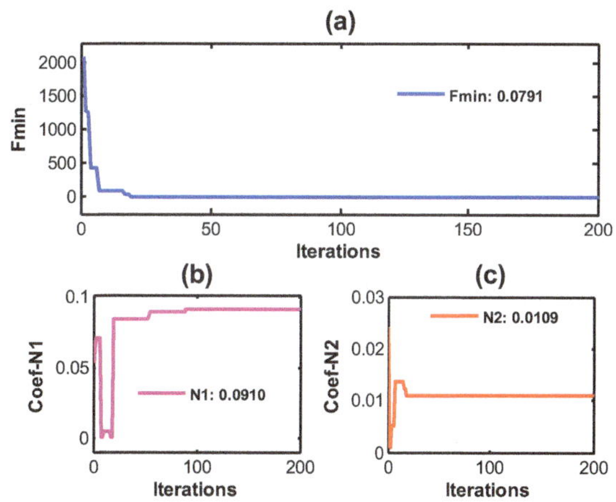

Figure 4. Performance of Cuckoo search algorithm for test system - 2 for (a) Plot of fitness function, (b) Variation of free coefficient N1, and (c) Variation of free coefficient N2

$$G(s)\Big|_{original} = \left[\frac{2s^5+3s^4+16s^3+20s^2+8s+1}{2s^6+33.6s^5+155.94s^4+209.46s^3+102.42s^2+18.3s+1}\right] \quad (19)$$

$$G_r(s) = \frac{N_1 s + N_2}{s^2+0.2012s+0.01099} \quad (20)$$

$$G(s)\Big|_{Proposed} = \left[\frac{0.0910s+0.0109}{s^2+0.2012s+0.01099}\right] \quad (21)$$

The other second order reduced models as typical Pade approximant [1] and Multi objective Harmony search algorithm [1] based are as in equation (22) and equation (23), respectively.

$$G(s)\Big|_{Pade} = \left[\frac{6.87815s+1.09228}{89.54625s^2+12.96860s+0.99732}\right] \quad (22)$$

$$G(s)\Big|_{Soloklo} = \left[\frac{7.80016s+0.81849}{87.58712s^2+12.43314s+0.81657}\right] \quad (23)$$

Table 2. Performance parameters of original test system-2 and reduced models

Para-meters	Original [1]	Proposed	Pade [1]	Soloklo[1]
Over-shoot	0	0.0193	7.46	7.61
Rise Time	22.7	21.4609	15.1	14.3
Settling Time	40	36.0165	50.4	53
ITAE	-	30.4949	585.0442	138.72
IAE	-	0.5987	12.7028	4.906
ISE	-	0.0052	2.8887	0.5377

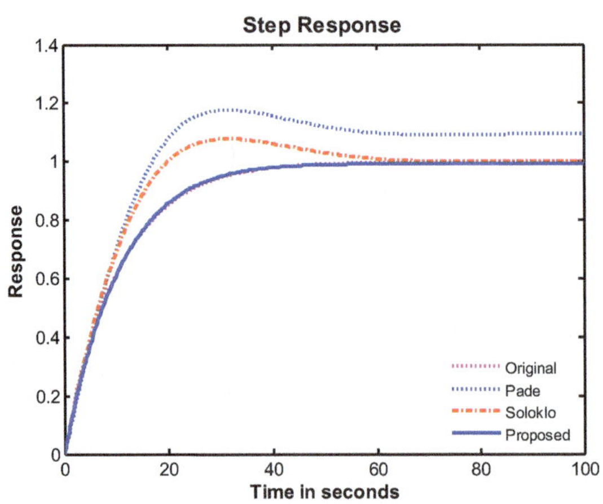

Figure 5. Step response of Original system, 2nd order reduced (Proposed) and other comparing reduced models as in [1]

The Step response of Original system, 2nd order reduced (Proposed) and other comparing reduced models of [1] as in equation (19) and equations (21) - (23) is shown in Figure 5. The step response information and PIs based performance is summarized in Table 2, where in, the proposed ROM possesses least value of the PI values resulting to superior performance.

4.3. Example-3: SISO System

Considering a test system as boiler system of order nine as in [47] and presented in equation (24). Here, the desired third order reduced model is represented by equation (25). The parameter bounds for free coefficients in CSA are as $120 \le N_1 \le 160$, $4000 \le N_2 \le 4500$ and $4400 \le N_3 \le 5000$ for 200 iterations and the denominator is determined by typical Routh approximation as in [41,46]. The performance of Cuckoo Search algorithm for determining free coefficients is shown in Figure 6(a) in terms of fitness function plot and coefficients in Figures 6(b) - 6(d). Using optimized parameters, the third order reduced model by CSA is shown in equation (26). The other third order model as in [1] as typical Pade and HSA based multi-objective are represented in equations (27) - (28), respectively.

The step signal based response of these systems is shown in Figure 7. The step response by proposed method is able to mimic the original system more appropriately as compared to others. The other comparing parameters as overshoot, rise time, settling time and performance indices as ISE (Integral square error), ITAE (Integral time multiplied absolute error) and IAE (Integral absolute error) of these step responses are determined and enlisted in Table 3. The performance parameters with proposed reduced model in terms of ISE, ITAE and IAE are having least values as compared to other methods, proving superior performance.

$$G(s) = \left[\frac{146.4s^8+9.81\times10^4 s^7+5.999\times10^7 s^6+3.206\times10^{10} s^5+3.582\times10^{12} s^4+1.113\times10^{14} s^3+1.154\times10^{15} s^2+3.971\times10^{15} s+3.063\times10^{15}}{s^9+659.8\,s^8+4.136\times10^5 s^7+2.13\times10^8 s^6+2.422\times10^{10} s^5+8.737\times10^{11} s^4+1.523\times10^{13} s^3+1.221\times10^{14} s^2+3.636\times10^{14} s+2.406\times10^{14}}\right] \quad (24)$$

$$G_r(s) = \left[\frac{N_1 s^2+N_2 s+N_3}{s^3+29.8634s^2+399.8925s+355.9565}\right] \quad (25)$$

$$G_r(s) = \left[\frac{144.50s^2+4119.30s+4531.50}{s^3+29.8634s^2+399.8925s+355.9565}\right] \quad (26)$$

$$G(s)|_{Pade} = \left[\frac{145.36242 s^2 + 431231031 s + 4701.85734}{s^3 + 23.23900\, s^2 + 420.38264 s + 371.29177} \right] \quad (27)$$

$$G(s)|_{Soloklo} = \left[\frac{148.12856 s^2 + 4398.96963 s + 4725.72521}{s^3 + 29.90996 s^2 + 429.17178 s + 371.99085} \right] \quad (28)$$

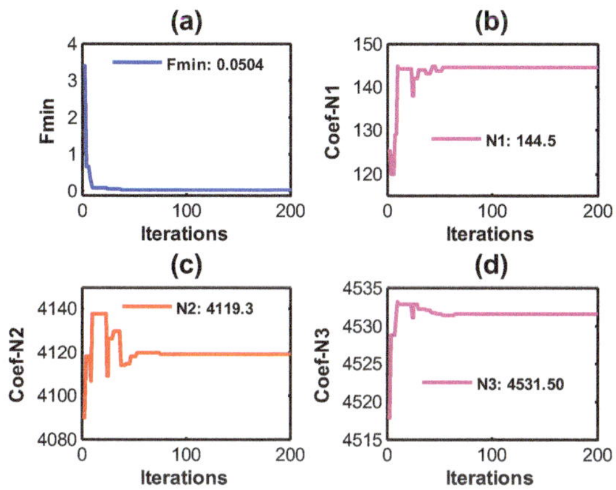

Figure 6. Performance of Cuckoo search algorithm for test system - 3 for (a) Plot of fitness function, (b) Variation of free coefficient N_1, (c) Variation of free coefficient N_2, and (c) Variation of free coefficient N_3

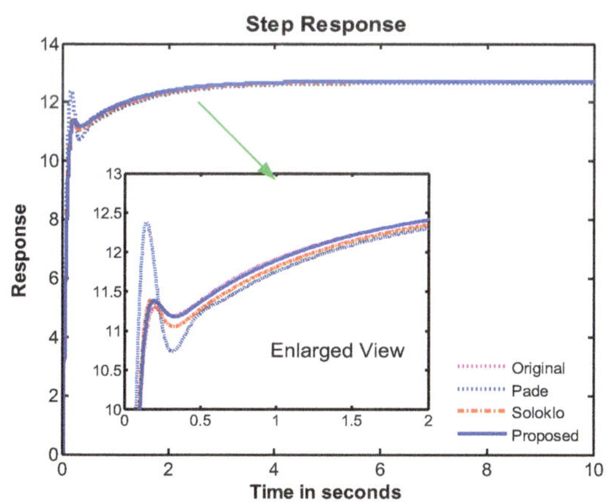

Figure 7. Step response of Original system, 2nd order reduced (Proposed) and other comparing reduced models as in [1]

Table 3. Performance parameters of original and reduced models

Para-meters	Original [47]	Proposed	Pade [1]	Soloklo [1]
Over-shoot	0	0	0	0
Rise Time	0.543	0.5613	0.0918	0.612
Settling Time	2.28	2.2520	2.39	2.36
ITAE	-	0.2374	53.9939	21.622
IAE	-	0.0719	3.0904	1.2432
ISE	-	0.0027	0.3561	0.0269

4.4. Example-4: MIMO system

Let us consider a Multi-Input Multi-Output (MIMO) system as in [41,48] of sixth order having the following transfer matrix as in equations (29) - (30). The system in sub-system format is represented in equations (31) - (34). The proposed second order reduced model is represented by equation (35).

$$G(s) = \begin{bmatrix} \frac{2(s+5)}{(s+1)(s+10)} & \frac{(s+4)}{(s+2)(s+5)} \\ \frac{(s+10)}{(s+1)(s+20)} & \frac{(s+6)}{(s+2)(s+3)} \end{bmatrix} \quad (29)$$

$$G(s) = \begin{bmatrix} G_{11}(s) & G_{12}(s) \\ G_{21}(s) & G_{22}(s) \end{bmatrix} \quad (30)$$

$$G_{11}(s) = \left[\frac{2s^5 + 70s^4 + 762s^3 + 3610s^2 + 7700s + 6000}{s^6 + 41s^5 + 571s^4 + 3491s^3 + 10060s^2 + 13100s + 6000} \right] \quad (31)$$

$$G_{12}(s) = \left[\frac{s^5 + 38s^4 + 459s^3 + 2182s^2 + 4160s + 2400}{s^6 + 41s^5 + 571s^4 + 3491s^3 + 10060s^2 + 13100s + 6000} \right] \quad (32)$$

$$G_{21}(s) = \left[\frac{s^5 + 30s^4 + 331s^3 + 1650s^2 + 3700s + 3000}{s^6 + 41s^5 + 571s^4 + 3491s^3 + 10060s^2 + 13100s + 6000} \right] \quad (33)$$

$$G_{22}(s) = \left[\frac{s^5 + 42s^4 + 601s^3 + 3660s^2 + 9100s + 6000}{s^6 + 41s^5 + 571s^4 + 3491s^3 + 10060s^2 + 13100s + 6000} \right] \quad (34)$$

$$G(s) = \frac{\begin{bmatrix} N_{11}(s) & N_{12}(s) \\ N_{21}(s) & N_{22}(s) \end{bmatrix}}{D_2(s)} \quad (35)$$

The 6th order denominator of the system is reduced to 2nd order model by applying Routh approximation method as in [41] and is being represented by equation (36).

$$D_2(s) = s^2 + 1.5480\, s + 0.7091 \quad (36)$$

Since, the denominator of equations (31) - (34) is same; therefore, the denominator would remain same. The numerators are different, would be determined by using Cuckoo search algorithm. The general form of numerator with free coefficient can be represented as $N_1 s + N_2$. The performance of CSA in terms of fitness function is shown in Figure 8 and the plot of the free coefficients for $N_{11}(s)$, $N_{12}(s)$, and $N_{21}(s)$ and $N_{22}(s)$ is shown in Figure 9. The optimized numerators are shown in equation (37) - (40) and the resulting reduced MIMO system is represented by equation (41).

$$N_{11}(s) = 0.0672\, s + 0.7086 \quad (37)$$

$$N_{12}(s) = 0.5223\, s + 0.2831 \quad (38)$$

$$N_{21}(s) = 0.4552\, s + 0.3543 \quad (39)$$

$$N_{22}(s) = 1.1398\, s + 0.7083 \quad (40)$$

$$G_2(s) = \frac{\begin{bmatrix} 0.0672\, s + 0.7086 & 0.5223\, s + 0.2831 \\ 0.4552\, s + 0.3543 & 1.1398\, s + 0.7083 \end{bmatrix}}{s^2 + 1.5480s + 0.7091} \quad (41)$$

The reduced second order transfer matrix by Prasad [48] is given as in equation (42).

$$R_2(s) = \frac{\begin{bmatrix} 1.18156s + 3.65079 & 1.04664s + 1.46031 \\ 0.49819s + 1.82539 & 1.6911s + 3.65079 \end{bmatrix}}{s^2 + 4.3374s + 3.65079} \quad (42)$$

The reduced second order transfer matrix by Desai [41] is given as in equation (43).

$$R_2(s) = \frac{\begin{bmatrix} 0.9475s + 0.7091 & 0.4892s + 0.2837 \\ 0.455s + 0.3546 & 1.126s + 0.7091 \end{bmatrix}}{s^2 + 1.548267s + 0.7091} \quad (43)$$

The reduced second order transfer matrix by Parmar [20] is given as in equation (44).

$$R_2(s) = \frac{\begin{bmatrix} \begin{pmatrix} 0.8503087s \\ +0.6171331 \end{pmatrix} & \begin{pmatrix} 0.4617562s \\ +0.2466069 \end{pmatrix} \\ \begin{pmatrix} 0.4093304s \\ +0.3086095 \end{pmatrix} & \begin{pmatrix} 0.9976611s \\ +0.6171125 \end{pmatrix} \end{bmatrix}}{s^2 + 1.34952s + 0.6181} \quad (44)$$

The step response for original system, proposed reduced model and other published reduced models [20,41,48] are shown in Figures 10 - 13 and the corresponding performance indices (ITAE, IAE and ISE) are shown in Table 4 – Table 7. The step response for G_{11} as in Figure 10, shown the best mimic operation as compared to others while in Figure 11 (for G_{12}, G_{21} and G_{22}) is Vishwakarma and Prasad. As the performance indices are quantitative representation of the responses and minimum value represents the best performance. By observation of Table 4 – Table 7, the all PIs (ITAE, IAE and ISE) found to be minimum with the proposed reduced model. It is found that the proposed method is able reduce manual calculative complexity as in the pole-clustering and other method.

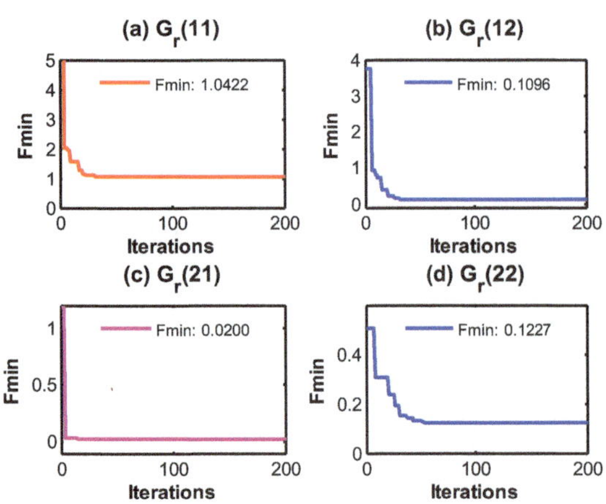

Figure 8. Fitness function for optimization of free coefficients using CSA for denominator by Routh approximation method

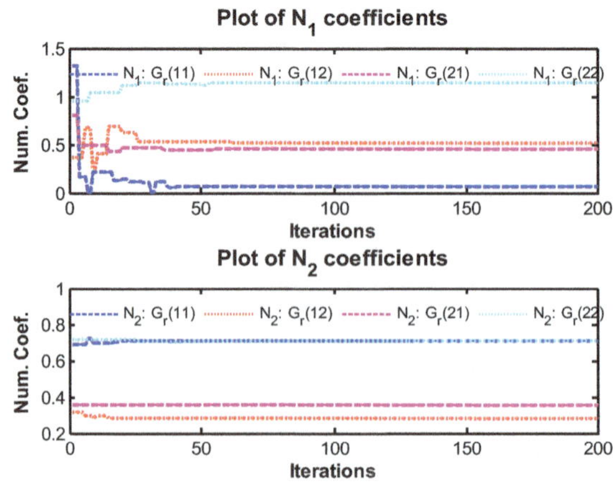

Figure 9. Plot of free coefficients N_1 and N_2 for $G_r(11)$, $G_r(12)$, $G_r(21)$ and $G_r(22)$ sub-systems of proposed reduced models using CSA.

Table 4. Performance indices (IAE, ITAE and ISE) for step response of different reduced models of $G_{11}(s)$

Methods	$R_{11}(s)$		
	IAE	ITAE	ISE
Proposed	0.35	0.5411	0.000825
Desai [41]	1.25	2.9822	0.00672
Prasad [48]	1.21	2.6621	0.001515
Parmar [20]	1.31	3.4544	0.014498

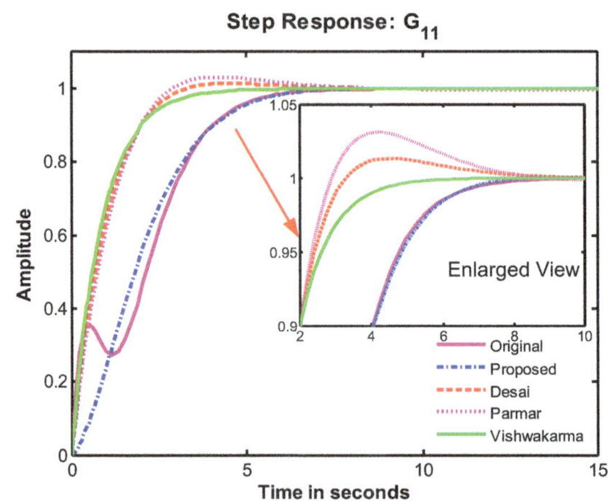

Figure 10. Step response of Original system, $G_{11}(s)$ and different reduced models.

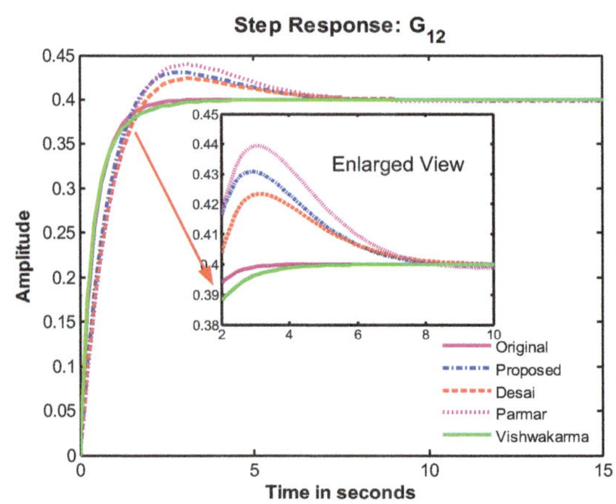

Figure 11. Step response of Original system, $G_{12}(s)$ and different reduced models

Table 5. Performance indices (IAE, ITAE and ISE) for step response of different reduced models of $G_{12}(s)$

Methods	$R_{12}(s)$		
	IAE	ITAE	ISE
Proposed	0.1665	0.5	6.006E-3
Desai [41]	0.1559	0.37	0.005918
Prasad [48]	0.0168	3E-2	7.845E-5
Parmar [20]	0.2169	0.70	0.008744

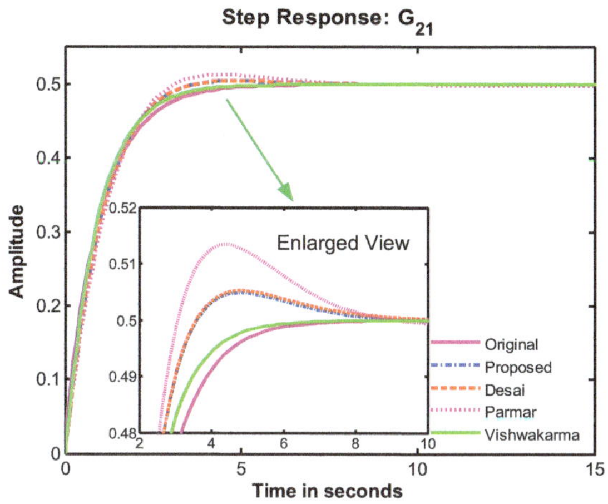

Figure 12. Step response of Original system, $G_{21}(s)$ and different reduced models.

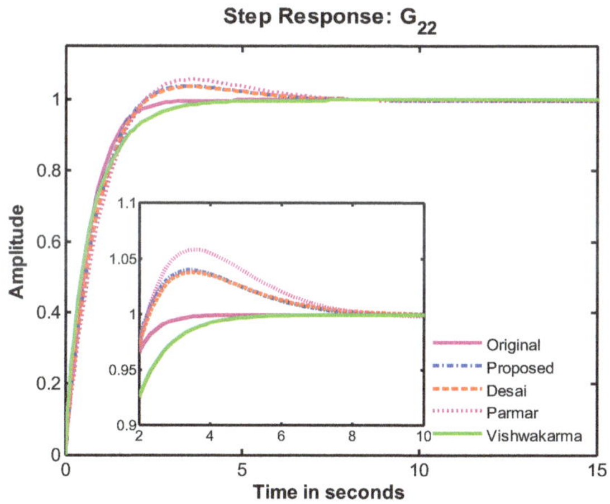

Figure 13. Step response of Original system, $G_{22}(s)$ and different reduced models

Table 6. Performance indices (IAE, ITAE and ISE) for step response of different reduced models of $G_{21}(s)$

Methods	$R_{21}(s)$		
	IAE	ITAE	ISE
Proposed	0.0763	0.2594	1.005E-3
Desai [41]	0.0766	0.2387	0.000958
Prasad [48]	0.0342	0.0717	0.000299
Parmar [20]	0.1288	0.4807	0.002538

Table 7. Performance indices (IAE, ITAE and ISE) for step response of different reduced models of $G_{22}(s)$

Methods	$R_{22}(s)$		
	IAE	ITAE	ISE
Proposed	0.196	0.7638	0.005123
Desai [41]	0.191	0.6182	0.005896
Prasad [48]	0.125	0.2207	0.004681
Parmar [20]	0.315	1.1704	0.015741

5. Conclusion

The Cuckoo Search algorithm based model order reduction is designed with minimization of ISE pertaining to a unit step input for optimization of numerator free coefficients. The denominator of the original systems for SISO and MIMO systems are reduced using Routh approximation method. The optimization process is for bounded constraints varying as per examples and the order of reduction. The performance of the Cuckoo Search algorithm based reduced order model for different systems is compared to Routh approximation and Big Bang Big Crunch, Dominat Pole clustering and Big Bang Big Crunch, Stability equation and Genetic algorithm, typical Pade and HSA based multi-objective ROM and outperforms in terms of performance indices as ITAE, IAE and ISE.

References

[1] H. N. Soloklo and M. M. Farsangi, "Multiobjective weighted sum approach model reduction by Routh-Pade approximation using harmony search," Turk J Elec Eng & Comp Sci, vol. 21, pp. 2283 - 2293, 30.10.2013 2013.

[2] V. Singh, D. Chandra, and H. Kar, "Optimal Routh approximants through integral squared error minimisation: computer-aided approach," IEE Proceedings - Control Theory and Applications, vol. 151, pp. 53-58, 2004.

[3] V. Singh, "Obtaining Routh-Pade approximants using the Luus-Jaakola algorithm," IEE Proceedings - Control Theory and Applications, vol. 152, pp. 129-132, 2005.

[4] V. Singh, D. Chandra, and H. Kar, "Improved Routh-Pade approximants: a computer-aided approach," IEEE Transactions on Automatic Control, vol. 49, pp. 292-296, 2004.

[5] N. K. Sinha and G. J. Lastman, "Reduced Order Models for Complex Systems-A Critical Survey," IETE Technical Review, vol. 7, pp. 33-40, 1990/01/01 1990.

[6] I. El-nahas, N. K. Sinha, and R. T. H. Alden, "Pade and Routh approximation in the time domain," in Decision and Control, 1983. The 22nd IEEE Conference on, 1983, pp. 243-246.

[7] C.-q. Gu and J. Yang, "Stable Routh-Padé-type approximation in model reduction of interval systems," Journal of Shanghai University (English Edition), vol. 14, pp. 369-373, 2010/10/01 2010.

[8] B.-W. Wan, "Linear model reduction using Mihailov criterion and Padé approximation technique," International Journal of Control, vol. 33, pp. 1073-1089, 1981/06/01 1981.

[9] D. K. Sambariya and R. Prasad, "Routh Approximation based Stable Reduced Model of Single Machine Infinite Bus System with Power System Stabilizer," in DRDO-CSIR Sponsered: IX Control Instrumentation System Conference (CISCON - 2012), Department of Instrumentation and Control Engineering, Manipal Institute of Technology (A Constitute Institute of Manipal University), Manipal-576104; ISBN 978-93-82338-26-0 | © 2012 Bonfring; CIS-162; , 2012, pp. 85-93.

[10] T. C. Chen, C. Y. Chang, and K. W. Han, "Model reduction using the stability-equation method and the continued-fraction method," International Journal of Control, vol. 32, pp. 81-94, 1980/07/01 1980.

[11] S. R. Desai and R. Prasad, "A new approach to order reduction using stability equation and big bang big crunch optimization," Systems Science & Control Engineering, vol. 1, pp. 20-27, 2013/12/01 2013.

[12] A. Sikander and R. Prasad, "A Novel Order Reduction Method Using Cuckoo Search Algorithm," IETE Journal of Research, vol. 61, pp. 83-90, 2015/03/04 2015.

[13] D. K. Sambariya and R. Prasad, "Routh Stability Array Method based reduced model of Single Machine Infinite Bus with Power System Stabilizer," in International Conference on Emerging Trends in Electrical, Communication and Information Technologies (ICECIT-2012), ISBN-9789351070504, at Srinivasa Ramanujan Institute of Technology, Rotarypuram (V), B K

Samudram (M), Anantapur (Dist). - 515701, Andhrapradesh, India., 2012, pp. 27-34.

[14] R. K. Appiah, "Linear model reduction using Hurwitz polynomial approximation," International Journal of Control, vol. 28, pp. 477-488, 1978/09/01 1978.

[15] Y. Shamash, "Model reduction using the Routh stability criterion and the Padé approximation technique," International Journal of Control, vol. 21, pp. 475-484, 1975/03/01 1975.

[16] D. K. Sambariya and G. Arvind, "High Order Diminution of LTI System Using Stability Equation Method," British Journal of Mathematics & Computer Science, vol. 13, pp. 1-15, December 29, 2015 2016.

[17] D. K. Sambariya and H. Manohar, "Model order reduction by differentiation equation method using with Routh array Method," in 10th International Conference on Intelligent Systems and Control (ISCO 2016), Karpagam College of Engineering, Coimbatore, Tamilnadu, India, 2016, pp. 341-346.

[18] D. K. Sambariya and H. Manohar, "Preservation of Stability for Reduced Order Model of Large Scale Systems Using Differentiation Method," British Journal of Mathematics & Computer Science, vol. 13, pp. 1-17, 2016.

[19] H. Manohar and D. K. Sambariya, "Model order reduction of MIMO system using differentiation method," in 10th International Conference on Intelligent Systems and Control (ISCO 2016), Karpagam College of Engineering, Coimbatore, Tamilnadu, India, 2016, pp. 347-351.

[20] G. Parmar, R. Prasad, and S. Mukherjee, "Order Reduction of Linear Dynamic Systems using Stability Equation Method and GA," International Journal of Computer, Information, and Systems Science, and Engineering, vol. 1, pp. 1, 26-32 2007.

[21] U. Salma and K. Vaisakh, "Reduced Order Modeling of Linear MIMO Systems Using Soft Computing Techniques," in Swarm, Evolutionary, and Memetic Computing. vol. 7077, B. Panigrahi, P. Suganthan, S. Das, and S. Satapathy, Eds., ed: Springer Berlin Heidelberg, 2011, pp. 278-286.

[22] I. Saaki, P. C. Babu, C. K. Rao, and D. S. Prasad, "Integral square error minimization technique for linear multi input and multi output systems," in Power and Energy Systems (ICPS), 2011 International Conference on, 2011, pp. 1-5.

[23] F. Wu and J. J. Jaramillo, "Computationally Efficient Algorithm For Frequency-Weighted Optimal H∞ Model Reduction," Asian Journal of Control, vol. 5, pp. 341-349, 2003.

[24] J. Shen and J. Lam, "H∞ Model Reduction for Positive Fractional Order Systems," Asian Journal of Control, vol. 16, pp. 441-450, March 2014 2014.

[25] Y. Ebihara, Y. Hirai, and T. Hagiwara, "On $\mathcal{H}\infty$ model reduction for discrete-time linear time-invariant systems using linear matrix inequalities," Asian Journal of Control, vol. 10, pp. 291-300, 2008.

[26] A. Ghafoor, V. Sreera, and R. Treasure, "Frequency weighted model reduction technique retaining Hankel singular values," Asian Journal of Control, vol. 9, pp. 50-56, 2007.

[27] L. Cao and H. M. Schwartz, "Reduced-order models for feedback stabilization of linear systems with a singular perturbation model," Asian Journal of Control, vol. 7, pp. 326-336, 2005.

[28] D. K. Sambariya and R. Prasad, "Stable reduced model of a single machine infinite bus power system with power system stabilizer," in International Conference on Advances in Technology and Engineering (ICATE-2013) 2013, pp. 1-10.

[29] D. K. Sambariya and R. Prasad, "Stable Reduction Methods of Linear Dynamic Systems in Frequency Domain," 17th National Power Systems Conference (NPSC-2012) at Indian Institute of Technology (BHU), Vranasi, pp. 549-558, December 12-14, 2012 2012.

[30] D. K. Sambariya and R. Prasad, "Stability Equation method based Stable Reduced Model of Single Machine Infinite Bus System with Power System Stabilizer," International Journal of Electronic and Electrical Engineering, vol. 5, pp. 101-106, September 29-30, 2012 2012.

[31] D. K. Sambariya and H. Manohar, "Model order reduction by integral squared error minimization using bat algorithm," in Proceedings of 2015 RAECS UIET Panjab University Chandigarh 21 – 22nd December 2015, Chandigarh, India, 2015, pp. 1-7.

[32] S. Walton, O. Hassan, K. Morgan, and M. R. Brown, "Modified cuckoo search: A new gradient free optimisation algorithm," Chaos, Solitons & Fractals, vol. 44, pp. 710-718, 2011.

[33] X.-S. Yang and S. Deb, "Multiobjective cuckoo search for design optimization," Computers & Operations Research, vol. 40, pp. 1616-1624, 2013.

[34] E. Valian, S. Tavakoli, S. Mohanna, and A. Haghi, "Improved cuckoo search for reliability optimization problems," Computers & Industrial Engineering, vol. 64, pp. 459-468, 2013.

[35] Y. Xin-She and S. Deb, "Cuckoo Search via Levy flights," in World Congress on Nature & Biologically Inspired Computing, 2009. NaBIC 2009., 2009, pp. 210-214.

[36] I. Pavlyukevich, "Levy Flights, Non-local Search and Simulated Annealing," J Comput Phys, vol. 226, pp. 1830-1844, 26 Jan 2007 2007.

[37] X.-S. Yang and S. Deb, "Engineering Optimisation by Cuckoo Search," International Journal of Mathematical Modelling and Numerical Optimisation, vol. 1, pp. 330-343, 23 Dec 2010 2010.

[38] P. Civicioglu and E. Besdok, "A conceptual comparison of the Cuckoo-search, particle swarm optimization, differential evolution and artificial bee colony algorithms," Artif. Intell. Rev., vol. 39, pp. 315-346, 2013.

[39] A. Gandomi, X.-S. Yang, and A. Alavi, "Cuckoo search algorithm: a metaheuristic approach to solve structural optimization problems," Engineering with Computers, vol. 29, pp. 17-35, 2013/01/01 2013.

[40] S. Mukherjee and R. N. Mishra, "Order reduction of linear systems using an error minimization technique," Journal of the Franklin Institute, vol. 323, pp. 23-32, 1987.

[41] S. R. Desai and R. Prasad, "A novel order diminution of LTI systems using Big Bang Big Crunch optimization and Routh Approximation," Applied Mathematical Modelling, vol. 37, pp. 8016-8028, 2013.

[42] B. Philip and J. Pal, "An evolutionary computation based approach for reduced order modelling of linear systems," in Computational Intelligence and Computing Research (ICCIC), 2010 IEEE International Conference on, 2010, pp. 1-8.

[43] D. K. Sambariya and R. Prasad, "Optimal Tuning of Fuzzy Logic Power System Stabilizer Using Harmony Search Algorithm," International Journal of Fuzzy Systems, vol. 17, pp. 457-470, September 2015 2015.

[44] D. K. Sambariya and R. Prasad, "Robust tuning of power system stabilizer for small signal stability enhancement using metaheuristic bat algorithm," International Journal of Electrical Power & Energy Systems, vol. 61, pp. 229-238, 2014.

[45] D. K. Sambariya and R. Prasad, "Design of Robust PID Power System Stabilizer for Multimachine Power System Using HS Algorithm," American Journal of Electrical and Electronic Engineering, vol. 3, pp. 75-82, July 16, 2015 2015.

[46] V. P. Singh and D. Chandra, "Routh-approximation based model reduction using series expansion of interval systems," in Power, Control and Embedded Systems (ICPCES), 2010 International Conference on, 2010, pp. 1-4.

[47] R. Salim and M. Bettayeb, "H2 and H∞ optimal model reduction using genetic algorithms," Journal of the Franklin Institute, vol. 348, pp. 1177-1191, 2011.

[48] C. B. Vishwakarma and R. Prasad, "MIMO System Reduction Using Modified Pole Clustering and Genetic Algorithm," Modelling and Simulation in Engineering, vol. 2009, 2009.

Locomotion of Snake Robot through the Pipe

František Trebuňa[1], Ivan Virgala[2,*], Michal Kelemen[2], Tomáš Lipták[2]

[1]Department of Applied Mechanics and Mechanical Engineering, Technical University of Košice, Faculty of Mechanical Engineering, Košice, Slovakia
[2]Department of Mechatronics, Technical University of Košice, Faculty of Mechanical Engineering, Košice, Slovakia
*Corresponding author: ivan.virgala@tuke.sk

Abstract The paper deals with concertina locomotion of snake robot. Mathematical model of robot locomotion is based on locomotion of natural snake. The kinematic model is derived by homogeneous transformation matrices. In the fourth section, the mechanical design of experimental snake robot LocoSnake is introduced. The robot can perform planar motion by means of segments with 2 DOF, one revolute and one prismatic joint. In the following section the simulation of concertina locomotion is done in Matlab. Simulation is compared with experiments and the conclusion are discussed.

Keywords: *locomotion, simulation, snake robot*

1. Introduction

Locomotion of snake robots is inspired by biological snakes. Only concertina or rectilinear locomotion can be used for inspections tasks in the pipes or channels. However, both locomotion patterns are not very effective, in many cases there is no other option for locomotion. One of the first qualitative research concerning the biological snake was done by J. Gray [1] and H. W. Lissmann [2]. The authors investigated the muscle activity on real snake (boa occidentalis) and their works are often use as basic for this field of research. Next great milestone in this field was work of professor Shiego Hirose. He was an author of the first snake-like robot – ACM III in 1972 [3]. From this point the research on snake robots increased, especially in last two decades.

Sound research and investigation of new approach of concertina locomotion is presented in [4]. The authors dealt with new curve describing concertina locomotion of snake robot. Shape of this curve rise from the shape of snake body during its locomotion. However, snake robot locomotion in this study does not consider real behavior of robot, especially its anchoring to the walls of confined space. In [5] the authors investigated the utilizing anchoring of snake robot in the confined space. In the study there is determined the minimum number of self-locking contact points for concertina locomotion. Next serious contribution to the field of snake robot concertina locomotion is presented in [6]. The authors investigated friction of snake body on the surface. In the paper the video sequences of snake concertina locomotion is showed. On the video sequences can be seen that snake changes its anchoring to the confined space based on confined space width.

Based on previous research our aim established locomotion pattern of snake robot moving in the confined space, what will be in our case pipe with squared shape. The paper is divided into following sections: at first the sequence of motion of snake robot is introduced. Next is determination of kinematic model of snake robot concertina locomotion. Subsequently, the kinematic model is simulated in software Matlab. At last, the concertina locomotion is verified by experimental snake robot LocoSnake in the pipe with squared shape. The results in the conclusion are discussed.

2. Design of Concertina Locomotion

A snake performs concertina locomotion especially in narrow spaces where it cannot use lateral undulation. This is typically for snakes with small cross-section of their body. Two features are characteristic for concertina locomotion. The first feature consists of anchoring the snake rear by pushing against the walls of confined space while rest of body moves forward. The second feature consists of anchoring front of snake by pushing against the walls of confined space and drawing the snake rear to forward. By repeating of this sequence the snake performs forward locomotion. Anchoring the body of snake can be performed by several shapes of body.

The study of prof. Hu [6] and his team shows the way of snake body anchoring in dependence on diameter of channel.

Above mentioned sequence of locomotion can be in the Figure 1 seen. In this figure the video sequence of biological snake locomotion is shown. In this study the snake moves through the different pipe diameters and ways of its locomotion were observed. From other video sequences in [6] can be seen that the narrower pipe diameter is, the more similar snake locomotion to "concertina" motion is. According to this locomotion (Figure 1) will be designed sequence of locomotion of snake robot consisting of eight identical links, see Figure 2. During locomotion the biological snake uses muscle activity by which it can reach bending of particular parts to the sides as well as it can reach lengthening and shortening of these parts. From this reason we replace biological snake body by several identical links, which can move to the sides as well as they can lengthen and shorten [9]. For motion to the sides the revolute joint is assumed and for lengthening and shortening the prismatic joint is assumed. Our analysis considers snake robot with eight identical links, but it may be extended by other links.

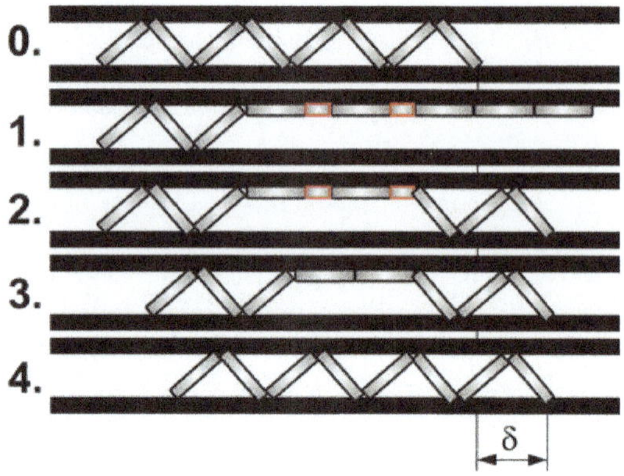

Figure 1. Sequence of snake robot locomotion

In the Figure 1 one locomotion cycle is shown. Three links are considered as static links and other five links move forward. This is the first phase of locomotion cycle. The 4th and 5th links lengthen by means of prismatic joint (red color). In the second phase the front of robot is anchored by pushing the first three links against the walls of pipe. During the third phase the last three links are slightly released in order to 4th and 5th links can be shortened. In the last fourth phase the rest of snake robot body only attract to the front part of the robot. By repeating of these four phases snake robot performs forward locomotion in the pipe.

The letter δ denotes the traveled distance in one locomotion cycle. This distance is dependent on the relation between length of snake robot link and diameter of the pipe.

The pipe diameter is of course important parameter. Locomotion according the Figure 2 can be performed when pipe diameter is lower than length of snake robot link. Locomotion according the Figure 2 would be able with pipe diameter higher than length of snake robot link only when robot has more links. However, snake robots are the robots with many identical links there can be added next links according to requirements.

Description of this locomotion can be expressed by following considerations:

- Pipe has rectangular cross-section
- Pipe diameter is lower than length of snake robot link
- One locomotion cycle consists of four phases. By repeating of these phases the snake robot performs forward locomotion with traveled distance δ during one cycle
- Snake robot uses for its locomotion revolute as well as prismatic joints
- Static links have to push against the walls of pipe with such torque in order to they stay static

For locomotion in the pipe can be used also different locomotion pattern like in [6,7] or [8,9]. The authors used for their study traveling wave locomotion of snake robot.

Uniqueness of our solution is in utilization of revolute as well as prismatic joint in one snake robot link. Snake robot with this kinematic configuration is on one hand more flexible and universal to the changes of environment. On the other hand utilization of prismatic joint in each link is more similar to a biological snake muscle activity. [10]

3. Kinematic Aspect of Concertina Locomotion

The kinematic model describes geometrical aspects of snake robot motion. The model of the investigated snake robot consists of 8 identical links and each of link has 2 degrees of freedom. Each link has one revolute joint and one prismatic joint. Link has length L_i and position of center of gravity CG_i. The reference frame (ground fixed frame) has denotation $O = \{x_0, y_0\}$. The position vector from the reference frame to the origin of particular link is $\boldsymbol{r}_i \in \mathbb{R}^2$, to end of the head link $\boldsymbol{r}_H \in \mathbb{R}^2$ and to origin of rear link $\boldsymbol{r}_R \in \mathbb{R}^2$. The protrusion of particular link by prismatic joint is l_i.

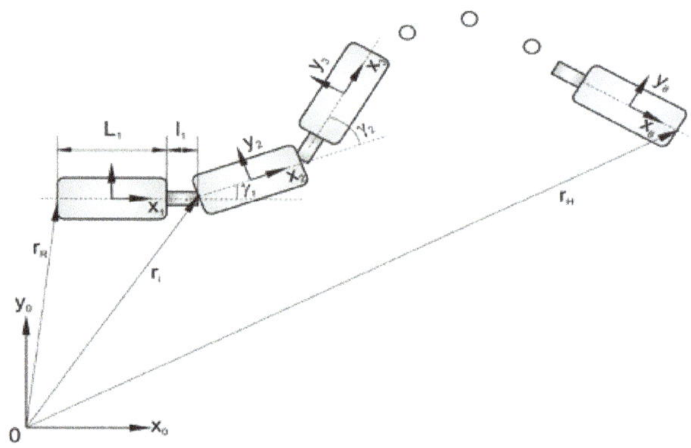

Figure 2. Kinematic aspect of investigated snake robot

Angle between two adjacent links is denoted as γ_i. Considering above mentioned kinematic convention, each link of snake robot is described by homogeneous transformation matrix

$$A_{i-1}^{i} = \begin{bmatrix} & & & r_{xi} \\ & R_{i-1}^{i} & & r_{yi} \\ & & & 0 \\ 0 & 0 & 0 & 1 \end{bmatrix} \tag{1}$$

Where R_{i-1}^{i} presents rotation matrix from coordinate system i to coordinate system i-1, r_{xi} and r_{yi} are coordinates of position vector of i-th link. Final transformation matrix from i-th link to the reference frame O is denoted as

$$T_0^i = \prod_{i=1}^{n} A_{i-1}^{i} \tag{2}$$

T_0^i represents transformation matrix of i-th link to reference frame O. Position vector of any link can be expressed by multiplication of corresponding transformation matrices according to equation (2). By transformation matrix of last link to reference frame can be also described workspace of snake robot.

Next important issue of snake robot kinematic is expressing of its traveled distance during one locomotion cycle. It is clear that the higher traveled distance of one locomotion cycle will be, the higher average velocity the robot will has. The analysis is based on relation between pipe diameter and length of snake robot link, see Figure 3.

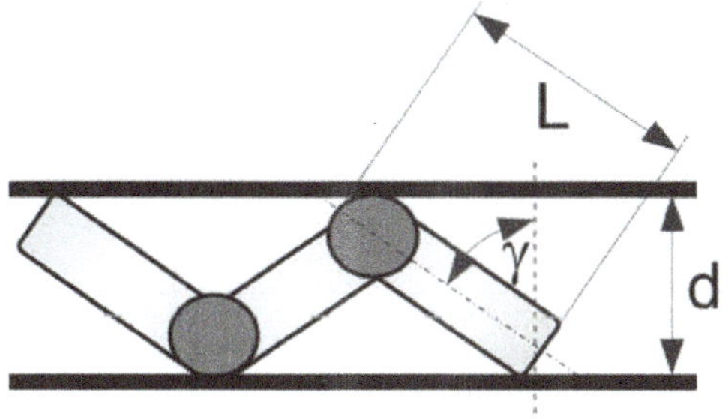

Figure 3. Relation between pipe diameter and length of snake robot link

According to Figure 1 the traveled distance δ can be derived. Let consider that angle between links is the same γ, see Figure 2 – 0th phase. Maximum protrusion of prismatic joint is l (red color), see Figure 1 – 1th phase. By considering all phases we can derive final traveled distance of one locomotion cycle denotes as δ by following relation

$$\delta = 2(L+l) - 2L\sin\left[\arccos\left(\frac{d}{L}\right)\right]. \quad (3)$$

From the equation (3) can be seen, that final traveled distance δ is function of pipe diameter d, length of snake robot link L and maximum protrusion of prismatic joint l_i. Graphic expression of this relation in the Figure 4 is shown.

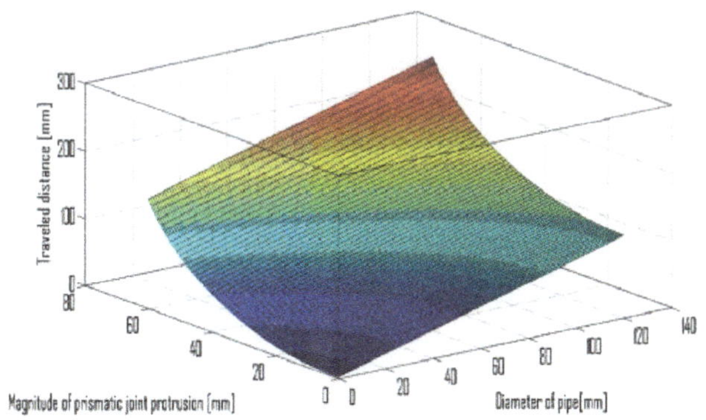

Figure 4. Dependence of traveled distance on changing pipe diameter and prismatic joint protrusion

4. Design of Snake Robot LocoSnake

For our study the experimental snake robot LocoSnake was designed. LocoSnake is of unique kinematic structure not used up to now yet. Each segment of the snake robot consists of two degrees of freedom. The one degree of freedom is represented by the revolute joint with the rotation range of ± 90° from the zero position. The next degree of freedom is represented by the prismatic joint with the range of 0 to 50 mm. Each end of the segment is equipped with the clutch enabling the connection of other parts and the signal cables and the power source cables transportation.

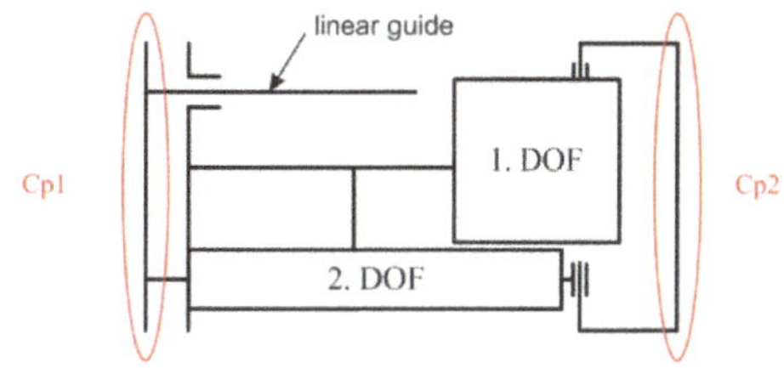

Figure 5. Kinematic configuration of LocoSnake

Based on the kinematic structure the model of snake inspired robot was designed by SolidWorks software, see Figure 6, Figure 7 and Figure 8. However, it was necessary to specify the dimension demands due to the accuracy of the model configuration. The dimensions of the action parts responsible for the segment dimensions were the most important criterion.

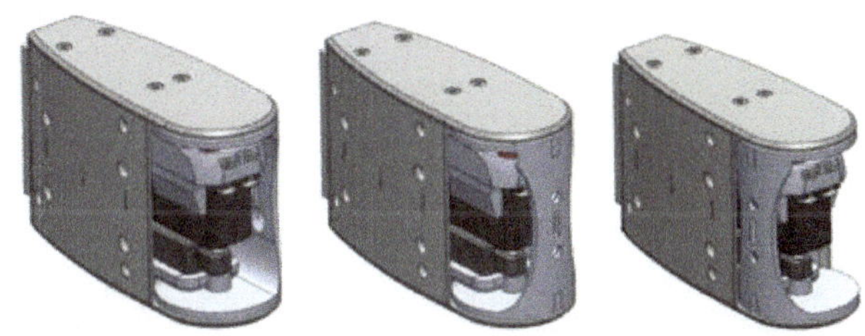

Figure 6. 3D model of LocoSnake link a) +90° b) 0° c) -90°

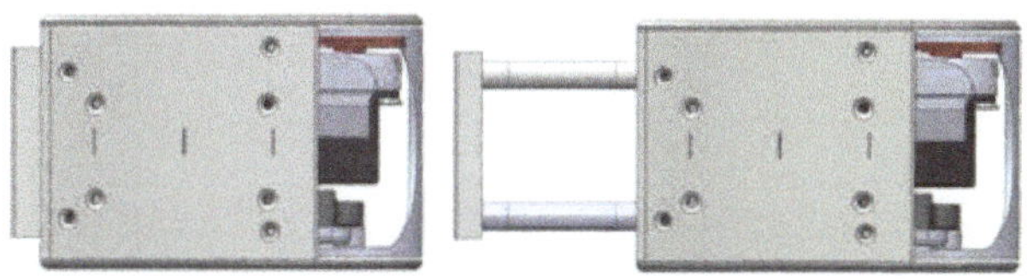

Figure 7. 3D model of LocoSnake link a) 0mm b) 50 mm

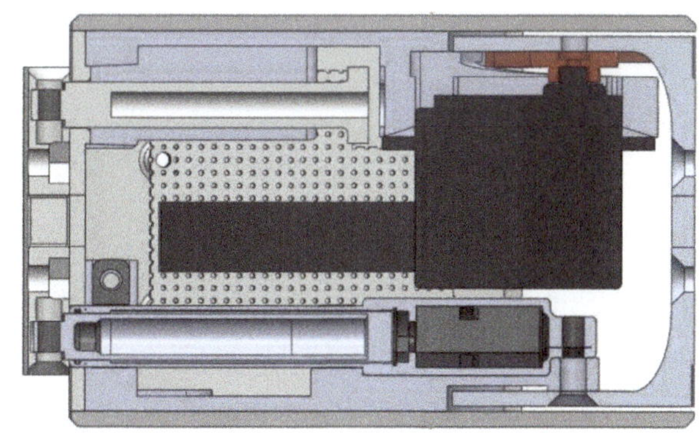

Figure 8. 3D model of LocoSnake link inside

The construction of LocoSnake is designed from parts made by the 3D printer were made. One link of LocoSnake weights 225 g with maximum dimensions: length 130 mm; width 80 mm; height 47 mm.

5. Simulation and Experimental Verification of Concertina Locomotion

In the section two the sequence of concertina locomotion for snake robot was designed and in the section three the direct kinematic model was derived. Based on these two sections the concertina locomotion can be simulated. Simulation is done in software Matlab.

The parameters of the simulation and experiment are: length of link $L_i = 0.13$ m, prismatic protrusion of link is $l_i = 0.05$ m, radius of the pipe (rectangle cross-section) is $d = 0.12$ m, number of links $n = 8$, weight of one link $w = 0.225$ kg.

Used approach is based on transformation matrices, described in the section 3. Actual position and orientation of each link can be obtained in arbitrary time.

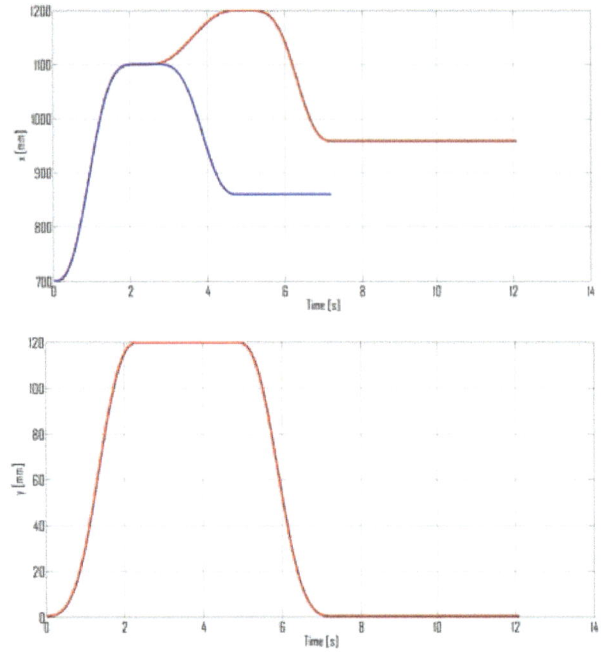

Figure 9. Position of head link during one locomotion cycle

In the Figure 9 the head position in x-axis and y-axis is shown during one locomotion cycle. Red color shows snake robot locomotion using prismatic joints with protrusion 50 mm and blue color shows locomotion without using prismatic joints.

Position of the head link in y-axis changes only in the range 0 – 120 mm, what corresponds with diameter of pipe. Position of the head link in x-axis changes from the initial position 700 mm to 958 mm (red color). Consequently, the final traveled distance of head link using prismatic joints (HLPJ) is 258 mm, what is the distance of one locomotion cycle during concertina locomotion. One locomotion cycle without using prismatic joints (HL) gives 160 mm of traveled distance what makes 98 mm difference between model with prismatic joints and without them. This considerable difference can be seen also in the Figure 5 as well.

While difference of traveled distance between HL and HLPJ is 98 mm, one locomotion cycle of HL is almost twofold faster than HLPJ. Traveled distance of HL can be higher than traveled distance of HLPJ after several cycles. Of course, this hypothesis is only theoretical and it depends on speed and dynamics of real actuators of snake robot. However, this study shows that by means of prismatic joints can be achieved considerably higher traveled distance of one locomotion cycle. By utilization of several prismatic joints (together with adding additional links to robot) this effect increases.

The course of entire simulation is animated in Matlab. Particular figures from the simulation are in the Figure 10 shown.

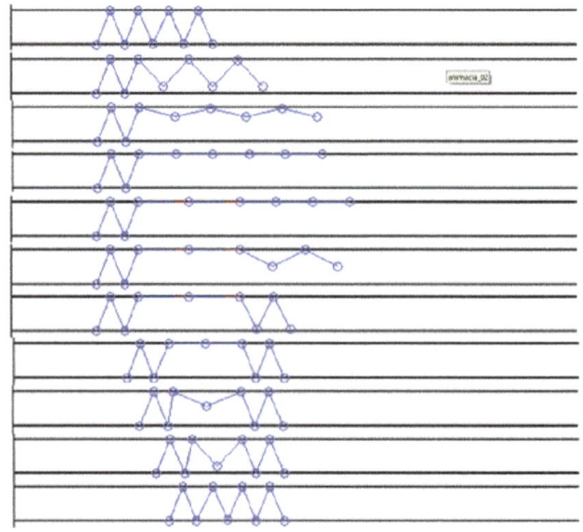

Figure 10. Position of head link during one locomotion cycle

In the Figure 10 only one locomotion cycle is shown. Prismatic joints protrusion is depicted by red color. Visualization only serves for verification purposes of kinematic model which is above mentioned.

Next point of our study is verification of achieved results by experimental analysis. For this purposes have been used snake robot - LocoSnake. Snake robot LocoSnake was designed and built in 2012 at Department of Mechatronics – Technical University of Košice, Slovakia. This robot was designed for experimental purposes of concertina locomotion analysis. Each segment of the robot has 2-DOF, one revolute servomechanism and one prismatic servomechanism. Length and weight of segment are the same like in simulation model. Each servomechanism has its own inner position controller and position of all segments are controlled by 32-bit microcontrollers BasicAtom Pro 28-M, 16 MHz. Supply voltage of the servomechanisms is 6 V DC. The speed of rotation of servomechanism without any load is approximately 60° per 0.2 seconds. Considering other segments as load the speed significantly decreases. The speed of prismatic servomechanism protrusion without any load is 23 mm/s. Considering other segments as load the speed of servomechanism decreases also.

Diameter of pipe is the same like in simulation. In the following shots the concertina locomotion of LocoSnake is shown.

Figure 11. Experiment with snake robot LocoSnake

The Figure 11 shows that LocoSnake can performs concertina locomotion according to computer simulation.

Considering that snake moves from left side to right side, mark the first segment from right side as 1 and the first segment from left side as 8. The most critical point of locomotion is in 2nd and 6th shot, during pushing static segments against walls of pipe. Let consider 2nd shot. Last three segments (6, 7, 8) push against the walls of pipe and this is base for forward motion of other segments. During the experiment there has to be increased the supply of electrical current in order to increasing of revolute joint performance. Next critical point is backlash rising between two adjoining segments what creates not quite pushing the segments against the walls of pipe.

6. Conclusion

A concertina locomotion from the view of biology is introduced. Based on knowledge about locomotion pattern, new pattern of snake robot is designed using revolute and prismatic joints as well. Assuming our locomotion pattern the relation for traveled distance is derived with dependence on diameter of pipe with rectangle cross section, and length of snake robot link. This relation shows signification of using prismatic joint for traveled distance in one locomotion cycle. The direct kinematic model of snake robot is derived using transformation matrix. Kinematic model for simulation in Matlab is used. There were done two simulations. The first with assuming prismatic joints and the second without them. By focusing on traveled distance the difference is considerable. Next, an experiment with snake robot LocoSnake was done.

The main aim of experiment was verification of concertina locomotion in the pipe. The most critical point of experiment was evolving required torque of first static segment (adjoining with first moving segment) in order to other segments can move forward. Next critical point is that junction between two adjoining segments is not too rigid and there arise backlash.

The results of experiment are different from simulation mainly because of differences of kinematic configuration between simulation and real model. Even though, the experiments shows that concertina locomotion is able with snake robot LocoSnake and expose new limitations and disadvantages of its design. The experiment also shows uniqueness of kinematic configuration using revolute as well as prismatic joints, what is for concertina locomotion significant.

Acknowledgement

The work has been accomplished under the research project APVV-0091-11 financed by the Ministry of Education. This contribution is also the result of the project implementation: Research of modules for intelligent robotic systems (ITMS: 26220220141) supported by the Research & Development Operational Programme funded by the ERDF.

References

[1] J. Gray, The mechanism of locomotion in snakes, in: Journal of Experimental Biology, 1946, Vol. 23, No. 2, pp. 101-120.

[2] H. W. Lissmann, Rectilinear locomotion in a snake, in: The Journal of Experimental Biology, 1949, pp. 368-379.

[3] S. Hirose, Biologically Inspired Robots: Snake-like Locomotors and Manipulators, Oxford University Press, 1993, Oxford.

[4] A. Akbarzadeh, Jal. Safehian, Jav. Safehian, H. Kalani, Generating Snake Robot Concertina Locomotion Using a New Dynamic Curve, in: International Journal of Modeling and Optimization, 2011, Vol. 1, No. 2, pp. 134-140.

[5] F. Barazandeh, B. Bahr, A. Moradi, How Self-Locking Reduces Actuators Torque in Climbing Snake Robots, IEEE International Conference on Advanced Intelligent Mechatronics, 2007, pp. 1-6.

[6] H. Marvi, D. L. Hu, Friction enhancement in concertina locomotion of snakes, in: Journal of the Royal Society, 2012, pp. 1-14.

[7] H. Kalani, A. Akbarzahed, J. Safehian, Traveling Wave Locomotion of Snake Robot along Symmetrical and Unsymmetrical Body Shapes. In: Robotics (ISR) - 6th German Conference on Robotics, pp. 62-68, 2010.

[8] Suzumori, K., Wakimoto, S., Takata, M., A miniature Inspection Robot Negotiating Pipes of Widely Varying Diameter. Proceedings of the 2003 IEEE International Conference on Robotics & Automation, Taiwan, pp. 2735-2740, 2003

[9] Božek, P., Chmelíková, G.. Implementation of robot offline programming. In Annals of DAAAM and Proceedings of DAAAM Symposium. s.0331-0332.

[10] Zubkova Y., Turygin Y., Experimental Investigation of Dynamic Error for Manipulator Output Link During Its Motion Path Tryout, Acta Mechanica Slovaca. Volume 19, Issue 3, Pages 34-37, ISSN 1335-2393.

SimMechanics and Forward and Inverse Problem of Dynamics

Darina Hroncová[1,*], Róbert Rákay[2], Tomáš Lipták[1]

[1]Department of Mechatronics, Faculty of Mechanical Engineering, Technical University of Košice, Letná 9, 042 00 Košice, Slovak republic

[2]Department of Automation, Control and Human Machine Interactions, Faculty of Mechanical Engineering, Technical University of Košice, Letná 9, 042 00 Košice, Slovak republic

*Corresponding author: darina.hroncova@tuke.sk

Abstract In the paper we show the possibilities of physical modeling in Matlab/SimMechanics on a simple mechanical model of robot arm with two links. We deal with its direct and inverse dynamics. Direct problem investigates the mechanical movement of the robot under the action of given forces and moments. In inverse problem we identify the forces which cause a prescribed motion of the effector. Both problems are solved using block diagrams in SimMechanics. Results are presented graphically.

Keywords: *forward dynamics, inverse dynamics, Matlab, Simulink, SimMechanics*

1. Introduction

Computer modeling is an effective tool to accelerate and improve the design of new mechanical systems. Matlab and SimMechanics [1] are appropriate software tools for creating computer models [7]. In dynamic analysis there are two basic tasks - inverse and direct analysis [5]. In inverse problem we identify the forces which cause a prescribed motion of the effector. Direct problem investigates the movement of the mechanical system under the action of given forces.

2. Description of the Model

We investigate the model of two link of manipulator which is a quite often analyses dynamic system (Figure 1a). Links are connected by two joints and there is a motor at each joint and joint angles are φ_1 and φ_2.

Location of the endpoint L of member 2 in the coordinate system O, x, y is given by [2]:

$$x_1 = l_1 \cos\phi_1 + l_1 \cos(\phi_1 + \phi_2), \tag{1}$$

$$x_2 = l_1 \sin\phi_1 + l_2 \sin(\phi_1 + \phi_2) \tag{2}$$

The arm is to start from rest at a known position and move to a desired position. The task is to investigate the necessary movement in kinematics pairs which would ensure the relocation of the end point L of member 2 from position L_0 (4.5, 0) to position L_1 (0, 3) (Figure 1b). At the beginning and end of the movement the velocity and acceleration of point L should be zero.

We try to find the time diagram of angular rotations φ_1 and φ_2 in the form of a fifth degree polynomials [2]:

$$\phi_1(t) = a_1 t^5 + a_2 t^4 + a_3 t^3 + a_4 t^2 + a_5 t + a_6, \tag{3}$$

$$\phi_2(t) = b_1 t^5 + b_2 t^4 + b_3 t^3 + b_4 t^2 + b_5 t + b_6 \tag{4}$$

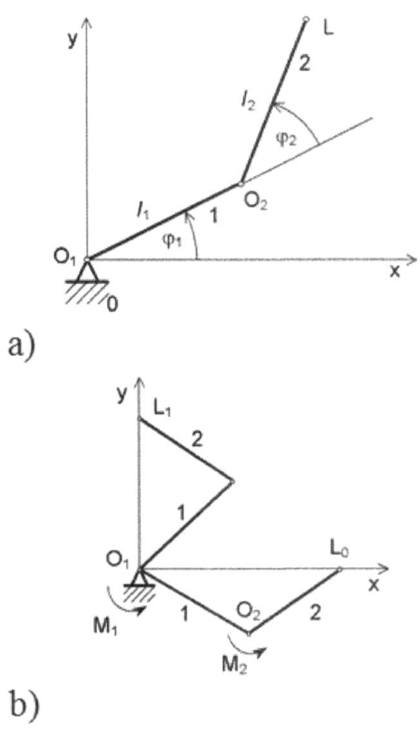

Figure 1. Model of manipulator: a) links are connected by two joints; b) starting and final position of two link of manipulator

where $a_6 = \varphi_1(0)$ and $b_6 = \varphi_2(0)$ are the initial condition at time $t = 0$ and coefficient vectors:

$$a = [a_1\ a_2\ a_3\ a_4\ a_5\ ]^T, \quad (5)$$

$$b = [b_1\ b_2\ b_3\ b_4\ b_5\ ]^T \quad (6)$$

are to be determined to provide the desired motion. The choice of the degree of the polynomials will be explained as the equations of motion are described. For desired initial coordinates (x_1, x_2) at time t=0 and desired final coordinates at time $t = t_f$,the required values for angles $\varphi_1(0), \varphi_2(0), \varphi_1(t_f), \varphi_2(t_f)$ can be found using trigonometry. For given values of $\varphi_1(0), \varphi_1(t_f)$ and t_f and for given values of $\varphi_2(0), \varphi_2(t_f)$ and t_f , matrix equations are to be set up and solved for coefficient vector ***a*** and coefficient vector ***b***. These results are to be used to plot the path of the hand.

The remaining constraints in the arm motion are that the velocity and acceleration of the links are to be zero at the known starting location and desired final location. This implies that the angular velocity and angular acceleration of the two angles are to be zero at time $t = 0$ and $t = t_f$. The angular velocity of the first link at time t is the derivative of the angle $\varphi_1(t)$ with respect to time [2]:

$$\dot{\varphi}_1(t) = 5a_1 t^4 + 4a_2 t^3 + 3a_3 t^2 + 2a_4 t + a_5 \quad (7)$$

The velocity at $t = 0$ is

$$\dot{\varphi}_1(0) = a_5 = 0 \quad (8)$$

and thus, coefficient $a_5 = 0$. The angular acceleration is the second derivative of the angle $\varphi_1(t)$ with respect to time:

$$\ddot{\varphi}_1(t) = 20a_1 t^3 + 12a_2 t^2 + 6a_3 t + 2a_4 \quad (9)$$

The acceleration at $t = 0$ is:

$$\ddot{\varphi}_1(0) = 2a_4 = 0 \quad (10)$$

and thus, coefficient $a_4 = 0$. Writing the three constraints on $\varphi_1(t)$ and its derivatives at time $t = t_f$ in matrix form:

$$\begin{bmatrix} t_f^5 & t_f^4 & t_f^3 \\ 5t_f^4 & 4t_f^3 & 3t_f^2 \\ 20t_f^3 & 12t_f^2 & 6t_f \end{bmatrix} \cdot \begin{bmatrix} a_1 \\ a_2 \\ a_3 \end{bmatrix} = \begin{bmatrix} \phi_1(t_f) - \phi_1(0) \\ 0 \\ 0 \end{bmatrix} \quad (11)$$

Similarly, for $\varphi_2(t)$:

$$\begin{bmatrix} t_f^5 & t_f^4 & t_f^3 \\ 5t_f^4 & 4t_f^3 & 3t_f^2 \\ 20t_f^3 & 12t_f^2 & 6t_f \end{bmatrix} \cdot \begin{bmatrix} b_1 \\ b_2 \\ b_3 \end{bmatrix} = \begin{bmatrix} \phi_2(t_f) - \phi_2(0) \\ 0 \\ 0 \end{bmatrix} \quad (12)$$

Note that there are three equations and three unknowns for each angle and its two derivatives. This is the reason for choosing a fifth-degree polynomial to control the motion. The lower degree three terms (t^2, t^1, t^0) have coefficients with value zero to meet the constraints at t=0. This leaves the higher degree three terms (t^5, t^4, t^3) and corresponding three unknown coefficients. Adding additional higher degree terms would require additional constraint equations to be able to compute the values of the corresponding coefficients [2].

Assuming the following initial and final values and link lengths:

t_f =2 s

$\varphi_1(0)$=-36,87°

$\varphi_2(0)$= 90°

$\varphi_1(t_f)$= 42,7785°

$\varphi_2(t_f)$= 125,3593°

l_1 =0,36 m

l_2 =0,27 m

which correspond to a starting hand location of (4.5, 0) and a destination location of (0,3), and plot the path of the end point L of the member 2 [6].

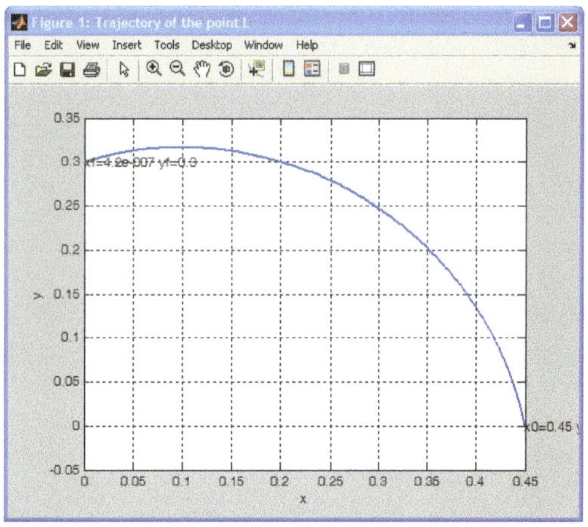

Figure 2. Window Figure 1 from Matlab simulation

Time diagram of the trajectory of the point L is shown in Figure 2 and Figure 3 from Matlab solution [8]:

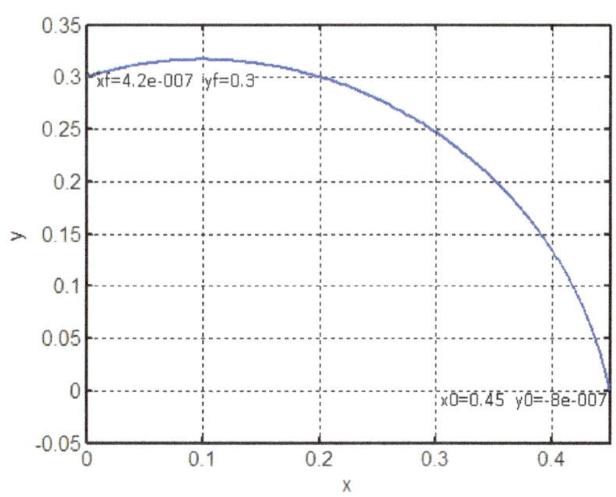

Figure 3. Calculated path of the point L of the robot hand

Time diagram of angular rotation, angular velocity and angular acceleration of the members is shown in Figure 4 [9].

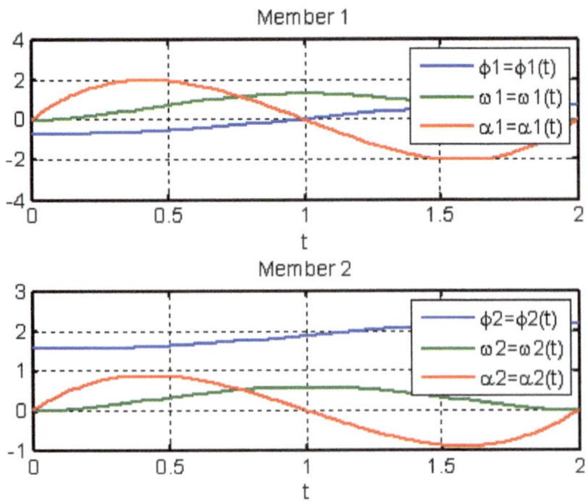

Figure 4. Time diagrams of the mechanical system with kinematical parameters

The constants in the polynomials are determined from the initial and end conditions of the endpoint L. The degree of these polynomials is chosen so that they are able to express all the desired conditions. After substituting the initial and final conditions we obtain the following values of their respective constants:

$a_1 = 0.2606 \qquad b_1 = 0.1157$

$a_2 = -1.3032 \qquad b_2 = -0.5786$

$a_3 = 1.7377 \qquad b_3 = 0.7714$

$a_4 = 0 \qquad b_4 = 0$

$a_5 = 0 \qquad b_5 = 0$

$a_6 = -36.87 * \pi / 180 = 0.6283$

$b_6 = 90 * \pi / 180 = 1.5708$

for $l_1 = 0.36$ and $l_2 = 0.27$.

3. Inverse Problem

While solving the inverse problem we determine the moments M_1 and M_2 required to obtain the prescribed movement of the mechanical system [3]. Inputs are the prescribed movements of its individual members. Member 2 is to be moved from position L_0 to the desired position L_1 along trajectory determined in the preceding paragraph. Block diagram of the mechanical system for solving the inverse problem [8] is in Figure 5.

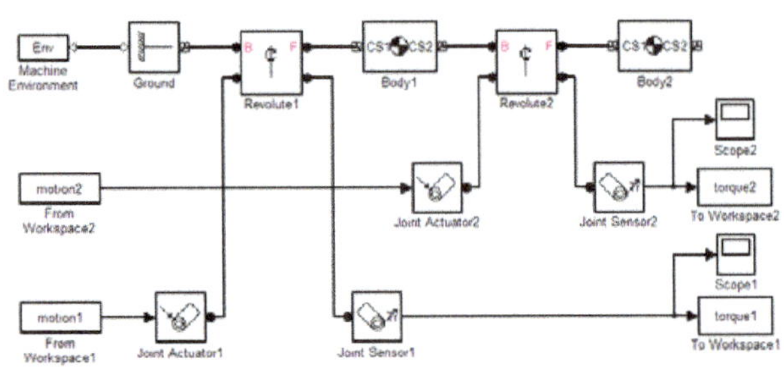

Figure 5. Block diagram of the mechanical system in SimMechanics (inverse dynamics)

The block of the Body1 (Figure 6) in SimMechanics scheme is shown [7]:

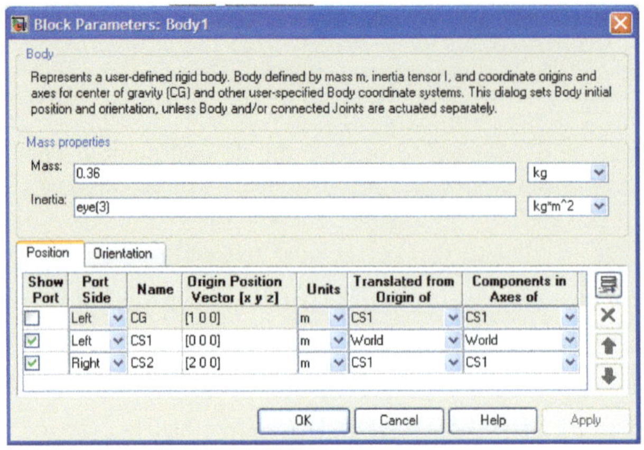

Figure 6. Block Parameters: Body1 of the mechanical system in SimMechanics (inverse dynamics)

The block of the Body2 (Figure 7) in SimMechanics scheme:

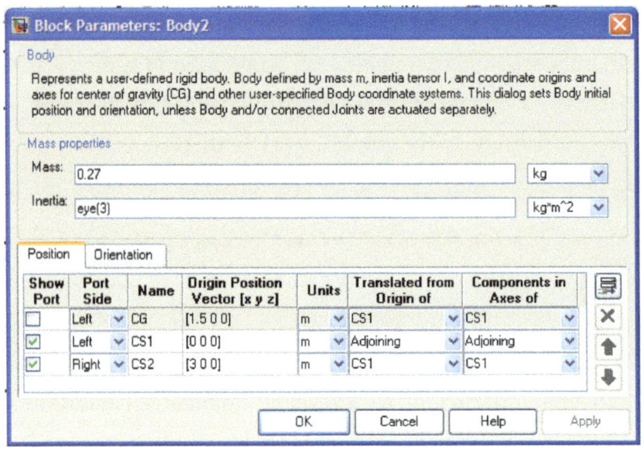

Figure 7. Block Parameters: Body2 of the mechanical system in SimMechanics (inverse dynamics)

Time diagrams of torques [9] in the kinematics pairs M_1 and M_2 obtained by solving the model in the above block diagram are shown in Figure 8.

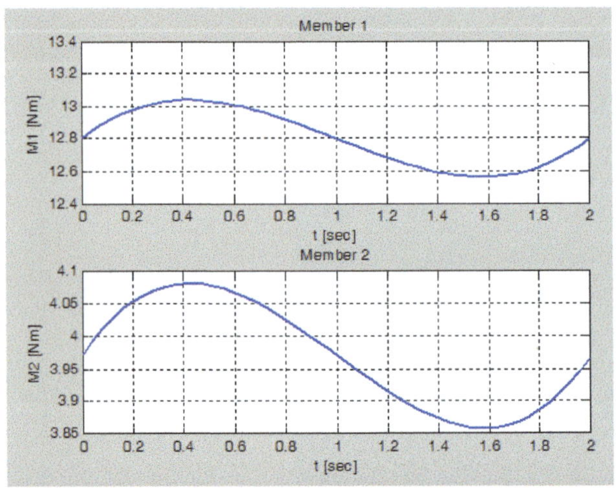

Figure 8. Time diagram of actuator forces of the mechanical system

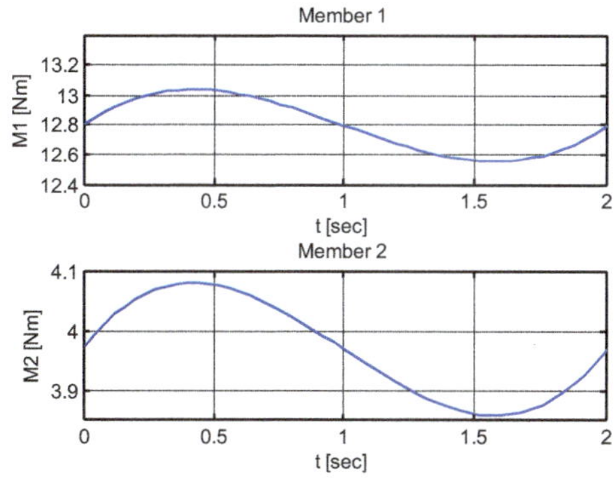

Figure 9. Time diagram of actuator forces of the mechanical system

4. Direct Problem

Direct problem investigates the movement of the mechanical system under the load of the forces obtained in the previous paragraph. Application of the results from the inverse problem on the model in Figure 10 should result in motion as we prescribed it at the beginning of the analysis [4].

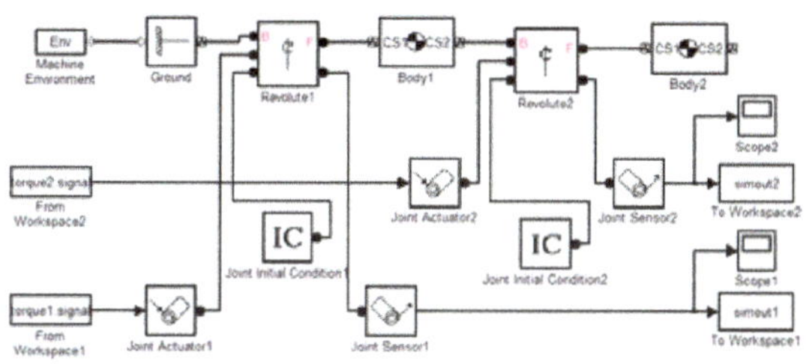

Figure 10. Block diagram of the mechanical system in SimMechanics (forward dynamics)

The block Joint Initial Condition for $\varphi_1(0)$ in SimMechanics scheme:

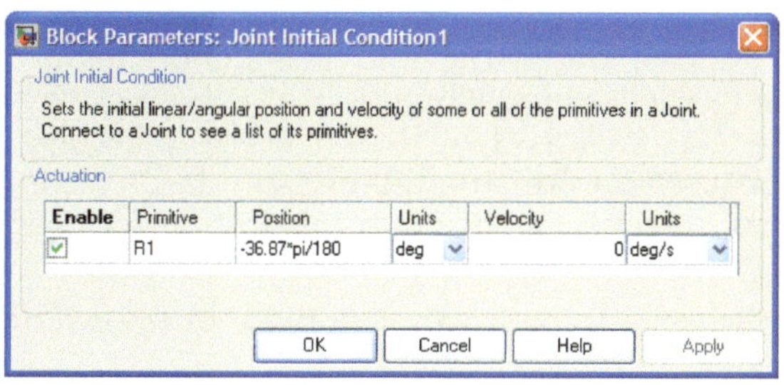

Figure 11. Block Joint Initial Condition in SimMechanics (forward dynamics)

The block Joint Initial Condition for $\varphi_2(0)$ in SimMechanics scheme [7]:

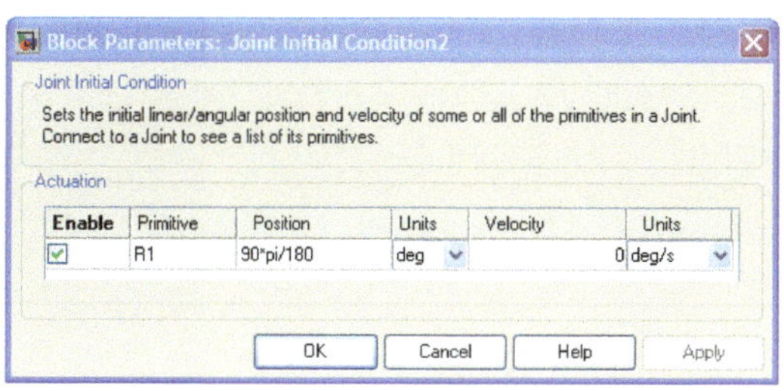

Figure 12. Block Joint Initial Condition in SimMechanics (forward dynamics).

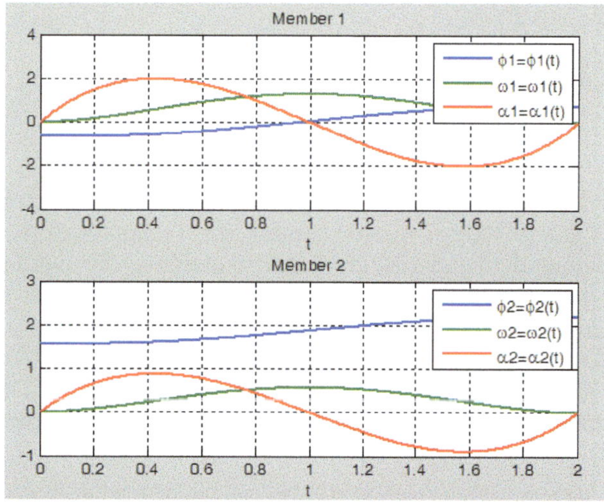

Figure 13. Time diagram of SimMechanics simulation

Results obtained from the simulation in Figure 13 correspond with the prescribed motion we determined from the desired location of the endpoint L that are in Figure 4.

5. Conclusion

In the work is shown a procedure for solving direct and inverse dynamic problem of the mechanism of double pendulum using modeling in SimMechanics. SimMechanics allows to simulate such mechanical systems which is a common problem in engineering practice. Results are obtained in form of time diagram of the desired variables. Tasks are solved numerically, model is compiled by using block diagrams. Mastering this methodology provides a suitable tool for solving problems of teaching and practice.

Acknowledgement

This work was supported by grant projects VEGA No. 1/1205/12 and grant projects VEGA No. 1/0937/12.

References

[1] The MathWorks,Inc. SimMechanics User's Guide, November 2002. http://www.mathworks.com.

[2] http://www.pwsztar.edu.pl/~r_wiatr/matlab/%28eBook%29%20%20Engineering%20-%20Matlab%20Programming.pdf

[3] Gmiterko, A., 2004. *Mechatronics*, Emilenia, Košice, Slovakia.

[4] Gmiterko, A., Šarga, P., Hroncová, D., 2012. *Mechatronika I*, Technická univerzita Košice, Košice, Slovakia.

[5] Hroncová, D., 2012. *Studies of compatible software tools for abstract and concrete design of mechatronic systems and linking them for the purpose of logical and physical modeling*, Dissertation thesis, Košice, Slovakia.

[6] Karban, P., 2006. *Výpočty a simulace v programech Matlab a Simulink*. Computer Press, Brno, Česko.

[7] Schlotter, M., 2003. *Multibody System Simulation with SimMechanics*. University of Canterbury.

[8] Virgala, I., Frankovský, P., Kenderová, M.: *Friction Effect Analysis of a DC Motor*. In: American Journal of Mechanical Engineering. 2013. Vol. 1, no. 1 (2013), p. 1-5

[9] Vittek, J., 1997. *Matlab pre elektrické pohony*, Nakladateľstvo Žilinskej univerzity, Žilina, Slovensko, p. 35-36.

Permissions

All chapters in this book were first published in AUTOMATION, by Science and Education Publishing; hereby published with permission under the Creative Commons Attribution License or equivalent. Every chapter published in this book has been scrutinized by our experts. Their significance has been extensively debated. The topics covered herein carry significant findings which will fuel the growth of the discipline. They may even be implemented as practical applications or may be referred to as a beginning point for another development.

The contributors of this book come from diverse backgrounds, making this book a truly international effort. This book will bring forth new frontiers with its revolutionizing research information and detailed analysis of the nascent developments around the world.

We would like to thank all the contributing authors for lending their expertise to make the book truly unique. They have played a crucial role in the development of this book. Without their invaluable contributions this book wouldn't have been possible. They have made vital efforts to compile up to date information on the varied aspects of this subject to make this book a valuable addition to the collection of many professionals and students.

This book was conceptualized with the vision of imparting up-to-date information and advanced data in this field. To ensure the same, a matchless editorial board was set up. Every individual on the board went through rigorous rounds of assessment to prove their worth. After which they invested a large part of their time researching and compiling the most relevant data for our readers.

The editorial board has been involved in producing this book since its inception. They have spent rigorous hours researching and exploring the diverse topics which have resulted in the successful publishing of this book. They have passed on their knowledge of decades through this book. To expedite this challenging task, the publisher supported the team at every step. A small team of assistant editors was also appointed to further simplify the editing procedure and attain best results for the readers.

Apart from the editorial board, the designing team has also invested a significant amount of their time in understanding the subject and creating the most relevant covers. They scrutinized every image to scout for the most suitable representation of the subject and create an appropriate cover for the book.

The publishing team has been an ardent support to the editorial, designing and production team. Their endless efforts to recruit the best for this project, has resulted in the accomplishment of this book. They are a veteran in the field of academics and their pool of knowledge is as vast as their experience in printing. Their expertise and guidance has proved useful at every step. Their uncompromising quality standards have made this book an exceptional effort. Their encouragement from time to time has been an inspiration for everyone.

The publisher and the editorial board hope that this book will prove to be a valuable piece of knowledge for researchers, students, practitioners and scholars across the globe.

List of Contributors

Mehdi Namdari, Hooshang Jazayeri-Rad and Seyed-Jalaladdin Hashemi
Department of Instrumentation and Automation, Petroleum University of Technology, Ahwaz, Iran

Lidong Wang
Department of Engineering Technology, Mississippi Valley State University, Itta Bena, USA

Masih Vafaee Ayouri and Mehdi Shahbazian
Department of Instrumentation and Automation Engineering, Petroleum University of Technology, Ahwaz, Iran

Bahman Moslemi
Department of Basic Science, Petroleum University of Technology, Ahwaz, Iran

Mahboobeh Taheri
Senior expert in R&D, Sarkhoon & Qeshm Gas treating Company, Bandar Abbas, Iran

Pooria Karimi and Hooshang Jazayeri-Rad
Department of Instrumentation and Automation, Petroleum University of Technology, Ahwaz, Iran

Michal Kelemen, Ivan Virgala, Ľubica Miková, Peter Frankovský, Tomáš Lipták and Milan Lörinc
Department of Mechatronics, Technical University of Kosice, Faculty of Mechanical Engineering, Kosice, Slovak Republic

Tatiana Kelemenová
Department of Biomedical Engineering and Measurement, Technical University of Kosice, Faculty of Mechanical Engineering, Kosice, Slovak Republic

Darina Hroncová, Alexander Gmiterko and Tomáš Lipták
Department of Mechatronics, Faculty of Mechanical Engineering, Technical University of Košice, Letná 9, 042 00 Košice, Slovak republic

Sajjad Aliesfahani and Mehdi Shahbazian
Department of Instrumentation and Automation Engineering, Petroleum University of Technology, Ahwaz, Iran

Satheesh G
Electrical and Electronics Engineering Department, G.Pulla Reddy Engineering College, Kurnool, India

T Bramhananda Reddy and CH Sai Babu
Electrical and Electronics Engineering Department, University college of Engineering Kakinada, JNTUK, Kakinada, India

Tatiana Kelemenová and Peter Sedlačko
Department of Mechatronics, Technical University of Kosice, Faculty of Mechanical Engineering, Kosice, Slovak Republic

Michal Kelemen, Ivan Virgala, Ľubica Miková, Peter Frankovský, Milan Lörinc and Tomáš Lipták
Department of Biomedical Engineering and Measurement, Technical University of Kosice, Faculty of Mechanical Engineering, Kosice, Slovak Republic

Rudolf Jánoš
Department of robotics, Technical university of Kosice, Kosice, Slovakia

M. Khodadadi and M. Shahbazian
Department of Instrumentation and Automation Engineering, Ahwaz Faculty of Petroleum, PUT, Ahwaz, Iran

M. Aghajani
Department of Chemical Engineering, Ahwaz Faculty of Petroleum, PUT, Ahwaz, Iran

Marek Moravič
Department of Construction, Automotive and Transport Engineering, Faculty of Mechanical Engineering, Technical University of Košice, Košice, Slovak republic

Oskar Ostertag
Department of Applied Mechanics and Mechanical Engineering, Faculty of Mechanical Engineering, Technical University of Košice, Košice, Slovak republic

Darina Hroncová
Department of Mechatronics, Faculty of Mechanical Engineering, Technical University of Košice, Košice, Slovak republic

Ľubica Miková, Michal Kelemen, Alexander Gmiterko and Lukáš Kačmár
Technical University of Košice, Faculty of Mechanical Engineering, Department of Mechatronics, Košice, Slovakia

Kamil Židek
Faculty of Manufacturing Technologies with a seat in Presov, Presov, Slovak Republic

Vladislav Maxim
Institute of Automation, Robotics and Mechatronics, FME, Technical University of Kosice, Slovak Republic

Ľubica Miková, Michal Kelemen and Alexander Gmiterko
Technical University of Košice, Faculty of Mechanical Engineering, Department of Mechatronics, Košice, Slovakia

Chii-Huei Yu
Department of Management and Information, Nan Jeon University of Science and Technology, Tainan City, Taiwan

Kamran Vafaee and Behdad Geranmehr
School of Mechanical Engineering, Iran University of Science and Technology (IUST), Tehran, IRAN

Alice J. Lin
Marshall University, Huntington,West Virginia, USA

Charles B. Chen
West VirginiaUniversity, Morgantown,West Virginia, USA

Navid Yazdanparast and Mehdi Shahbazian
Department of Instrumentation and Automation Engineering, Petroleum University of Technology, Ahwaz, Iran

Masoud Aghajani
Department of Chemical Engineering, Petroleum University of Technology, Ahwaz, Iran

Saeed Pour Abed
Senior expert in Instrumentation Unit, Sarkhoon&Qeshm Gas treating Company, Bandar Abbas, Iran

M. Shahbazian, H. Jazayerirad and M. Ebnali
Department of Instrumentation and Automation Engineering, Ahwaz Faculty of Petroleum Engineering, Ahwaz, Iran

Lozhkin A.G., Arkhipov I.O. and Sultanov R.O.
1Departments "Software", M.T. Kalashnikov Izhevsk State Technical University, Izhevsk, Russia

Abramov I.V. and Nikitin Yu.R.
Mechatronic Systems department, M.T. Kalashnikov Izhevsk State Technical University, Izhevsk, Russia

Alexander Gmiterko, Ivan Virgala, Lubica Miková, Peter Frankovský, Tatiana Kelemenová and Michal Kelemen
Department of Mechatronics, Technical University of Kosice, Faculty of Mechanical Engineering, Kosice, Slovak Republic

Dusan Simsik, Alena Galajdova and Robert Rakay
Department of Automation, Control and Human-Machines Interactions, Mechanical Engineering Faculty, Technical University of Kosice, Slovakia

Daniela Onofrejova
Department of Industrial Engineering and Management, Mechanical Engineering Faculty, Technical University of Kosice, Slovakia

Hatem Elaydi
Electrical Engineering Department, Islamic University of Gaza, Gaza, Palestine

Mohammed Wadi
Electrical Engineering Department, Yildiz Technical University, Istanbul, Turkey

Kamil Židek
Faculty of Manufacturing Technologies with a seat in Presov, Presov, Slovak Republic

Vladislav Maxim
Institute of Automation, Robotics and Mechatronics, FME, Technical University of Kosice, Slovak Republic

Sławomir Kłos and Justyna Patalas-Maliszewska
University of Zielona Góra, Faculty of Mechanical Engineering, Licealna 9, 65-417 Zielona Góra, Poland

Peter Trebuna
Technical University of Kosice, Faculty of Mechanical Engineering

Mohammad Khodadadi and Mahdi Shahbazian
Department of Instrumentation and Automation Engineering, Ahwaz Faculty of Petroleum, PUT, Ahwaz, Iran

Masoud Aghajani
Department of Chemical Engineering, Ahwaz Faculty of Petroleum, PUT, Ahwaz, Iran

Ivan Virgala, Michal Kelemen, Alexander Gmiterko and Tomáš Lipták
Department of Mechatronics, Technical University of Košice, Faculty of Mechanical Engineering, Košice, Slovakia

Róbert Rákay, Martin Višňovský, Alena Galajdová and Dušan Šimšík
Department of Automation, Control and Human Machine Interactions, Technical University of Košice, Košice, Slovak Republic

Eva Ostertagová
Department of Mathematics and Theoretical Informatics, Faculty of Electrical Engineering and Informatics, Technical University of Košice, Nemcovej 32, 042 00 Košice, Slovak Republic

Oskar Ostertag
Department of Applied Mechanics and Mechatronics, Faculty of Mechanical Engineering, Technical University of Košice, Letná 9, 042 00 Košice, Slovak Republic

Marek Vagaš, Vladimír Baláž and Lucia Koukolová
Department of Robotics, Faculty of mechanical engineering, Technical University of Košice, Slovakia

Jozef Putala
Matador Industries, a. s., General engineering & Automatization, Dubnica n. Váhom, Slovakia

Sonali Mayuresh Akolkar and Bansidhar Eknath Kushare
Electrical Engineering Department, K.K.W.I.E.E. & R, Nashik, India

Yogesh Misra
Mewar University, Chittorgarh, Rajasthan, India

H R Kamath
Malwa Institute of Technology, Indore, Madhya Pradesh, India

Vasiliy Kazak, Dmitry Shevchuk and Nataliia Tymoshenko
National Aviation University, Kiev, Ukraine

Alena Galajdová and Dušan Šimšík
Department of Automation, Control and Human Machine Interactions, Technical University of Košice, Slovakia

Snejana Yordanova and Yavor Sivchev
Faculty of Automation, Technical University of Sofia, Sofia, Bulgaria

Peter Sedlačko, RadovanHudák, Tatiana Kelemenova and Jozef Živčák
Department of Biomedical Engineering and Measurement, Faculty of Mechanical Engineering, Technical University of Košice, Košice 042 00, Slovakia

Marek Penhaker
Department of Cybernetics Biomedical Engineering, Faculty of Electrical Engineering and Computer Science, VŠB- Technical University of Ostrava, Ostrava 70800, Czech Republic

D. K. Sambariya and Omveer Sharma
Department of Electrical Engineering, Rajasthan Technical University, Kota, India

František Trebuňa
Department of Applied Mechanics and Mechanical Engineering, Technical University of Košice, Faculty of Mechanical Engineering, Košice, Slovakia

Ivan Virgala, Michal Kelemen and Tomáš Lipták
Department of Mechatronics, Technical University of Košice, Faculty of Mechanical Engineering, Košice, Slovakia

Darina Hroncová and Tomáš Lipták
Department of Mechatronics, Faculty of Mechanical Engineering, Technical University of Košice, Letná 9, 042 00 Košice, Slovak republic

Róbert Rákay
Department of Automation, Control and Human Machine Interactions, Faculty of Mechanical Engineering, Technical University of Košice, Letná 9, 042 00 Košice, Slovak republic

Index